GEOM[illegible]RMS

ELLIS HORWOOD SERIES IN MATHEMATICS AND ITS APPLICATIONS

Series Editor: Professor G. M. BELL, Chelsea College, University of London
(and within the same series)
Statistics and Operational Research
Editor: B. W. CONOLLY, Chelsea College, University of London

Baldock, G. R. & Bridgeman, T.	**Mathematical Theory of Wave Motion**
de Barra, G.	**Measure Theory and Integration**
Beaumont, G. P.	**Introductory Applied Probability**
Burghes, D. N. & Borrie, M.	**Modelling with Differential Equations**
Burghes, D. N. & Downs, A. M.	**Modern Introduction to Classical Mechanics and Control**
Burghes, D. N. & Graham, A.	**Introduction to Control Theory, including Optimal Control**
Burghes, D. N., Huntley, I. & McDonald, J.	**Applying Mathematics**
Burghes, D. N. & Wood, A. D.	**Mathematical Models in the Social, Management and Life Sciences**
Butkovskiy, A. G.	**Green's Functions and Transfer Functions Handbook**
Butkovskiy, A. G.	**Structure of Distributed Systems**
Chorlton, F.	**Textbook of Dynamics, 2nd Edition**
Chorlton, F.	**Vector and Tensor Methods**
Conolly, B.	**Techniques in Operational Research Vol. 1: Queueing Systems Vol. 2: Models, Search, Randomization**
Dunning-Davies, J.	**Mathematical Methods for Mathematicians, Physical Scientists and Engineers**
Eason, G., Coles, C. W., Gettinby, G.	**Mathematics and Statistics for the Bio-sciences**
Exton, H.	**Handbook of Hypergeometric Integrals**
Exton, H.	**Multiple Hypergeometric Functions and Applications**
Exton, H.	**Q-Hypergeometric Functions**
Faux, I. D. & Pratt, M. J.	**Computational Geometry for Design and Manufacture**
Firby, P. A. & Gardiner, C. F.	**Surface Topology**
Gardiner, C. F.	**Modern Algebra**
Goodbody, A. M.	**Cartesian Tensors**
Goult, R. J.	**Applied Linear Algebra**
Graham, A.	**Kronecker Products and Matrix Calculus: with Applications**
Graham, A.	**Matrix Theory and Applications for Engineers and Mathematicians**
Griffel, D. H.	**Applied Functional Analysis**
Hoskins, R. F.	**Generalised Functions**
Hunter, S. C.	**Mechanics of Continuous Media, 2nd (Revised) Edition**
Huntley, I. & Johnson, R. M.	**Linear and Nonlinear Differential Equations**
Jaswon, M. A. & Rose, M. A.	**Crystal Symmetry: The Theory of Colour Crystallography**
Jones, A. J.	**Game Theory**
Kemp, K. W.	**Introductory Statistics**
Kim, K. H. and Roush, F. W.	**Applied Abstract Algebra**
Kosinski, W.	**Field Singularities and Wave Analysis in Continuum Mechanics**
Marichev, O. I.	**Integral Transforms of Higher Transcendental Functions**
Meek, B. L. & Fairthorne, S.	**Using Computers**
Muller-Pfeiffer, E.	**Spectral Theory of Ordinary Differential Operators**
Nonweiler, T. R. F.	**Computational Mathematics: An Introduction to Numerical Analysis**
Oldknow, A. & Smith, D.	**Learning Mathematics with Micros**
Oliviera-Pinto, F.	**Simulation Concepts in Mathematical Modelling**
Oliviera-Pinto, F. & Conolly, B. W.	**Applicable Mathematics of Non-physical Phenomena**
Rosser, W. G. V.	**An Introduction to Statistical Physics**
Scorer, R. S.	**Environmental Aerodynamics**
Smith, D. K.	**Network Optimisation Practice: A Computational Guide**
Srivastava, H. M. and Manocha, H. L.	**A Treatise on Generating Functions**
Stoodley, K. D. C., Lewis, T. & Stainton, C. L. S.	**Applied Statistical Techniques**
Sweet, M. V.	**Algebra, Geometry and Trigonometry for Science Students**
Temperley, H. N. V. & Trevena, D. H.	**Liquids and Their Properties**
Temperley, H. N. V.	**Graph Theory and Applications**
Twizell, E. H.	**Computational Methods for Partial Differential Equations in Biomedicine**
Wheeler, R. F.	**Rethinking Mathematical Concepts**
Whitehead, J. R.	**The Design and Analysis of Sequential Clinical Trials**
Willmore, T. J.	**Total Curvature in Riemannian Geometry**

GEOMETRY OF SPATIAL FORMS

PETER C. GASSON, D.I.C., C.Eng., M.I.Mech.E., F.I.E.D.
Senior Research Engineer, Department of Aeronautics
Imperial College of Science and Technology
University of London

ELLIS HORWOOD LIMITED
Publishers · Chichester

Halsted Press: a division of
JOHN WILEY & SONS
New York · Brisbane · Chichester · Ontario

First published in 1983 by
ELLIS HORWOOD LIMITED
Market Cross House, Cooper Street, Chichester, West Sussex, PO19 1EB, England

The publisher's colophon is reproduced from James Gillison's drawing of the ancient Market Cross, Chichester.

Distributors:

Australia, New Zealand, South-east Asia:
Jacaranda-Wiley Ltd., Jacaranda Press,
JOHN WILEY & SONS INC.,
G.P.O. Box 859, Brisbane, Queensland 40001, Australia

Canada:
JOHN WILEY & SONS CANADA LIMITED
22 Worcester Road, Rexdale, Ontario, Canada.

Europe, Africa:
JOHN WILEY & SONS LIMITED
Baffins Lane, Chichester, West Sussex, England.

North and South America and the rest of the world:
Halsted Press: a division of
JOHN WILEY & SONS
605 Third Avenue, New York, N.Y. 10016, U.S.A.

British Library Cataloguing in Publication Data
Gasson, Peter C.
Geometry of spatial forms. – (Ellis Horwood series in mathematics and its applications)
1. Geometry, Descriptive
I. Title
516'.6 QA501

Library of Congress Card No. 83-12893

ISBN 0-85312-620-8 (Ellis Horwood Ltd. – Library Edition)
ISBN 0-470-20011-1 (Halsted Press – Library Edition)
ISBN 0-85312-652-6 (Ellis Horwood Ltd. – Student Edition)
ISBN 0-470-20009-X (Halsted Press – Paperback Edition)

Printed in Great Britain by Unwin Brothers of Woking.

Table of Contents

DEDICATION

To my parents

The whole of physics, that is the whole of the philosophy of nature is nothing but geometry.

Réné Descartes (1596–1650)

PREFACE

Geometry is that branch of mathematics which treats of spatial magnitudes and examines the configuration of spatial forms.

The word spatial relates to space, and space is the plenum or void in which all material bodies have their being.

Configuration relates to the structure of form, and form is the mode of arrangement best suited to fitness and efficiency of purpose.

The subject we are about to embark on is called the Geometry of Spatial Forms.

The geometry of spatial forms is an applicable subject which spans several separately taught domains.

Analytic, transformational and topological geometry are topics taught by the Mathematics department. Descriptive geometry is taught by the Drawing or Graphics department and computer-aided methods come from yet another source. It is, therefore, most important to weigh and assimilate the content of these separately taught courses and the role of this book is to present all the most important branches of applicable geometry, in a form best suited to the needs of shape design architects and engineers.

Having been intimately concerned with the practice, and to a lesser extent, with the teaching of geometry for many years, it has been a pleasure for me to further research my subject: to write and illustrate a book of these proportions.

My grateful thanks are due: to Carol Willis who typed the entire script so precisely, with very few errors. To Maureen, my wife, who prepared numerous drafts, letters and amendments towards the end. To the editorial and production staff at Ellis Horwood Ltd. To Professor G. M. Bell, series editor, and M. J. Pratt who reviewed my early draft and made several helpful suggestions.

It is a great privilege for me, an engineer, to contribute to this very fine mathematical series and I extend my most sincere thanks to Ellis Horwood for accepting and publishing my work.

Peter C. Gasson: April 1983

INTRODUCTION

The work of Architects, Designers and Engineers is inseparable from the collective laws and geometrical elements of idealised 3-space. Idealised 3-space is the abstract representational notion we have of the void or plenum in which we live and in which everything known to us has its being.

In manufacture the transmutation of forms is achieved by the application of appreciable force, and entails the expenditure of immense amounts of energy: but in design the geometrical constraints of idealised 3-space are so strong that they are immutable and independent of whatever forces or energies are brought to bear.

These geometrical constraints are the invariant universals on which all real design depends.

The existence of N.C. and C.N.C. machine tools and computer graphic facilities means that geometrical modelling is a directly applicable subject. Every N.C. machine tool in the land takes its motion commands in the form of digital coordinate data, obtained from an orthographic layout drawing, or from a spatial model, or is derived from the appropriate geometrical equation. The availability of multi-axis machine tools makes it possible to consider the automated or semi-automated production of truly three-dimensional sculptured forms, of a complexity that could not previously have been considered. The need for three-dimensional skills has never been greater.

It is, for instance, widely recognised that few people are born with a natural capacity for three-dimensional design and thinking. People who experience no great difficulty in visualising, conceptualising and calculating 2D forms in 2-space, frequently suffer a mind block when forms of a higher spatial order are considered.

The needs, ways, means and ends of the Architect, Designer and Engineer are different from those of the pure mathematician; nevertheless the fluent facility to visualise spatial relationships and the ability to calculate the extent and disposition of forms in 3-space, are assets that no free thinking nor practising Architect, Designer or Engineer can do without.

Architects and industrial designers have a sound conception of form but often lack the necessary mathematical expertise. Professional Engineers though often less sensitive to form than Architects and Industrial Designers, are generally the better equipped to solve the many analytic problems associated with 3-space. Technician engineers are best qualified, from a practical point of view, to undertake work in the automated part programming field, but frequently lack the necessary geometrical and mathematical primary background, and this inhibits their work and projects in numerous ways.

Three dimensionality

In terms of performance, the superiority of three-dimensional configurations is widely acknowledged and everywhere apparent in nature.

The main reason(s) why the third dimension has not been more widely evoked lies in the fact that many would-be users are unacquainted with, or are inept at applying, 3D methods.

Until quite recently the design and manufacture of 3D (as opposed to $2\frac{1}{2}$ D) forms (components, moulds, tools, structures, mechanisms, etc.) has been inhibited by the two-axis composition and manual operation of traditional machine tools; however, the continuing development and wide availability of multi-axis N.C. and C.N.C. machines, plus the increasing use of high-powered computational and graphical facilities, means that three-dimensional design and manufacture of engineering and architectural components is now an attractive proposition.

It is recognised that three-dimensionality is not the inevitable answer in every instance. No more than the search for minimum weight is the be-all and end-all of structural design. The on-going search for minimum weight does, however, serve to focus the designer's attention and leads directly to improved and more efficient designs. The study of three-dimensionality is here proposed in much the same spirit, and a broadly philosophical, mathematical, methodological approach is pursued throughout.

The essence of structure

Structural pattern is present in all things. It is to be observed on a microscopic scale, in crystals: on the macroscopic scale, in cosmos. The presence of structure is obvious in a girder bridge – in a geodesic dome – less apparent in a balloon. The study of geometrical structure is universal and all-important. Consider the following instances.

The economy of hexagonal packing, as in the two-dimensional close packing of a bees' honeycomb, is known to many: the manner in which the mid-plane of each cell is fashioned from three canted rhombic faces, set as in the rhombic dodecahedron, is perhaps less well known.

The manner in which oranges may be stably stacked in a tetrahedral cluster is akin to the way in which equi-spheres close pack in three dimensions about a centre nuclear sphere with twelve, forty-two, ninety-two surrounding spheres in successive layers is in turn akin to the isotropic vector matrix on which the highly efficient 'octet' space grid is based.

A good example of the inflexible nature of geometrical constraints imposed on structure in 3-space, is provided by one of today's multi-patched, multi-coloured beach balls. Anyone who has such a ball to hand may check that the faceted spherical surface is largely composed of hexagonal patches, with twelve pentagonal patches interspersed at preordained positions. It may be shown mathematically that it is, in fact, impossible to construct a faceted spherical or part spherical surface (a dome for instance) if one has but hexagonal panels at one's disposal. There are always twelve pentagonal panels in a completely spherical ball and a set quanta of hexagons, the number depending on the mathematical concept of frequency.

The apparently random structure of detergent soap froth, in the quasi-static state, has been identified as a conglomerate of all-space-filling (Kelvin) tetra-kai-deca-hera. A knowledge of this fact enabled the author to propose that for the purpose of a finite element analysis the alveoli of the human lung could be modelled as a close-packed cluster of open-ended truncated octahedra.

Practical applications

Elemental problems, akin to those first faced by early masons (desirous of cutting stones at compound angle abound in 3D design. Many of these problems can be solved, to a sufficient degree of accuracy, using the traditional descriptive graphic method, but an ever-increasing number of problems require solutions to a greater accuracy than the descriptive method can provide. Note that the descriptive method is, however, still an indispensable stage in the process.

The calculation and call-up of countless skew lines, planes, intersections and surfaces is an inescapable consequence of complex three-dimensional design and the methods of analytic geometry are indispensable to this end.

The design of shells of exact mathematical type and form; for example the layout of hyperbolic paraboloid, conoid, spherical, elliptical, geodesic cylindrical domes, roofs and vaults is entirely mathematically based but rendered practical by graphic description.

The production of special-purpose mathematical lofts: such as those required for wings, hulls, auto bodies, turbine blades, bottle moulds and shoe lasts is an area of current consideration.

Further applications include the study of articulation and mobility problems as represented by the operation–retraction–stowage, of an aircraft undercarriage, by the design of three-dimensional robotic manipulators, by the design of correctly articulated biomechanical replacement limbs, hip, knee, foot, and hand joints, and by the resolution of spatial aesthetic and ergonomic problems in general.

Virtue of the 3D approach

Although the conceptualisation of 3-space forms is inherently more difficult and more demanding than broadly equivalent work in the $2\frac{1}{2}$D field, and the calculation of stresses, dynamic responses and flows more complex: the analysis of 3D geometrical forms is often more direct than is the plane geometrical method.

The present educational system

The present educational approach to analytic and descriptive geometry and structural form design is low-key and fragmentary; and the scope and interrelationship between these three separately taught subjects is scarcely ever realised.

2D analytic geometry is taught as part of pure mathematics, at A level, plus a little 3D analytic geometry when the syllabus is covered fully. The problems considered at school are always abstract and the average student never finds practical application for the methods he has only partially learned.

The very basic elements of 3D (Mongean) descriptive geometry are learned only by a few (by those who elect to take this option). No more than 25% of students entering Aero Department, Imperial College, have any previous knowledge of the subject.

The undergraduate teaching of 3D analytic geometry falls within the province of the mathematics department and whilst taught to mathematicians is not formally studied in depth by Architects and Engineers.

The undergraduate teaching of 3D (Mongean) descriptive geometry is undertaken by the drawing or graphics department and in recent years the subject has been taught less widely than was previously the case.

T.E.C. students study engineering drawing at various levels and this is important but they do not study descriptive geometry or analytic geometry in depth.

According to Booker there are many practising draughtsmen and hence designers who are unaware of the role to be played by Monge's axioms.

Structural configuration is usefully studied in schools by way of model building (tesselation and polyhedral models) and these important structures are studied in greater depth by Architectural students and by some Structural Engineers, but no formal training in the theory of structural form is commonly given to engineers, though shape and form relevant to particular processes and functions is of course considered.

With regard to private study there are, no doubt, many interested parties who find it difficult and time-consuming to track-down, learn-up and assimilate subjects that were not, or are not, presented to them as a unit, and in the case of some students the mere fact that subjects are segregated into units often presents an impasse. There are also others, older in years, who need help with the basic elements before proceeding to new postgraduate/professional texts, such as

Computational Geometry for Design and Manufacture by Faux and Pratt and *The Analysis of Mechanisms and Robot Manipulators* by Duffy, for example.

Concluding Statement

At the time of writing the in-words are: CAD, CAM, CADM and CAE and it would appear that now is a time of great expansion and expectation in these fields. It is a healthy sign that the young cannot wait to obtain hands-on experience in computer-aided design and manufacture and it is no doubt right for them to do so; but the success of their mission and the extent of their future success undoubtedly depends on a sound understanding of the basic geometrical elements. Many people eager to enter, indeed many already in, the computer-aided field are thin on basic geometrical understanding.

As with any subject, the effect of an inadequate understanding of first principles is to stunt further learning and thence through the agents of frustration becomes the ultimate cause, reason and excuse for 'packing it in'.

What better reason could there be, for all interested parties, to engage themselves in this important study?

1 Descriptive geometry

They seized upon stakes and stones for their purposes. They hammered great stones into sharper shape and when the sparks flew into the dry leaves amidst which they squatted and the red flower of fire appeared, it appeared so mild and familiar that they were not dismayed.

H. G. Wells

1.1 INTRODUCTION

The need to picture one's thoughts, the need to develop these thoughts by means of diagrams and the need to transmit one's design aspirations by means of formal drawings, before cutting wood, stone, metal or plastic has been a self-evident truth for a long time.

The proverbial sketch on the back of an envelope may serve to provide a much-needed instant impression but the value of explicit diagrams, drawn truthfully to scale, should never be underrated. Readers of this book should learn to make their own diagrams and use them both as a means of defining the problem and as a means to its solution. (N.B. The present is a time of transition from traditional (manual) to computer-aided techniques and both these approaches will be considered, the latter techniques in Chapter 11.)

Technical graphics is thus a subject of immense importance to the architect and the engineer. There are of the order of 100,000 drawings required to specify the shape and mode of working of a moderately large aircraft and a similar though less extensive role is played by drawings in other fields. Drawings of the orthographic type are used extensively by all architects and engineers. In addition to their use in a purely communicative role, the production of a near full set of orthographic descriptive drawings remains the principal means, of evolving and solving spatial problems. Descriptive geometry is thus a problem-solving procedure, whereby the plan(s) and elevation(s) of 3D objects are formally delineated and represented in 2-space: on plane sheets of paper or on computer graphics viewing screens.

The principle of aligning the widths of plans and the heights of elevations and locating other features by means of parallel projection is known to have been practised by the painter and engraver Albrecht Durer (1471–1528) who was probably the first to write a book on the subject. The rationale of descriptive geometry is however generally ascribed to the French military engineer and academician, Gaspard Monge (1746–1818).

Monge observed that two orthogonal views (traces) of a 3D object suffice to

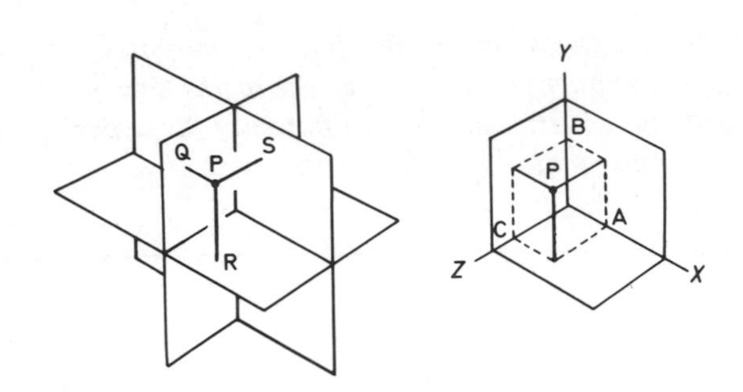
Q
P
S
R
Y
B
P
A
C
Z
X

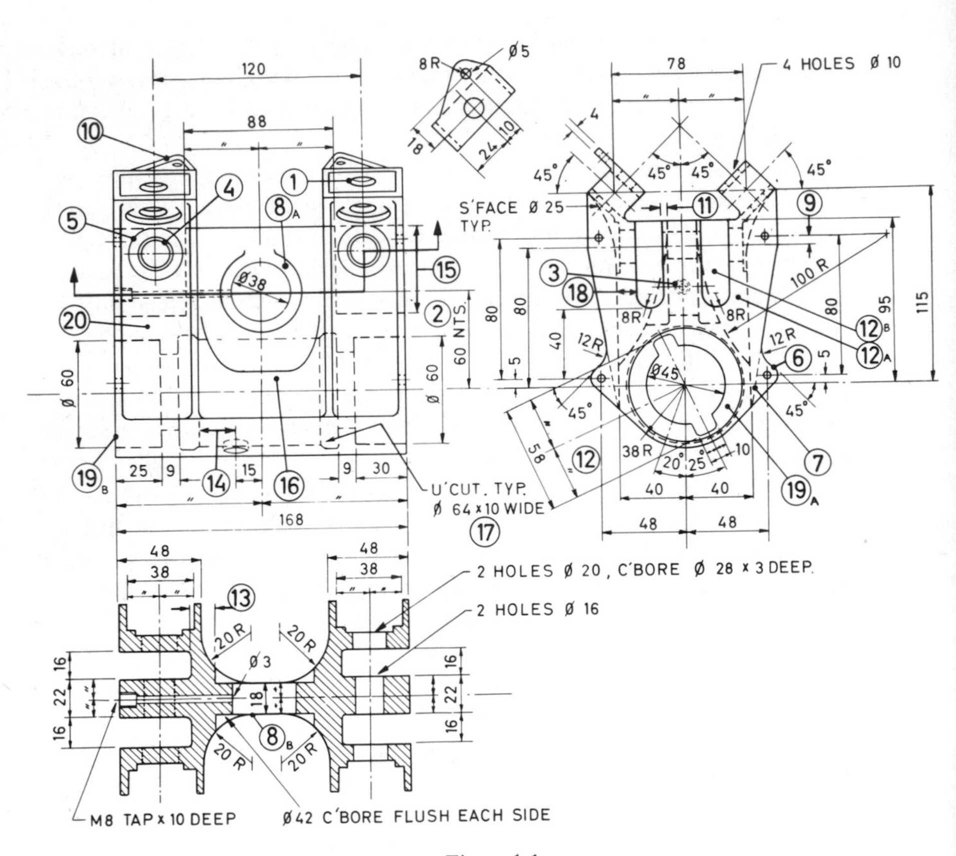
120
88
8R
Ø5
18
24
10
78
4 HOLES Ø 10
4
45°
45°
45°
45°
S'FACE Ø 25
TYP.
Ø38
60 NTS.
Ø 60
Ø 60
80
80
40
5
100 R
80
95
115
12R
12R
8R
8R
Ø45
45°
45°
58
38 R
20°
25°
10
40
40
48
48
25
9
15
9
30
168
U'CUT. TYP.
Ø 64 x 10 WIDE
48
38
48
38
2 HOLES Ø 20, C'BORE Ø 28 x 3 DEEP.
2 HOLES Ø 16
20R
20R
20R
20R
Ø 3
18
16
22
16
16
22
16
M8 TAP x 10 DEEP
Ø42 C'BORE FLUSH EACH SIDE

Figure 1.1

delineate its form. Monge's development of the subject is based on the use of two mutually perpendicular picture planes. A vertical plane on which elevations are drawn and a horizontal plane on which plans are drawn.

In the method proposed by Monge the imaginary or real object (as held in the mind's eye) is placed between the observer's eye and the planes on which the traces are drawn on paper or recorded on a graphics screen. The projectors (that is the lines of sight between eye, object and trace) are deemed to be parallel and this implies that the observer views the object from an infinite distance and hence there is no apparent diminution of size with distance nor distortion of angle.

Simple prismical objects (these sometimes described as being $2\frac{1}{2}$D) can be completely defined (delineated) by the information contained in two mutually perpendicular projections. The delineation of more complex objects which may include planes of various shapes which lie obliquely to the principal lines of vision may necessitate the use of one or more auxiliary planes of projection.

In order to draw two-dimensional traces of three-dimensional objects on two-dimensional paper, or to display these traces on a two-dimensional computer graphics viewing screen, it is necessary to release the 3-space hold on the three mutually perpendicular picture planes.

The system of projection described above is essentially that developed by Gaspard Monge. Monge set up his elevations in a vertical plane (called the VP) and drew his plans in the horizontal plane (called the HP). The essence of the Mongian method is that points on the object may be transferred, by means of parallel projectors, onto two (or three) mutually perpendicular planes of reference, henceforth called picture planes. Monge summarised the power of his orthogonal or orthographic method by stating that planes (by which he meant objects of all kinds) may be treated by their traces. Ruled surfaces treated by the projections of their rulings, surfaces of revolution by means of concentric circles in plan and a single median in elevation.

If we view a point on the object, called the object point P, Fig. 1.1, from an infinite distance, then the line of sight and hence the projectors PS, PR, PQ will in fact translate the space point P to the so-called trace positions S, R, Q, when the point P is viewed in mutually perpendicular directions. It follows that QP = BS = OA, RP = AS = OB, SP = BQ = OC. (N.B. OA, OB, OC, are the Cartesian coordinates of the point P.)

1.2 FIRST ANGLE ORTHOGRAPHIC PROJECTION

In the method described above the real or imaginary object is placed between the observer's standing point (eye) and the picture planes onto which the traces of all object points are drawn and this way of viewing an object is known as first angle orthographic projection.

If as in Fig. 1.3, the point P represents the front near-side head lamp and the point Q represents the rear off-side stop lamp of a motor vehicle we see that points P and Q are uniquely defined by their traces and the circuits made by their respective projectors should be carefully observed.

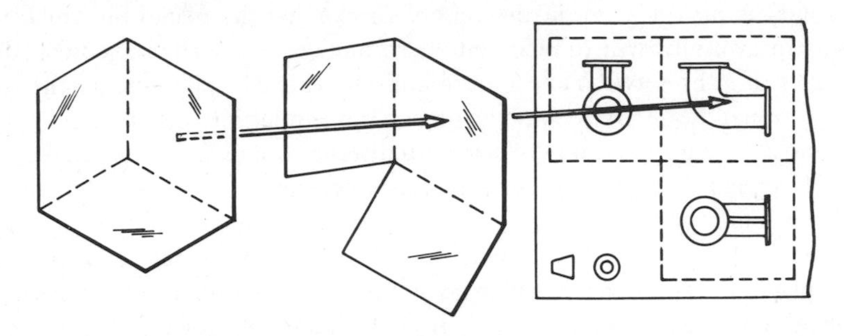

Figure 1.2

If the plane of the drawing 2-space is to contain the origin O, then in first angle projection the three mutually perpendicular picture planes, Fig. 1.2, hinge down away from the eye until they lie flat in the plane of the paper. When the front of the car in side elevation is to the left, as shown in Fig. 1.3, the frontal view of the car is drawn on the right and conversely the rear of the car is drawn on the left. A bird's eye-view showing the top of the car is drawn below the side elevation as shown in the figure.

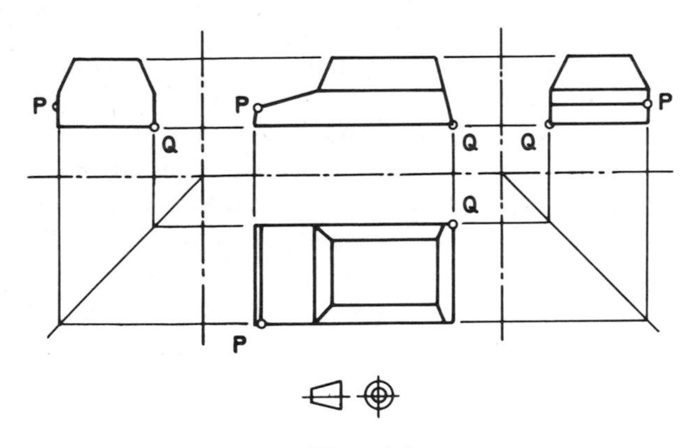

Figure 1.3

The British Standard BS 308 defines first angle projection as follows:

BS 308 definition of first angle projection: In first angle projection each view shows what would be seen by looking on the far side of an adjacent view. The BS symbol for first angle projection is as shown in the figure.

1.3 THIRD ANGLE PROJECTION

When using the third angle system we place the object in the diagonally opposite (hidden) octant of Fig. 1.1, and in order to view the object when in this octant of space we deem the planes of projection to be transparent. The third angle system thus contrasts with the first angle system in that the picture planes are placed between the observer and the object.

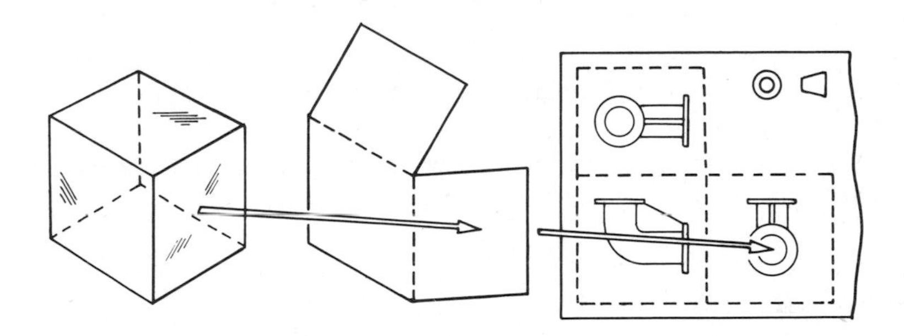

Figure 1.4

If the plane of the drawing 2-space is to contain the origin O, then the mutually perpendicular planes of reference hinge up towards the eye.

As seen from Fig. 1.5, the spatial position of P is located in third angle projection by exactly the same procedure used in first angle projection. Thus when the front of the car in side elevation is to the left, as in Fig. 1.5, the frontal view of the car is drawn on the left (adjacent to the feature to which it refers). The rear view and the bird's eye plan view, likewise appear adjacent to the features to which they refer.

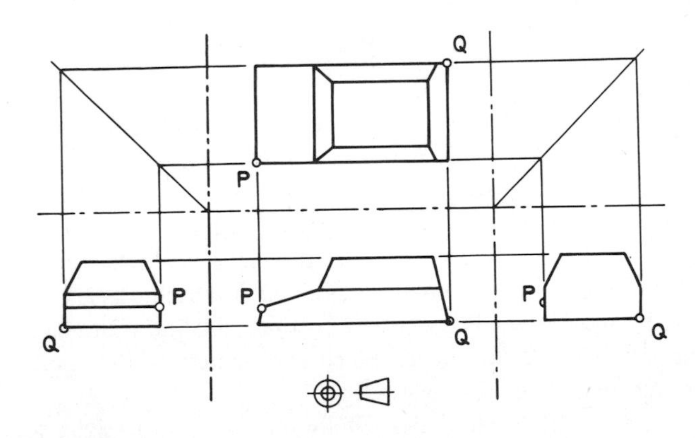

Figure 1.5

The British Standard BS 308 defines third angle projection as follows:

BS 308 definition of third angle projection: In third angle projection each view shows what would be seen by looking on the near side of an adjacent view. The BS symbol for third angle projection is as shown in the figure.

1.4 VISIBILITY IN ORTHOGRAPHIC DRAWINGS

It is customary in both first and third angle projection to draw the side elevation of an object the right way up, that is the roof of a house is towards the top of the paper.

In first angle projection a bird's eye-view of the roof is drawn below the side elevation – this bird's eye-view is called the plan.

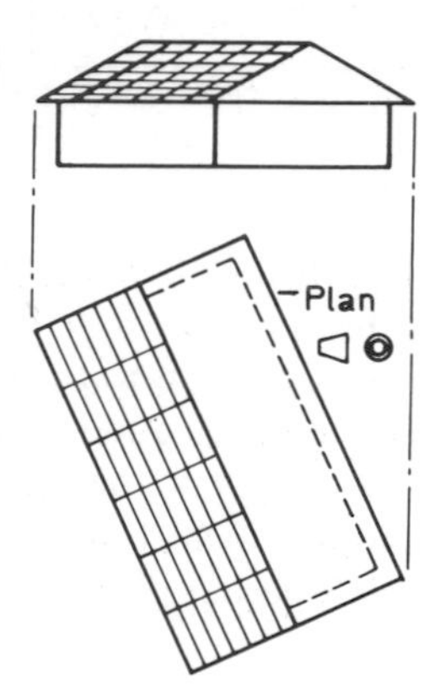

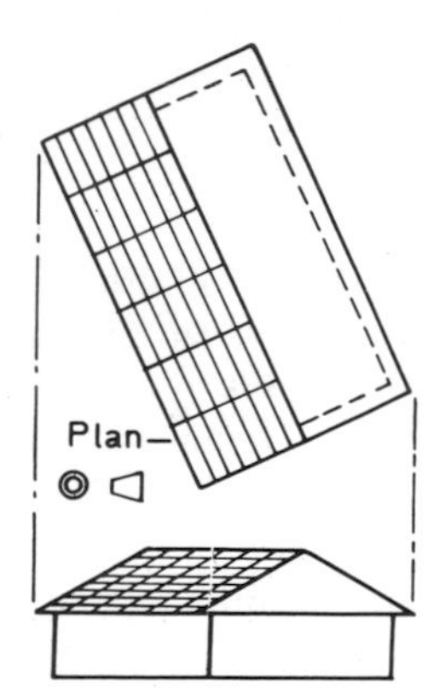

Figure 1.6

In third angle projection a bird's eye-view of the roof is drawn above the side elevation and this bird's eye-view is also called the plan.

We observe from Fig. 1.6, that side view and plan are identical in either case, only their respective positions (on the paper) are different.

Referring now to Fig. 1.7, we see that the circle is above the square in side elevation and when viewed from above, the circle obscures part of the square. The part of the square which is exposed to the sight line is drawn in a full continuous line. The part of the square which is obscured by the circle is indicated by means of a broken line, as shown in Fig. 1.7.

Any geometrical detail which lies wholly exposed to the sight line is called a visible feature. Any geometrical detail which is obscured by the presence of

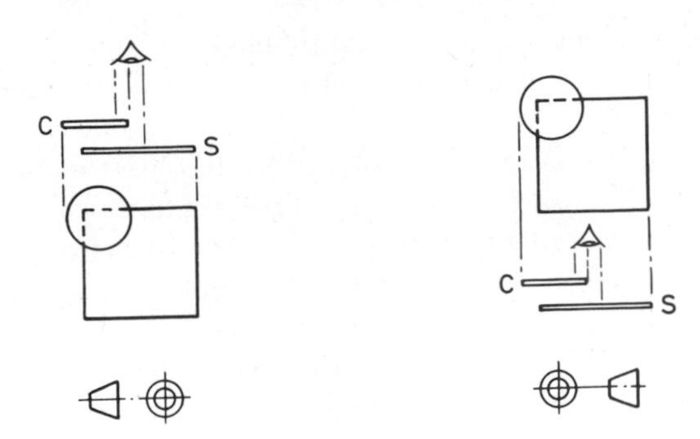

Figure 1.7

another detail (and is not therefore directly visible) is called a hidden detail and is indicated where necessary by a broken line.

The rule for depicting visible and hidden detail is a full line when visible and broken line when obscured.

1.5 SECTION PLANES

In many complex objects the amount of hidden detail is considerable and to include it all would lead only to confusion. In practice the amount of hidden

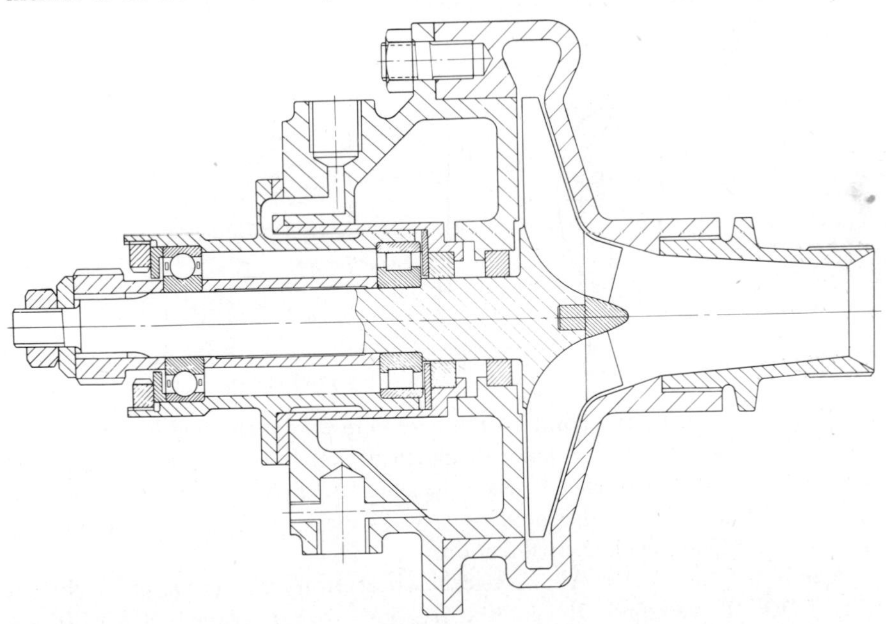

Figure 1.8

detail shown on any one drawing or view should be limited to that which is considered essential. In cases where the build-up of hidden detail cannot be avoided section planes are used.

A section plane is akin to the act of cutting an object in two, removing that piece of the object which is obstructing the eye and viewing the cross-section revealed by the cut. In order to distinguish section planes from outside views, all cross-sections are hatched as shown in Fig. 1.8, preferably with hatching lines at 45°. When two adjacent mating parts are sectioned, as would be the case in an assembly, the hatching lines associated with adjacent parts should slope, as far as is possible, in opposite directions. A single example, Fig. 1.8, should clarify the issue.

1.6 PROPORTIONAL DEVELOPMENT

Given two cross-sections of predetermined shape, the method of proportional development enables intermediate cross-sections to be developed in accord with the rules of linear interpolation.

If we consider the case of an elliptical based cone then one end cross-section is a point, the other end a given ellipse. All lines from points on the ellipse which home on the apex are straight lines, called generators and all intermediate cross-sections are ellipses of similar aspect ratio. (NB. Readers not conversant with orthographic notation are referred to the summary article 1.8.1.)

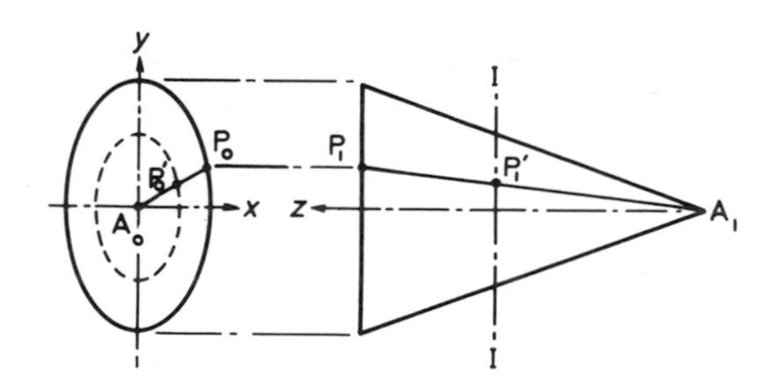

Figure 1.9

If, in Fig. 1.9, P_0 is a point on the base ellipse then the line P_0A is the end view of a generator P_1A_1. A point P_1' may legitimately be regarded as a point on the cross-section II and since P_1' is on the generator P_1A_1 it projects to the point P_0' on the line P_0A_0. Further points such as P_1' may thus be used to build up the profile of any intermediate cross-section.

Given the base curve AB we select two arbitrary points O′, O″ such that OO′ = OO″. If we require the contour (section) corresponding to $a/A = b/B = \frac{1}{3}$, then we make oa = OA/3 and ob = OB/3.

Next we choose any arbitrary point P_0 on the base curve AB and draw in the line OP_0. We project P_0 onto the line AO′, onto the line aO′, onto the line OP_0, as shown by the arrows. We may also (as a check) project P_0 onto the line BO″ onto the line bo″ onto the line OP_0, but this check may not always be necessary.

If we make OO′ and OO″ of different lengths then we have the means of producing a progressive but different rate of thinning on either axis. The method may also be applied to curved profiles as shown in Fig. 1.10.

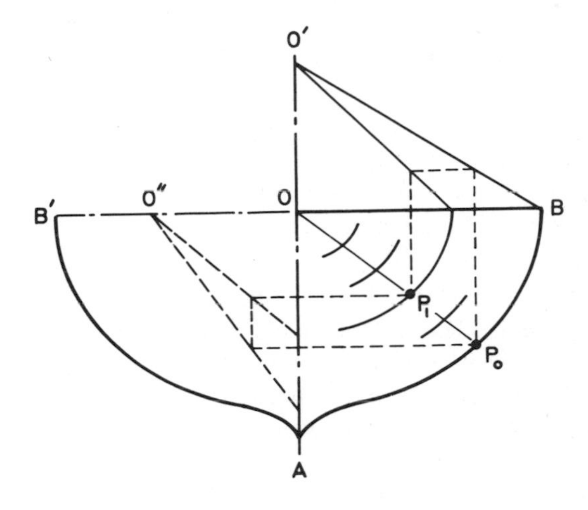

Figure 1.10

Prior to the evolution of computer-aided methods, proportional development was widely used in the shape design of ships, motor vehicles and aircraft.

1.7 TRUE SHAPE AND AUXILIARY PLANES OF PROJECTION

The need to introduce auxiliary planes of projection on which to draw the true lengths of lines and the true shapes of planes which appear inclined in both plan and elevation, becomes apparent on asking the question, what is the true shape of the hatch back on the simplified motor vehicle, Fig. 1.11. The only way we can view the flat hatch back surface, true shape, is to look in a direction normal to its surface and in order to do this we must set up an auxiliary plane parallel to the flat surface, the true shape of which is required. The true shape of an inclined flat surface may thus be obtained by direct projection.

1.7.1 Cams – a further instance of true shape layout

Cam profiles shaped in such a way that they impart a prescribed motion to a following element are widely featured in machines of all kinds. A simple instance

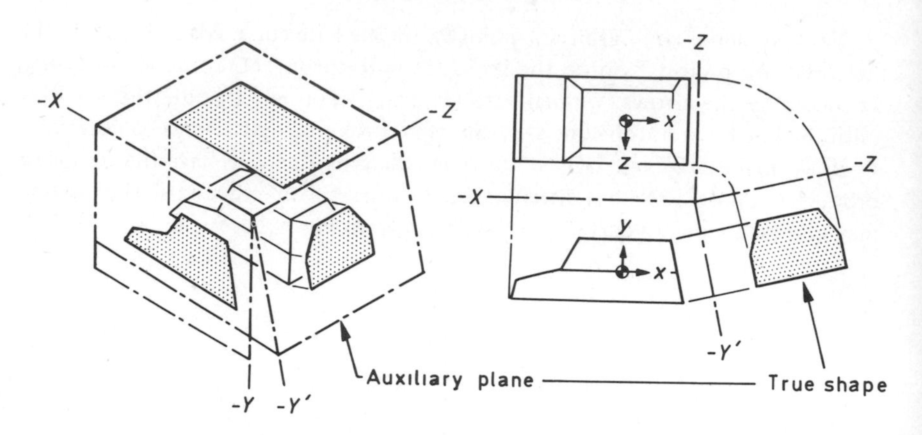

Figure 1.11

would be a cam so shaped that it imparts simple harmonic reciprocating motion to an on-centre stem follower. A point is said to move with simple harmonic motion when its acceleration is proportional to its displacement. Reciprocating S.H.M. is modelled by considering the linear motion of the projection of a point which revolves about a centre at a constant radius and at a constant speed. In this way the motion of the follower is thus defined. The problem is then to translate the pre-defined motion of the follower (the pre-defined motion need not be simple harmonic) into a practical cam profile which will do the job. Leaving aside kinematic considerations which in any particular problem influence the maximum speed, total lift and minimum radius, etc., the problem of translating a rectalinear motion into a rotary cam profile is solved as explained below.

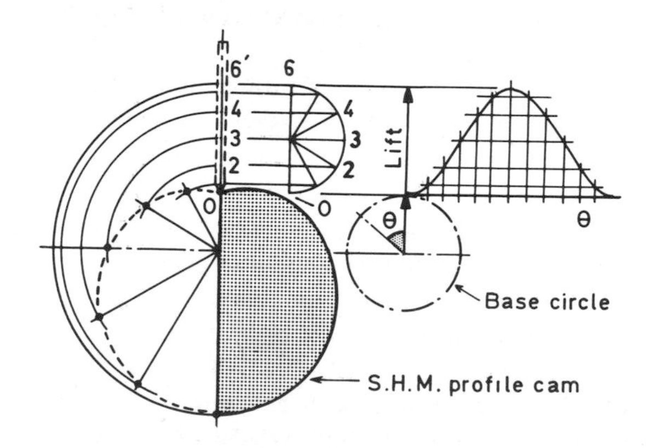

Figure 1.12

With reference to Fig. 1.12, we recall our definition of simple harmonic motion. We draw a half circle of diameter 0–6 equal to the total lift. We divide the half circle into six (or more) equal angular divisions. The projections 0, 1′, 2′, 3′, 4′, 5′, 6′ then represent the follower lift at equal increments of rotation. We next decide on a suitable minimum cam radius and set up the cam centre at C, vertically below F_0. Using the line CF_0 as zero datum we draw in twelve radial lines CF_1, CF_2, CF_3, etc., at equal angles. The intersection(s) of these lines with the arcs CF'_1, CF'_2, CF'_3, etc., yield points on the profile required.

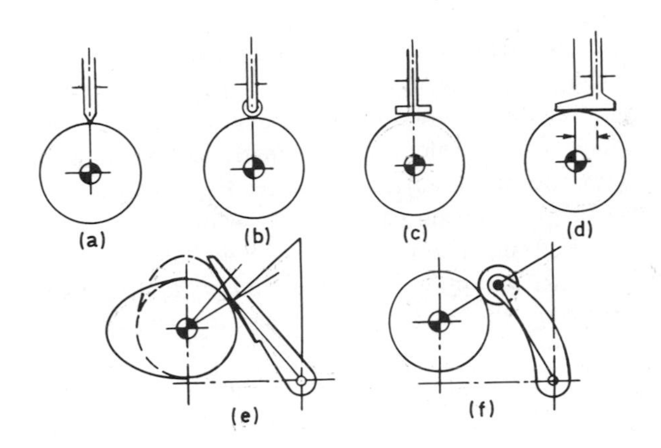

Figure 1.13

1.8 PURE DESCRIPTIVE GEOMETRY

The problem-solving procedure known as descriptive geometry enables the multiplicity of true lengths, true angles, true shapes, lines of intersection and lines of interpenetration, which occur in complex three dimensional forms to be displayed and their magnitudes determined by graphical means.

1.8.1 Orthographic notation

A formal notation whereby points, lines and planes can be readily identified in different orthographic views is often found useful.

1. A point in 3-space is identified by an upright Roman unscripted capital letter. Thus the letters A, B, C, . . ., P denote the actual positions of real object points or features in space.
2. The parallel projection of P onto the first plane of projection may be identified by either P_1 or p_1.

3. The parallel projection of P onto the second plane of projection may be identified by either P_2 or p_2.
4. The parallel projection of P onto the third plane of projection may be identified by either P_3 or p_3.
5. The call up of points, projected into auxiliary planes, follows an ascending order.
6. The line which separates one projection (picture) plane from another is called a ground line. The ground line between plane 1 and plane 2 is commonly referred to as ground line 1, 2.

1.8.2 Summary of the projective properties of lines and planes in 3-space

(a) Any line which lies parallel to a projection plane is seen true length in the adjacent view, see Fig. 1.14(a) unless it is viewed end on.

(b) Any line which is viewed along its true length is seen end-on and projects to a single point or dot in the adjacent view, see Fig. 1.14(b).

(c) Parallel lines in space remain parallel in all orthographic views except when the lines are viewed end-on in which case they are seen as two separate points or dots. The other exception is when the lines are viewed so that one line lies directly behind the other, in which case a single line only is seen, see Fig. 1.14(c).

(d) The 90° angle between two perpendicular lines will be seen in every view in which at least one of the lines appears true length. The sole exception to

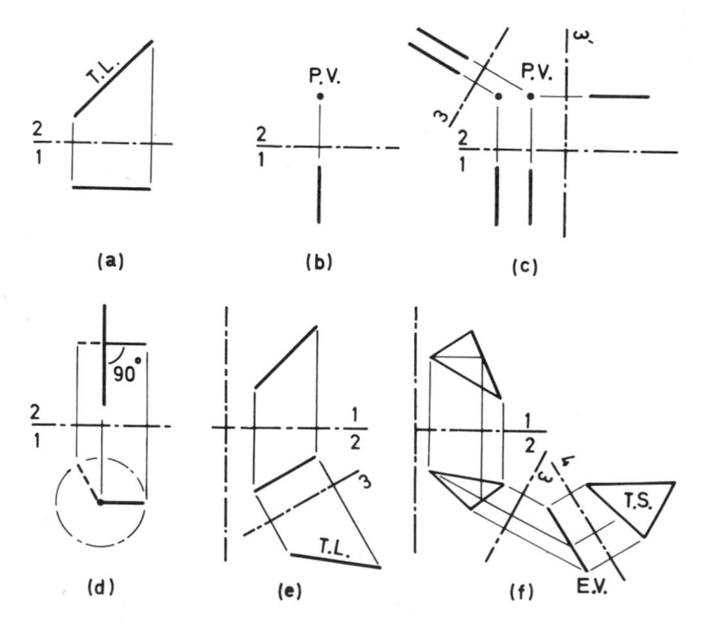

Figure 1.14

this rule is when one of the lines is true length the other a point view, see Fig. 1.14(d).

(e) Any line which lies obliquely to two mutually perpendicular picture planes may be viewed true length by setting up an auxiliary picture plane which lies parallel to the line in either of the given orthographic views. The rule case (a) then applies. See Fig. 1.14(e).

(f) Any plane which lies obliquely to two mutually perpendicular picture planes may be viewed true shape by projecting, first an auxiliary edge view and then projecting a second auxiliary view (normal to the edge view) which yields the true shape of the plane required. The method and truth of (f) will be demonstrated after a few relatively simple examples have been considered.

Example

The true shape and several edge views of a circle, the true shape of which lies parallel to a projection plane is shown in Fig. 1.15. The following points shall be observed: (1) the constant distance of all edge views from their accompanying ground lines, (2) the projective relationship of a typical point P on the circumference and its orthographic traces.

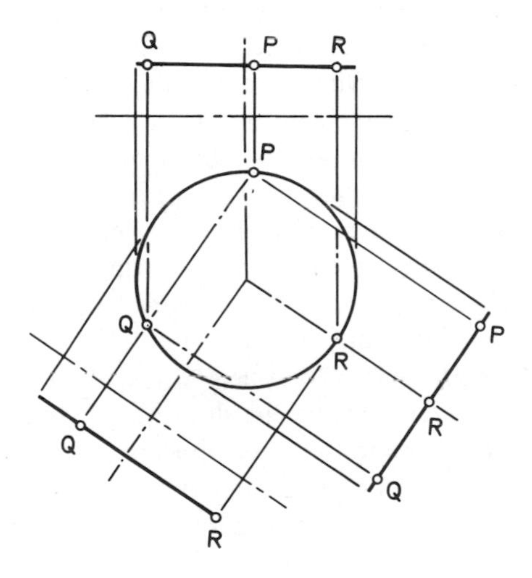

Figure 1.15

Example

In Fig. 1.16, the edge view of a circle of known diameter is inclined to a picture plane and the problem is to derive its projected shape in the adjacent view.

We know that the given line is the edge view of a circular disc, such as a coin for example. We know from experience, or by trial, that when we look at a circular disc in a direction normal to its surface, we see its true shape circular

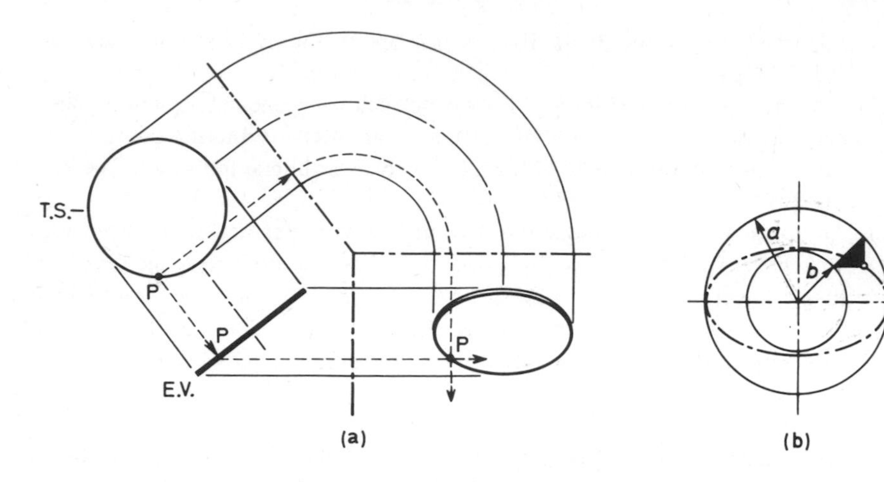

Figure 1.16

profile. As we tilt the circle (top towards or away from the eye) the apparent height of the circle or coin becomes less and less as we tilt it but the width never changes. The apparent shape of the surface is transformed from a full circle, through a continuous series of slimmer and slimmer elliptical stages, ultimately becoming an edge view before re-emergence as an inverted obverse.

We have seen that any point on an object may be projected to a unique position and it is a matter of pure convenience that we choose to divide the periphery of the true shape circle (which is projected in a direction normal to the given edge view) into twelve equal sectors giving twelve points which lie on the circumference. These twelve points may then be projected, along with their corresponding edge view positions (points) to obtain the apparent shape as seen in the direction of the arrow A. The projected shape is, of course, an ellipse, the minor axis of which is determined by the slope of the given edge line. The major axis lies square-on to the eye and therefore appears true length.

The method employed may be applied to any shape or solid providing sufficient information is given, and there are a wide range of recurrent problems the solution to which follows a more or less standard procedure. We consider a number of the most important cases.

1.8.3 The true length of a line which is parallel to a projection plane

The true length of a line may only be viewed true length when the length of the line is set normal to the line of vision. A line in an orthographic projection will project to a true length, if and only if, it lies parallel to the picture plane onto which it is projected.

Consider the much simplified space capsule shown in Fig. 1.17, in which the views 1, 2 and 3 are given. The antenna AB is clearly shown in all three views but in no view is its true length given (and the same is true of the sloping front-screen and hatch back posts of the motor vehicle, Fig. 1.11).

The true length of the antenna AB may be obtained by setting up an auxiliary plane parallel to one orthographic view of the antenna. The line A_3B_3, Fig. 1.17, is the true length in this instance.

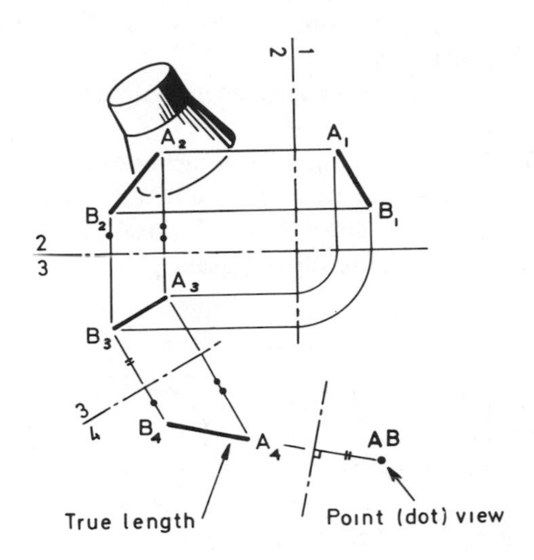

Figure 1.17

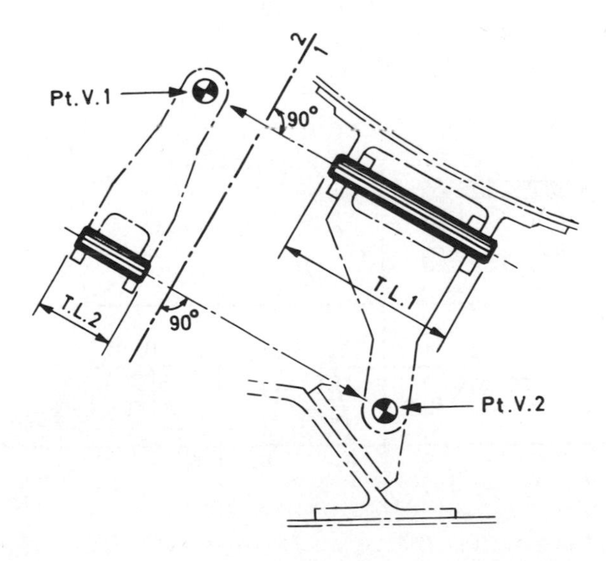

Figure 1.18

1.8.4 The end view of a line is a dot

The dot, or point view, of a line is seen when the line is viewed along its length.

A line in an orthographic projection will be seen as a dot, if and only if the line lies normal to the plane onto which it is projected. See Fig. 1.18.

1.8.5 The true length of an oblique line by rotation

When a line in an orthographic projection lies skew to two reference planes its true length may be obtained by rotation.

The principle of rotation was enunciated by Gaspard Monge and is employed as follows.

When a line such as $A_1 B_1$, in Fig. 1.19, is inclined to two adjacent projection planes its trace in one view may be rotated so that it (the trace) becomes $A_1 B_1'$ parallel to the common groundline. The orthogonal projection of the rotated trace projects to true length in the adjacent plane.

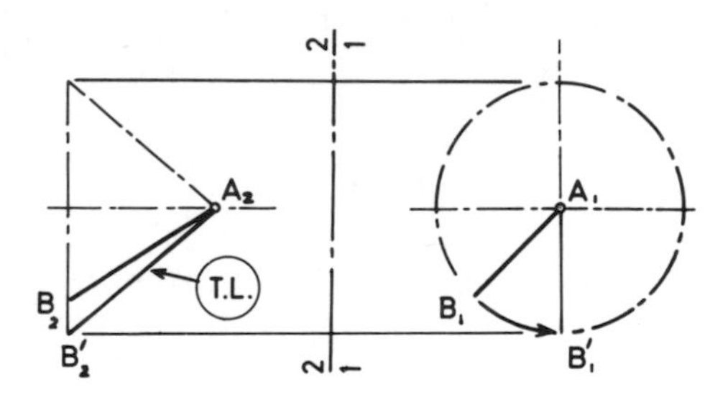

Figure 1.19

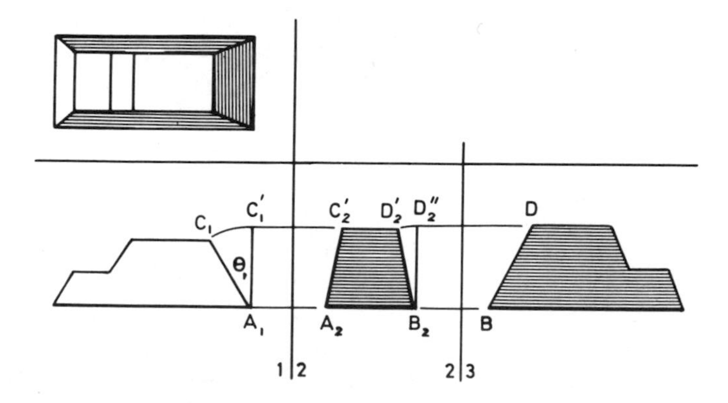

Figure 1.20

1.8.6 Practical application of rotation principle

With reference to Fig. 1.20 we see that the hatchback of the vehicle is represented by the line A_1C_1. When A_1C_1 is rotated through an angle θ_1, it becomes $A_1C'_1$, and since $A_1C'_1$ lies parallel to the ground line, its orthogonal projection into the adjacent plane yields the line $A_2C'_2$, which is the true length of AC. The true shape of the hatchback is therefore as shown.

The true shape of the side panel is obtained by rotating $A_2C'_2$ or $B_2D'_2$ through the angle θ_2. The true shape of the side panel is shown in Fig. 1.20.

1.8.7 The edge view of a plane

Whilst a plane may be of any shape, three points suffice to define the lay of a plane in 3 space. These three points may be joined by three lines to form a triangle and any orthographic projection of this triangle serves to identify the plane to which these three points belong.

The true shape of a plane is only visible when the plane is viewed in a direction normal to its surface.

The three points A, B, C, Fig. 1.21, clearly define a plane which lies obliquely to the three principal planes of reference.

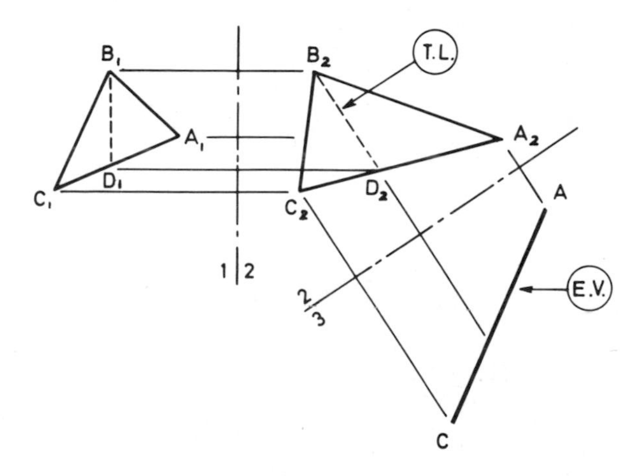

Figure 1.21

We solve this problem by recalling that a line which lies parallel to a reference plane appears true length in the adjacent view. We therefore introduce a line M_1N_1 into the orthographic plane π_1, which is parallel to the adjacent ground line. If the point M_1 lies on the line A_1B_1 and the point N_1 lies on the line B_1C_1, the line M_1N_1 lies in the plane defined by the points A_1, B_1, C_1. The projected points M_2N_2 therefore lie on the lines A_2B_2, B_2C_2 and hence the line M_2N_2 is the true length of the line MN, and since MN lies in the plane of the triangle ABC, we may obtain the edge view of the plane ABC by viewing the true length

M_2N_2 end on. We use the same procedure employed to obtain the point view in article 8.4 but in this instance, we repeat the procedure three times over: once for each cardinal point. If accurately draughted the points A_3, B_3, C_3, Fig. 1.21, should lie on a straight line and this straight line is the required edge view of the plane ABC.

1.8.8 The true shape of a plane

Having obtained the edge view of a plane, using the procedure of article 8.7, we may obtain the true shape of the plane by viewing its edge view at right angles. The procedure for erecting auxiliary planes and setting up ground lines should now be clear and the true shape of the plane defined by the three points A, B, C, is as shown by the projected points A_4, B_4, C_4, Fig. 1.22.

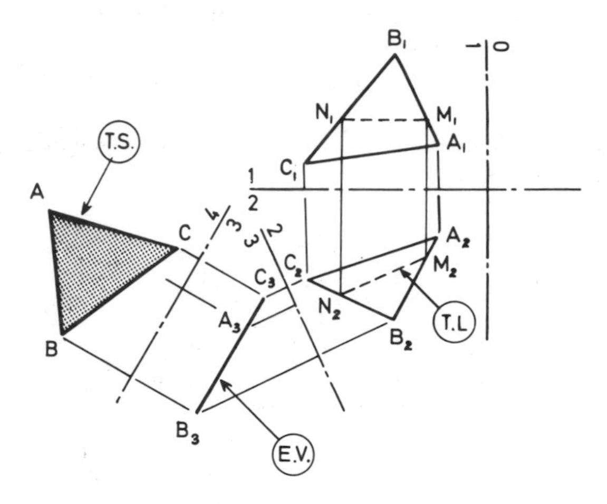

Figure 1.22

1.8.9 The intersection of a line and a plane

When the direction of a skew line and a skew plane are defined by their orthographic projections (traces), the point at which the line pierces the plane may be determined as follows:

Let a line AB be defined by its orthographic projections and a skew plane be defined by three points R, S, T. Using the procedure of article 1.8.7, we prepare an edge view and thence using the procedure of article 8.8 produce the true shape of the plane. We then superimpose the orthographic projections of the line as shown in Fig. 1.23.

We note the apparent intersection between the line and the plane at the point P_1, but do not know (at this stage) whether the line AB lies in front, behind or through the limited plane RST. We therefore assume that point P_1 lies on the line and we obtain its orthographic position P_2, by projection. We see from

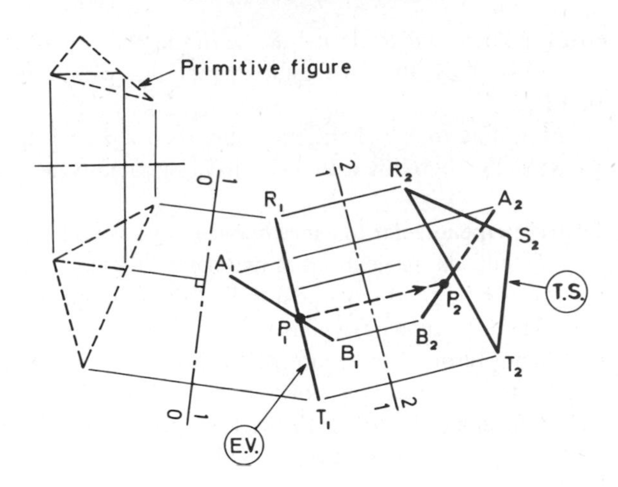

Figure 1.23

Fig. 1.23, that whilst P_2 is on the line A_2B_2 (as it should be) the point P_2 lies outside the triangle $R_2S_2T_2$, and therefore the line AB does not pierce the limited plane defined by RST.

If the points RST represented the corner points of a triangular bulk head and the line AB was a control cable or pipe line, then as shown above the cable or pipe line would be clear of the bulk head and no provision to pass through the bulk head would have to be provided. If, however, the three points RST were

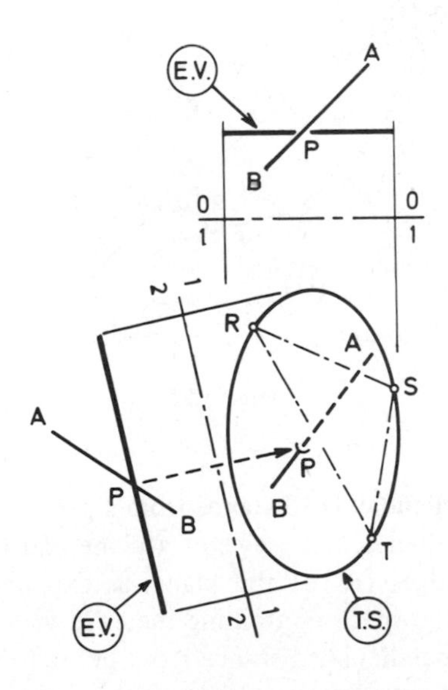

Figure 1.24

purely geometrical points used to define the lie of the plane of an elliptical bulk head, as shown in Fig. 1.24, then the cable or pipe line would pierce the bulk head at the point P_2.

We conclude by noting that had the projection of P_1 placed the point P_2 inside the triangle $R_2S_2T_2$, then this would also have signified penetration.

1.8.10 The shortest perpendicular distance between two skew lines

Given mutually perpendicular projections of two skew lines we set up an auxiliary plane (ground line) parallel to one of them so that the subsequent projection (in which the second line also appears) yields the first line true length. We set up a second auxiliary plane (ground line) normal to the true length of the first line and thereby obtain its point view as shown in Fig. 1.25. A line from the point view which makes an angle of 90° with the final projection of the other line, gives the shortest distance (or clearance) required.

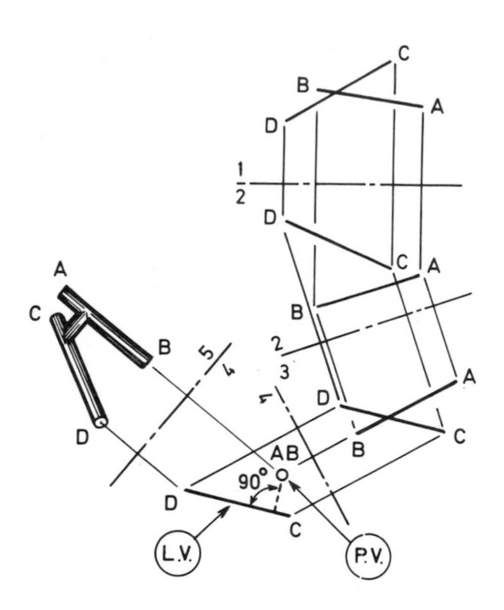

Figure 1.25

1.8.11 The shortest perpendicular distance from a point to a plane

Given two mutually perpendicular views of a plane and the position of a given point we obtain the edge view of the plane, as explained in article 8.7, and project the point into the space containing the edge view of the plane as shown in figure 1.26. The perpendicular distance from point to plane is the line which passes through the point and cuts the plane at right angles.

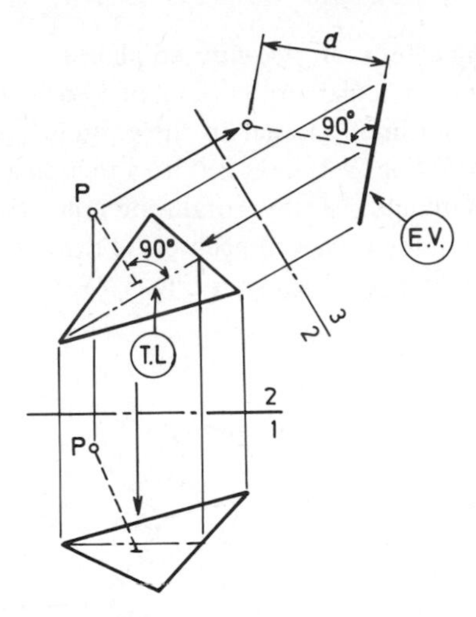

Figure 1.26

1.8.12 The angle between a line and a plane

Given two mutually perpendicular views of the given plane and line, we use the method of article 8.7 to obtain an edge view of the plane and the method of article 8.8 to obtain its true shape (by which we mean the shape inferred by its three defining points). We project the line into each of these auxiliary views as shown in Fig. 1.27, and then set up a new auxiliary plane parallel to the line (as it appears in the true shape projection) and produce the final view. The angle between the true length line and edge view is the true angle required.

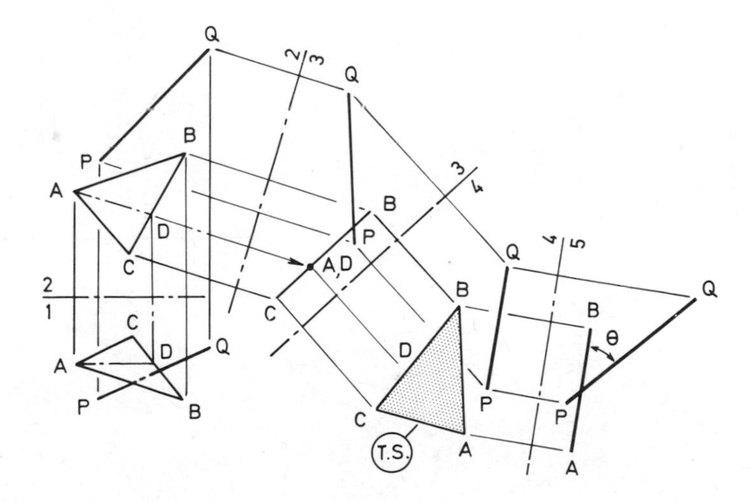

Figure 1.27

1.8.13 The dihedral angle between two limited planes

Given two mutually perpendicular projections of two inclined planes, we set up an auxiliary plane (ground line) parallel to the given intersection and obtain the true length of the intersection. We, next, set up a second auxiliary plane (ground line) normal to true length and thereby obtain the point view of the intersection. We project the three defining points of each plane throughout and the final view presents the true dihedral angle as shown in Fig. 1.28.

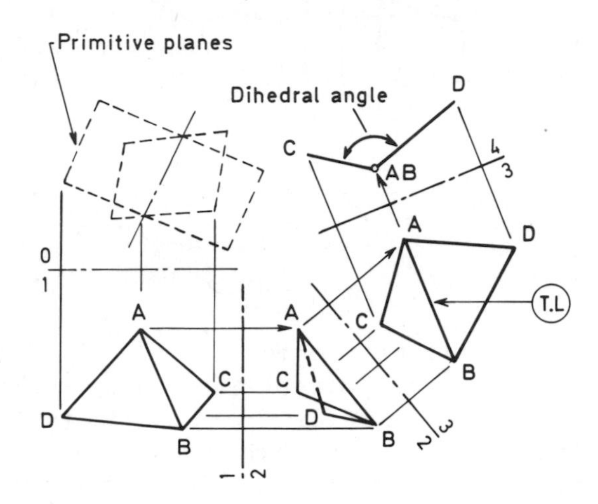

Figure 1.28

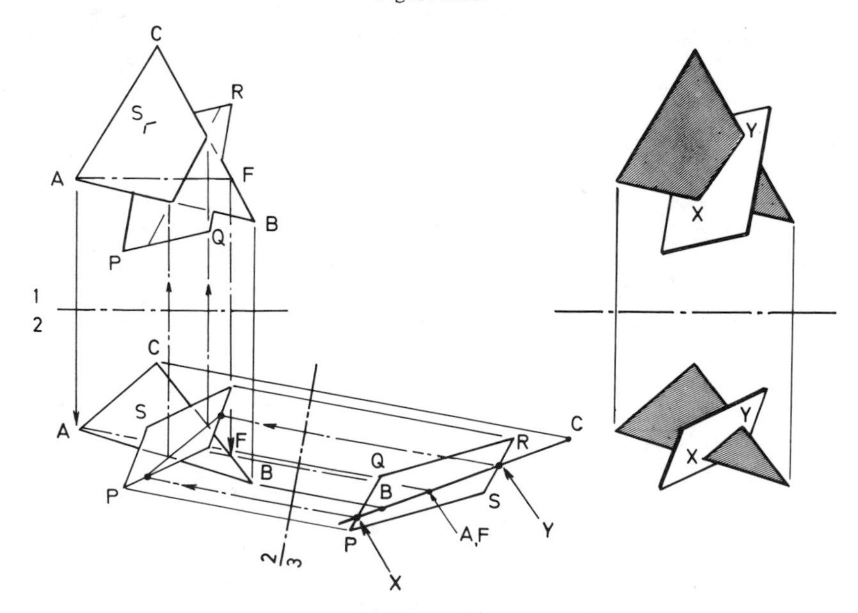

Figure 1.29

1.8.14 The line of intersection between two limited planes

Given planes ABC and PQRS as shown in Fig. 1.29, the line of intersection may be determined (if it exists) as follows.

Draw a frontal line A_1F_1 in plane π_1 parallel to the common ground line. The projection of A_1F_1 appears true length in plan. Set up an auxiliary plane normal to A_2F_2 and thereby obtain the edge view of the plane π_1 (the points A_3F_3 are coincident).

Project plane π_2 into the auxiliary plane and mark the two points X_1, Y_1 (if they exist). The line of intersection between planes π_1 and π_2 is thus obtained by projecting the points X_3, Y_3 back into the given plan on elevation as shown in Fig. 1.29.

1.8.15 Affine properties, an aid to pictorial presentation

The affine (similarity) properties of a regular pentagon are shown in Fig. 1.30(a). The dimensional properties of the same pentagon are shown at (b).

When the three plane pentagons shown at (c) are folded to form a typical trihedral apex they constitute part of a dodecahedron, see Chapter 6. The vertex A′ merges with the vertex A and the vertex F′ merges with the vertex F. Assuming the edge DE remains parallel to the picture plane the face dimension A′F′ remains a true length dimension. A′F′ is equal to the face dimension AC, where $AC = 2s \cos 36 = 1.618035$ and the face dimension $BN = s \sin 36 + s \sin 72 = 1.53884s$.

The orthographic representation of a dodecahedron may be produced by starting from a true shape pentagon face $A_1B_1C_1D_1E_1$ and from this face an edge view B_2N_2 representing the face dimension B_1N_1 may be projected directly, as shown in Fig. 1.30(c).

The pentagonal face $D'_1E'_1F'_1G'_1H'_1$ projects to the edge view $N'_2G'_2$. In a dodecahedron the face containing N_2G_2 lies at an angle as shown in the figure. The angle ψ is determined by the rotation of the edge line N_2G_2, and its intersection with the line A_2F_2. The length of A_2F_2 is equal to AC and hence the angle ψ is determined.

Applying the cosine rule to the triangle $A_2N_2F_2$:

$$4\,s^2 \cos^2 36 = 2\,s^2 \sin^2 72\,(1 - \cos\psi)$$

whence

$$\cos\psi = 1 - \frac{2\cos^2 36}{\sin^2 72}$$

$$= -0.44721$$

$$\psi = 116.56482°$$

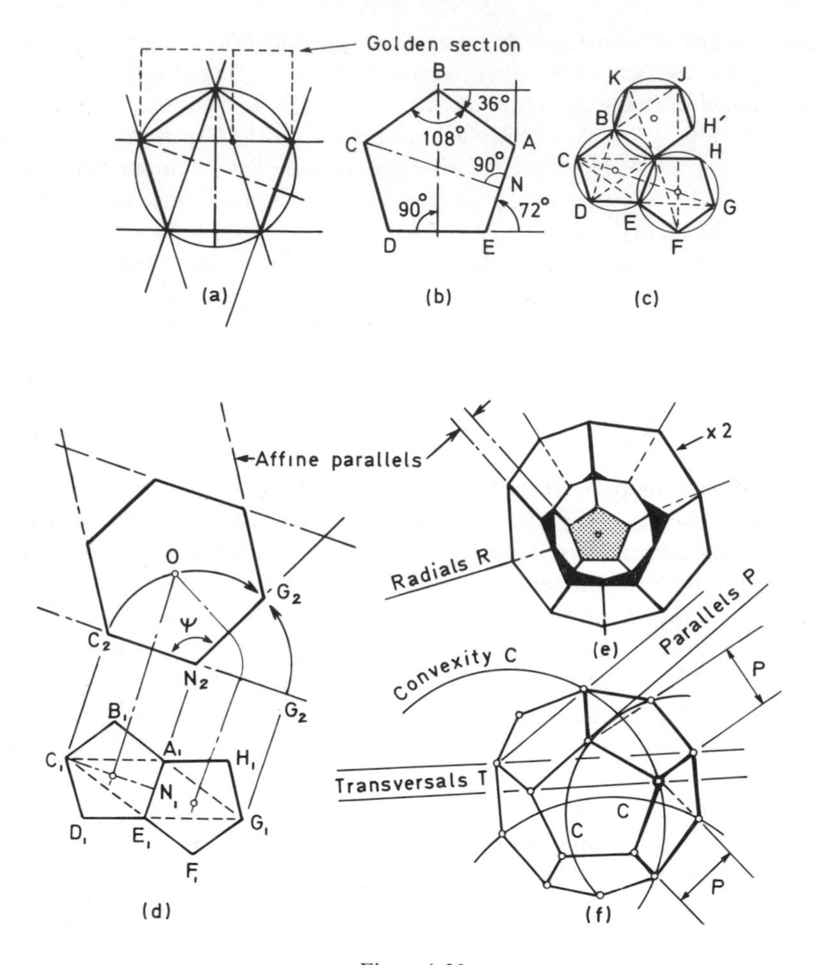

Figure 1.30

The lines A_2F_2 and B_2G_2 are normal to the line N_2P_2 which bisects the angle.

The centre O, projects to the centre O_2, where O_2 lies on the line N_2P_2.

The line B_2K_2 passes through O_2 and since segments B_2O_2, O_2K_2, are equal, the position of K_2 and the edge length G_2K_2 is determined. The remainder of the figure is easily completed by virtue of the affine property which dictates that opposite faces and edges are parallel.

A view of the image $\pi 2$ may be projected in any direction to produce the image $\pi 3$. Thus the angle ω is an arbitrary angle of tilt in Fig. 1.30.

The points $(A_3B_3C_3D_3E_3F_3G_3H_3)$ lie on projectors perpendicular to the ground line. The width dimensions A_3C_3, E_3D_3, F_3H_3, lie parallel to the picture plane and therefore appear true length. The twin faced profile $(A_3B_3C_3D_3H_3G_3F_3E_3)$ is thus completely determined.

Since the line containing D_2G_2 makes an angle ω with the ground line the projected face $(D_3E_3F_3G_3H_3)$ is not a true shape.

The remaining visible faces may be completed by using the affine requirement that all edges are parallel to their opposite diagonals, and all radial edges pass through the centre O_3. The diagonal C_3I_3 is thus parallel to the edge D_3H_3 and other affine properties (indicated by dashed lines) may be used to check the accuracy of the figure, and hidden detail may similarly be included.

Affine properties may thus be used to aid the construction of the regular and semi-regular figures and affine properties may similarly be used as an aid to the graphical representation of various types of segmented structures.

1.8.16 The tangent plane to a sphere

The traces of a plane which is tangential to a given sphere may be obtained by the following construction.

If the plan and elevation of a sphere are as shown in Fig. 1.31, and P is any point through which the tangent plane is to be directed, then we proceed as follows. Draw a radial line through the points P_1 and P_2 in plan and elevation. Draw a line P_1Q_1 perpendicular to the radial line in plan, to intersect the ground line at Q. Erect a perpendicular Q_1Q_2. Draw a line through P_2 (in elevation) parallel to the ground line to intersect the perpendicular Q_1Q_2 at Q_2. A line perpendicular to C_2P_2 drawn through Q_2 intersects the ground line in F. The line through FQ_2 and the line FH parallel to P_1Q_1 (in plan) are the vertical and horizontal traces of the tangent plane through the point P. The true angle of this plane may be obtained by taking an auxiliary view as usual.

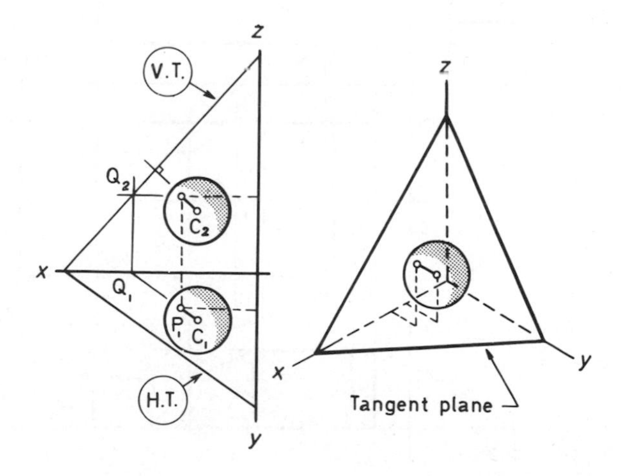

Figure 1.31

1.9. SURFACE DEVELOPMENTS

When making three-dimensional forms from folded or rolled flat sheet, it is necessary to determine all the true lengths, true angles, true shapes involved. The layout of these true shapes is called a surface development. Except for a small bending allowance which takes count of the slight local plastic deformation which necessarily occurs when a material is bent, the surface shapes of the development are exact replicas of those on the three dimensional form.

Example

If we refer to the model auto vehicle, Fig. 1.11, we observe that whilst the roof and lower side panels are delineated true shape in plan and side elevation and the true shape of the hatch back is as shown by the auxiliary view, the upper side panels as also other areas are nowhere drawn true shape and to this end additional auxiliary planes of projection could be created.

A point of general interest is that although the edge line ABC appears as a single line in side elevation it is in truth a double line segment as shown by the auxiliary view. The true angle ABC is as shown in the true shape of the hatch back. The line AB is thus the true length of the edge line AB and the true depth of the upper side panel is as indicated by the rotation construction, Fig. 1.32.

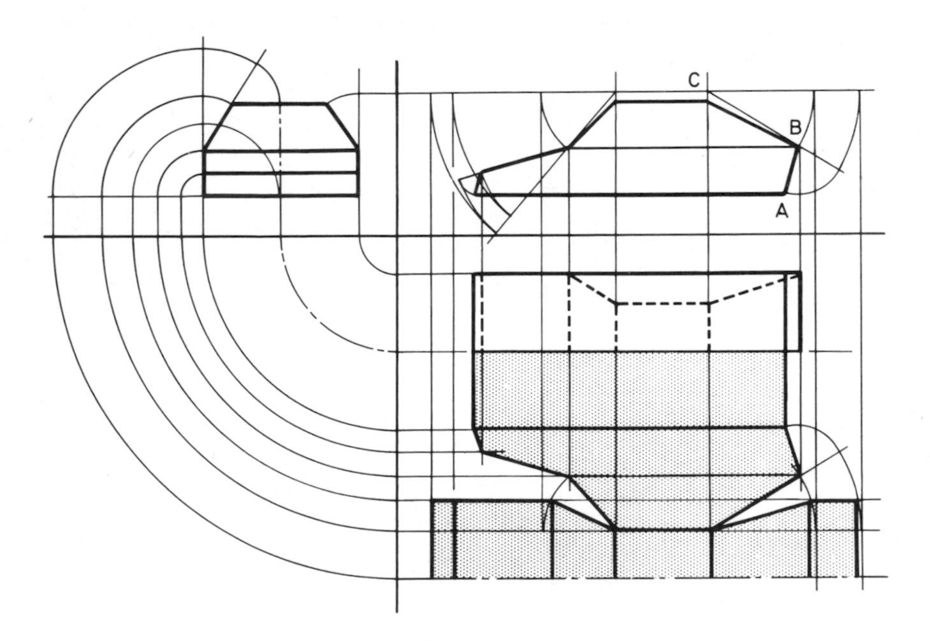

Figure 1.32

Example

As a simple example of how a curved surface may be developed we consider rolling the shell of the space capsule, Fig. 1.33, from an initially flat sheet of metal. The two orthographic views indicate that the central shell is a frustum of a cone, which for the purpose of this example has been extended in both front and side elevation to include the apex of the parent cone.

We note that when the circumferential points 0–12 are joined to the apex, in plan, the lines 01, 02, 03, etc., are all generators which lie on the surface of the cone. When transferred to front and side elevations these lines lie inclined to the axis, as shown in figure 1.33.

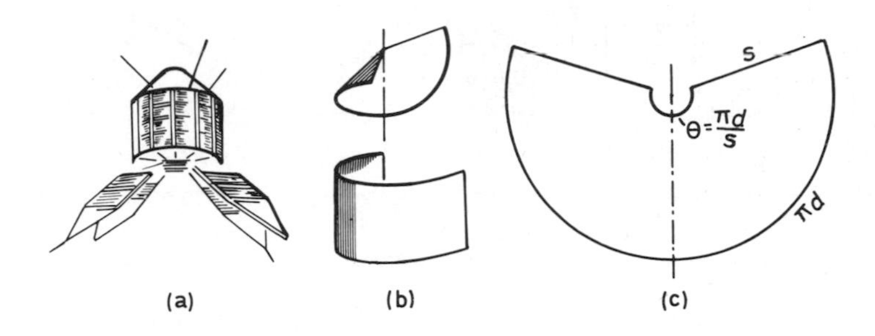

Figure 1.33

If the base circle circumference is $2\pi R$, the perpendicular height of the full cone H and the slope height of the full cone S. It is found that the curved surface may be rolled from a flat sheet of material cut to the shape of a sector.

We observe that the slope height $S = \sqrt{R^2 + H^2}$, where $R = D/2$ and since arc length = $\theta^c R$

$$\theta^c = \frac{\pi D}{S}, \quad \theta^\circ = \left(\frac{\pi D}{S}\right)\left(\frac{180}{\pi}\right)$$

whence

$$\theta^\circ = \frac{D}{S} \cdot 180^\circ.$$

If $R = 3, H = 4, S = 5$ and $D = 6$, whence

$$\theta^\circ = \frac{6}{5} \cdot 180 = 216^\circ$$

and the development for the frustum is as shown in the figure.

1.10 INTERPENETRATION OF INTERSECTING SOLIDS

It is easy to visualise that if two circular cylinders (pipes) are to merge to form a right-angled piece then this is only possible if the merging end of each cylinder is cut at 45° as shown in Fig. 1.34. When the intersection is not a right angle the common line of interpenetration is always the bisector of the angle between the two equidiameter cylindrical elements.

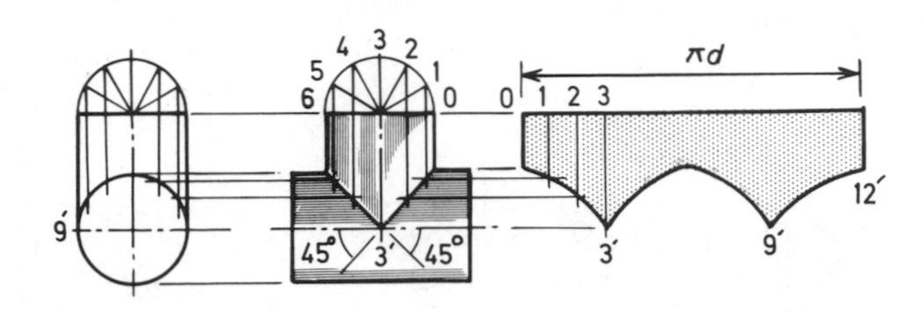

Figure 1.34

In the case of a non-symmetrical tee piece formed by two pipes of different diameters, the line of penetration is slightly more complex. The reader who understands article 1.9 should experience no undue difficulty in following Fig. 1.35, on which the developments of the two separate pieces are also shown.

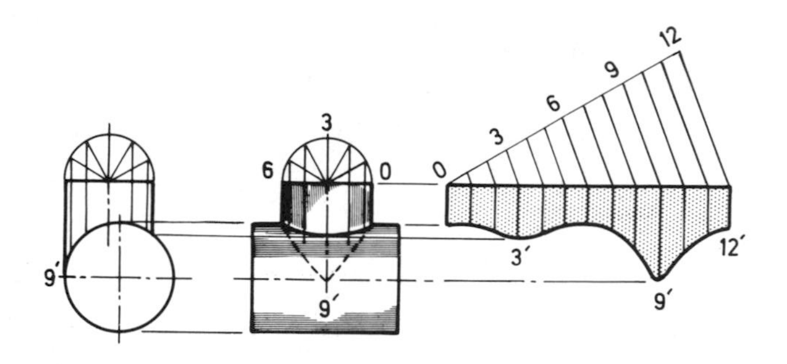

Figure 1.35

The development and interpenetration curves for a more complex three dimensional manifold are also shown in Fig. 1.36.

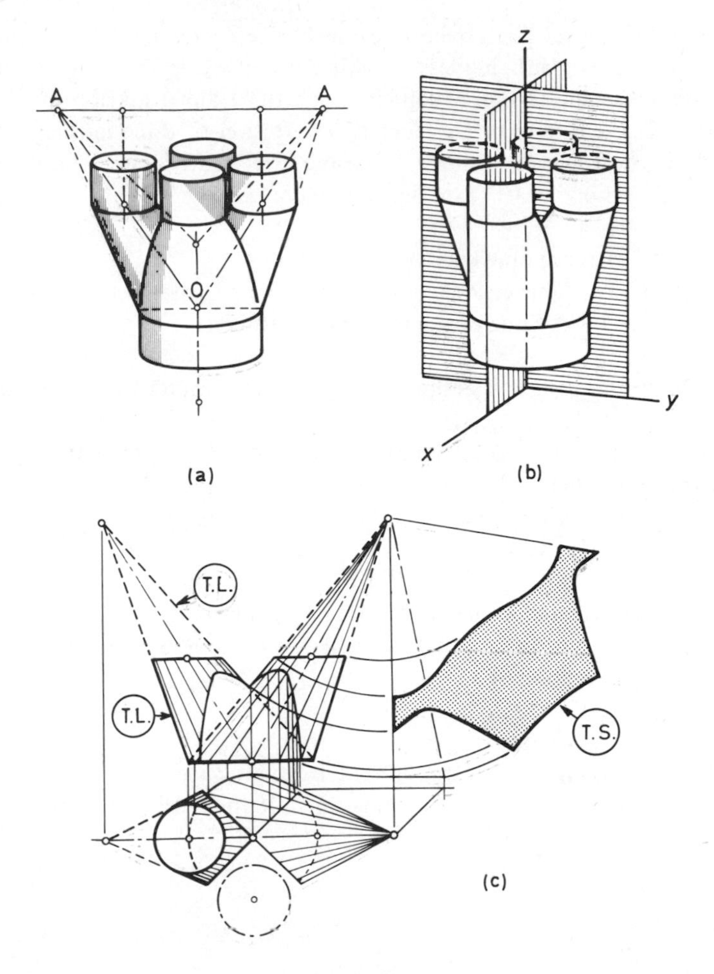

Figure 1.36

1.11 PARALLEL PICTORIAL PROJECTION

The method of orthographic projection enables the designer draughtsman to present three dimensional objects square on to the eye so that all relevant details are displayed true shape, and herein lies both the strength and the weakness of the method. From a manufacturing point of view, where it is imperative to give precise details of true shape and size, no better method has yet been discovered, for clearly a full tabulation of all the coordinates of all the points of an object is no substitute for a fully dimensional orthographic delineation. The weakness of the orthographic method is that to many eyes two dimensional plans, elevations and end views fail to transmit the impression of substance.

There is, therefore, good reason to seek alternative methods of pictorial presentation that whilst being less suitable for conveying metrical information, give a better visual impression of solidarity and three dimensional form. Our first illustration, Fig. 1.1, is in fact a pictorial impression of three mutually perpendicular planes in 3-space and the fact that the three right angles are not drawn as right angles does not detract from this impression.

1.11.1 Parallel pictorial drawings

Reasonable pictorial impressions based on the use of parallel projection can be readily produced by any one of the following methods.

1. Isometric single scale projection, using one scale factor applied to all three axes.
2. Dimetric double scale projection, using two different scale factors applied to two axes, the third axis being equal to one of them.
3. Trimetric triple scale projection, using three scale factors applied one to each axis.

We seek a rapid method of drawing reasonably authentic pictorial impressions and the clue to how this end may be achieved is suggested by considering the orthographic presentation of a cube.

1.11.2 Isometric projection

We observe from Fig. 1.37, that in order to produce a reasonable pictorial likeness of a real cube we need to rotate and tilt the square-on cube through at least two nonorthogonal angles.

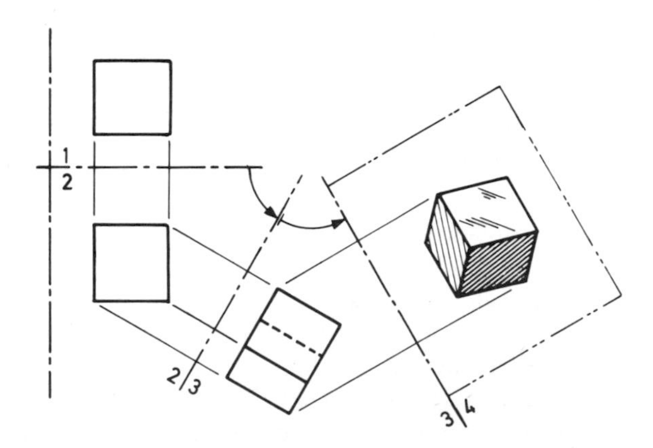

Figure 1.37

A simple procedure for producing pictorial impressions was proposed by the Rev. William Farish (born 1759). The method is known as isometric drawing as it is not a projective technique. As may be seen from Fig. 1.38, a cube with axes OX, OY, OZ attached to three mutually perpendicular edges may be rotated and tilted so that the foreshortened lengths of OX, OY, OZ are the same. The single condition for which this statement is true occurs when a rotation of 45° about one axis, is followed by a tilt of 35°16′ about the same axis after it has been rotated. When viewed in the appropriate direction each axis of the trihedral system then slopes away from or towards the eye at the same real angle. It is therefore to be expected that the three axes in an isometric projection should be separated one from the other by 120°. Thus if one axis is vertical, the other two axes are inclined to the horizontal at ±30°. It is therefore an easy matter to quickly produce isometric drawings directly using a 30° set square.

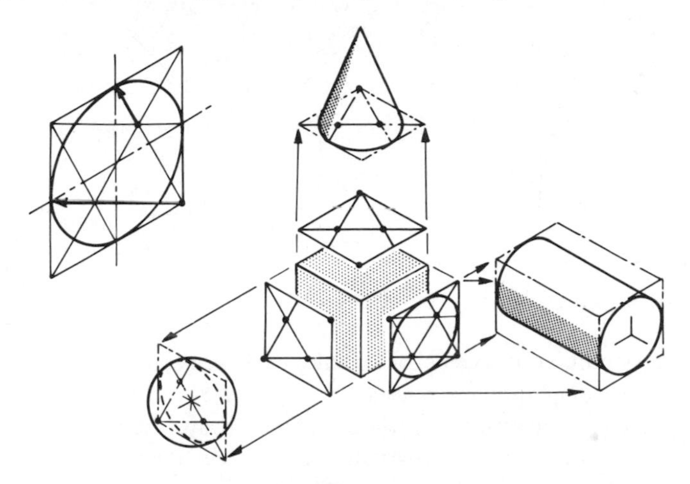

Figure 1.38

1.11.3 The isometric ellipse

Our frequent need to draw circles and cylinders in isometric which lie parallel to one or other of the three principal planes has led to the following approximate method.

1. construct a square (rhombus) circumscribing the circle, in the plane required;
2. draw in the longest diagonal;
3. complete the W construction shown in the figure;
4. draw in the enclosed approximate ellipse, by using a compass set to the two radii shown.

(N.B. The construction is approximate since a real ellipse is a continuous curve and not a piecewise make-up of four circular arcs.)

1.11.4 Isometric scale

The principal advantage of isometric drawing is that all three scales are equal and for many practical purposes it is acceptable to make lengths in the isometric view equal to corresponding lengths in the orthographic view, even though all lengths ought to be foreshortened. It is easily proved mathematically or checked graphically, by Fig. 1.39, that the isometric reduction factor, or scale is equal to $\sqrt{2}/\sqrt{3} = 0.816$ times the true length. When this scale factor is used the isometric drawing is identical to the isometric as produced by orthographic projection.

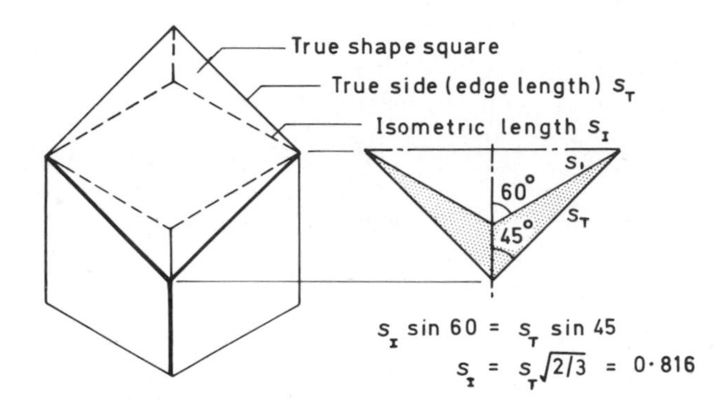

Figure 1.39

1.11.5 Dimetric, trimetric, axometric or axiometric projection

Whilst the isometric method provides a very rapid means of constructing parallel 'perspectives', the symmetrical orientation is not always the best view-point to use. In axiometric projection we may choose the view point to suit the features of a particular object, but having done so, have then to ascertain the extent of foreshortening. In a dimetric drawing there are two different scales, in a trimetric drawing there are three different scales. We therefore need to consider the dual problem of dimetric and trimetric scales.

1. to determine the scales of a pictorial projection, given the directions of axes,
2. to determine the direction of axes given three convenient scales.

We have the choice between, convenient easily measured angles and less convenient scales, and less easily measured angles with more convenient integral scales.

1.11.6 Axiometric scale when the direction of axes is given

Given the trimetric axes OX, OY, OZ, Fig. 1.40, we draw AB normal to OZ, AC normal to OY, BC normal to OX. We visualise the centre as lifted towards the eye, such that OABC is in effect a tetrahedron. Because the axes OX, OY, OZ, are mutually perpendicular the tetrahedral edge OB is perpendicular to the face OAC, and similarly for other corresponding faces and edges. An auxiliary view parallel to the axis OY, yields the edge line $B_1 O_1$ and the altitude of the sloping face as the line $\alpha_1 O_1$.

The angle $B_1 O_1 D_1$ is a right angle and this may be confirmed by drawing a semicircle on the line $B_1 D_1$. The angle β shows the true angle of the edge OB. The edges OA and OC similarly converge on O_1 and the distances O″A and O″C may be obtained by rotation as shown in the diagram.

The lengths O″A, O″B, O″C, represent the scale distances to be used in respect of the trimetric axes OX, OY, OZ.

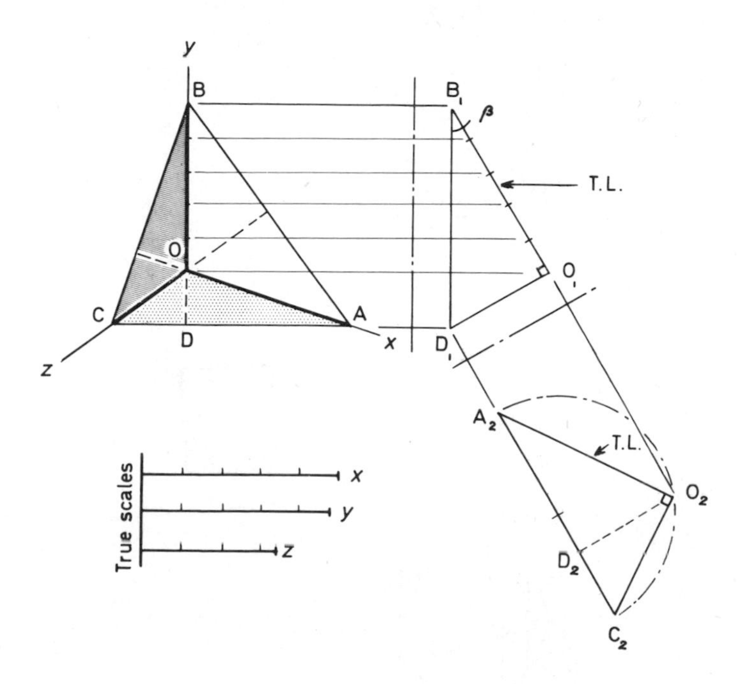

Figure 1.40

1.11.7 Axiometric scale when the scales are chosen

This is the converse of the problem above. We use this alternative approach when we wish to operate with easy scales.

1.11.8 Direct trimetric projection

Given a suitable orthographic plan and elevation, a trimetric projection may be produced using the direct trimetric projection technique originated by Rudolf

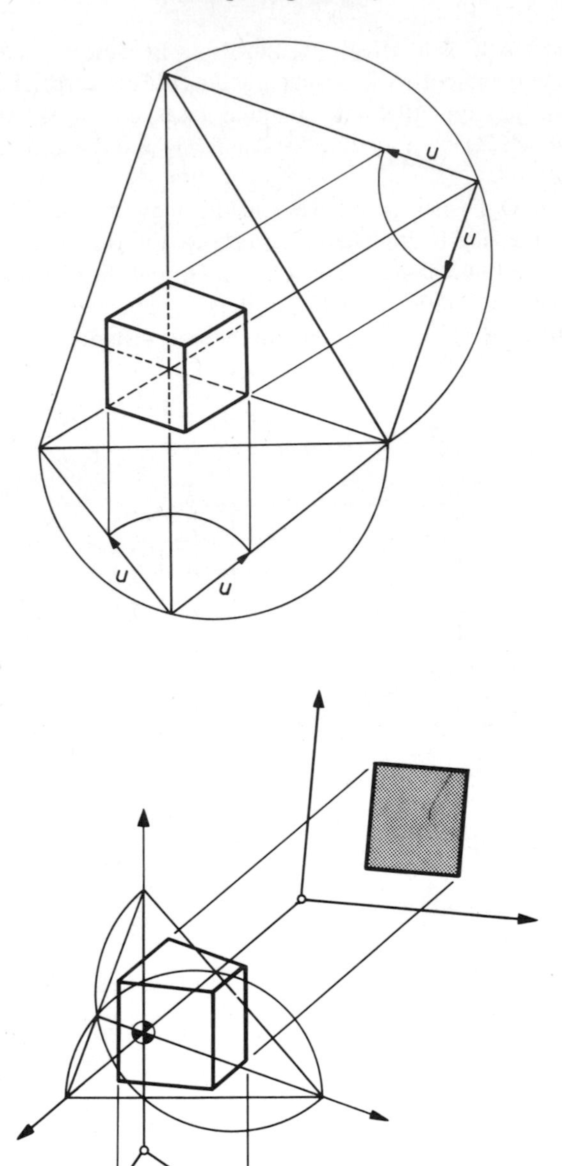

Figure 1.41

Schuessler in 1905 and developed by Schmid and Eckhart. The given plan and elevation are transferred (graphically or by cutting out and fixing to a backing sheet) to any convenient inclined position(s) as shown on Fig. 1.41.

The projection, Fig. 1.41, is self-explanatory but we note the superimposed semi-circles and their points of intersection with the reference axes. The axes O_1X_1, O_1Y_1, O_1Z_1 and the intersections O_1A, O_1B, O_1C, are equivalent to the axes and apex intersection, discussed in the previous article.

1.12 TRUE PERSPECTIVE PROJECTION

Design engineers who are mainly engaged in the design and manufacture of engineering type components rarely need to make use of the relatively more time-consuming 'true to life' perspective methods. The architect and the industrial designer are, however, more directly concerned with creating a true to life aesthetic impression and they rightly make good use of perspective. In view of the relatively high level of draughting skill and the time needed to produce complex perspectives, manually, so called, one, two and three point methods have been developed. (N.B. Perspective projections can now be made quite easily using computer-aided methods which entail little if any more work than do the corresponding parallel projections).

The terms one, two and three point perspective refer to the number of vanishing points on any one drawing.

1.12.1 Definition of terms used in perspective drawing

1. Standing point. The point from which the object is viewed.
   (N.B. 5 feet above the ground line from any distance unless otherwise stated).
2. Centre line of vision. The line from the observer's eye to the centre of object interest.
3. Cone of vision. The full cone angle formed by the extreme spread of vision. For the purpose of perspective drawings the maximum included angle is generally taken to be 90°, but 60° generally leads to clearer viewing.
4. Picture plane. The imaginary plane (screen) which intercepts the diverging cone of vision on which the picture is drawn.
5. Vanishing points. Any two parallel lines on an object appear to converge in a perspective view. The point at which they meet is called a vanishing point. The maximum number of vanishing points is three and all lines on an object home to one of them. Simple (quasi) perspectives may be drawn using one, or two vanishing points only.

1.12.2 Single point perspective

This is the simplest and least realistic of the three methods. As commonly arranged a single point perspective contains horizontal and vertical lines in the frontal plane and depth lines converging to a single vanishing point on the horizon, as shown in Fig. 1.42.

A commonly used, single point internal perspective is shown in Fig. 1.43.

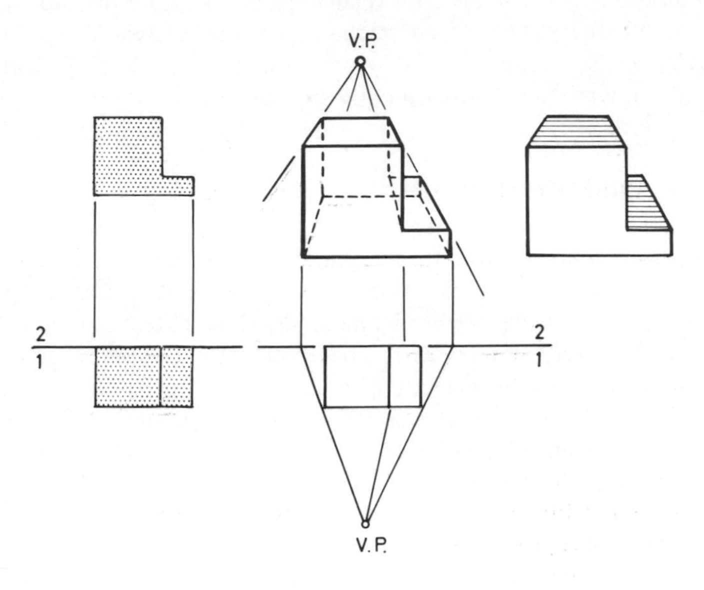

Figure 1.42

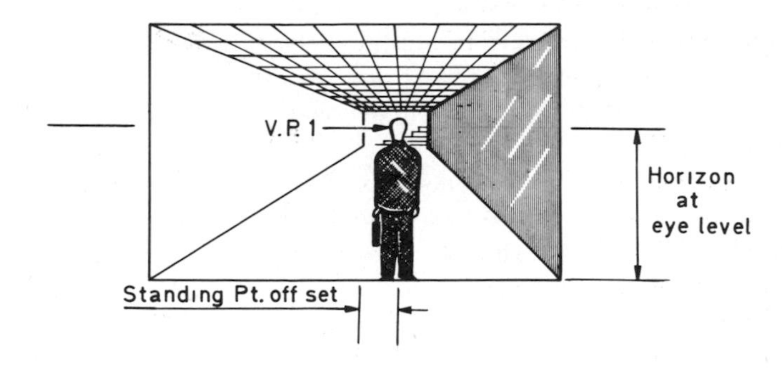

Figure 1.43

1.12.3 Two point perspective

Here a greater dynamic realism is introduced by showing, for instance, all three sides of a cube. Lines in the first plane converge to vanishing point 1. Lines in the second plane converge to vanishing point 2. Lines in the third plane converge to vanishing point 1 or 2 depending on their direction. A two point perspective is shown in Fig. 1.44.

We observe that in order to view three sides of a prismical block the centre line of vision and the face of the block must be at an angle.

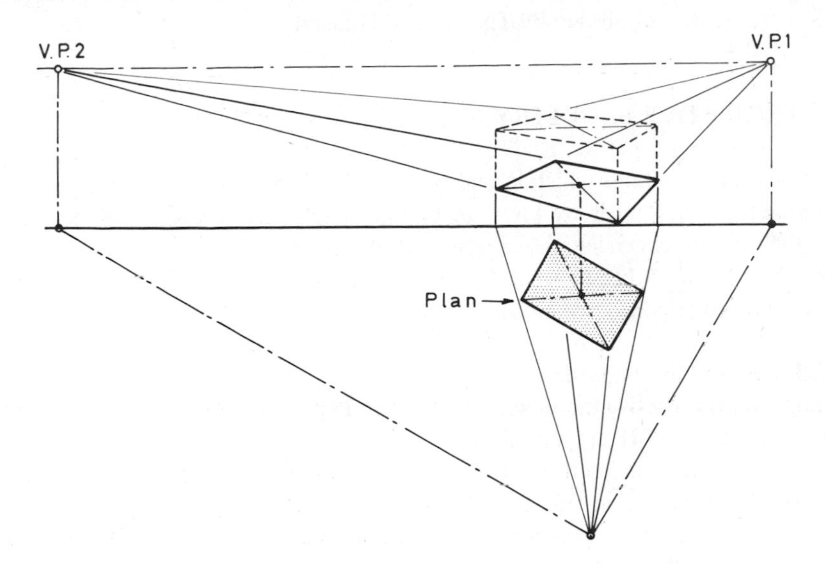

Figure 1.44

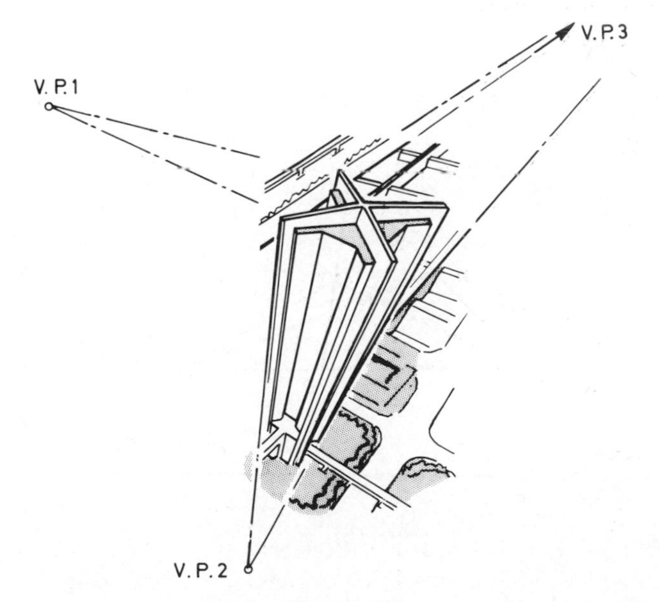

Figure 1.45

1.12.4 Three point perspective

The method of drawing which yields the most dynamic impression is three point perspective, in which three vanishing points are used.

A three point perspective layout is shown for reference in Fig. 1.45. For a more detailed understanding the reader is recommended to read *Basic Perspective* by Robert W. Gill, published by Thames and Hudson.

1.13 PROJECTIVE GEOMETRY

We have seen that lines which in reality are parallel are non-parallel in a perspective presentation. Angles we know to be right angles are drawn more acute and even the ratios between lengths are changed. Perspective drawings do nevertheless give a better, more life-like, impression of realism, than do drawings produced by parallel projection.

1.13.1 Cross ratio

An important projective theorem, accredited to Pappas and Menelaus (c. AD 300) states that if AC, BD are the diagonals of a quadrilateral ABCD, in which the sides are produced in pairs to intersect at P and Q, then a line drawn from P through the intersection of AC, BD cuts the line AB in R, and similarly a line drawn from Q intersects AD in S, such that AR: AB: AQ and AS: AD: AP are in harmonic progression. (N.B. The word harmonic links the harmonic progression of lengths to a musical scale).

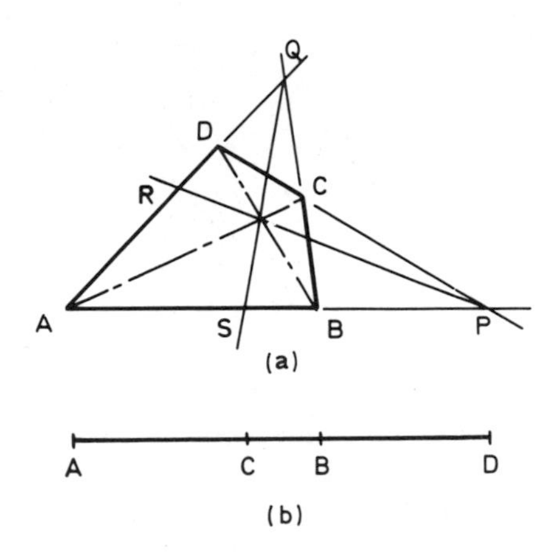

Figure 1.46

If four points A, C, B, D, (note the order) are on a straight line, as in Fig. 1.46 then the ratio

$$\frac{AC \cdot BD}{AD \cdot BC}$$

is an harmonic or cross ratio of the range A, B, C, D, in which we observe the convention that AC = –CA and write (ABCD) to indicate the order in which the points are taken. The numerator is written with the points in given order while the denominator is written with the second and fourth points interchanged.

Similarly if the four lines OA, OB, OC, OD are concurrent then the cross ratio is

$$\frac{\sin AOC \cdot \sin BOD}{\sin AOD \cdot \sin BOC}$$

and it follows that if OA = a, OB = b, OC = c, OD = d then

$$\frac{AC \cdot BD}{AD \cdot BC} = \frac{(c-a)(d-b)}{(d-a)(c-b)}$$

where BC in Fig. 1.46, is negative.

Cross ratios are thus seen to be projective and when the cross ratio equals minus 1 we have the particular formulation of an harmonic range.

For convenience in labelling diagrams the natural sequence A, B, C, D may be preferred and when this natural order is adopted the cross ratio is

$$\frac{(a-b)(c-d)}{(b-c)(d-a)}$$

The importance of the cross ratio is that it is a property common to both real object and picture. For instance, if we consider the rectangle ABCD, Fig. 1.47, to be the plan of a real object, then the quadrilateral A′B′C′D′ with sides projected in pairs (as in Fig. 1.47) with PQ as the vanishing points in its perspective transformation.

Example

If a three span bridge is supported on equi-spaced piers, determine the cross ratio that defines their spacing.

Solution

If we take our datum at the first pier $a = 0, b = 1, c = 2, d = 3$ and the cross ratio is

$$\frac{(a-b)(c-d)}{(b-c)(d-a)} = \frac{(-1)(-1)}{(-1)(+1)} = -\frac{1}{3}$$

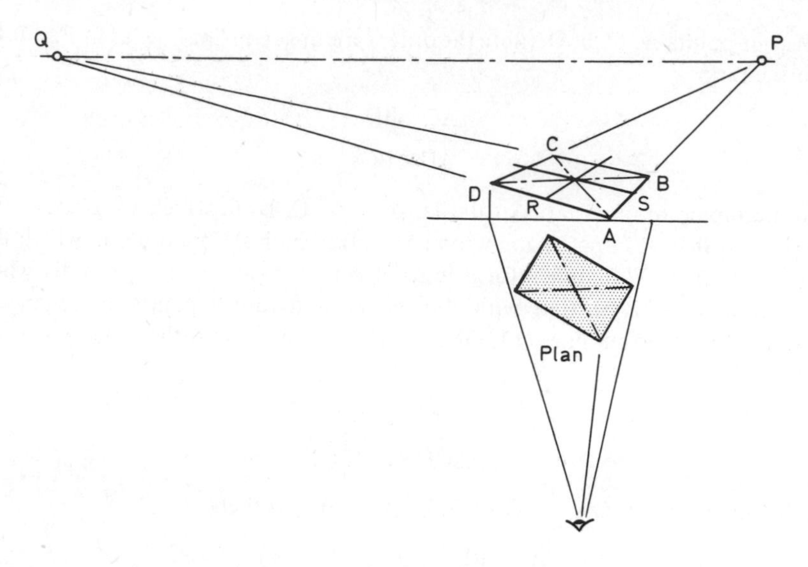

Figure 1.47

1.13.2 Pappas duality theorem

If A, B, C, and D, E, F, are point sets on any two straight lines, then we may join these points as in Fig. 1.48, to create intersections P, Q, R, such that no matter how the point sets A, B, C, D, E, F, are arranged P, Q, R, will always lie on a straight line. We observe that there are nine points connected by nine lines as shown in the figure.

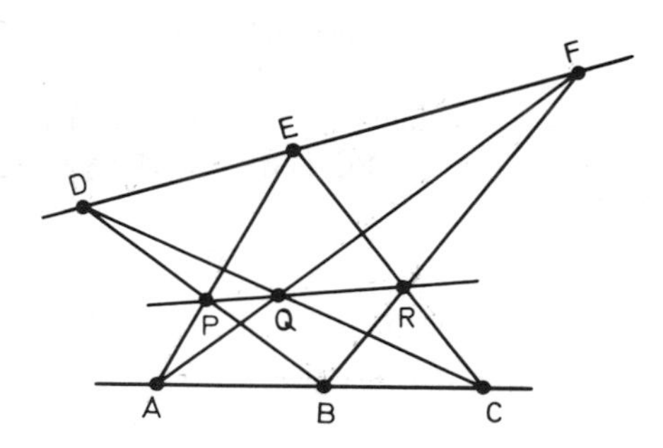

Figure 1.48

1.13.3 Desargues theorem

The theorem states that if two triangles ABC and A′B′C′ lie in a plane and situated so that straight lines drawn through corresponding vertices meet in a single point O. Then the extensions of corresponding sides intersect in three co-linear points P.Q.R. The plane figure of Desargues (1593–1662) may be visualised as a three dimensional pyramid by adding the embellishments, Fig. 1.49.

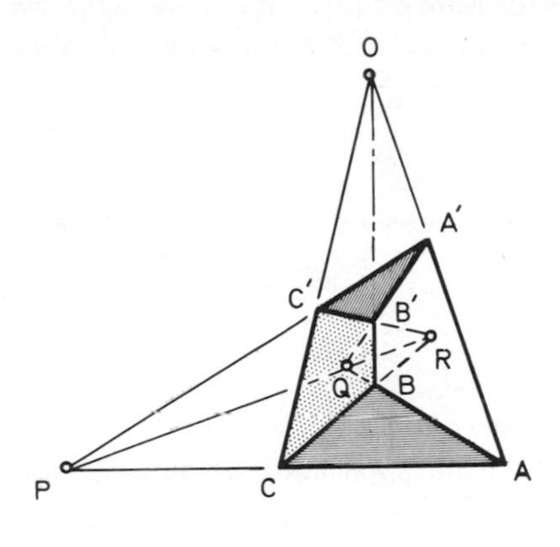

Figure 1.49

Two triangles are said to be perspective (from a point) when each pair of vertices are joined by two concurrent lines.

2 Analytic geometry in two dimensions

When we consider a figure of three sides, we form the idea of a triangle, and we afterwards make use of it as the universal to represent to our mind other figures of three sides.

Réné Descartes

2.1 INTRODUCTION

Plane analytic geometry is the application of algebraic principles to the geometry of points, lines, planes and curves, the positions and shapes of which are allowed to vary in two dimensions only. The abstract geometrical concepts point, line and plane are fundamental entities of increasing dimensionality. By this, we mean that a mathematical point is a size-less dot, the function of which is to indicate position. In practice all drawn dots are of necessity, made large enough to be visible to the naked eye, but are in principle sizeless. The function of all drawn dots is to indicate a purely dimensionless position. A mathematical point is thus a position devoid of extent (size) and said to be of dimensionality zero.

Many different lines may be drawn through two separate points but two points suffice to define a straight line. A straight line is thus defined as the shortest distance between two given points and the distance between two points is called the length of the line. Length is a one dimensional measure and since lines have neither breadth nor thickness lines are said to be one-dimensional.

A curved and/or twisted surface is two dimensional.

The geometrical entities point, line and plane, exhaust the scope of two-dimensional geometry. Two-dimensional methods are, however, widely applicable to three-dimensional problems and use is made of two dimensional concepts in many practical design applications. The study of two dimensional analytic geometry was founded by the French mathematicians René Descartes (1596-1650) and Pierre Fermat (1601-1665) and known as the Cartesian coordinate system.

2.2 POINT-POSITION CARTESIAN COORDINATES

The position of every point in a plane is completely defined by two mutually perpendicular Cartesian coordinate dimensions. When plotting a point on a two-dimensional graph we are supposed to lay down a distance OA = a along the ox

axis and then a distance OB = b along the oy axis and then project two lines AP and BP parallel to the oy and ox axes respectively. The point at which these two projection lines intersect is the two-dimensional location of the point P.

In practice, however, we more often lay down a distance OA = a along the axis ox and add to it a distance AP = b in a direction parallel to the axis oy. The point at which the line AP terminates is the point location of P. (N.B. We observe that OA = BP, OB = AP.)

The position of the point P is said to be located in terms of its Cartesian coordinates (a, b) and the point P is referred to as the point P(a, b). The x coordinate is always listed first, the y coordinate second and in three-dimensional problems the z coordinate third. To distinguish coordinate dimensions from other entries we enclose them in brackets as shown.

2.2.1 Point position polar coordinates

Another approach is to lay down a known length OP = r along the ox axis and swing this known length r through an angle θ, about the origin O. The position of the point P is thus located in terms of its polar coordinates (r, θ) and the point P is referred to as the point P(r, θ).

The two alternative ways of locating a point P are as illustrated in Fig. 2.1.

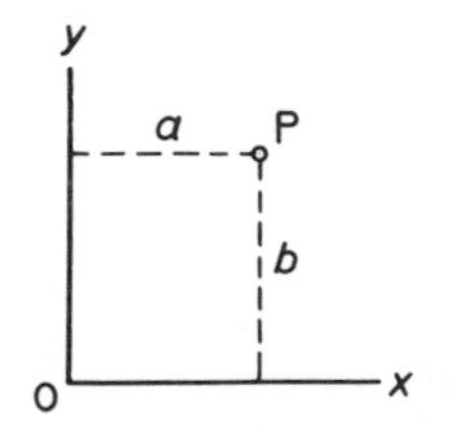

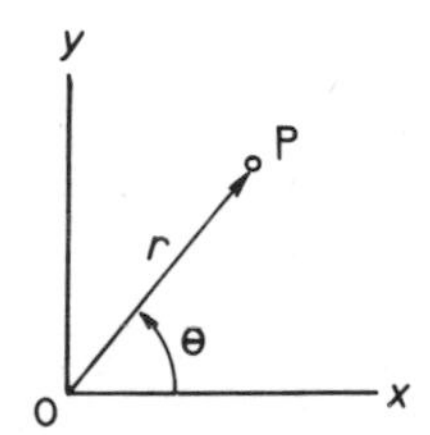

Figure 2.1

2.3 THE LOCUS OF A POINT

The graph of a point which moves in accord with a predefined mathematical law is called its locus. The plural of locus is loci. We shall in the pages to come, write the equations of a wide range of loci, each of which has its own characteristic.

2.4 TWO-DIMENSIONAL GRAPHS AND SIGN CONVENTIONS

When plotting a two-dimensional graph, we adopt the universal convention that; x to the right is positive, x to the left is negative, y up is positive, y down is negative. The two-dimensional space defined by the positive and negative axes of

xoy comprises four quadrants known as the first, second, third and fourth quadrants, which one always takes in anticlockwise order as shown in Fig. 2.2.

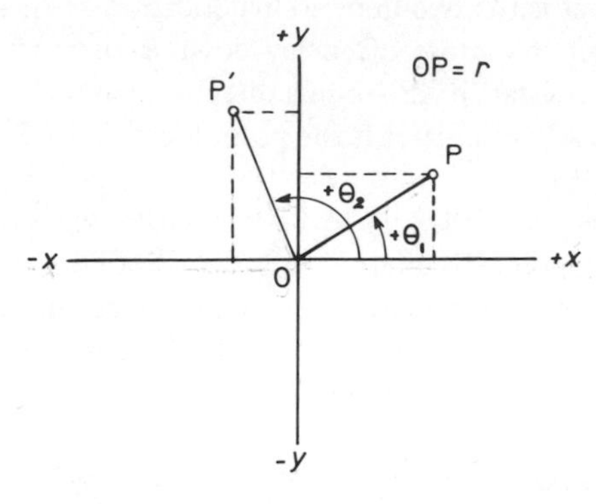

Figure 2.2

The two coordinates $(+x, +y)$ define a point P and hence an angle θ_1 in the first quadrant. The three coordinates (x, y) and θ_1 are clearly related as follows

$$\frac{y}{x} = \tan\theta_1$$

$$\frac{y}{r} = \sin\theta_1$$

$$\frac{x}{r} = \cos\theta_1$$

$$r^2 = x^2 + y^2$$

The two coordinates $(-x, +y)$ define a point and hence an angle θ_2 in the second quadrant and the sign allotted to $\tan\theta_2$, $\sin\theta_2$ and $\cos\theta_2$ is to be observed.

$$\frac{+y}{-x} = -\tan\theta_2$$

$$\frac{+y}{+r} = +\sin\theta_2$$

$$\frac{-x}{+r} = -\cos\theta_2$$

and similarly for quadrants three and four. We observe that all three trigonometric functions are positive in the first quadrant. Sine alone is positive in the second quadrant. Tangent alone is positive in the third quadrant and cosine alone is positive in the fourth quadrant. Hence the jingle: all, sine, tan, cos.

The need for this rule is clearly due to the cyclic rise and fall of the trigonometric functions. The variation of which is as shown in Fig. 2.3.

The radius $OP = r$ and the angle θ measured counter-clockwise are always positive.

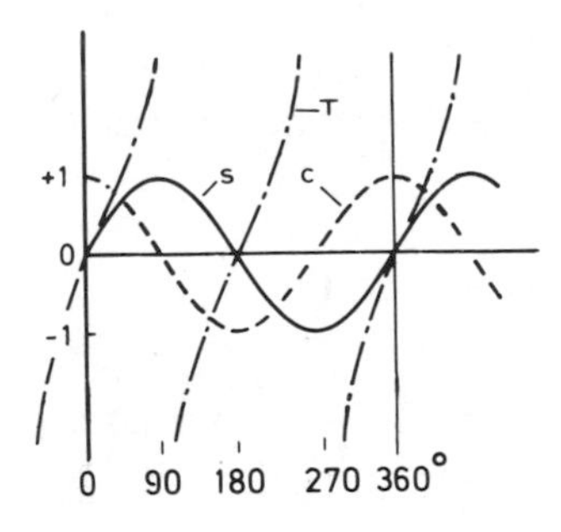

Figure 2.3

2.5 THE EQUATIONS TO A STRAIGHT LINE

There are numerous ways in which a straight line may be defined and these will now be considered.

2.5.1 A straight line which lies parallel to an axis

Since all points on a line parallel to the oy axis lie at a constant distance a from the axis oy and all points on a line parallel to the axis ox lie at a constant distance b from the axis ox, the equations to the two lines are

$$x = a, y = b \tag{2.1}$$

The equations $x' = x - a$, $y' = y - b$ would mean that the points (x', y') were displaced by amounts a and b from the points (x, y).

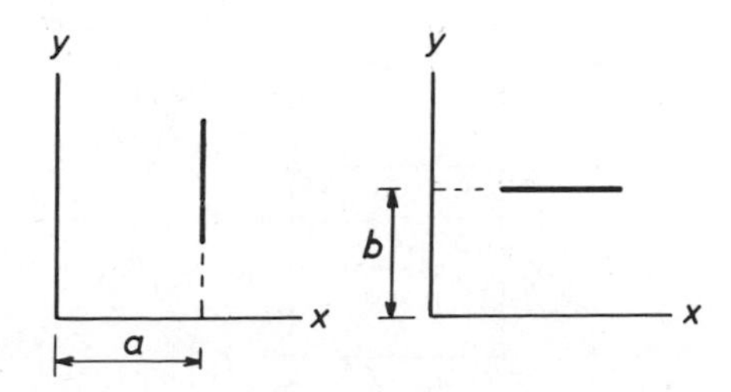

Figure 2.4

2.5.2 A straight line which lies inclined to an axis

Given the coordinates of a point $(x_1 y_1)$ on the line and the angle θ, the co-ordinates of a (second) general point (x, y) are clearly $(x_1 + L \cos \theta)$, $(y_1 + L \sin \theta)$ and since

$$\frac{y - y_1}{\sin \theta} = \frac{x - x_1}{\cos \theta} = L$$

$$(y - y_1) = (x - x_1) \tan \theta$$

whence

$$(y - y_1) = (x - x_1) m \tag{2.2}$$

where $m = \tan \theta$.

If the line produced intersects the oy axis at a point $y = b$, the equation which defines the line at every point is

$$y = mx + b \tag{2.3}$$

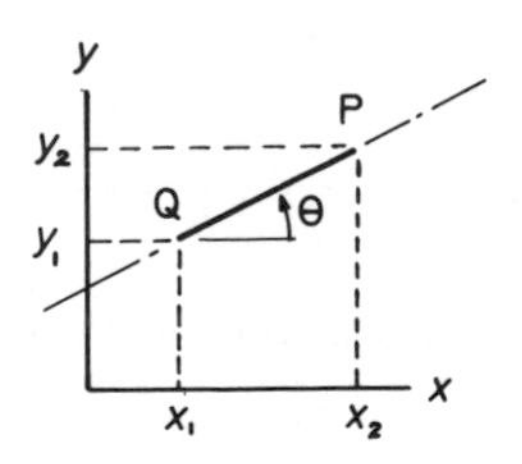

Figure 2.5

2.5.3 A straight line in terms of two given points

If the two given points are $A(x_1, y_1)$, $B(x_2, y_2)$ and $P(x, y)$ is a general point on the line as shown in Fig. 2.6: the triangles MPA and NBP are similar. Hence the required equation is

$$\frac{x - x_1}{x_2 - x_1} = \frac{y - y_1}{y_2 - y_1} \tag{2.4}$$

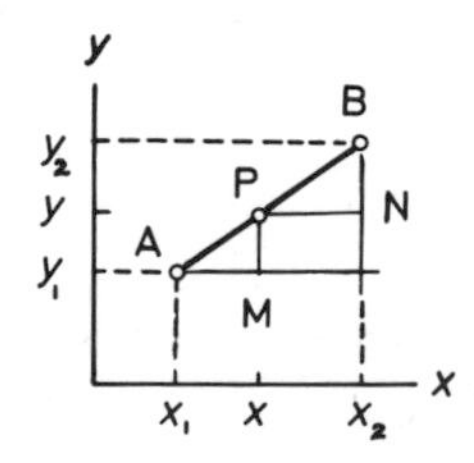

Figure 2.6

2.5.4 A straight line in terms of its intercepts

Let the intercepts of the line on the axes ox, oy, Fig. 2.7, equal a and b. Let P(x, y) be any point on the line. The required equation may then be obtained by equating the area AOB to the sum of the areas AOP, POB.

$$\tfrac{1}{2}bx + \tfrac{1}{2}ay = \tfrac{1}{2}ab$$

hence

$$\frac{x}{a} + \frac{y}{b} = 1 \tag{2.5}$$

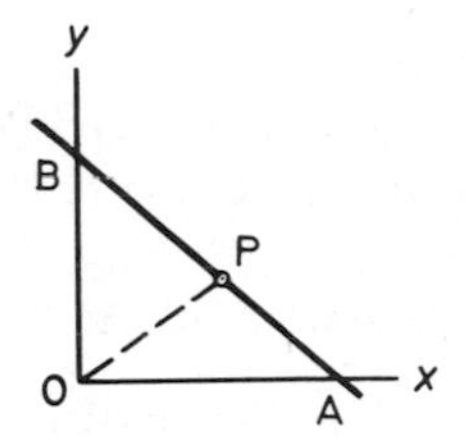

Figure 2.7

2.5.5 A straight line in perpendicular form

Let the perpendicular distance from origin to line equal p and the angle of the perpendicular relative to the ox axis equal α, then the required equation to the line is obtained directly from an inspection of Fig. 2.8.

$$x \cos\alpha + y \sin\alpha = p \tag{2.6}$$

This equation is called the perpendicular equation to a straight line.

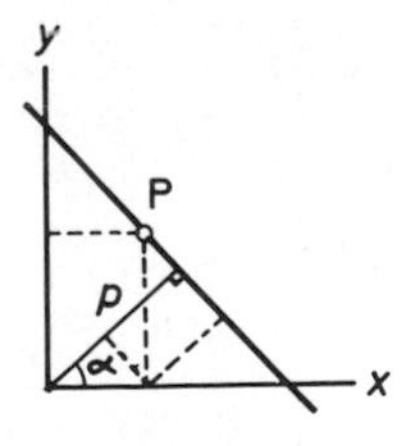

Figure 2.8

2.5.6 A straight line in polar form 1

If P(r, θ) is any point on the line and (p, α) are the polar coordinates of its perpendicular, as shown in Fig. 2.9, then the required result is obtained by inspection.

$$r \cos(\theta - \alpha) = p \tag{2.7}$$

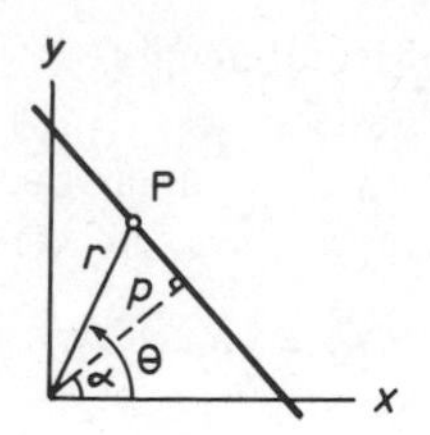

Figure 2.9

2.5.7 A straight line in polar form 2

If $A(r_1\theta_1)$, $B(r_2\theta_2)$ are the two given points and $P(r, \theta)$ is a general point on the line, then we may equate the area APB, Fig. 2.10, to zero and thereby obtain the equation required.

We observe that the elemental areas AOB, AOP, POB, are

$$\text{Area } AOB = \tfrac{1}{2}r_1r_2\sin(\theta_2 - \theta_1)$$
$$\text{Area } AOP = \tfrac{1}{2}r_1r\sin(\theta - \theta_1)$$
$$\text{Area } POB = \tfrac{1}{2}r_2r\sin(\theta_2 - \theta)$$

and since $AOB + AOP + POB = APB = 0$

$$r_1r_2\sin(\theta_2 - \theta_1) - r_1r\sin(\theta - \theta_1) - r_2r\sin(\theta_2 - \theta) = 0$$

and

$$r_1r_2\sin(\theta_2 - \theta_1) + r_1r\sin(\theta_1 - \theta) + r_2r\sin(\theta - \theta_2) = 0$$

hence

$$\frac{\sin(\theta_2 - \theta_1)}{r} + \frac{\sin(\theta_1 - \theta)}{r_2} + \frac{\sin(\theta - \theta_2)}{r_1}$$

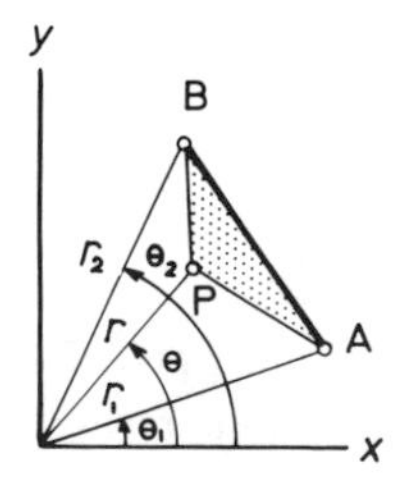

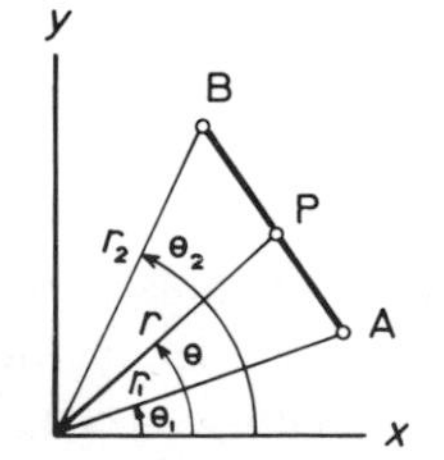

Figure 2.10

2.5.8 General equation to a straight line

(1) By writing $m = -$ A/B and $b = -$ C/B we may easily prove that the equation $Ax + By + C = 0$ is equivalent to the equation $y = mx + b$. The so-called general equation to a line $Ax + By + C = 0$ is thus another way of writing equation (2.3). Since the general equation contains three constants A, B and C, three independent equations suffice to eliminate the A, B, and C. Given the three equations:

$$\begin{aligned} Ax_1 + By_1 + C &= 0 \\ Ax_2 + By_2 + C &= 0 \\ Ax + By + C &= 0 \end{aligned} \tag{2.8}$$

where (x, y) is the general point.

By equating the first and third equation we obtain

$$\frac{A}{C}(x - x_1) + \frac{B}{C}(y - y_1) = 0$$

and by equating the first and second equation

$$\frac{A}{C}(x_2 - x_1) + \frac{B}{C}(y_2 - y_1) = 0$$

Hence

$$\frac{x - x_1}{x_2 - x_1} = \frac{y - y_1}{y_2 - y_1}$$

The equivalence between equation (2.4) and equation (2.8) is therefore proved. (N.B. This equation is also true if oblique coordinates are used, see article 15.3).

(2) By writing the general equation $Ax + By + C = 0$ in the form:

$$\frac{Ax}{C} + \frac{By}{C} + 1 = 0$$

it is a straightforward step to prove that equation (2.5) and equation (2.8) are equivalent.

Thus we may write:

$$-\frac{A}{C}x - \frac{B}{C}y = 1$$

whence

$$\frac{x}{-C/A} + \frac{y}{-C/B} = 1$$

and if we put; $-$ C/A $= a$, and $-$C/B $= b$, we see that equation (2.5) and equation (2.8) are equivalent.

(3) The general equation $Ax + By + C = 0$ may be reduced to perpendicular form, $x \cos \alpha + y \sin \alpha = p$, by dividing throughout by $\sqrt{A^2 + B^2}$.

Thus we write:

$$\frac{Ax}{\sqrt{A^2 + B^2}} + \frac{By}{\sqrt{A^2 + B^2}} = \frac{C}{\sqrt{A^2 + B^2}} = p$$

hence

$$x \cos \alpha + y \sin \alpha = p$$

when C is positive the equation above becomes

$$\frac{Ax}{\sqrt{A^2 + B^2}} - \frac{By}{\sqrt{A^2 + B^2}} = \frac{C}{\sqrt{A^2 + B^2}}$$

2.5.9 The perpendicular distance from a point to a line

Draw a line parallel to the given line which passes through the point $P(x, y)$ so that the y intercept of this new line is either $(b + d \cos \theta)$ or $(b - d \cos \theta)$, according to whether the point $P(x, y)$ is above or below the given line. If the point $P(x, y)$ is above the given line, d is positive and the equation of the line through P is:

$$y = mx + b + \frac{d}{\cos \theta}$$

where

$$\cos \theta = \frac{1}{\sqrt{m^2 + 1}}$$

Hence the perpendicular distance from a point to a straight line is

$$d = \frac{y - mx - b}{\sqrt{m^2 + 1}} \text{ or } \frac{mx - y + b}{\sqrt{m^2 + 1}} \tag{2.9}$$

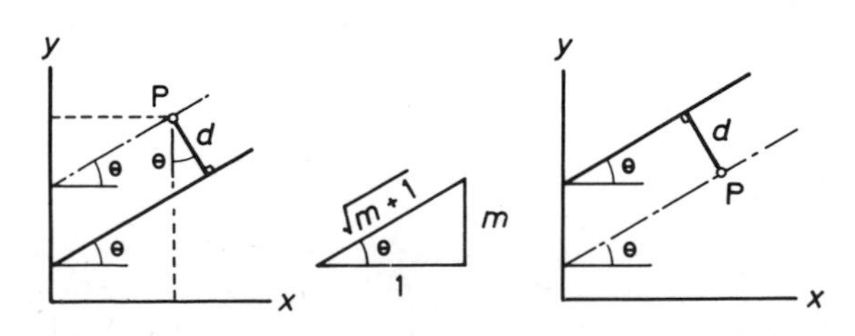

Figure 2.11

2.5.10 The intersection of two straight lines

The intersection between two lines $A_1x + B_1y + C_1 = 0$ and $A_2x + b_2y + C_2 = 0$ is obtained by eliminating y and x from each equation in turn. Suffice to record that the coordinates of intersection are

$$x = \frac{B_1C_2 - B_2C_1}{A_1B_2 - A_2B_1}$$
$$y = \frac{C_1A_2 - C_2A_1}{A_1B_2 - A_2B_1} \qquad (2.10)$$

2.5.11 The angle between two straight lines

If two straight lines $y = m_1x + b_1$ and $y = m_2x + b_2$ make angles θ_1 and θ_2 with the *ox* axis and $\phi = (\theta_1 - \theta_2)$ is the angle between them. See Fig. 2.12.

Then $\tan\phi = \tan(\theta_2 - \theta_1)$ expands to

$$\tan\phi = \frac{\tan\theta_2 - \tan\theta_1}{1 + \tan\theta_2 \tan\theta_1}$$

but $\tan\theta_1 = m_1$ and $\tan\theta_2 = m_2$ hence

$$\tan\phi = \frac{m_2 - m_1}{1 + m_2m_1} \qquad (2.11)$$

when $\tan\phi = 0$ the lines are parallel and $m_1 = m_2$ and conversely when the lines are perpendicular $\tan\phi = \infty$, $m_1m_2 = -1$.

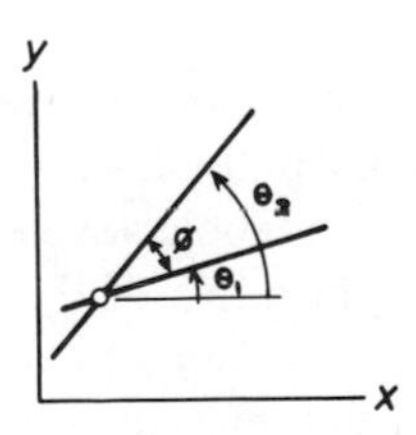

Figure 2.12

2.6 PLANE CURVES

There are many different curves of interest to the designer and we will consider the catenary, involute, lemniscate, cycloid, epicycloid, hypocycloid, nephroid, cardioid, astroid, deltroid and various spiral curves in due course. We begin by considering polynomials and conic sections.

2.6.1 Polynomial curves

The importance of polynomial curves of the form $y = a_0 + a_1x + a_2x^2 \ldots a_nx^n$ is partly due to the ease with which such curves may be differentiated, and partly due to the fact that a polynomial of degree n may be directed through $n + 1$ predefined points, simply by allotting appropriate values to the coefficients a_0, $a_1 \ldots a_n$.

The equation $y = a_0 + a_1x$ is, of course, a straight line. The equation $y = a_0 + a_1x + a_2x^2$ is a quadratic. The equation $y = a_0 + a_1x + a_2x^2 + a_3x^3$, a cubic, and such equations are differentiable term by term. For instance the differential of

$$y = a_0 + a_1x + a_2x^2 + a_3x^3 + \ldots a_nx^n$$

is

$$\frac{dy}{dx} = a_1 + 2a_2x + 3a_3x^2 \ldots na_nx^{n-1}.$$

When working with cubic polynomial equations of the type $y = a_0 + a_1x + a_2x^2 + a_3x^3$ it is sometimes found beneficial to regroup coefficients in accord with Horner's rule.

$$y = [(a_3u + a_2)u + a_1] + a_0$$

N.B. Horner's rule reduces the number of multiplications required to evaluate a cubic.

2.6.2 The conic sections

We may best begin our study of curves by considering that very important class of curves, known since the time of Apollonius (263–200 BC) as the conic sections. The circle, ellipse, hyperbola and parabola are all sections of a right circular cone and their shape may be defined in terms of the Cartesian coordinate system as follows.

The classical treatment defines these curves in terms of the locus of a mobile point in relation to a fixed point (the focus) and a fixed straight line (the directrix).

The locus (that is the path) of a point which moves so that its distance from a fixed point, the focus, bears a constant ratio, the eccentricity, to its distance from a fixed straight line, the directrix, is a conic.

If the eccentricity is zero, the conic is a circle, if the eccentricity is unity the conic is a parabola, if the eccentricity is less than unity but greater than zero, the conic is an ellipse, if the eccentricity is greater than unity, the conic is a hyperbola.

The Cartesian equations for these four conics will now be considered.

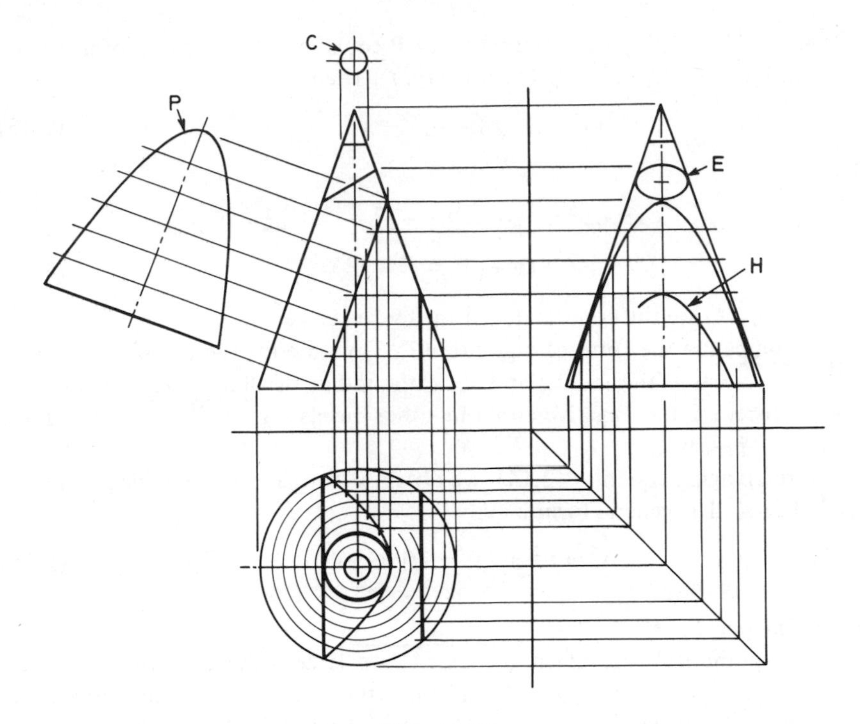

Figure 2.13

2.6.3 The equation to a circle

The simplest Cartesian equation of a circle is obtained by applying the rule of Pythagoras to a centralised circle.

As shown by Fig. 2.14, OA = x, OB = y, OP = r, and hence

$$x^2 + y^2 = r^2$$

The equation of a circle with its centre off-set from the origin of axis is readily obtained as follows:

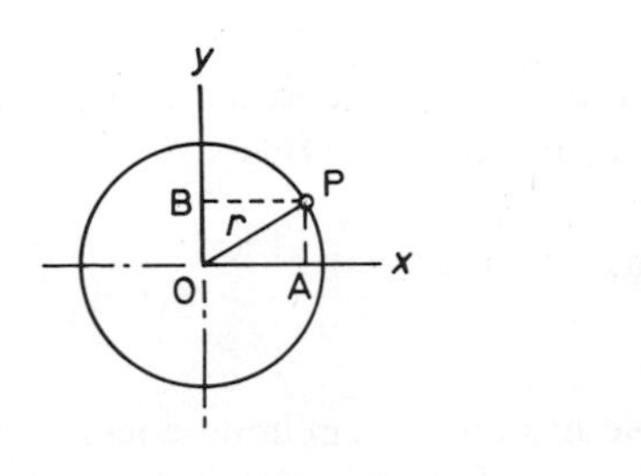

Figure 2.14

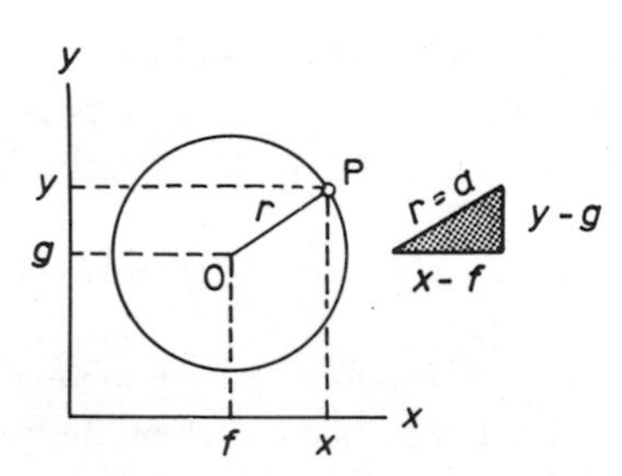

Figure 2.15

Let the coordinates of a general point P equal (x, y). Let the coordinates of the centre of the circle, Fig. 2.15, equal (g, f). Then

$$(x - g)^2 + (y - f)^2 = a^2 \tag{2.13}$$

and hence

$$x^2 + y^2 - 2gx - 2fy + g^2 + f^2 - a^2 = 0$$

$$(x + g)^2 + (y + f)^2 = g^2 + f^2 - c \tag{2.14}$$

where $(-g, -f)$ are the coordinates of the circle centre.

On comparing the form of equation (2.13) with equation (2.14) we see that $\sqrt{g^2 + f^2 - c} = a$ (the radius) and since the radius of all real circles is finite we always disregard the negative root. In other words $\sqrt{g^2 + f^2 - c} = a$ and a is necessarily positive.

On rearranging equation (2.14), we obtain the equation to a circle in its most general form. The general form is

$$x^2 + y^2 + 2gx + 2fy + c = 0 \tag{2.15}$$

2.6.4 The point circle

In the hypothetical case when $\sqrt{g^2 + f^2 - c}$ is zero the equation to a circle becomes $(x + g)^2 + (y + f)^2 = 0$ and since the square of a number cannot be negative, each bracketed term must be zero. That is $(x + g) = 0$, $(y + f) = 0$ whence $x = -g$, $y = -f$. The point $(-g, -f)$ is of zero size and is known as a point circle.

2.6.5 Three point definition of a circle

Three coplanar points not on a straight line suffice to define a circle.

The general equation to a circle is:

$$x^2 + y^2 + 2gx + 2fy + c = 0$$

and hence the constants g, f and c may be evaluated, by substituting the co-ordinates of three known points and solving the three equations simultaneously.

2.6.6 Polar definition of a circle

Let the polar coordinates of a point P equal (r, θ) and the coordinates of its centre C equal (p, α) whence we obtain from the cosine rule

$$r^2 - 2rp\cos(\theta - \alpha) + p^2 - a^2 = 0 \tag{2.16}$$

This is the polar equation to a circle.

A special case occurs when the pole lies on the circumference, as shown in Fig. 2.16, and in this case $r = 2a\cos(\theta - \alpha)$ or if OA passes through the circle centre, $r = 2a\cos\theta$.

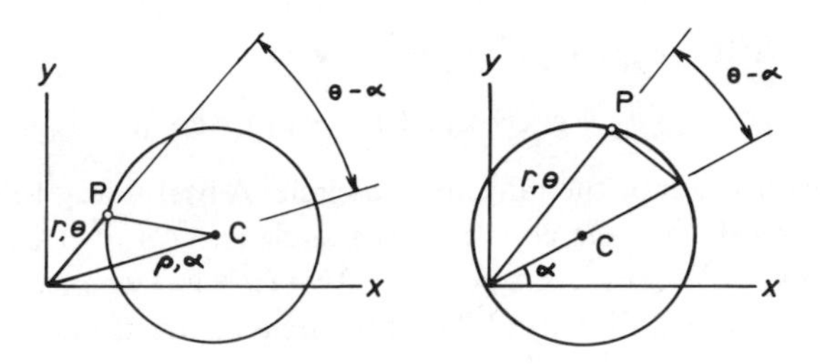

Figure 2.16

2.6.7 The tangent to a circle

Let $P(x_1y_1)$, $Q(x_2y_2)$ be two points near to each other on a circle.

The equation of the line PQ may be written

$$\frac{y - y_1}{x - y_1} = \frac{y_2 - y_1}{x_2 - x_1}$$

and since both points P and Q lie on the circle

$$x_1^2 + y_1^2 + 2gx_1 + 2fy_1 + c = 0$$
$$x_2^2 + y_2^2 + 2gx_2 + 2fy_2 + c = 0$$

On subtracting and rearranging

$$(x_2 - x_1)(x_1 + x_2 + 2g) + (y_2 - y_1)(y_1 + y_2 + 2f) = 0$$

whence

$$\frac{y_2 - y_1}{x_2 - x_1} = -\frac{x_1 + x_2 + 2g}{y_1 + y_2 + 2f}$$

is the equation to the chord PQ. If we let point Q approach point P, x_2 approaches x_1 and y_2 approaches y_1 and when P and Q are co-incident, $x_2 = x_1$ and $y_2 = y_1$. By putting $x_2 = x_1$ and $y_2 = y_1$ in the equation for the chord PQ we obtain the required equation for the tangent.

$$(x - x_1)(x_1 + g) + (y - y_1)(y_1 + f) = 0$$

$$x(x_1 + g) + y(y_1 + f) = x_1^2 + y_1^2 + gx_1 + fy_1$$

But the equation to a circle is

$$x_1^2 + y_1^2 + 2gx_1 + 2fy_1 + c = 0$$

and hence

$$x(x_1 + g) + y(y_1 + f) + gx_1 + fy_1 + c = 0$$

This equation may be re-arranged to give

$$xx_1 + yy_1 + g(x + x_1) + f(y + y_1) + c = 0 \tag{2.17}$$

and this is the equation of the tangent to a circle. A useful way to remember this equation is to recall that the equation to a circle is: $x^2 + y^2 + 2gx + 2fy + c = 0$ and put $x^2 = xx_1$, $y^2 = yy_1$, $2x = (x + x_1)$, $2y = (y + y_1)$ whence the equation to the tangent is obtained directly. When the circle centre is co-incident with the origin of co-ordinates $f = 0$ and $g = 0$ hence the equation to the tangent at the point $(x_1 y_1)$ reduces to

$$xx_1 + yy_1 = r^2$$

The equation to the tangent may alternatively be derived by differentiating the general equation. If the equation to a circle is

$$x^2 + y^2 + 2gx + 2fy + c = 0$$

$$2x + 2y\frac{dy}{dx} + 2g + 2f\frac{dy}{dx} = 0$$

whence the slope at the point $(x_1 y_1)$ is

$$\frac{dy}{dx} = -\left(\frac{x_1 + g}{y_1 + f}\right)$$

and the equation to the tangent through $(x_1 y_1)$ is

$$y - y_1 = -\left(\frac{x_1 + g}{y_1 + f}\right)\left(x - x_1\right)$$

2.6.8 The normal to a circle

The normal to a circle is the straight line which is perpendicular to the tangent and passes through the centre point (f, g) and the point $(x_1 y_1)$ on the circumference. The slope of the normal with respect to the axes xoy is $m = (y_1 - g)/(x - f)$ and the equation of the normal is

$$\frac{x - x_1}{x_1 + g} = \frac{y - y_1}{y_1 + f} \tag{2.18}$$

or

$$y(x_1 + g) - x(y_1 + f) + fx_1 - gy_1 = 0$$

whichever form is preferred.

2.6.9 The equations to an ellipse

An ellipse is the locus of a point P(x, y) which moves so that its distance from a fixed point F (the focus) is ϵ times its distance from a fixed straight line (the directrix). Where ϵ the eccentricity is less than 1.

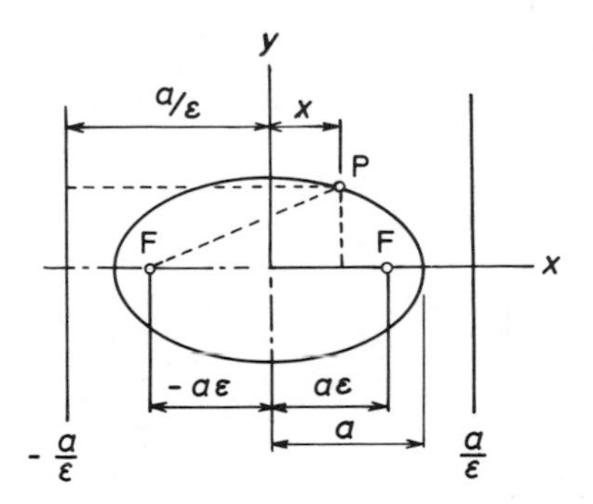

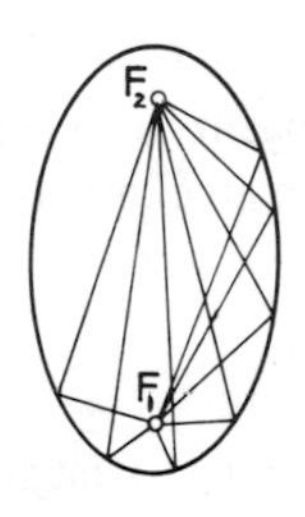

Figure 2.17

The equation to an ellipse assumes its simplest form when the focus is at ($-a$, o) and the directrix at ($-a/\epsilon$, o) as shown in Fig. 2.17.

$$FP = \epsilon PM \text{ by definition}$$

with reference to Fig. 2.17 $$\epsilon^2 \left(\frac{a}{\epsilon} + x\right)^2 = (x + a\epsilon)^2 + y^2$$

$$a^2(1 - \epsilon^2) = x^2(1 - \epsilon^2) + y^2$$

$$\frac{x^2}{a^2} + \frac{y^2}{a^2(1 - \epsilon^2)} = 1$$

or by writing $$a^2(1 - \epsilon^2) = b^2$$

$$\frac{x^2}{a^2} + \frac{y^2}{b^2} = 1 \tag{2.19}$$

This is the equation to an ellipse symmetrically disposed about the axes xoy, when the two foci of the ellipse are located at ($-a\epsilon$, o) and ($a\epsilon$, o) the major axis is aligned with the axis ox. When the two foci are at (o, $-a\epsilon$) and (o, $a\epsilon$) the major axis is aligned with the axis oy.

The alignment of an ellipse may be transformed to any other orientation by means of the appropriate transformation chapter 5.

It is easily proved that the sum of the focal distances of any point on an ellipse is equal to the major axis.

The length of the major axis is $2a$, the length of the minor axis is $2b$. Hence

$$F_1P + PF_2 = 2a$$

This equation enables an ellipse to be drawn using two pins, a pencil and a single piece of string.

2.6.10 The tangent to an ellipse

The equation to an ellipse is

$$\frac{x^2}{a^2} + \frac{y^2}{b^2} = 1$$

and the equation to a straight line

$$y = mx + b$$

The local slope of an ellipse may be obtained by differentiation, the local slope dy/dx equals m.

On differentiation equation (2.19), becomes

$$\frac{2x}{a^2} + \frac{2y}{b^2}\frac{dy}{dx} = 0$$

and hence

$$\frac{dy}{dx} = -\frac{b^2x}{a^2y}$$

The tangent at the point (x_1y_1) is equal to $-b^2x_1/a^2y_1$.

The tangent to an ellipse is the straight line

$$y - y_1 = -\frac{b^2x_1}{a^2y_1}(x - x_1)$$

whence

$$b^2xx_1 + a^2yy_1 = b^2x_1^2 + a^2y_1^2$$

and the equation of the tangent is:

$$\frac{xx_1}{a^2} + \frac{yy_1}{b^2} = 1 \qquad (2.20)$$

We note that this equation may be obtained by substituting $x_1 = x$, $y_1 = y$ in the equation for an ellipse.

2.6.11 The normal to an ellipse

The normal is the line which passes through the point P(x, y) and is perpendicular to the tangent at that point. The slope of the tangent is

$$-\frac{b^2 x_1}{a^2 y_1}$$

Hence the slope of the normal is

$$+\frac{a^2 y_1}{b^2 x_1}$$

The equation to the normal at the point P$(x_1 y_1)$ is

$$y - y_1 = \frac{a^2 y_1}{b^2 x_1}(x - x_1)$$

and this equation may be written as follows:

$$\frac{x - x_1}{x_1/a^2} = \frac{y - y_1}{y_1/b^2} \qquad (2.21)$$

This is the equation to the normal of an ellipse.

2.6.12 The Polar and tangent polar equation to an ellipse

We obtain the polar equation from the Cartesian equation by making the substitution $x = r\cos\theta$, $y = r\sin\theta$.

If

$$\frac{x^2}{a^2} + \frac{y^2}{b^2} = 1$$

then

$$\frac{\cos^2\theta}{a^2} + \frac{\sin^2\theta}{b^2} = \frac{1}{r^2} \qquad (2.22)$$

and this is the polar equation to the ellipse.

The tangent polar definition of a curve is a particularly useful way of specifying the trajectory of moving points (loci). In problems such as the flight of a rocket in which the propulsive and drag forces act tangentially to the flight path and lift and centripetal forces act normal to it, the tangent polar definition is an ideal form.

Any point on a curve may be defined in terms of its polar coordinates (r, θ) and the perpendicular distance p of its tangent.

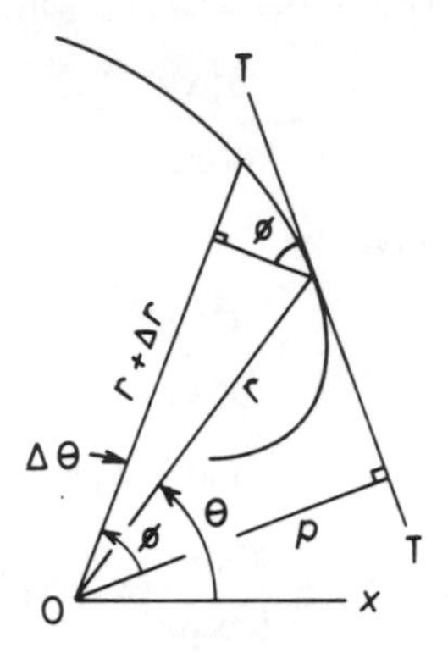

Figure 2.18

$$\cos \phi = \lim \theta \to 0 \quad \frac{r\Delta\theta}{(r^2(\Delta\theta)^2 + (\Delta r)^2)^{\frac{1}{2}}}$$

and on dividing the R.H.S. top and bottom by $\Delta\theta$

$$\cos \phi = \frac{r}{\left(r^2 + \left(\frac{\Delta r}{\Delta\theta}\right)^2\right)^{\frac{1}{2}}}$$

and hence in the limit as $\Delta\theta \to 0$

$$\cos \phi = \frac{r}{\left(r^2 + \left(\frac{\partial r}{\partial\theta}\right)^2\right)^{\frac{1}{2}}}$$

and since $\cos \theta = p/r$

$$\frac{1}{p^2} = \frac{1}{r^2} + \frac{1}{r^4}\left(\frac{\mathrm{d}r}{\mathrm{d}\theta}\right)^2$$

which is the polar tangent equation to a curve.

2.6.13 A practical application of the ellipse

Whilst not wishing to open old wounds it is instructive to recall the Comet disasters of thirty years ago. Suffice to say that the shape design of the windows was partly to blame. For the purpose of the present appraisal it is sufficient to consider the effect of window cut-outs in a thin walled pressurised cylinder. If we consider only the pressurisation loads in a circular cylindrical aircraft fuselage (and such an assumption gives a fair representation of the stresses in the forward fuselage of Concorde) it is easily proved that the circumferential (hoop) stresses are twice the longitudinal stresses and on this basis, a moderately good window can be designed. As a first approximation and for the purposes of this discussion we ignore the effect of curvature. That is, we assume that the window cut-outs are small in comparison with the fuselage radius (the windows in Concorde are indeed quite small and again this, for our purpose, is a fair assumption).

If we begin by considering the effect of a slit (crack) in the fuselage skin we can easily see that a crack of given length would be less likely to propagate if orientated as in Fig. 2.19(a) than if orientated as in Fig. 2.19(b).

This is because the effect of the crack, Fig. 2.19(a), has hardly any effect on the distribution of force (stress) lines in the hoop direction, whereas the same crack when broad-side on would cause a considerable deviation of the force (stress) lines as shown in Fig. 2.19(b). N.B. In this simple explanation we have assumed that the maximum nominal stress occurs in the hoop direction and have ignored the effect of the longitudinal stresses, due to fuselage bending. We also bear in mind that in situations where a fatigue cycle is expected (i.e. repeated cycles of pressurisation and depressurisation in this case), it is important to make all structural transitions as smooth and as gentle as possible. Now suppose we transform our slit (crack) into a very slender ellipse and then fatten it up, with a view to making a sensible window opening. It may be shown without too much difficulty that the best shape for a window in a flat sheet loaded by a 2:1 stress field, is an ellipse with major and minor axes in the ratio 2:1, and with major axis aligned with the direction of the largest stress. It may also be shown that if the elliptical window is set the right way round, the local stress concentration factor is $K = 1.5$. If on the other hand, the elliptical window is set the wrong way round the local stress concentration factor is $K = 4.5$. A window set the wrong way round would thus be far more prone to fatigue.

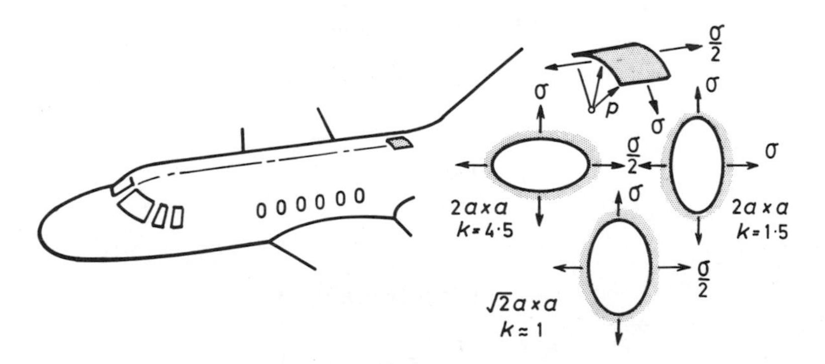

Figure 2.19

In practice the designer would need to form a frame around the window opening and with the aid of the stress man would shape and proportion the frame in such a way that the level of local stresses be further reduced.

A mathematical solution to this idealised problem was supplied by Mansfield in 1953 and we observe that this theoretical solution has a theoretical stress concentration factor of $K = 1.0$. The advantage of the Mansfield neutral hole is that it does not disturb the 2:1 stress pattern produced by pressurisation and therefore improves the structure against fatigue. We concluded by observing that the optimum proportions of a reinforced ellipse is not 2:1 but $2{:}\sqrt{2}$. Such is the finesse of geometrical shape design.

2.6.14 The equations to a hyperbola

An hyperbola is the locus of a point $P(x, y)$ which moves so that its distance from a fixed point F (the focus) is ϵ times its distance from a fixed straight line (the directrix), where ϵ the eccentricity is greater than 1.

When the focus is at the point $(-a\epsilon, o)$ and the directrix is the line $x = -a/\epsilon$, the equation to a hyperbola is obtained as follows:

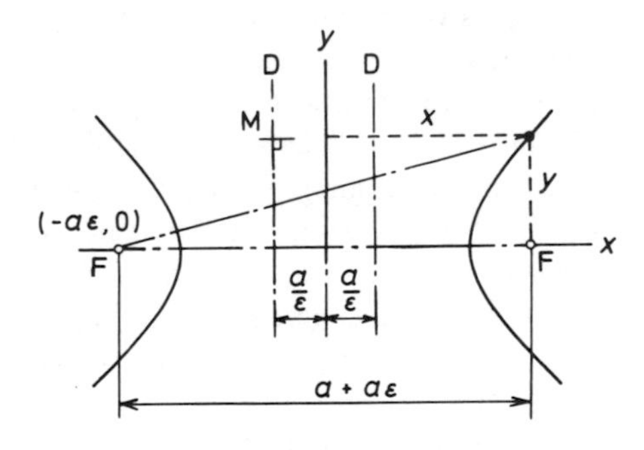

Figure 2.20

With reference to Fig. 2.20,

$$FP = \epsilon PM \text{ (by definition)}$$

$$FP^2 = \epsilon^2 \, PM^2$$

Now

$$PM = \left(x + \frac{a}{\epsilon}\right)$$

and hence

$$FP^2 = (x + a\epsilon)^2 + y^2$$

$$\epsilon^2 \left(x + \frac{a}{\epsilon}\right)^2 = (x + a\epsilon)^2 + y^2$$

whence

$$\frac{x^2}{a^2} - \frac{y^2}{a^2(\epsilon^2 - 1)} = 1$$

and by writing

$$b^2 = a^2(\epsilon^2 - 1)$$

$$\frac{x^2}{a^2} - \frac{y^2}{b^2} = 1 \tag{2.23}$$

This is the equation to an hyperbola.

2.6.15 The tangent of a hyperbola

The equation to an hyperbola is:

$$\frac{x^2}{a^2} - \frac{y^2}{b^2} = 1$$

The equation to a straight line is:

$$y = mx + b$$

The local slope of an hyperbola may be obtained by differentiation; the local slope dy/dx equals m.

On differentiating equation (2.23), we obtain

$$\frac{2x}{a^2} - \frac{2y}{b^2}\frac{dy}{dx} = 0$$

whence

$$\frac{dy}{dx} = + \frac{b^2 x}{a^2 y}$$

The tangent at the point $(x_1 y_1)$ is $+b^2 x_1/a^2 y_1$.

The equation to the tangent at the point $(x_1 y_1)$ is

$$y - y_1 = + \frac{b^2 x_1}{a^2 y_1}(x - x)$$

from which

$$a^2 y y_1 - a^2 y_1^2 = bxx_1 = b^2 x_1^2$$

$$\frac{yy_1}{b^2} - \frac{y_1^2}{b^2} = \frac{xx_1}{a^2} + \frac{x_1^2}{a^2}$$

and hence

$$\frac{xx_1}{a^2} - \frac{yy_1}{b^2} = \frac{x_1^2}{a^2} - \frac{y_1^2}{b^2} = 1$$

whence

$$\frac{xx_1}{a^2} - \frac{yy_1}{b^2} = 1 \tag{2.24}$$

We note that this equation may be obtained by substituting $x_1 = x$, $y_1 = y$, in the equation for an hyperbola.

2.6.16 The normal to an hyperbola

The normal is the line which passes through the point $P(x_1 y_1)$ and is perpendicular to the tangent at that point.

The slope of the tangent is:

$$+\frac{b^2 x_1}{a^2 y_1}$$

Hence the slope of the normal is:

$$-\frac{a^2 y_1}{b^2 x_1}$$

The equation to the normal at the point $P(x_1 y_1)$ is:

$$y - y_1 = -\frac{a^2 y_1}{b^2 x_1}(x - x_1)$$

and this equation may be rewritten as follows:

$$\frac{x - x_1}{x_1/a^2} = \frac{y - y_1}{y_1/-b^2} = -\frac{(y - y_1)}{y_1/b^2} \tag{2.25}$$

2.6.17 The rectangular hyperbola

A special case of the hyperbola is that which results when the axes of coordinates and the asymptotes coincide: With reference to Fig. 2.21,

$$\begin{aligned} OK &= OL + LK \\ &= PM \cos 45 + OM \cos 45 \\ PK &= PN - NK = PN - LM \\ &= PM \sin 45 - OM \sin 45 \end{aligned}$$

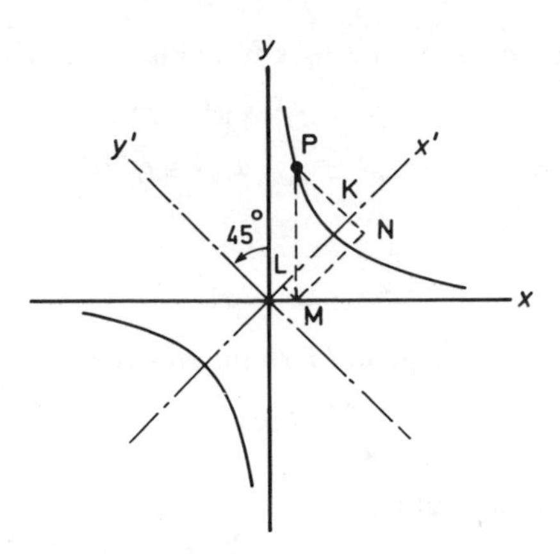

Figure 2.21

If the point P is referred to the ox, oy, axes OM = x, MP = y. By substituting into the equation above we obtain:

$$\mathrm{X} = \frac{(y+x)}{\sqrt{2}}, \quad \mathrm{Y} = \frac{(y-x)}{\sqrt{2}}$$

hence

$$\frac{(y+x)^2}{2} - \frac{(y-x)^2}{2} = a^2$$

and

$$xy = \frac{a^2}{2}$$

If

$$c^2 = \frac{a^2}{2}$$

$$xy = c^2 \qquad (2.26)$$

and this is the equation to a rectangular hyperbola.

2.6.18 The equation to a parabola

A parabola is the path of a point which moves in such a way that its distance d from a fixed focus F is equal to the distance s from a fixed straight line, the directrix.

Since $d = s = x$ (by definition) we see from Fig. 2.22 that

$$(2a - x)^2 + y^2 = x^2$$

$$4a^2 - 4ax + y^2 = 0$$

whence

$$y^2 = 4a(x - a) \tag{2.27}$$

If the origin is placed at the point (a, o) the equation above reduces to the form

$$y^2 = 4ax. \tag{2.28}$$

2.6.19 The tangent to a parabola

If we differentiate the equation to a parabola we obtain the measure of its local slope $dy/dx = m$ and can express the tangent in straight line form. Now, on differentiating equation (2.27), we obtain $2y\, dy/dx = 4a$. Whence

$$\left(\frac{dy}{dx}\right) = \frac{2a}{y_1}$$

$$y - y_1 = \frac{2a}{y_1}(x - x_1)$$

and the equation to the tangent at the point $(x_1 y_1)$ is

$$yy_1 - y_1^2 = 2a(x - x_1) \tag{2.29}$$

2.6.20 The normal to a parabola

Since the normal is perpendicular to the tangent the slope of the normal is equal to $-1/m = -dx/dy$.

Hence the equation to the normal is written

$$y - y_1 = \frac{y_1}{2a}(x - x_1)$$

$$\frac{y - y_1}{y_1} = \frac{x - x_1}{2a} \tag{2.30}$$

2.6.21 Arch bridges, a practical application of the parabola

A small structure which carries load(s) which are transverse to its length is called a beam. Beams of rectangular or I cross-section carry these loads in bending and their span is strictly limited by the tensile/compressive strength of the materials of which they are made. In order to carry transverse loads over a significant span it is necessary to develop the concepts of a simple beam into a bridge.

If we consider the simplest possible case of a uniformly distributed load we find that the bending moment (the principal cause of stress and deflection in long beams and arch bridges) varies at various sections across the span as follows.

The reaction to a total uniformly distributed load wl is $R = wl/2$. The shear force due to a small element of length of loaded structure is wx. The moment of this loaded element is $\left(wx.\frac{x}{2}\right)$. If the sign of the reaction R is positive the sign of the elemental load is negative and hence the bending moment due to a small element of length is

$$m = \frac{wl}{2} - \frac{wx^2}{2}$$

The build-up and decay of bending moment across the span is thus of a parabolic form, and since the bending moment is the principal source of stress and strain in a beam and bridge, and since the bending moment varies according to a parabolic law, and since beams and bridges of uniform width are usually desirable, the effective structural depth of the bridge is varied in such a way that it everywhere matches the bending moment (in practice the depth near to the abutments needs to be sufficient to carry shear loads but apart from this the arch of a uniformly loaded bridge corresponds to the sag of a suspension cable, both are of parabolic form).

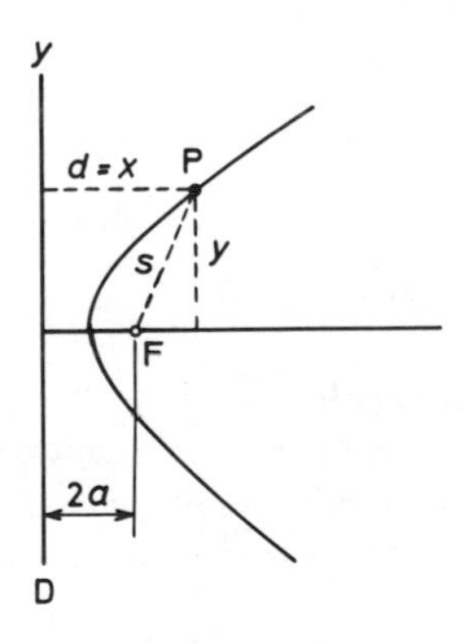

Figure 2.22

2.7 PARAMETRIC EQUATIONS TO CONICS

The form of the equations so far considered is such that, regardless of whether the equation is in the explicit or implicit form, we find it necessary to allocate a value to either x or y and thence determine the corresponding value of its partner. We may, however, avoid this difficulty by rewriting the equation in parametric form. One important advantage of using a parametric format is that by defining a parameter t, say, such that both x and y are functions of t, we are able to generate points on a curve directly, without limiting the outcome to a particular value of x or y.

2.7.1 Parametric equation to a straight line

Given the co-ordinates of the end points of a straight line of unit length, we may readily derive its parametric equations. If $t = \text{AP}$ is a linear measure of the position of a general point P along the line AB, where AB is of unit length, then the coordinates of P may be expressed as follows:

$$x' = t \cos \theta \text{ and } y' = t \sin \theta$$

where x' and y' are the coordinates of P relative to the origin A. However,

$$\cos \theta = \frac{(x_2 - x_1)}{l} \text{ and } \sin \theta = \frac{(y_2 - y_1)}{l}$$

hence when l is of unit length

$$x' = t(x_2 - x_1) \text{ and } y' = t(y_2 - y_1)$$

To obtain the coordinates of the point P, referred to the original axes xoy we merely add the x and y coordinates of the point A to the x' and y' coordinates above.

$$\begin{aligned} x &= x_1 + t(x_2 - x_1) \\ y &= y_1 + t(y_2 - y_1) \end{aligned} \tag{2.31}$$

are thus the parametric equations required.

2.7.2 Parametric equation to a circle

Consider the case of a circle of unit radius with centre at (0, 0).

As shown in Fig. 2.23, the coordinates of a general point P, on the unit circle are $x = \cos \theta = t$ say, $y = \sin \theta = \sqrt{1 - t^2}$ hence $x = t$, $y = \sqrt{1 - t^2}$ are the para-

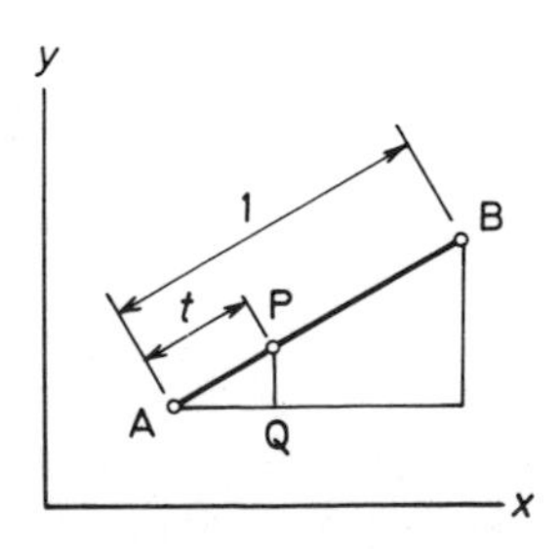

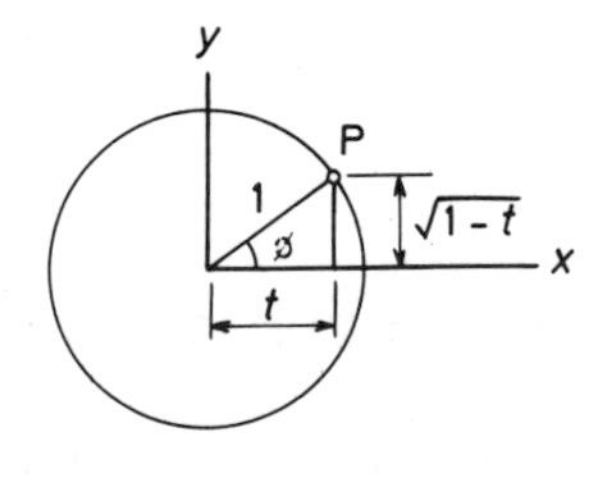

Figure 2.23

metric equations of an on-centre circle. One advantage of this approach is that any number of (x, y) coordinate pairs may be obtained by specifying a single variable t. In the particular case of a unit circle t must clearly lie between the limits of $\cos\theta$, that is t must not lie outside the range ± 1.

A few values of t, x, y, are listed in Table 2.1.

Table 2.1

t	−1.0	−0.8	−0.6	0	0.6	0.8	1.0
x	−1.0	−0.8	−0.6	0	0.6	0.8	1.0
y	0	−0.6	−0.8	1.0	0.8	0.6	0

The x, y coordinates, Table 2.1, apply to a circle of unit radius, that is to the circle $x^2 + y^2 - 1 = 0$. If the x, y coordinates appropriate to some other radius are needed, we simply multiply the coordinates above by the radius of the circle required. That is we write:

$$\begin{aligned} x &= r\cos\phi, \quad y = r\sin\phi \\ x &= rt, \qquad y = r\sqrt{1-t^2} \end{aligned} \tag{2.32}$$

2.7.3 Parametric equations of an ellipse

An ellipse may be considered as an oblique view of a circle. If $2a$ and $2b$ denote the major and minor axes, the parametric equation is written:

$$\begin{aligned} x &= a\cos\phi, y = b\sin\phi \\ x &= at, \qquad y = b\sqrt{1-t^2} \end{aligned} \tag{2.33}$$

we see that:

$$\frac{dx}{d\phi} = -a\sin\phi \quad \text{and} \quad \frac{dy}{d\phi} = b\cos\phi$$

and hence

$$\frac{dy}{dx} = -\frac{b}{a}\cot\phi = \frac{-b}{a}\,\frac{t}{\sqrt{1-t^2}}$$

The tangent is thus the line through the point $(a\cos\phi, b\sin\phi)$, and the equation to the tangent is

$$y - b\sin\phi = -\frac{b}{a}\cot\phi(x - a\cos\phi)$$

and this equation may be rewritten to give:

$$\frac{x}{a}\cos\phi + \frac{y}{b}\sin\phi = 1$$

$$\frac{xt}{a} + \frac{y}{b}\sqrt{1-t^2} = 1 \tag{2.34}$$

The normal is the line through the point (a cos θ, b sin θ) is the line for which

$$\frac{dy}{dx} = \frac{a}{b}\tan\phi = \frac{a}{b}\,\frac{(1-t^2)}{t}$$

and hence the equation to the normal may be written as:

$$y - b\sin\phi = \frac{a}{b}\tan\phi\,(x - a\cos\phi)$$

and this equation may be simplified to give:

$$ax\sec\phi - by\,\mathrm{cosec}\,\phi = a^2 - b^2$$

$$\frac{ax}{t} - \frac{by}{\sqrt{1-t^2}} = a^2 - b^2 \tag{2.35}$$

2.7.4 Parametric equation to a hyperbola

Although a hyperbola does not have a real angle which directly corresponds with the angle ϕ as demonstrated for the circle and the ellipse, it is neverthless possible to write parametric equations which describe the hyperbola completely.

We observe that $\sec^2\phi = 1 + \tan^2\phi$ and hence $\sec\phi = \sqrt{1 + \tan^2\phi}$, whence

$$x = a\sec\phi,\ y = b\tan\phi \tag{2.36}$$

and this statement may be proved by direct substitution into the equation

$$\frac{x^2}{a^2} - \frac{y^2}{b^2} = 1$$

We also observe that

$$\frac{dx}{d\phi} = a\sec\phi\tan\phi \quad \text{and} \quad \frac{dy}{d\phi} = b\sec^2\phi$$

The tangent through the point (a sec ϕ, b tan ϕ) with slope b sec ϕ/a tan ϕ is

$$y - b\tan\phi = \frac{b\sec\phi}{a\tan\phi}(x - a\sec\phi)$$

and this reduces to

$$\frac{x}{a}\sec\phi - \frac{y}{b}\tan\phi = 1$$

$$\frac{x}{at} - \frac{y\sqrt{1-t^2}}{bt} = 1 \tag{2.37}$$

The normal through the point (a sec ϕ, b tan ϕ) with slope $-a$ tan ϕ/b sec ϕ is the line

$$y - b\tan\phi = \frac{-a\tan\phi}{b\sec\phi}(x - a\sec\phi)$$

and this reduces to

$$ax\sin\phi + by = (a^2 + b^2)\tan\phi.$$

$$ax + by\sqrt{1-t^2} = (a^2 + b^2)\ \frac{1-t^2}{t} \tag{2.38}$$

2.7.5 Parametric equation to a rectangular hyperbola

The Cartesian equation to a rectangular hyperbola is $xy = c^2$. The parametric equations are:

$$x = ct \quad y = \frac{c}{t}$$

$$\frac{dx}{dt} = c, \quad \frac{dy}{dt} = \frac{-c}{t^2} \tag{2.39}$$

$$\frac{dy}{dx} = \frac{-1}{t^2}$$

and the equation to the tangent is:

$$y - \frac{c}{t} = \frac{-1}{t^2}(x - ct) \tag{2.40}$$

from which

$$y = \frac{-x + 2ct}{t^2}$$

The equation to the normal is:

$$y - \frac{c}{t} = t^2(x - ct)$$

whence

$$y = t^2(x - ct) + c. \tag{2.41}$$

2.7.6 Parametric equation to a parabola

It is easy to verify, by substitution, that the parametric equations:

$$x = at^2, y = 2at \tag{2.42}$$

satisfy the equation:

$$y^2 = 4ax$$

and hence the parabola is completely defined by the parametric equations given.

$$\frac{dx}{dt} = 2at \quad \text{and} \quad \frac{dy}{dt} = 2a$$

hence

$$\frac{dy}{dx} = \frac{1}{t}$$

The equation to the tangent line through the point $(at^2, 2at)$ is:

$$y - 2at = \frac{1}{t}(x - at^2) \tag{2.43}$$

and the equation to the normal through the point $(at^2, 2at)$ is:

$$y + tx = 2at + at^3 \tag{2.44}$$

Example

Use the parametric definition of a parabola to prove that the angle of incidence and the angle of refraction are equal.

The coordinates of P are $(at^2, 2at)$. The coordinates of S are (a, o). We see from Fig. 2.24, that the length of the incident ray is SP where

$$SP^2 = (at^2 - a)^2 + (2at - o)^2 = a^2(t^2 + 1)^2$$

The equation to the tangent TQ is

$$x - ty + at^2 = 0$$

which intersects the ox axis at $y = 0$.

Hence the parametric coordinates of the point Q are $(-at^2, 0)$.

The length $OQ = at^2$ and the length $OS = a$ and hence the length $QS = a(t^2 + 1)$ and we see that QS = SP, that is the triangle QSP is isosceles, which means that $\angle PQS = \angle SPQ$. If PR (the reflected ray) is parallel to ox $\angle TPR = \angle PQS = \angle SPQ$.

Hence the incident ray SP and the reflected ray PR are equally inclined to the tangent through P. It follows that the angles SPN and NPR are also equal and hence the angle of incidence equals the angle of reflection.

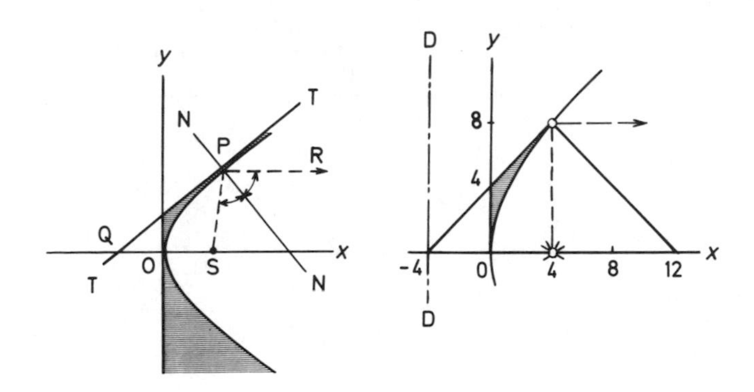

Figure 2.24

2.8 MORE LOCI

A locus is the path of a point which moves in accord with some fixed mathematical rule-in mechanisms imposed by mechanical constraint. The word loci is the plural of locus. The conic sections are loci, and could well have been considered under this heading. The simplest of all loci is perhaps a circular arc, which is the path mapped out by a point constrained to move at a constant distance from a fixed centre. A slightly more complex case is that of an ellipse, in which a point is constrained by two points to move in such a manner that the sum of its distances from these points, the foci is constant.

2.8.1 The catenary

The curve adopted by a completely flexible chain of uniform density, that hangs from its ends in a vertical plane is a catenary. Galileo Gallilei (1599) seems to have been the first to speak of this curve but mistook it for a parabola. The curve was, however, correctly identified by James Bernoulli (1691).

Since the shape of a catenary is the natural shape adopted by a flexible chain when hanging in a vertical gravitational field, its shape is derived by considering the equilibrium of forces.

If, in Fig. 2.25, the weight of the chain per unit length is w , then a length of chain s weighs ws and this downward force is supported by the horizontal tension

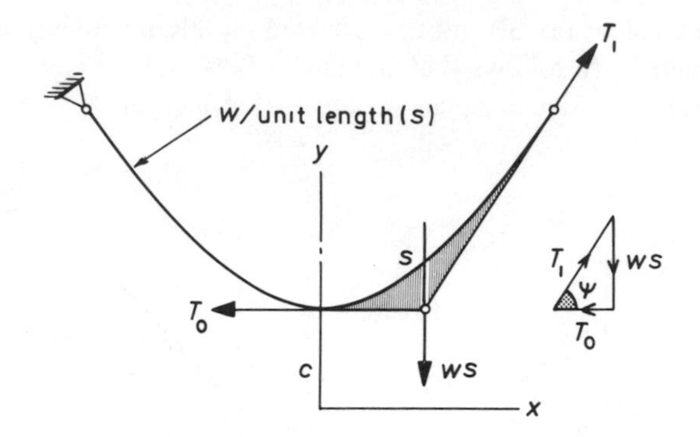

Figure 2.25

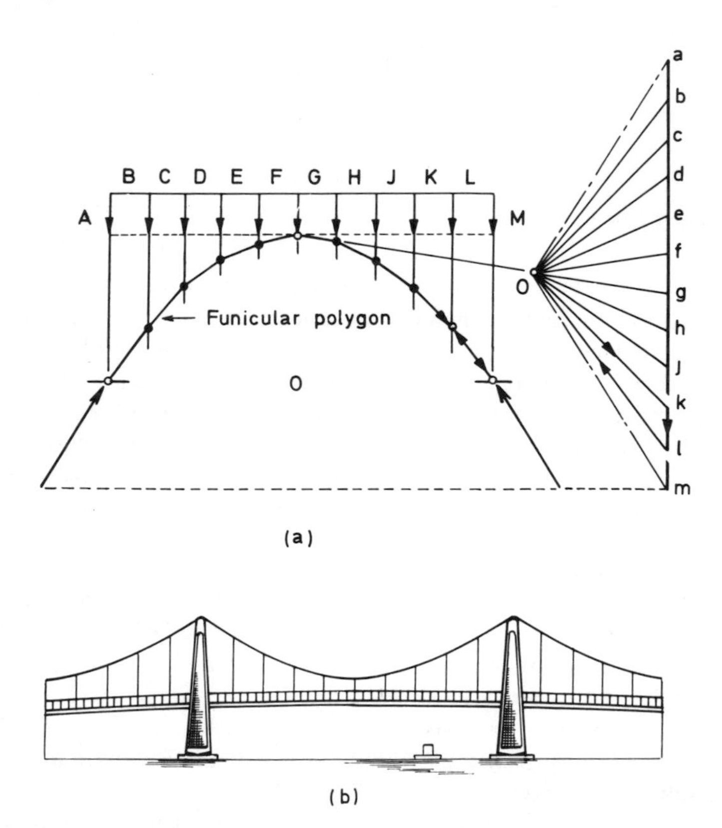

Figure 2.26

T_0 and the inclined tension T_1 as shown in Fig. 2.25. If the inclination of T_1 is ψ to the horizontal

$$T_1 \cos \psi = T_0 \text{ and } T_1 \sin \psi = ws$$

whence

$$T_0 = wC$$

where $C = s/\tan \psi$.

The derivation of the x and y coordinates of the catenary curve is lengthy and the reader is referred to the relevant mathematical texts on the subject. We merely quote the coordinates here.

$$\begin{aligned} x &= C \log (\sec \psi + \tan \psi) \\ y &= C \sec \psi . \end{aligned} \tag{2.45}$$

As mentioned above, it is a common slip to mistake a catenary for a parabola and the difference between the two should be noted. The catenary is the curve adopted by a chain or string of uniform density. The parabola is the natural shape of a cable used to support a uniformly distributed horizontal load, a condition approached in some suspension bridges.

2.8.2 The involute

The path taken by the free end of a taut string, as it unwinds from a uniform circular cylinder is the common involute. Involutes were first studied by Huygens (1695).

The straight lines (P_1T_1), (P_2T_2), (P_3T_3), Fig. 2.27(a), represent the taut string as it unwinds. The lines (P_1T_1), (P_2T_2) are tangential to the base circular cylinder at points T_1 and T_2 and normal to the curve through the points P_1 and P_2.

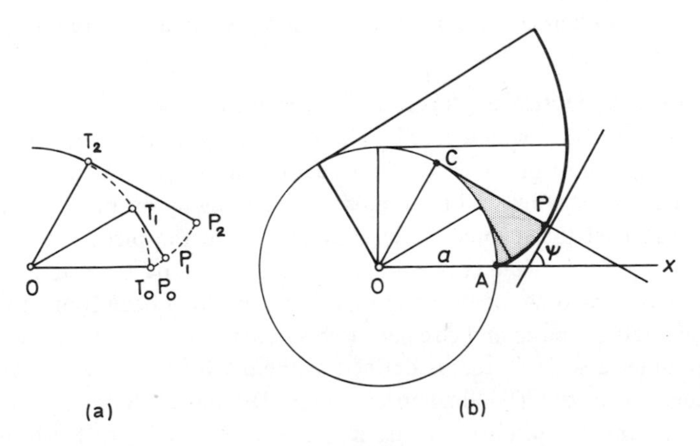

Figure 2.27

The lengths of the lines (P_1T_1), (P_2T_2) are clearly equal to the unwrapped arc lengths (T_0T_1), (T_0T_2), respectively.

Since the lines (P_1T_1), (P_2T_2) are normal to the involute, the points T_1, T_2, T_3, etc., lie on the involute.

It is also clear that every point on the string scribes its own involute and all involutes which have the same generating circle are identical curves. We may regard these involutes as being either parallel to, or displaced from, each other.

The intrinsic equation of the involute may be derived as follows:

Let the radius of the circle or cylinder equal OA $= a$. Let the radius of curvature at the point P equal, CP $= \rho$. Let the tangent to the involute at P make an angle ψ with the *ox* axis.

$$CP = AC$$

and since OC is parallel to the tangent TT

$$CP = AC = a\psi$$

$$CP = \rho = \frac{ds}{d\psi}$$

where s is the arc length AP

$$\frac{ds}{d\psi} = a\psi$$

We have only to integrate this expression to obtain the intrinsic equation to the involute. Thus:

$$s = \frac{a\psi^2}{2} \tag{2.46}$$

(N.B. Since $s = 0$ when $\psi = 0$, no constant of integration is required.

2.8.3 Gearing–a practical application of the involute curve

The involute curve is widely used as a practical gear tooth profile. From a manufacturing point of view involute teeth can be generated by means of a straight flanked cutting tool (as opposed to shaping by means of a form tool), and in this respect the involute form is preferable to the piecewise coupling of epicycloid and hypocycloid curves, once much favoured by the watch and clock-making trades. The form of involute gear teeth has long been standardised. The module, diametrical pitch and circular pitch systems are in common use.

The module (a scale factor) is defined as the pitch circle diameter divided by the number of teeth. The diametrical pitch DP being the reciprocal of the module. The circular pitch is the arc length between adjacent teeth, measured along the pitch circle circumference.

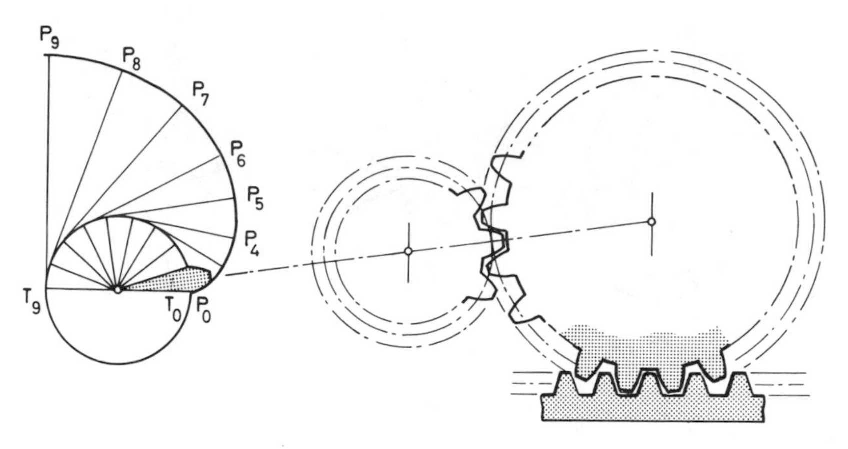

Figure 2.28

The virtue of these systems is that any two gears which have the same scale factor and the same pressure angle may be meshed together so as to form an efficient means of transmitting power. The meshing of two gears of differing diameters may thus be used to effect a speed ratio between a driving and a driven shaft.

Within reasonable limits any tooth profile may be chosen for one gear and the other profile designed to suit. In the interests of interchangeability, however, it is obviously advantageous to have the same profile on all gears of a given tooth size. Hence the system described above. There are also practical considerations which tend to favour one particular shape rather than another. Let us consider, however, the general requirements of shape.

From the angular velocity ratio theorem we know that the respective velocities of a driver and a driven element are inversely proportional to the perpendiculars drawn from the centres of rotation to the line of transmission. The truth of this statement is, of course, self-evident from Fig. 2.29, but readers requiring a mathematical proof are referred to the relevant texts on the theory of machines. It is, however, sufficient to say here that:

$$\frac{\omega_1}{\omega_2} = \frac{Q_2B}{Q_1A}$$

The instantaneous motion is thus identical to that which would be imparted by the pressure contact of two friction discs of diameters O_1A and O_2B which just touched at P. Notice, therefore, that if the ratio of speeds ω_1/ω_2 is to remain constant for all subsequent motions, then the common normal to the point of contact must necessarily pass through P. This, in fact, is the fundamental condition on which the form design of constant ratio toothed gears is based.

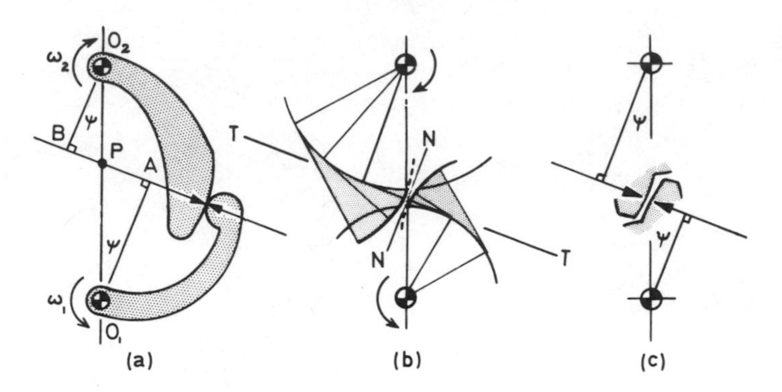

Figure 2.29

The design of disc gears which are required to impart a constant velocity ratio is thus governed by the following condition. The common normal at the point of contact C must pass through a fixed point P on a line joining the centres O_1 and O_2. The angle ψ is called the pressure angle and for modern gears is invariably equal to 20°.

We notice, from Fig. 2.29, that two tangent circles have been drawn to touch the line of action at the points A and B as defined by the previous discussion. These two circles are known as base circles. If we now imagine a cord (or belt) wrapped round the base circles as shown at (b) then by rotating one wheel the other may be induced to rotate. If we cut the cord at some arbitrary distance along the line AB, then we have two free ends of cord each of which emerges from its anchor disc in a direction tangential to its circumference. If, keeping the free length of cord taut, we map out the path of its end as we wind it on and off its associated disc, then we find that the path traced out (relative to the anchor disc) is an involute curve. The same is, of course, true of the other free end.

It follows from the observations made above that if we arrange to carry teeth on our discs, the profile of which is identical to the involute curves just described, then since tooth contact would be restricted to the point Q and since point Q lies on the line of action AB, which line by definition passes through P.

We have a tooth form which meets our requirement of trasmitting rotation at a constant speed ratio

$$\frac{\omega_1}{\omega_2} = \frac{R_2}{R_1}$$

For continuous rotation we must choose the position of Q so that a whole number of teeth are contained within each circumference. We must also ensure that before contact between one pair of mating teeth is lost the next pair of teeth have established contact and a satisfactory condition of working is usually ensured if the contact ratio between mating gears is greater than 1.4. In practical terms this means that contact must be maintained between two mating pairs of teeth for at least 40% of the time that a particular pair of teeth are in contact. N.B. The arc of action is the pitch circle arc through which a tooth moves from the beginning to the end of contact with a mating tooth. The contact ratio is the ratio of the arc of action to the circular pitch.

2.8.4 The lemniscate

Jacques Bernoulli (1695) first described the lemniscate as a mathematical function and James Watt (1736-1819) used a portion of a lemniscate curve, Fig. 2.30, as a means of obtaining approximate straight line motion.

The polar equation of a lemniscate is of the form:

$$r^2 = a^2 \cos 2\theta \text{ or } r^2 = a^2 \sin 2\theta$$

whence

$$(x^2 + y^2)^2 = a^2(x^2 - y^2)(x^2 + y^2) = 2a^2 xy$$

The radius of curvature is given by the equation:

$$\rho = \frac{a^2}{3r} = \frac{a}{3(\cos 2\theta)^{1/2}}$$

when AD = BC = a, AB = CD = $a\sqrt{2}$ and P is the mid-point of DC.

$$r^2 = a^2 \cos 2\theta \tag{2.47}$$

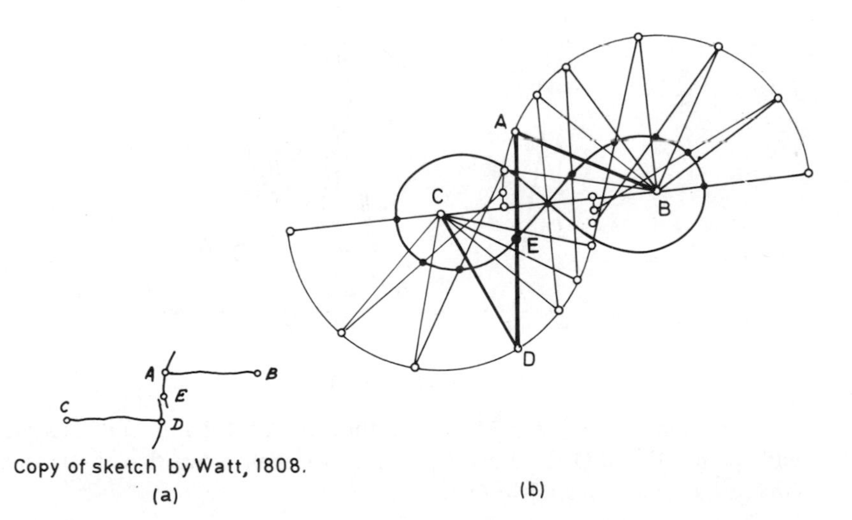

Figure 2.30

When OA = AB = a and BC = CP = OC = $a/\sqrt{2}$ and angle BOP

$$r^2 = \text{BP}^2 = \text{OB}^2$$
$$= 2a^2 - 4a^2 \sin\theta$$

whence

$$r^2 = 2a^2 \cos\theta \tag{2.48}$$

2.9 LOCI AND LINK MECHANISMS

The study of loci is of immense importance in the design of link mechanisms and is directly relevant to the current development of robots.

Prior to the advent of the electronic digital computer the synthesis of curves traced out by various points on a mechanism was an undertaking that depended on manual draughting skill. Today a far wider variety of mechanisms may be synthesised and checked out on the computer.

2.9.1 The Peaucllier straight line motion

The eight link planar mechanism shown in Fig. 2.31, was devised by Peaucllier (1864) and is the first of its type. N.B. In counting the links in a mechanism we always include the ground link as first recommended by Franz Realeaux (1839-1905).

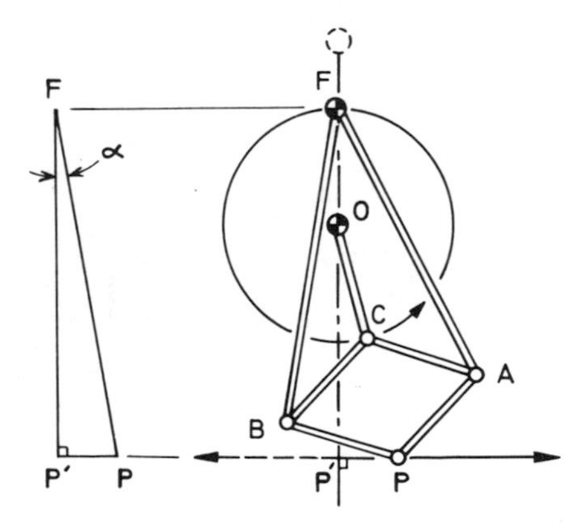

Figure 2.31

With reference to Fig. 2.31, we see that if FA = FB and APBC is a rhombus, with points F and O fixed pivot positions we shall prove that the motion of P (that is its locus) is a straight line.

$$\cos\alpha = \frac{\text{FP}'}{\text{FP}} = \frac{\text{FC}}{\text{FC}'}$$

whence

$$FP \cdot FC = FP' \cdot FC'$$

and by Pythagoras

$$\begin{aligned} FB^2 - BP^2 &= FM^2 - MP^2 \\ &= (FM + MP)(FM - MP) \\ &= FP \cdot FC = FP' \cdot FC' \end{aligned}$$

Now since FB, BP and FC′ are constant FP′ must be constant and P′ the projection of P onto the perpendicular through FO is always the same point. That is, the point P moves in a straight line in a direction perpendicular to the line through FO. The locus P′P is a straight line providing FA = FB and APBC is a rhombus, otherwise the locus P′P is a curve.

(N.B. This eight link mechanism (see article 7.1.7) was the first straight line linkage invented.

2.9.2 The pantograph

A well known mechanism widely used as a scaling device is the pantograph, several examples of which are shown in Fig. 2.32.

Pantographs have been used for many years as a means of producing reduced scale drawings and engravings. They are still used in many workshops where traditional copy form work is produced but are giving way to the introduction of N.C. machine tool methods.

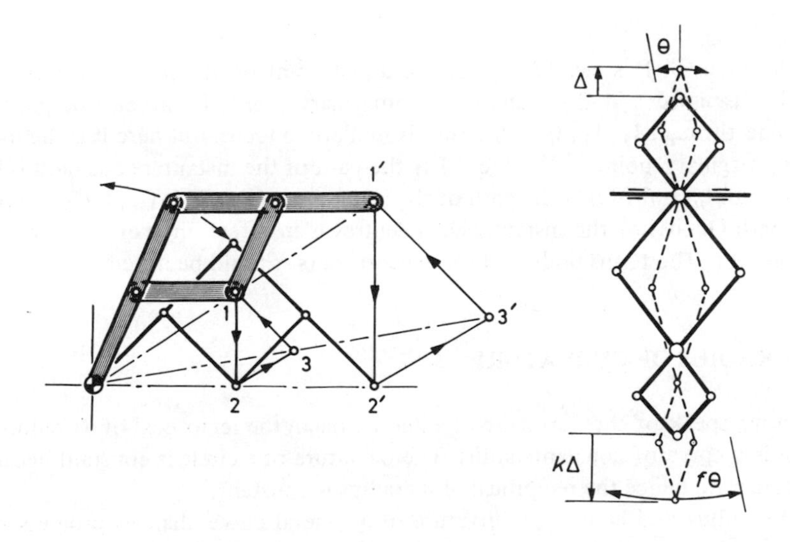

Figure 2.32

2.9.3 Instant centres

An inspection of the four bar chain, Fig. 2.33, shows that if the lines AB and DC (not the links) are produced they intersect at a point I, where I, may be regarded as the instantaneous centre of rotation. By this we mean that the instantaneous rotation of link AB may be regarded as being about the centre I, for the instant the motion is frozen.

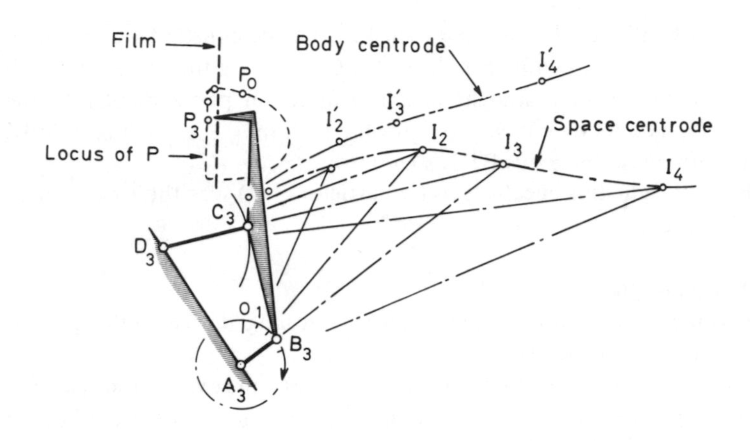

Figure 2.33

The locus of P is clearly the path of a real point on the mechanism as summarily discussed and the locus of the imaginary points I may also be plotted. The line through I_1, I_2, I_i. . .I_n, etc., is similarly a locus, but here it is the locus of an imaginary point I. The locus I is the path of the instantaneous radii $I_1 B_1$, $I_2 B_2$. . .$B_n I_n$ and is also the path of the instantaneous radii $I_1 C_1$, $I_2 C_2$. . .$I_n C_n$. The path (locus) of the instantaneous centres is in effect an evolute curve. See article 2.11. The terms body and space centrode(s) should be noted.

2.10 RADIUS OF CURVATURE

When we speak of the curvature of a line we mean the reciprocal of its radius. A circle is a curve of constant radius. The curvature of a circle is constant because its radius and hence the reciprocal of its radius is constant.

The radius and hence the curvature of a general curve changes progressively from point to point. The curvature of a curve is defined as the rate at which the

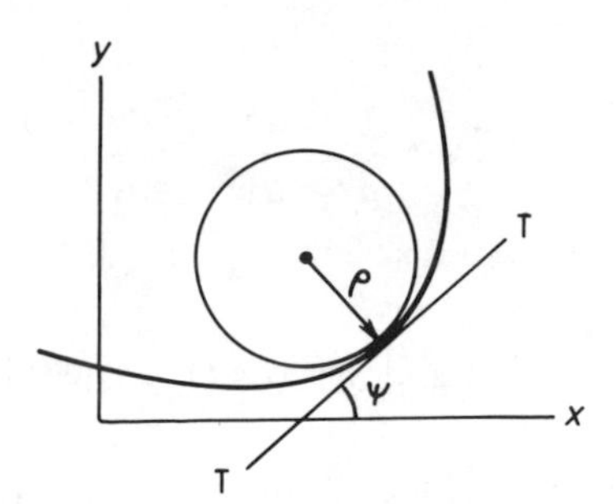

Figure 2.34

tangent turns with respect to the arc length s. The curvature p is commonly written:

$$\frac{1}{\rho} = \frac{d\psi}{ds} \quad \text{or} \quad \rho = \frac{ds}{d\psi}$$

where ψ is the angle the tangent to the curve makes with the ox axis. We observe that:

$$\tan \psi = \frac{dy}{dx} = \dot{y}$$

and use $\dot{y}$ to symbolise the first differential and $\ddot{y}$ to symbolise the second differential.

When

$$\tan \psi = \dot{y}$$

$$\sin \psi = \frac{\dot{y}}{\sqrt{1 + \dot{y}^2}}$$

$$\cos \psi = \frac{1}{\sqrt{1 + \dot{y}^2}}$$

We may obtain a general formula for the radius of curvature by differentiating $\tan \psi$ with respect to s.

The differential of $\tan \psi$ is $\sec^2 \psi$ and hence when differentiated with respect to s, $\tan \psi$ becomes

$$\sec^2 \psi \frac{d\psi}{ds} = \frac{d}{ds}\left(\frac{dy}{dx}\right) = \frac{d}{dx}\left(\frac{dy}{dx}\right) \cdot \frac{dx}{ds}$$

whence

$$\sec^2\psi \cdot \frac{1}{\rho} = \frac{d^2y}{dx^2} \cdot \frac{dx}{ds}$$

but

$$\frac{dx}{ds} = \cos\psi \text{ and } \sec^2\psi = (1 - \tan^2\psi)$$

from which we conclude, the radius of curvature ρ is given by the expression:

$$\rho = \frac{\left\{1 + \left(\frac{dy}{dx}\right)^2\right\}^{3/2}}{\frac{d^2y}{dx^2}} \tag{2.49}$$

N.B. ρ is positive when the curve is concave up, negative when concave down.

2.10.1 The length of an arc

The length of a curved line may be obtained by representing an increment of arc by a succession of infinitesimal Pythagorean triangles and summing arc lengths over the range of the curve required. Thus, in Fig. 2.35, dx, dy, ds, is a right-angled triangle with ds an infinitesimal part of the curve $y = f(x)$.

$$ds^2 = dx^2 + dy^2$$

$$\frac{ds^2}{dx^2} = 1 + \frac{dy^2}{dx^2}$$

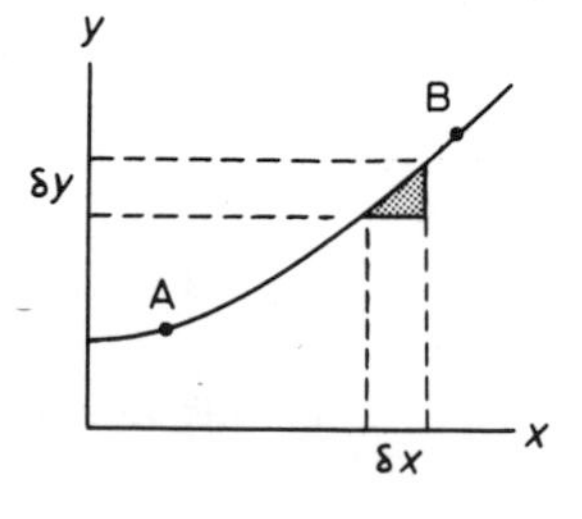

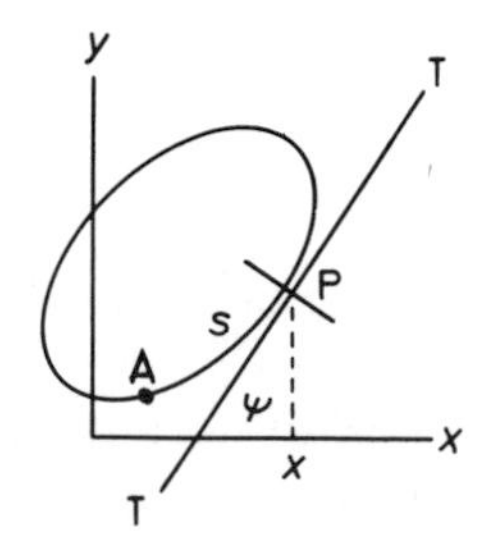

Figure 2.35

whence

$$\frac{ds}{dx} = \sqrt{1 + \left(\frac{dy}{dx}\right)^2}$$

$$\dot{s}^2 = \dot{x}^2 + \dot{y}^2$$

and the length of the arc from $t = t_1$ to $t = t_2$ for $t_2 > t_1$ is given by the equation

$$\text{Arc length} = \int_{t_1}^{t_2} \sqrt{\dot{x}^2 + \dot{y}^2}\, dt \qquad (2.50)$$

2.11 EVOLUTES

The evolute of a curve first considered by Apollonius as early as 200 BC is the locus of its centres of curvature.

A circle has a single radius struck from a single centre and hence a circle has no evolute curve. The parabola, ellipse, and hyperbola, however, have evolutes, as shown in Fig. 2.36.

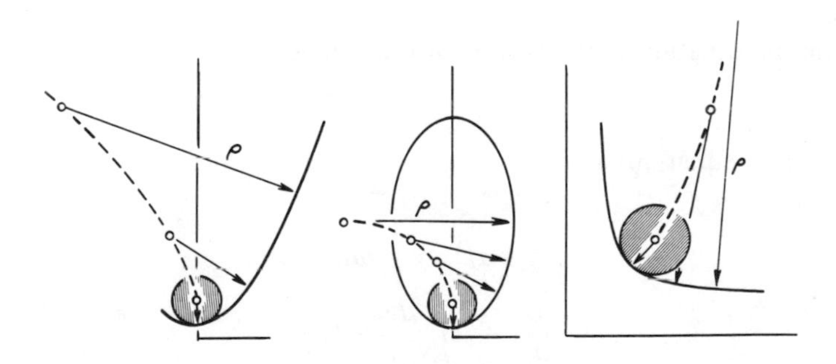

Figure 2.36

The coordinates of the centre of curvature may be written in the form:

$$X = x - \rho \sin \psi, \quad Y = y + \rho \cos \psi$$

where ψ is the angle defined in Fig. 2.36, and ρ is the radius of curvature.

It has also been shown that the radius of curvature is given by:

$$\rho = \frac{(1 + \dot{y}^2)^{3/2}}{\ddot{y}}$$

where

$$\dot{y} = \frac{dy}{dx} \quad \text{and} \quad \ddot{y} = \frac{d^2y}{dx^2}$$

$$\frac{dy}{dx} = \dot{y} = \tan\psi$$

whence

$$\sin\psi = \frac{\dot{y}}{\sqrt{1+\dot{y}^2}}, \quad \cos\psi = \frac{1}{\sqrt{1+\dot{y}^2}}$$

and hence the coordinates of the centre of curvature are

$$\begin{aligned} X &= x - \frac{\dot{y}(1+\dot{y}^2)}{\ddot{y}}, \\ Y &= y + \frac{1+\dot{y}^2}{\ddot{y}} \end{aligned} \tag{2.51}$$

The coordinates X and Y are the coordinates of the evolute curve.

Example
Determine the equation to the evolute of the parabola $y^2 = 4ax$.

Solution
The parametric equations are

$$x = at^2, \; y = 2at,$$

$$\frac{dx}{dt} = 2at, \; \frac{dy}{dt} = 2a$$

from which

$$\frac{dy}{dx} = \frac{1}{t}$$

$$\frac{d^2y}{dx^2} = \frac{1}{t}\,\frac{dt}{dx} = -\frac{1}{2at^3}$$

The coordinates of the evolute are therefore:

$$X = at^2 + \frac{1}{t}\left(1+\frac{1}{t^2}\right)\Big/\frac{1}{2at^3}$$

whence

$$X = 3at^2 + 2a$$

$$Y = 2at - \left(1 + \frac{1}{t^2}\right) \Big/ \frac{1}{2at^3}$$

whence

$$Y = -2at^3 .$$

We may now obtain the Cartesian equation to the parabola $y^2 = 4ax$ by eliminating the parametric variable t.

We know that

$$X = 3at^2 + 2a, \; Y = -2at^3$$

and it follows that:

$$\left(\frac{X - 2a}{3a}\right)^{3/2} = t^3 = -\frac{Y}{2a}$$

whence the equation to the evolute of the parabola $y^2 = 4ax$ is:

$$(X - 2a)^3 = \frac{27}{4} aY^2$$

2.12 ROULETTES

The path of a point or envelope of a line which is attached to the plane of a curve that rolls on a fixed curve is called a roulette. Roulettes were mentioned by Durer (1471-1528) and Daniel Bernoulli (c, 1710) and by Besant (1869).

2.12.1 The cycloid

The curve known as a cycloid is the locus of a point on the periphery of a circular disc, that rolls without slip along a straight line. The cycloid was first mentioned by Galileo (1599).

Consider a circle to roll without slip from 0 to N, as shown in Fig. 2.37. The arc length NP equals the distance rolled ON and ON = $r\theta$.

If (x, y) are the coordinates of the point P (the point P being originally at 0),

$$\begin{aligned} x &= \text{ON} - \text{PC} \sin\theta = r(\theta - \sin\theta) \\ y &= \text{CN} - \text{PC} \cos\theta = r(1 - \cos\theta) \end{aligned} \tag{2.52}$$

and these are the parametric equations of the cycloid.

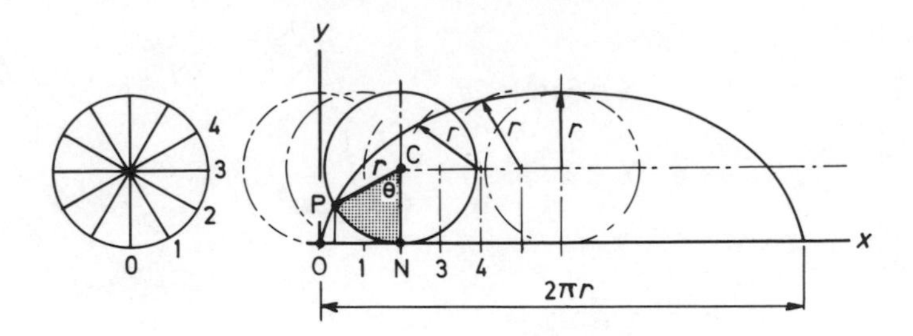

Figure 2.37

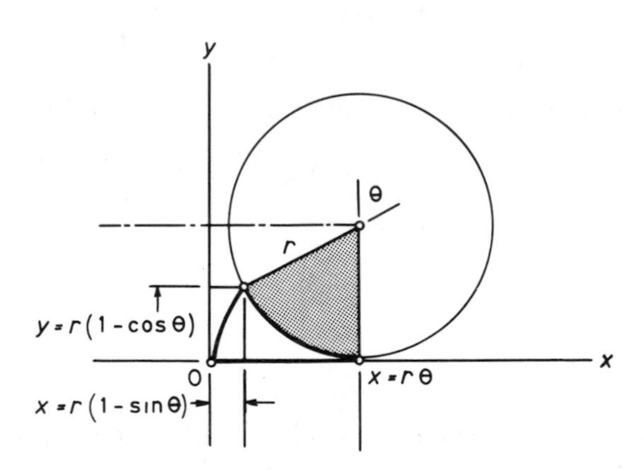

Figure 2.38

2.12.2 The epicycloid

The locus of a point on the periphery of a circle that rolls without slip around the outside of a fixed circle is called an epicycloid.

Let the radii of the two circles equal a and b. Let the circle of radius b roll around the fixed circle of radius a. The point $P(x, y)$ is thus the current position of a point on the periphery which started out from A.

In the absence of slip, the length of the arc AB equals the length of the arc PB. That is, $a\theta = b\phi$, as illustrated by Fig. 2.39.

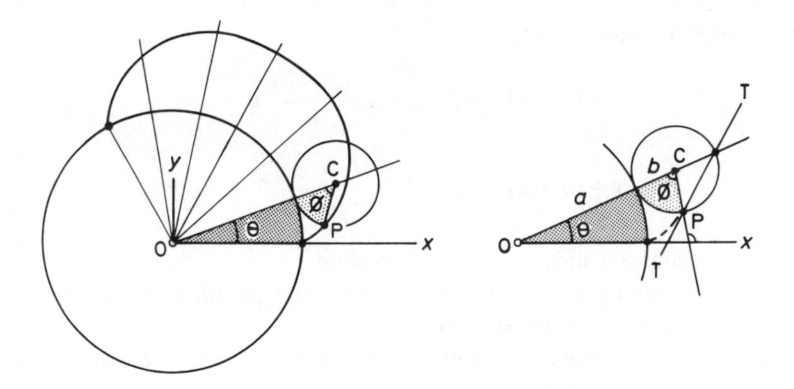

Figure 2.39

We observe that

$$\psi = (\theta + \phi) = \frac{(a+b)}{b}\,\theta$$

$$x = \mathrm{OC}\cos\theta + \mathrm{CP}\cos(180 - \psi)$$

$$= (a+b)\cos\theta = b\cos\frac{(a+b)}{b}\,\theta$$

$$y = \mathrm{OC}\sin\theta = \mathrm{CP}\sin(180 - \psi)$$

$$= (a+b)\sin\theta - b\sin\frac{(a+b)}{b}\,\theta$$

The (x, y) coordinates of the epicycloid are thus given by the expressions:

$$x = (a+b)\cos\theta - b\cos\frac{(a+b)}{b}\,\theta$$
$$y = (a+b)\sin\theta - b\sin\frac{(a+b)}{b}\,\theta \qquad (2.53)$$

2.12.3 The hypocycloid

The locus of a point on the periphery of a circle that rolls without slip around the inside of a fixed circle is called a hypocycloid.

The (x, y) coordinates of the hypocycloid may be obtained directly by writing $-b$ for b in the expression(s) above.

Thus the two expressions:

$$x = (a - b)\cos\theta + b\cos\frac{(a - b)}{b}\theta$$
$$y = (a - b)\sin\theta + b\sin\frac{(a - b)}{b}\theta \tag{2.54}$$

give the coordinates x and y for the hypocycloid.

The corresponding curves of points not on the periphery are called epitrochoids and hypotrochoids respectively.

(N.B. The term roulette is used to describe any loci of a point which is attached to a curve which rolls about another).

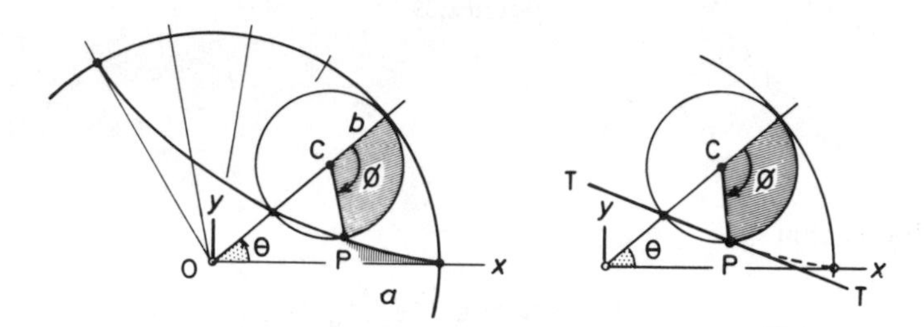

Figure 2.40

2.12.4 The nephroid

A nephroid curve is a two-cusped epicycloid, the rolling circle of which is related to the fixed circle by either of the ratios $a = 2b$ or $3a = b$. The nephroid was first studied by Huygens Gascharnhausen late in the fifteenth century.

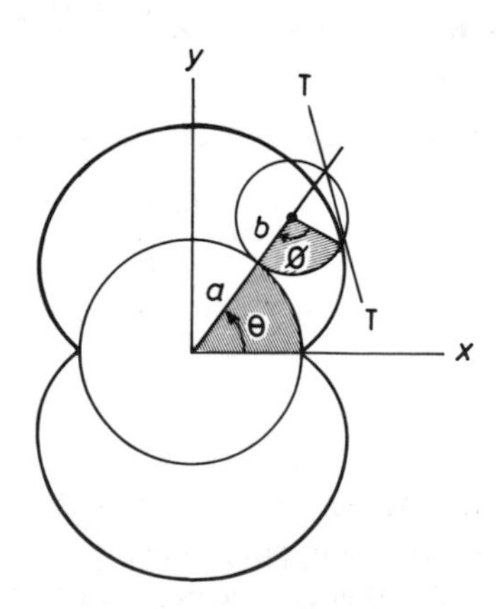

Figure 2.41

When $a = 2b$, the coordinates of the nephroid are:

$$x = b(3\cos\theta - \cos 3\theta) \qquad (2.55)$$
$$y = b(3\sin\theta - \sin 3\theta)$$

whence

$$(x^2 + y^2 - 4a^2)^3 = 108\, a^4 y^2$$

2.12.5 The cardioid and limacon

The equation $r = a\cos\theta$ is a circle, the equation $r = b + a\cos\theta$ is a limacon. When $a = b$, the equation becomes $r = a(1 + \cos\theta)$ and this is one form of cardioid.

A particular example of an epicycloid is the heart-shaped cardioid first evolved by Roemer (1674).

We consider the curve in terms of polar coordinates, in which the radius vector varies according to the relationship $r = a(1 + \cos\theta)$. We observe that when $\theta = 0, \cos\theta = 1$ and $r_0 = 2a$. When $\theta = 90°$, $\cos\theta = 0$ and $r_{90} = a$. When $\theta = 180°$, $\cos\theta = -1$ and $r_{180} = 0$. The cardioid curve is sketched in Fig. 2.42.

Equations of the cardioid include:

$$r = a(1 \pm \cos\theta),\ r = a(1 \pm \sin\theta)$$
$$x = a(2\cos\theta - \cos 2\theta) \qquad (2.56)$$
$$y = a(2\sin\theta - \sin 2\theta)$$

and

$$(x^2 + y^2 \pm 2ax)^2 = 4a^2(x^2 + y^2)$$

(N.B. All these equations site the cusp at the origin).

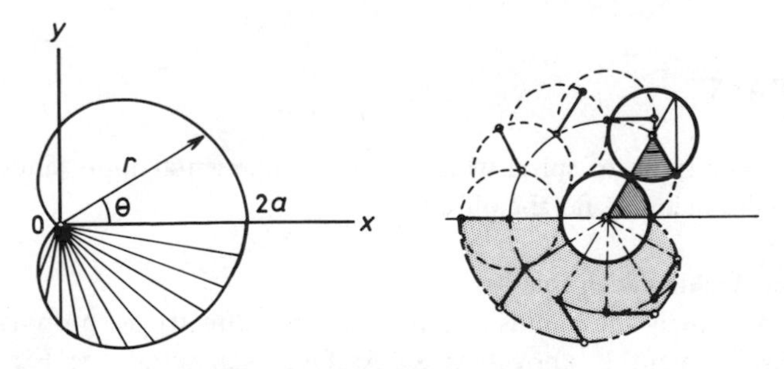

Figure 2.42

2.12.6 The astroid and deltroid

The astroid was first described by Roemer (1674) and the deltroid by Euler (1745).

The astroid and deltroid curves are both particular instances of the hypocycloid. The astroid is the locus of a point on the periphery of a rolling circle, the diameter of which is one quarter that of the fixed guiding circle. The deltoid is the loci of a point on the periphery of a guiding circle of radius a that rolls about a guiding circle of radius b where $a = 3b$ or alternatively $2a = 3b$. The two curves are shown in Fig. 2.43.

When $a = 3b$, the coordinates of the deltroids are:

$$\begin{aligned} x &= b(2 \cos \theta + \cos 2\theta) \\ y &= b(2 \sin \theta - \sin 2\theta) \end{aligned} \tag{2.57}$$

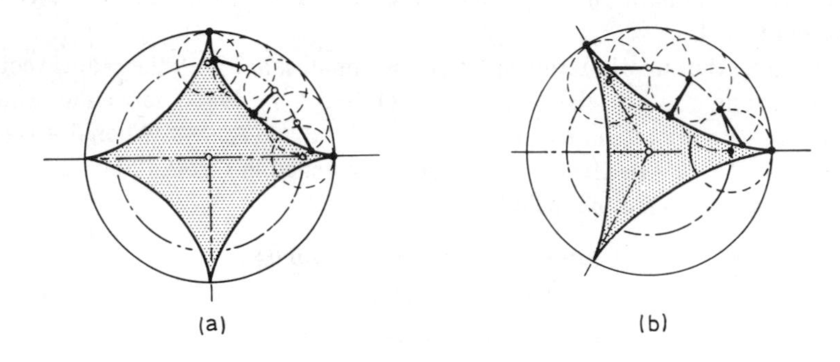

Figure 2.43

2.13 SPIRALS

There are three types of spiral which are of fundamental importance: the spiral of Archimedes being by far the oldest.

2.13.1 The Archimedean spiral

A spiral for which the radius vector increases directly in proportion to the angular displacement is known as an Archimedean spiral, see Fig. 2.44. The equation to an Archimedean spiral is:

$$r = a\theta \tag{2.58}$$

and we see that if the increase in θ is in arithmetic progression so too is the radius r.

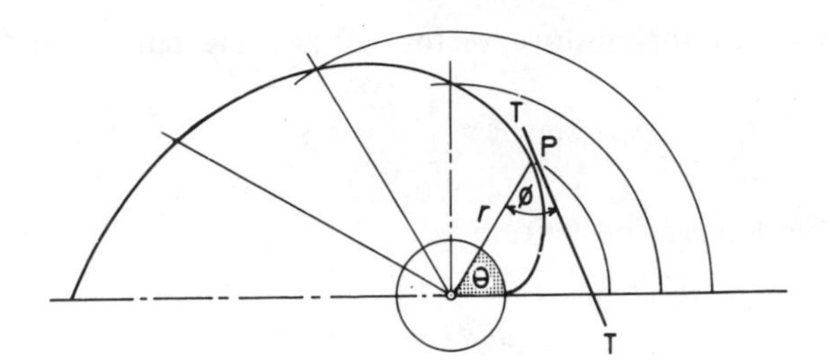

Figure 2.44

2.13.2 The hyperbolic spiral

The curve for which the radius vector varies inversely as the angular displacement is known as a hyperbolic spiral, see Fig. 2.45. The equation to an hyperbolic spiral is

$$r\theta - K = 0.$$

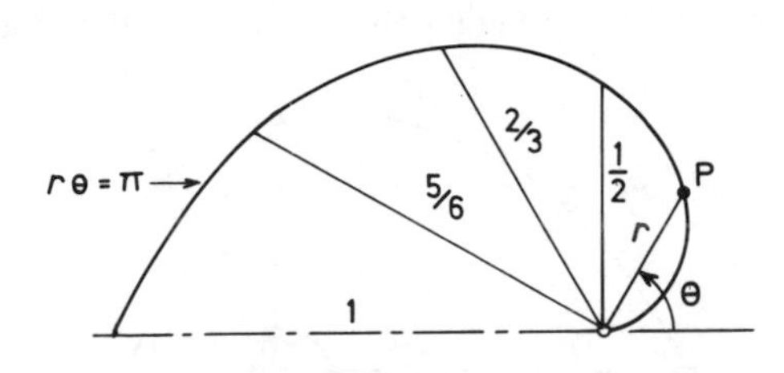

Figure 2.45

2.13.3 The logarithmic spiral

The logarithmic spiral (also known as the equi-angle spiral) is one in which equal increments of angle, i.e. increments in arithmetic progression, produce changes in the radius r which are in geometrical progression. The equation to a logarithmic spiral is obtained as follows.

The logarithmic spiral was first studied by Descartes who defined the curve as

$$r = a\mu^\theta$$

whence

$$\frac{dr}{d\theta} = r \log u$$

The angle between the positive vector OP and the tangent at P is therefore

$$\cot \phi = \frac{1}{r}\frac{dr}{d\theta} = \log u$$

which by definition is a constant.

Since

$$\mu^\theta = e^{\theta \log u} = e^{\theta \cot \phi}$$

the classical form of the equation to a logarithmic spiral is

$$r = ae^{\theta \cot \phi} \tag{2.59}$$

whence

$$\frac{dr}{ds} = \cos \phi$$

and since $(r - s \cos \phi)$ is constant, the arc length in terms of the radius vectors r_1 and r_2 is:

$$\text{arc length} = (r_2 - r_1) \sec \phi$$

or simply $r \sec \phi$ when r is the general point referred to the origin zero.

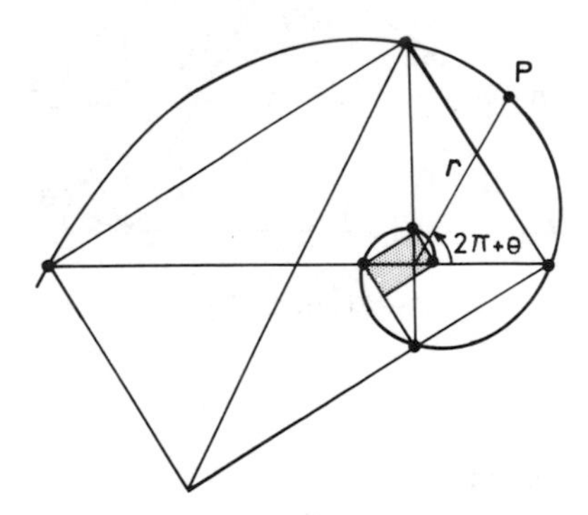

Figure 2.46

2.13.4 A practical application of spirals

The jaws of a self-centering chuck as used on a lathe are moved by means of a rotating scroll plate geared to an input turn key.

2.14 ENVELOPE CURVES

Curves, called envelopes, were first investigated by Leibnitz (1694), Taylor (1715) and Legrange (1774). We allow the coefficients of a given curve to depend on a variable parameter t, such that when t is given a definite value, a particular curve is produced. By giving the variable parameter t a series of different values, we create a family of curves all of a given type.

The equation $y = tx + a$ contains three variables x, y and t. The variable t clearly controls the slope of the (straight) line.

The equation $y = tx + a/t$ is another parametric equation to a straight line but here we are able to vary both the slope and the intercept; simply by specifying different values of the one parameter t.

Suppose we put $a = 6$ in the equation above. Then

$$y = tx + \frac{6}{t}$$

for all values of t.

When $t = 0.5$ we have the straight line:

$$y = 0.5x + 12$$

when $t = 1.0$

$$y = x + 6$$

when $t = 1.5$

$$y = 1.5x + 4$$

when $t = 6.0$

$$y = 6x + 1$$

and these four straight lines appear as plotted, in Fig. 2.47 from which the trend of their respective intersections is clearly visible.

2.14.1 Ultimate intersections

We have seen that different values of the parameter t give rise to a series of intersection as shown in Fig. 2.47.

The distance between these intersections can be reduced by taking smaller and smaller incremental values of t. When the difference between t_1 and t_2 is very, very small we approach the condition known as an ultimate intersection.

Consider the equation $\mu^2 x - \mu y + a = 0$, where μ is the variable parameter. Two members of the family of curves (straight lines) are thus described by the equations:

$$\mu_1^2 s - \mu_1 y + a = 0, \mu_2^2 x - \mu_2 y + a = 0$$

The point of intersection is obtained by subtracting these two equations. Whence the point of intersection satisfies the equation:

$$(\mu_1^2 - \mu_2^2)x - (\mu_1 - \mu_2)y = 0$$

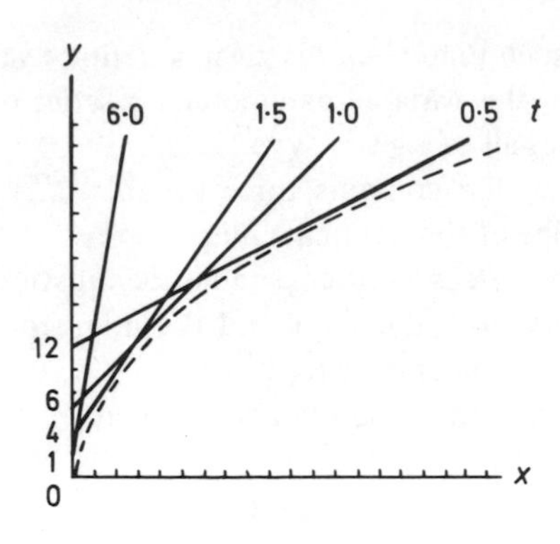

Figure 2.47

and hence the equation:

$$(\mu_1 + \mu_2)x - y = 0$$

If we allow μ_2 to approach μ_1, it ultimately becomes equal to μ_1, whence in the limit

$$2\mu_1 x - y = 0$$

We may now eliminate the variable parameter μ_1 by substituting $\mu_1 = y/2x$ into the original equation, for since

$$\mu_1^2 x - \mu_1 y + a = 0$$

$$\frac{y^2}{4x^2} - \frac{y \cdot y}{2x} + a = 0$$

and hence

$$y^2 - 4ax = 0$$

The equation to the envelope of the family of lines, $\mu^2 x - \mu y + a = 0$, is thus a parabola.

2.14.2 A general rule for envelopes

If the parameter μ or t appears as a power term it is theoretically possible to determine the envelope using the rule that two roots of μ are equal.

We have seen that the ultimate intersection of the line $\mu^2 \text{P} + \mu\text{Q} + \text{R} = 0$ may be expressed in the form:

$$\mu = -\frac{\text{Q}}{2\text{P}}$$

and that when μ is eliminated:

$$\frac{Q^2 \cdot P}{4P^2} - \frac{Q \cdot Q}{2P} + R = 0$$

whence

$$Q^2 = 4PR.$$

The condition of equal roots may be taken as a general rule.

Example

Suppose we seek the envelope of the curve

$$P \cos \theta + Q \sin \theta = R$$

where θ is the variable parameter

Solution

If we put $t = \tan \theta/2$ then $\cos \theta = (1 - t^2)/(1 + t^2)$, $\sin \theta = 2t/(1 + t^2)$, whence

$$P \frac{1 - t^2}{1 + t^2} + Q \frac{2t}{1 + t^2} = R$$

$$t^2(R + P) - 2tQ + (R - P) = 0$$

and

$$t = \frac{Q \pm \sqrt{Q^2 - R^2 + P^2}}{R + P}$$

We assume the rule and make the roots equal, that is we put $Q^2 = (R + P)(R - P)$, whence $t = Q/R + P$ and

$$R^2 = P^2 + Q^2$$

We conclude the envelope is a circle, which it most certainly is.

2.14.3 The hyperbolic envelope

The rectangular hyperbola may be described as an envelope, produced by a sliding line. The line AB, Fig. 2.48, may be defined in terms of its intercepts. a and b.

The intercept equation to the line AB is:

$$\frac{x}{a} + \frac{y}{b} = 1$$

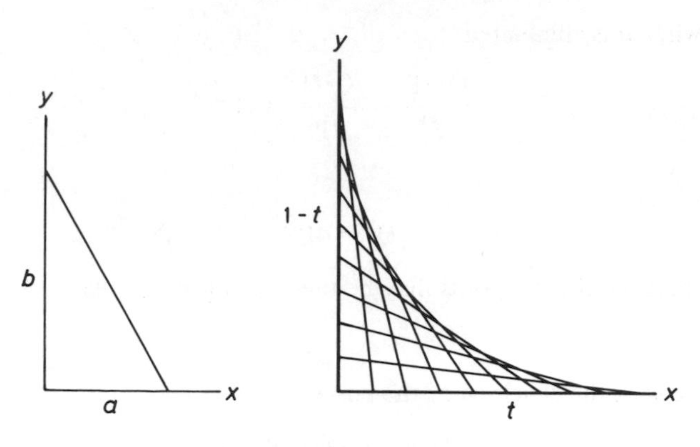

Figure 2.48

Now suppose we put $a = t$ and $b = (1 - t)$, where t is a variable parameter, free to take any fractional value between 0 and 1. Then

$$\frac{x}{t} + \frac{y}{1-t} = 1$$

and

$$ty + (1 - t)x - t(1 - t) = 0 \qquad (2.60)$$

We see that this equation is a function of x, y and t, which we may write in the following form:

$$f(x, y, t) = ty + (1 - t)x - t(1 - t) = 0$$

and

$$\frac{df}{dt} = y - x - 1 + 2t = 0$$

We note that df/dt varies depending on the choice of the parameter t. If we use the equation $f(x, y, t) = 0$, as given, and multiply the slope equation by t we are able to obtain x and y, as a function of the variable t.

We know that:

$$ty + (1 - t)x - t(1 - t) = 0$$

and multiplying the slope equation by t:

$$ty - tx - t(1 - 2t) = 0$$

we obtain by subtraction:

$$x = t^2$$

and by substituting t^2 for x, in the first equation:

$$y = (t - 1)^2 = (1 - t)^2$$

Having obtained the value of t in terms of x and y, the equation to the envelope is obtained by eliminating t.

$$t = \frac{x - y + 1}{2}$$

We substitute the value of t into the $f(xyt) = 0$ equation, whence

$$2y(x - y + 1) + 2x(1 - x + y) - (x - y + 1)(1 - x + y) = 0$$

and by careful reduction in terms we obtain the equation:

$$y^2 - 2xy + x^2 - 2y - 2x + 1 = 0 \qquad (2.60)$$

This is the implicit equation to the envelope, which we recognise as a hyperbola.

2.14.4 Caustic curves

The envelope of reflected or refracted light rays from a point source produced by a given curve is known as a caustic curve. A caustic curve is shown in Fig. 2.49.

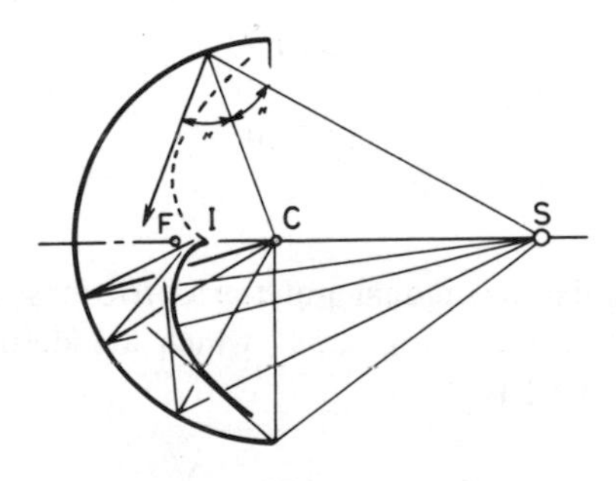

Figure 2.49

2.15 SYSTEMS OF LINES

There are many practical reasons why we may need to generate a family of lines and/or curves. An almost trivial instance is our frequent need to drill regular two dimensional arrays of holes in planar sheets. Another more sophisticated requirement is the generation of meshes required by the finite element computational technique.

2.15.1 An orthogonal mesh of straight lines

When using an N.C. machine tool to drill a two-dimensional array of holes, as in Fig. 2.50, it is necessary to instruct the machine tool to drill holes at all positions required. In point of fact this is done by writing a programme in A.P.T. or some other computer language. (N.B. A.P.T. stands for Automatically Programmed Tool).

In the first instance, however, we need to conceive the action of stepping.

The equation of a line V_1 parallel to the *oy* axis is $x = a$, and the equation of a line U, parallel to the *ox* axis is $y = b$, whence the point of intersection between lines U_1 and V_1 is the point $(a_1 b_1)$. Other lines and hence points on the mesh may be obtained by stepping the lines U and V by increments of p and q. The equation of the line U_2 is $x = a_1 + p = a_2$, and that of the line V_2 is $y = b_1 + q = b_2$. It is in practice relatively easy to tell an N.C. machine tool to "step, drill, step drill", etc.

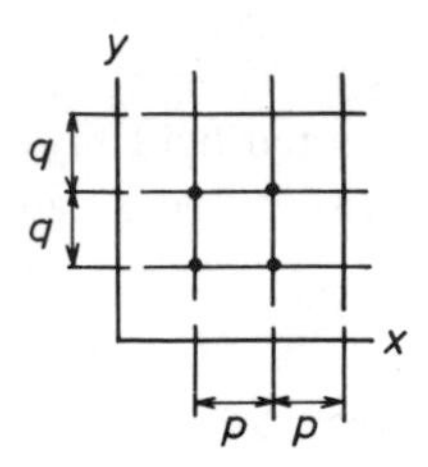

Figure 2.50

2.15.2 Diamond triangular, hexagonal and concentric meshes

Consider the case of two lines. Fig. 2.51, which are defined in perpendicular form. The equation to line 1 is:

$$x \cos 30 + y \sin 30 - \frac{\sqrt{3}}{2} = 0$$

whence

$$x\sqrt{3} + y - \sqrt{3} = 0$$

The equation to line 2 is:

$$x \cos 30 - y \sin 30 - \frac{\sqrt{3}}{2} = 0$$

whence

$$x\sqrt{3} - y - \sqrt{3} = 0$$

(N.B. The negative sin 30 in the equation to the second line is due to the angle of the perpendicular being in the fourth quadrant).

The point of intersection of these two lines is given by the condition:

$$x\sqrt{3} + y - \sqrt{3} = x\sqrt{3} - y - \sqrt{3}$$

whence

$$y = 0 \text{ and hence } x = 1.$$

A diamond grid, as shown at (c), may thus be obtained by telling control to step each line a distance (nx) in the ox direction, or by telling control to increase the perpendicular distance p by suitable increments.

A regular triangular, or semi-regular hexagonal tessellation may likewise be called by introducing a third line and displacing it parallel to ox in suitable increments, as shown in Fig. 2.51.

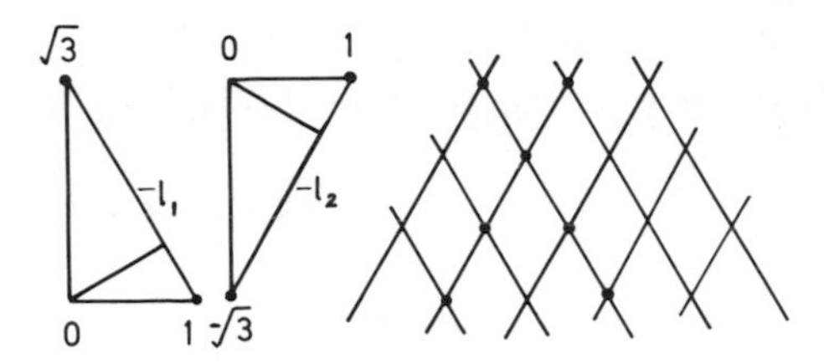

Figure 2.51

2.15.3 Oblique coordinates

In certain problems it may prove expedient to employ oblique (as opposed to orthogonal) coordinate axes. If the angle xoy is ω, and the angle PQR = θ, the angle QPR = $(\omega - \theta)$

$$\frac{\text{PR}}{\text{QR}} = \frac{\sin\theta}{\sin(\omega - \theta)}$$

and if we put $m = \sin\theta/\sin(\omega - \theta)$ it can be proved that

$$\tan\theta = \frac{m\sin\omega}{1 + m\cos\omega}$$

The equations above are as for rectangular axes except that m becomes $\sin\theta/\sin(\omega - \theta)$. The equation for a circle referred to oblique coordinates takes the form

$$(x - f)^2 + (y - g)^2 + 2(x - f) + 2(y - g)\cos\omega = a^2 \tag{2.61}$$

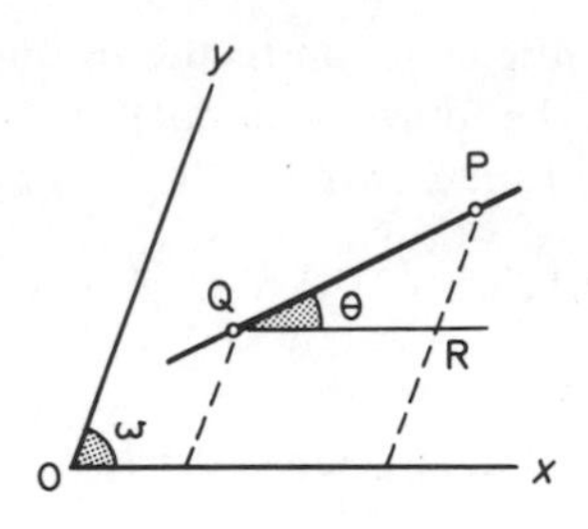

Figure 2.52

2.16 ORTHOGONAL CURVES

Systems of orthogonal curves arise in the study of magnetism, photo-elasticity and in the design of minimum weight structures.

2.16.1 Confocal conics

A system of confocal conics is shown in Fig. 2.53, from which it is seen that ellipse and hyperbola curves are super-imposed (overlaid) in such a way that the foci of each coincide.

A pair of conics which have the same foci, have the same centre and the two equations:

$$\frac{x^2}{a^2} + \frac{y^2}{b^2} = 1$$

and

$$\frac{x^2}{a^2 + \lambda} + \frac{y^2}{b^2 + \lambda} = 1$$

represent a system of confocal conics.

If a is greater than b, the conic:

$$\frac{x^2}{a^2 + \lambda} + \frac{y^2}{b^2 + \lambda} = 1$$

represents either an ellipse or a hyperbola, according to the magnitude of λ. The following rules apply:

i) the conic is an ellipse for all positive values of λ and some negative values, those for which λ is numerically greater than $-b^2$.
ii) the conic is the line $y = 0$, when $\lambda = -b^2$.
iii) the conic is a hyperbola when λ is less than b^2, but greater than a^2.
iv) the conic is the line $x = 0$, when $\lambda = -a^2$.
v) the conic is an imaginary ellipse when $\lambda < -a^2$.

Of the two confocal conics which pass through a given point, the one is always an ellipse, the other a hyperbola, and the intersection between the two curves is a right angle.

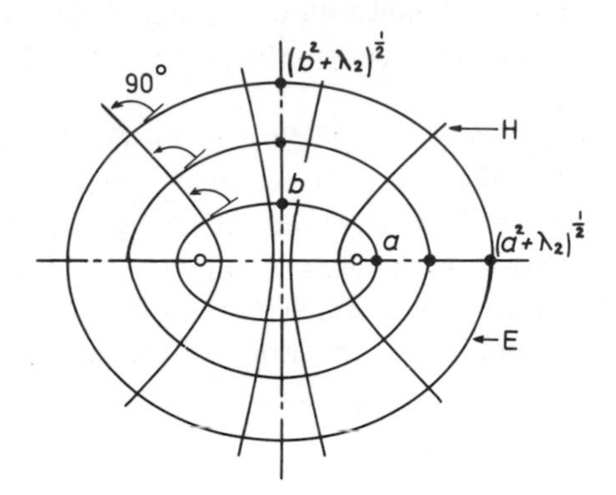

Figure 2.53

2.16.2 Orthogonal stress trajectories

The direction of principal stresses in beams is mapped by a system of orthogonal curves, the theory of which is fully explained in books on the Theory of Elasticity. Suffice to observe that the two principal stress curves, Fig. 2.54(a), cross the neutral axis at +45° and −45° respectively and that each set of curves crosses the boundary of the beam at right angles, or as an asymptote, as may be. A minimum weight structure based on this principal is shown at (b). It is called a Michell structure. See Hemp, Optimum Structures.

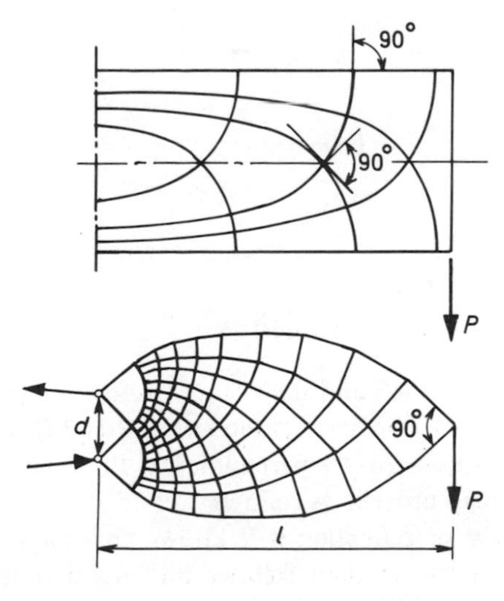

Figure 2.54

2.17 SYNTHESIS OF SHAPE IN TWO DIMENSIONS

The name of Louis Sullivan (1856–1924), architect, is often associated with the recommendation that form should follow function. Though this be an obvious first consideration in functional design the end result is, in many cases, by no means easy to achieve.

There are, however, instances in which it is possible to derive the ideal shape mathematically. The so-called ideal shape, that is the shape best suited to the prime functional objective must clearly be derived by way of the laws of physics and/or if these are unknown on the basis of experimental findings.

To take a very simple instance in which it is possible to define the entire shape of an object to meet a predefined function, we consider the synthesis of (1) a parabolic reflector; (2) a prismatic channel.

Example

Consider the requirement to synthesize a two dimensional reflecting surface the form of which is such that it everywhere converts the light from a divergent point source (placed on the axis of symmetry) into a parallel beam. Such is the requirement of many optical and heat projecting devices.

It is acknowledged that many readers of this book will already know the form of curve required. It is to be observed, however, that once it is recognised that the angle of incidence equals the angle of reflection, a fact known to be true from the findings of experimental physics, the derivation given below makes no assumption which cannot be directly deduced from the functional need.

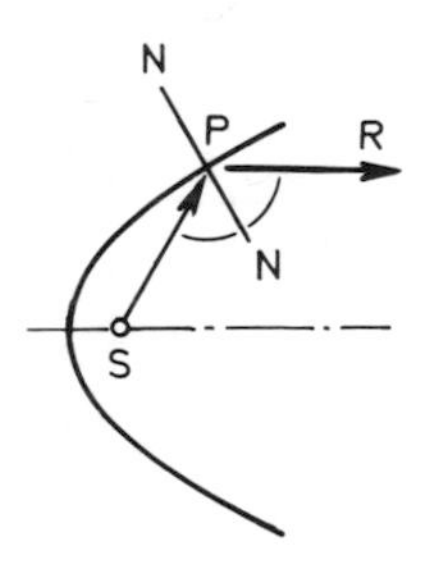

Figure 2.55

If the source is denoted by S and the point of reflection (of a typical ray) by P, then the local tangent at P may be extended to R and Q, where Q lies on the axis of symmetry. If we now drop a perpendicular PB, where B lies on the axis of symmetry, then we may proceed as follows:

According to the law of reflection it is known that angle RPT equals angle QPS. Let this angle equal α. It then follows directly that the angle PQS also equals α and hence the angle PSB equals 2α.

If we now erect a perpendicular SC and drop a further perpendicular CA, such that A is on the axis of symmetry, we also notice that angle ACS is also equal to α, which completes the construction necessary to the derivation of the required mathematical curve.

Using our powers of mathematical deduction we now set about deriving the mathematical equation to the curve we require. Before so doing, however, it is as well to recall that to work out the motions of the planets, treating the Earth as origin, is a very formidable problem indeed. If, however, we place the Sun at the centre then the problem, though still difficult, may be much more readily solved. So it is with the present problem. Thus a few minutes work would be sufficient to show that Q is probably not the best point to choose, and the same applies to point B.

It is sufficient to say that if one is working in polar coordinates then point S is the expedient point to choose. If, on the other hand, the equation is required in Cartesian form then the point A is to be preferred.

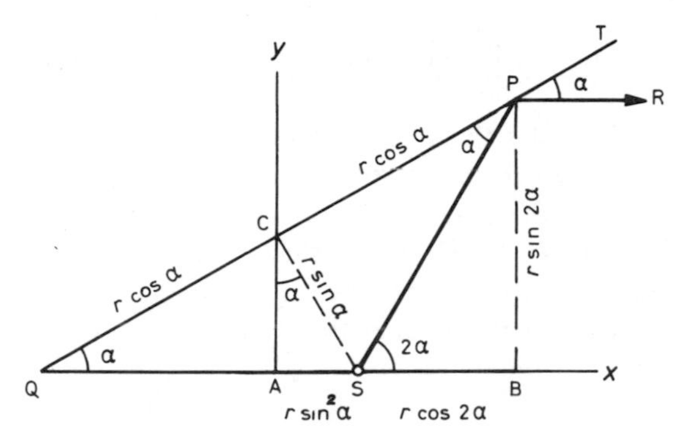

Figure 2.56

Let us tackle the problem on the assumption that point A is the origin of our rectangular axes x and y.

If we put SP = r then we may express all other lengths in terms of r.

Thus from an inspection of Fig. 2.56,

$$x = r \sin^2 \alpha + r \cos 2\alpha$$

Using the identity $\cos 2\alpha = \cos^2 \alpha - \sin^2 \alpha$

$$x = r \cos^2 \alpha$$

Similarly,

$$y = r \sin 2\alpha$$

Using the identity $\sin 2\alpha = 2 \sin\alpha \cos \alpha$,

$$y = 2r \sin \alpha \cos \alpha$$

and hence

$$y^2 = 4r^2 \sin^2\alpha \cos^2\alpha$$

But since it has already been shown that

$$x = r\cos^2\alpha$$

$$y^2 = 4xr\sin^2\alpha$$

Notice, however, that the length AS $= r\sin^2\alpha = a$, say.

Then we may write

$$y^2 = 4ax$$

which is the equation to a parabola.

The parabola is thus the curve required. The dimension (a) is called the focus and is the distance at which the light source must be placed.

Example

As a final example relating to the mathematical synthesis of shape, consider the following problem.

It is required to synthesise the cross-sectional shape of a straight prismatic channel (as frequently used for conveying liquids with suspended materials in process plant). Given that the channel section appropriate to minimum flow is defined by the parameters A_0, P_0. Derive the shape of channel required to ensure that at all flows greater than the minimum the mean velocity of flow will not be changed. It is found by experiment that the flow V is proportional to $(A/P)^n$ where n is a constant of the order of 0.5.

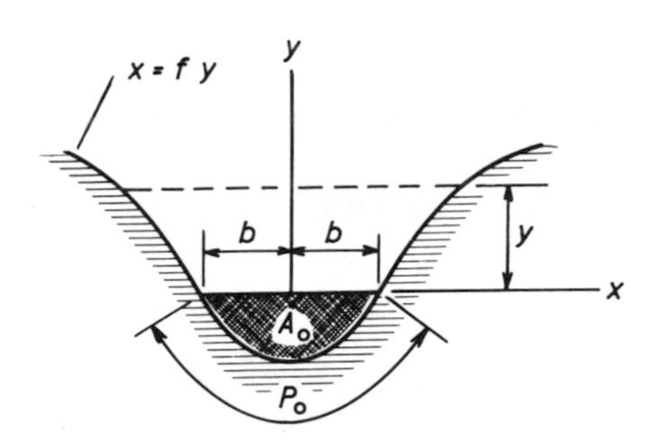

Figure 2.57

The problem is thus to derive the shape for which the ratio A/P is equal to A_0/P_0 and the same for all depths.

If it is stipulated that for all distances y above the minimum level x is a function of y, then the cross-sectional area A and the perimeter P may be expressed mathematically as follows:

$$A = A_0 + 2\int_0^y x(y)\mathrm{d}y$$

$$P = P_0 + 2\int_0^y \sqrt{1 + \left(\frac{\mathrm{d}x}{\mathrm{d}y}\right)^2}\mathrm{d}y$$

Since the problem requires that $A/P = A_0/P_0$

$$A_0 + 2\int_0^y \mathrm{d}(x)\mathrm{d}y = \frac{A_0}{P_0}\left[P_0 + 2\int_0^y \sqrt{1 + \left(\frac{\mathrm{d}x}{\mathrm{d}y}\right)^2}\mathrm{d}y\right]$$

Noting that both sides are functions of y we have by differentiating with respect to y:

$$x = \frac{A_0}{P_0}\sqrt{1 + \left(\frac{\mathrm{d}x}{\mathrm{d}y}\right)^2}$$

putting $A_0/P_0 = a$ and rearranging terms

$$\frac{\mathrm{d}x}{\mathrm{d}y} = \sqrt{\left(\frac{x}{a}\right)^2 - 1}$$

now by multiplying throughout by a and rearranging to isolate $\mathrm{d}y$ the equation becomes:

$$\frac{\mathrm{d}y}{a} = \frac{\mathrm{d}x}{\sqrt{x^2 - a^2}}$$

and this is a standard integral form, the solution to which may be written directly:

$$\pm\frac{y}{a} = \cosh^{-1}\frac{x}{a}$$

and hence by observing that $x = b$ at $y = 0$.

$$\pm\frac{P_0}{A_0}y = \cosh^{-1}\left(\frac{P_0 x}{A_0}\right) - \cosh^{-1}\left(\frac{P_0 b}{A_0}\right)$$

Readers acquainted with mathematics will also recall that since

$$\cosh^{-1}x = \pm\log_e\left(x + \sqrt{x^2 - 1}\right).$$

The above equation may be written in the following form:

$$\frac{P_0 y}{A_0} = \log_e \frac{x + \sqrt{x^2 - (A_0/P_0)^2}}{b + \sqrt{b^2 - (A_0/P_0)^2}}$$

Hence given the minimum flow parameters A_0 and P_0 the shape of the required channel is completely defined.

2.18 CENTROID OF PLANE FIGURES

The centroid (centre) of area of a plane figure is independent of its position and hence does not depend on one's choice of coordinate axes. The centroid of a circle is always its centre and the centroid of an ellipse always lies at the intersection of the major and minor axes. The centroid of a plane figure is akin to the centre of gravity of a real mass lamina, and is the point about which the sum of the first moments of all elemental areas is zero.

If a_1, a_2, a_3 are elements of area at distances x_1, x_2, x_3 and y_1, y_2, y_3 from an arbitrary datum then the centroid lies at $(\bar{x}, \bar{y})$ where

$$\bar{x} = \frac{a_1 x_2 + a_2 x_2 + a_3 x_3}{a_1 + a_2 + a_3}, \quad \bar{y} = \frac{a_1 y_1 + a_2 y_2 + a_3 y_3}{a_1 + a_2 + a_3}$$

that is

$$A\bar{x} = \Sigma\, ax \quad \text{and} \quad A\bar{y} = \Sigma\, ay$$

The above expressions allow the centroid to be calculated by means of Simpson's strip rule, for instance. Alternatively we may obtain the centroid of an area by evaluating the double integrals

$$\bar{x} = \frac{1}{A} \iint_A x dA, \quad \bar{y} = \frac{1}{A} \iint_A y dA$$

(N.B. The centroids of many common shapes are given in engineering handbooks).

2.18.1 Second moment of area

The quantity ydA is known as the first moment of area. The quantity $y^2 dA$ is known as the second moment of area and is akin to the second moment of inertia when mass is present. The second moment of area is an important geometrical property which determines the magnitude of stress and strain in beams and shafts.

The second moments of area about the axes ox and oy are defined as

$$I_x = \int_A y^2 dA, \quad I_y = \int_A x^2 dA, \quad I_{xy} = \int_A xy dA$$

and the second moment of area about the mutually perpendicular centroidal axis *oz*, termed the polar moment, is defined as

$$I_p = I_z = \int_A r^2 dA$$

where r is the radius of the elemental ring of width dr and area dA.

The polar moment of an ellipse is

$$I_p = I_x + I_y$$

$$= \frac{\pi ab^3}{4} + \frac{\pi ba^3}{4}$$

If the second moment of area about a parallel axis distance d from a centroidal axis is required it is given by

$$I' = I + Ad^2$$

The moments of inertia about inclined may be obtained by applying the appropriate transformation chapter 5.

2.19 PROPORTION

Although size may impress, the essence of all form is proportion. Early in our schooling, indeed, in our learning, teaching and practice we become increasingly aware that certain geometric proportions, offer topological, economical, constructional and/or aesthetic advantages over other alternative forms.

Triangles the sides of which are integral multiples of a specified unit length have a particular appeal to geometers. Right triangles in which all the sides are of integral length are a most useful class. See The Treasury of Mathematics.

The 3, 4, 5. . .5, 12, 13 and 7, 24, 25, triangles which follow the form $c - b = 1$ are perhaps the most well known and the triangle 33, 56, 65 is representative of another class. (N.B. a collection of clay tablets from Old Babylonia c. 1900–1600 BC list a total of seventeen integer triangles. (Pirates)

Unfortunately for geometers right triangles generally contain at least one irrational side and even the diagonal of the perfect square ($\sqrt{2}$) cannot be expressed, nor measured exactly.

The practical use and fitness of the unit square is obvious and the geometrical properties of the root two rectangle will now be explained.

2.19.1 The root two rectangle

The root two rectangle, in which the sides are in the same ratio as the side to diagonal of the unit square is a particularly useful form. The root two rectangle is used as the logical basis for sizing the continental papers A0, A1, A2, A3, A4. . .B0, B1, B2, B3, B4. . .C0, C1, C2, C3, C4. . .etc. see fig. 2.58.

The point about sizing paper in accord with this mathematically irrational proportion is that when the longer side is cut or folded in half, once, twice, three times – no matter how many times – the same 'irrational' proportional remains. The value of this way of thinking will be appreciated by all who have experienced the frustration of trying to cram a folded letter of old English proportions into a non-matching old English envelope. Happily the merit of using the root two rectangle as the basis for sizing paper is now quite widely known.

The designer who is left uninspired by the simple process of dividing equally in two's, three's and five's may, however, care to consider the use of a more dynamic rule.

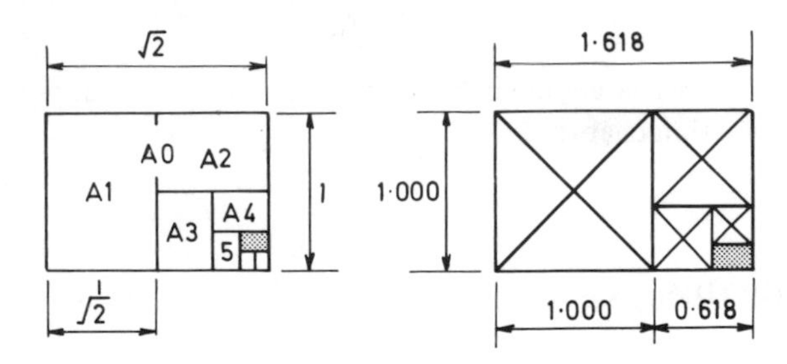

Figure 2.58

2.19.2 The golden section

The golden mean is a proportion (derived directly from Euclid proposition 11, Book II and proposition 30, Book VI) the presence of which Herodotus observed in the Great Pyramid has been widely used as a measure of proportion since c 300 BC. Not only does the golden rule provide the basis for a logical system of modular construction, it provides an aesthetically pleasing, some claim the most aesthetically pleasing layouts of all. See The Parthenon, Fig. 2.59, the proportions of which are typical of other Greek temples.

Taking a line of length $a + b$, the golden mean ratio or section is defined as the ratio a/b, where $a{:}b{::}b{:}a + b$.

By solving the relevant quadratic equation it is easy to show that

$$\frac{a}{b} = \frac{-1 \pm \sqrt{5}}{2} = 0.618034 \text{ or } -1.618034$$

Although at first sight this may not seem a particularly elegant result a number of interesting relationships can be directly attributed to it.

2.19.3 The Fibonacci series

It is, for example, an interesting fact that the reciprocal of 0.618034 equals 1.618034 and conversely. The square of 1.618034 equals 2.618034, which equals (1 + 1.618034) and this is one half the perimeter of the golden figure, in which the shorter side is of unit length. The numbers 0.618034, 1.618034/2, 1.000000 are in Arithmetic progression, the numbers 0.618034, 2/2.618034, 1.000000 are in Harmonic progression, while 1.000000, 1.618034, 2.618034 are in Geometric progression. The product of 0.618034 and 1.618034 equals 1.000000 and the sums of their two squares equals 3.000000. The number 0.618034 is the interval ratio to which the terms of the Fibonacci series tends, and the Fibonacci series has the property that each term is the sum of the two terms that proceeded it. Starting from zero the series reads:

$$0, 1, 1, 2, 3, 5, 8, 13, 21, 34, 55, 89 \ldots$$

and the interval ratio quickly converges.

$$\tfrac{0}{1} \to \tfrac{1}{1} \to \tfrac{1}{2} \to \tfrac{2}{3} \to \tfrac{3}{5} \to \tfrac{5}{8} \to \tfrac{8}{13} \to \tfrac{13}{21} \to \tfrac{21}{34} \to \tfrac{34}{55} \to \tfrac{55}{89} \ldots 0.6180$$

A further interesting property of the Fibonacci series is that if any three successive terms are taken at random the product of the first and third minus the square of the second is equal to plus or minus one alternately.

Right triangles with sides 1, 1. 1, 2. 2, 3. 3, 5. 5, 8. etc., have hypotenuse equal $\sqrt{2}$, $\sqrt{5}$, $\sqrt{13}$, $\sqrt{34}$, $\sqrt{89}$, etc., and the numbers under the radical sign are themselves Fibonacci numbers.

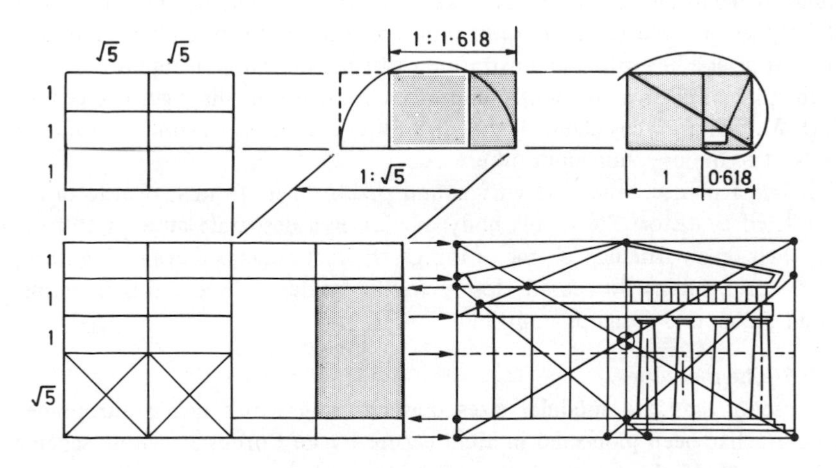

Figure 2.59

Although it has been stated by Holt, in his book Mathematics in Art, that 'The golden ratio is to the eye indistinguishable from the square root of two'. This view is not upheld by tests on pure perception. According to test results reported by Rowland, and reproduced in my own book Theory of Design the golden rectangle can not only be clearly distinguished from its root two rival but is preferred to it, by at least twice the number of people. Whereas the 21:34 approximation judged on its inherent aesthetic appeal typically collects between 33 and 35% of the vote, the $1:\sqrt{2}$ rectangle rarely collects more than 10%. The first Fibonacci approximation is the square and this collects on average less than 5%. The second Fibonacci approximation has sides in the ratio of 1:2 and this collects on average 8%. The third approximation is 2:3 and this collects on average 26%.

Whereas halving the longer side of the root two rectangle always produces two smaller rectangles of root two proportion: Dividing the longer side of the golden rectangle in accord with the golden rule always produces a square plus a smaller golden rectangle, the area of which is always 1/2.618034 times that of the original figure.

In art, evidence of the use, conscious or otherwise, of both the golden and root two rectangle abound. The works of Leonardo, Bellows, Mondrian and Seurat being, perhaps, the most well known.

2.19.4 Bionic sources of form

Evidence of Nature's apparent sympathy for the Fibonacci series is to be found in the spiral distribution of leaves on plant stems and in the number of petals in Aster, Daisy and Sunflower heads. The logarithmic spiral mapped by the radius vectors generated by the nesting of the golden rectangle is the same class of spiral found in snail-shell structures. The skin formation of onions, the leaf formation of lettuces and the spiral layup of pine cones are other notable examples, and Coxeter gives a beautiful demonstration of its presence in pineapples.

In spite of all this apparently impressive evidence the distinguished geneticist C. H. Waddington, has claimed 'that the Fibonacci series has nothing to do with any sort of biology, human or otherwise'.

It is, of course, true that writers and practitioners from Leonardo onwards have tried to distort the supple body of man in a desperate attempt to fit man into their own particular scheme of things, the fact remains that man's anthropometric proportions do accord closely to the golden rule and this in terms of sound ergonomic design is important.

2.19.5 The modulor

The use of standard modular sizes derived from man's own anthropometric structure has been pioneered in architecture by Le Corbusier, whose scheme is fully described in his book, *The Modulor.*

The basis of Le Corbusier's scheme is the adherence to a set of proportions derived from a two dimensional matrix grid, Fig. 2.60, in which the two co-ordinate dimensions grow in accordance with the golden rule.

Taking first the squares on the diagonal it is clear that square N can be filled

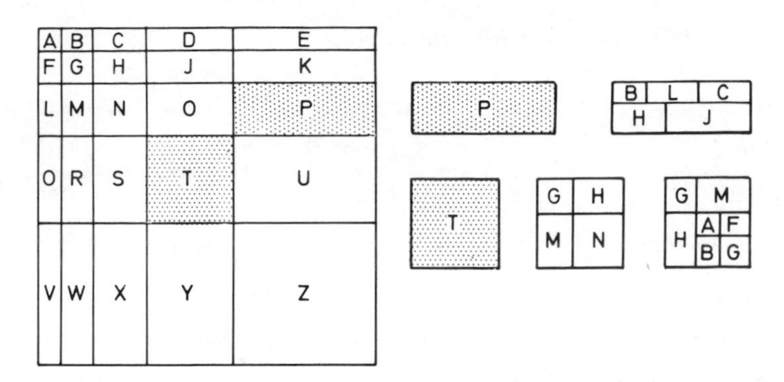

Figure 2.60

or divided by the divisions A, B, F, G, and similarly, divisions G, H, M, N, fill square T exactly. Notice, however, that square T may alternatively be compounded from the two rectangles R and S or by the three rectangles J, M, N, or equally well by rectangles J and O.

This most useful property is not confined to the filling and/or the division of squares, but applies equally well to the filling and/or division of all the rectangles, providing the matrix is suitably extended. The inner rectangles may, of course, be filled or divided as the matrix stands.

Thus, for example, rectangles C and D constitute the rectangle E, and the rectangle E plus the rectangle K together make up the rectangle P. The square Z plus the golden rectangle Y may be filled by O, P, T, U, while O may be replaced by M, N, and T, may be replaced by R, S. Division P may be replaced by E, K, and division U may be quartered by H, J, N, O, and so on. The variations and combinations are not far fromsbeing endless.

The area of each adjacent square increases or decreases in proportion to the square of the golden ratio, while adjacent rectangles increase in linear proportion. The area of corner rectangles adjacent to any square is equal to the area of the square itself.

Summary of Elementary Trig. Identities Table

$\cos^2 A + \sin^2 A = 1, \qquad \sec^2 A - \tan^2 A = 1, \qquad \operatorname{cosec}^2 A - \cot^2 A = 1$

$\sin A = \dfrac{2\tan\frac{A}{2}}{1+\tan^2\frac{A}{2}}, \qquad \cos A = \dfrac{1-\tan^2\frac{A}{2}}{1+\tan^2\frac{A}{2}}, \qquad \tan A = \dfrac{2\tan\frac{A}{2}}{1-\tan^2\frac{A}{2}}$

$2\cot 2A = \cot A - \tan A, \qquad 2\operatorname{cosec} 2A = \cot A + \tan A$

$\tan\frac{A}{2} = \operatorname{cosec} A - \cot A, \qquad \cot\frac{A}{2} = \operatorname{cosec} A + \cot A$

$\sin(A+B) = \sin A\cos B + \cos A\sin B, \qquad \sin(A-B) = \sin A\cos B - \cos A\sin B$

$\cos(A+B) = \cos A\cos B - \sin A\sin B, \qquad \cos(A-B) = \cos A\cos B + \sin A\sin B$

$\tan(A+B) = \dfrac{\tan A + \tan B}{1-\tan A\tan B}, \qquad \tan(A-B) = \dfrac{\tan A - \tan B}{1+\tan A\tan B}$

$\sin 2A = 2\sin A\cos A, \qquad \cos 2A = \cos^2 A - \sin^2 A, \qquad \tan 2A = \dfrac{2\tan A}{1-\tan^2 A}$

$$\cos 2A = 2\cos^2 A - 1 = 1 - 2\sin^2 A$$

$\sin 3A = 3\sin A - 4\sin^3 A, \quad \cos 3A = 4\cos^3 A - 3\cos A, \quad \tan 3A = \dfrac{3\tan A - \tan^3 A}{1-3\tan^2 A}$

$2\sin A\cos B = \sin(A+B) + \sin(A-B), \qquad 2\cos A\sin B = \sin(A+B) - \sin(A-B)$

$2\cos A\cos B = \cos(A+B) + \cos(A-B), \qquad 2\sin A\sin B = -\cos(A+B) + \cos(A-B)$

$\sin A + \sin B = 2\sin\left(\dfrac{A+B}{2}\right)\cos\left(\dfrac{A-B}{2}\right), \qquad \sin A - \sin B = 2\cos\left(\dfrac{A+B}{2}\right)\sin\left(\dfrac{A-B}{2}\right)$

$\cos A + \cos B = 2\cos\left(\dfrac{A+B}{2}\right)\cos\left(\dfrac{A-B}{2}\right), \qquad \cos A - \cos B = -2\sin\left(\dfrac{A+B}{2}\right)\sin\left(\dfrac{A -}{2}\right.$

3 Analytic geometry in three dimensions

This space may be thought of as a container that may be filled with any substance whatsoever. The same space that contains a basketful of grain may be filled with a basketful of sand. On the basis of this idea, geometry becomes the study of the space occupied by physical objects.

Irving Adler

3.1 THE CARTESIAN COORDINATE SYSTEM IN 3-SPACE

The spatial positions of points, lines, planes, curved surfaces and three-dimensional objects of all kinds, are customarily defined relative to three mutually perpendicular planes of reference. These three planes meet at a point (termed the origin) and produce three mutually perpendicular lines of intersection ox, oy, oz, to which all coordinate dimensions are referred. If, as in Fig. 3.1, the origin O lies at one corner of a rectangular prism, three mutually perpendicular edges aligned with the reference axes ox, oy, oz, then the most general object point P lies along the long diagonal at the corner farthest from O.

The spatial location of an object point P is conveniently specified relative to the origin O in terms of three ordered numbers (a, b, c) measured along, and projected from, the mutually perpendicular axes ox, oy, oz.

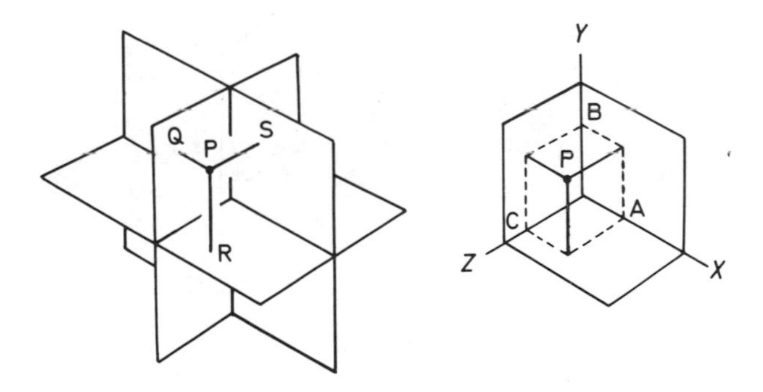

Figure 3.1

3.1.1 Absolute right- and left-handed axes

A right-handed system of axes is one in which the positive axial directions of ox, oy, oz and the positive rotational directions correspond to the motions of a right-handed screw. A right-handed screw is a screw for which clockwise turning, viewed from the tail, causes the head to translate axially in a direction away from the observer's eye and for which anticlockwise rotation causes the head to translate towards the eye.

Right-handed axes are used for all traditional mathematical calculations and when defining object points in the manner already explained.

Left-handed axes (those akin to a left-handed screw) are used in computer graphics but only in the role of eye coordinates see chapter 11.

The two distinct conventions are compared in Fig. 3.2, from which it is clear that changing the direction of one axis (in this case the z axis) changes the hand relative to the other two.

Except where the use of eye coordinates is specifically stated right-handed axes are used throughout.

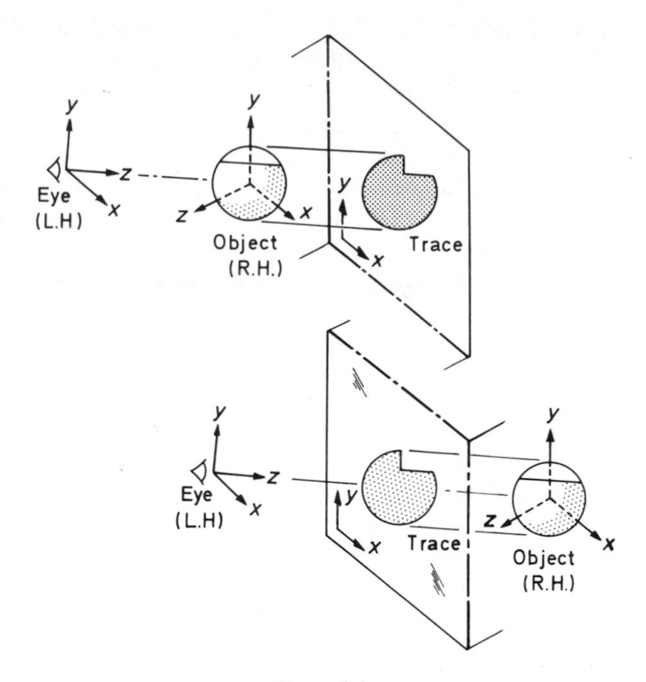

Figure 3.2

3.2 POINTS AND LINES IN 3-SPACE

A characteristic feature of three-dimensional analytical geometry is the frequent occurrence of direction ratios and direction cosines. We consider these two new concepts in depth.

3.2.1 The absolute position of an object point

An object point located by coordinates OA = a, OB = b, OC = c, is referred to as the point P (a, b, c) and the points P(ab), P(bc), P(ca) are its orthogonal projections (trace points) on the xoy, yoz, zox planes of reference.

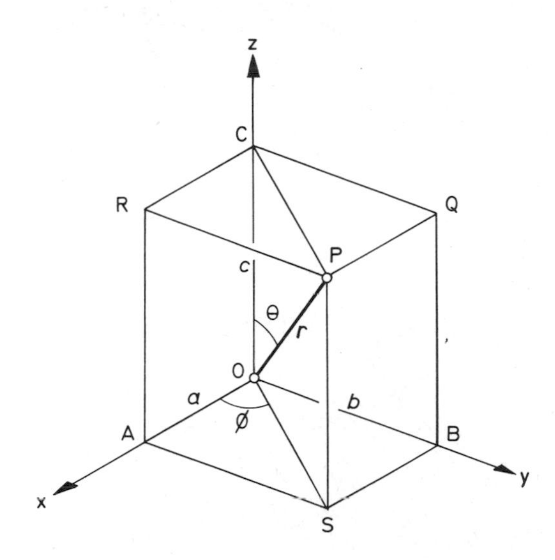

Figure 3.3

3.2.2 Direction ratios of a line

Any set of three ordered and corresponding increments in x, y and z, taken in ratio, serve to define the inclination (slope) of a line in 3-space and the three mutually perpendicular distances OA = a, AS = OB = b, SP = OC = c, Fig. 3.4, are the direction ratios of the line OP. In this particular instance the increments are taken directly from the origin and the direction ratios of the line OP and the coordinates of the point P are one and the same. In order to distinguish between coordinate dimensions and direction ratios we adopt the convention of writing coordinates in (x, y, z) spaced by a comma, and direction ratios (x: y: z) spaced by a colon. The two formulations are: (a, b, c) and (a: b: c) respectively. If P(x_1, y_1, z_1) and P(x_2, y_2, z_2) are any two points on a straight line then $(x_2 - x_1):(y_2 - y_1):(z_2 - z_1)$ are the direction ratios and since the inclination of a straight line is the same throughout its length any two points on the line suffice to define its inclination. The direction ratios:

$$(2:4:7),\ (4:8:14),\ (\tfrac{2}{7}:\tfrac{4}{7}:1)$$

all represent the same line, and indeed all lines parallel to it.

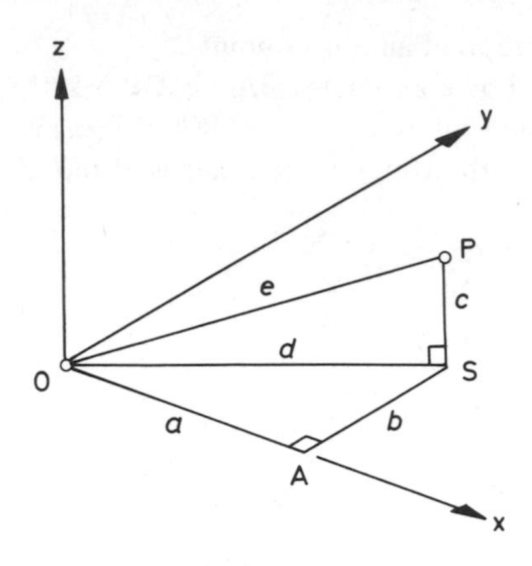

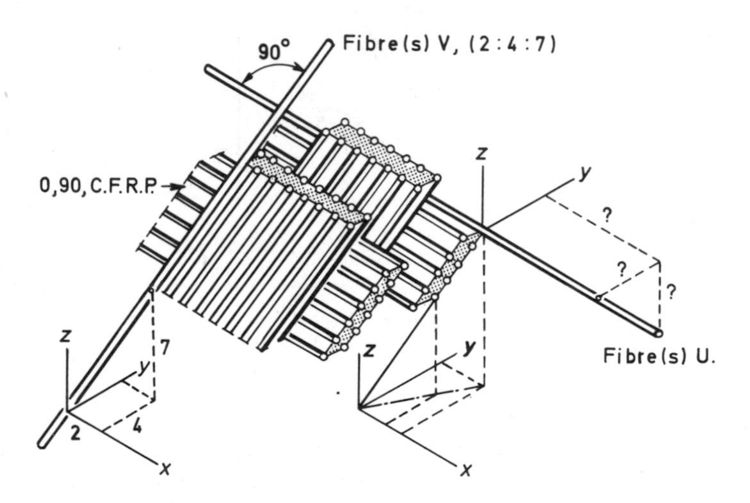

Figure 3.4

3.2.3 The true length of a line segment

Since the point A, Fig. 3.4, projects parallel to the axis *oy* to meet the orthogonal projection of the point B in the point S and the point C projects parallel to the axis *oy* to meet the orthogonal projection of the point B in the point Q and since the points R, Q and S project to the point P in similar fashion, the indirect line path (O → A → S → P) equals the direct line path (OP), and since the angles OAS, ASP are right angles, the length of the direct line path OP is equal to the square root of the Pythagoran sum of the indirect line segments OA = a, AS = OB = b, SP = OC = c.

$$\text{(TL)} \quad \text{OP} = \sqrt{a^2 + b^2 + c^2}$$

The orientation of a line which does not pass through the origin is completely defined by the coordinates of any two points on the line and the true length between these two points may be calculated as previously. That is the true length of a space line $P_1P_2 = L_{12}$ is:

$$(TL) \quad P_1P_2 = L_{12} = \sqrt{(x_2 - x_1)^2 + (y_2 - y_1)^2 + (z_2 - z_1)^2}$$

(N.B. The quantities $(x_2 - x_1):(y_2 - y_1):(z_2 - z_1)$ are the direction ratios of the line L_{12}.

Example

If P_1 is the point $(2, 0, 2)$ and P_2 is the point $(5, 4, 14)$ and all dimensions are in metres, find the true length of the line P_1P_2.

Solution

$$(x_2 - x_1) = (5 - 2) = 3$$
$$(y_2 - y_1) = (4 - 0) = 4$$
$$(z_2 - z_1) = (14 - 2) = 12$$

and the true length of the line P_1P_2 is:

$$(TL) \quad P_1P_2 = \sqrt{3^2 + 4^2 + 12^2} = 13 \text{ metres}$$

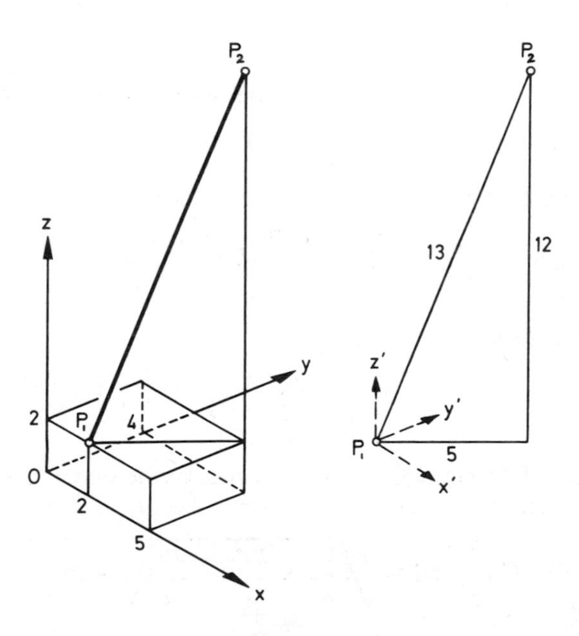

Figure 3.5

3.2.4 Local axes

We observe from the preceding discussion that the use of the direction ratios

$$(x_2 - x_1):(y_2 - y_1):(z_2 - z_1)$$

effectively relates the line P_1P_2 to a new set of local axes sited at P_1; whereas the use of the direction ratios

$$(x_1 - x_2):(y_1 - y_2):(z_1 - z_2)$$

effectively relates the line P_2P_1 to a new set of local axes sited at P_2.

If x_2 is greater than x_1 then the sign of $(x_2 - x_1)$ is positive and the sign of $(x_1 - x_2)$ negative, and vice versa. We must likewise observe the rules of positive angle. See Fig. 3.5 for a visual interpretation.

Example

An element from a torsionally stiff space frame of vertical height h is formed with an equilateral base of side $2h$ and a parallel equilateral top of side h, as shown in Fig. 3.6. Determine the true length of a typical inclined edge member.

Solution

We site the axes xoy as shown in the figure, whence the coordinates of P are:

$$(h, 0, 0)$$

The coordinates of Q are:

$$\left(\frac{h}{2}, \frac{h\sqrt{3}}{2}, h\right)$$

The direction ratios are:

$$\left(\frac{-h}{2} : \frac{h\sqrt{3}}{2} : h\right)$$

The true length of the line PQ is the square root of the sum of the squares of its direction ratios. Hence

$$\text{(TL)}\quad \text{PQ} = \sqrt{\frac{h^2}{4} + \frac{3h^2}{4} + \frac{4h^2}{4}} = h\sqrt{2}$$

and this result is confirmed by the accompanying descriptive method.

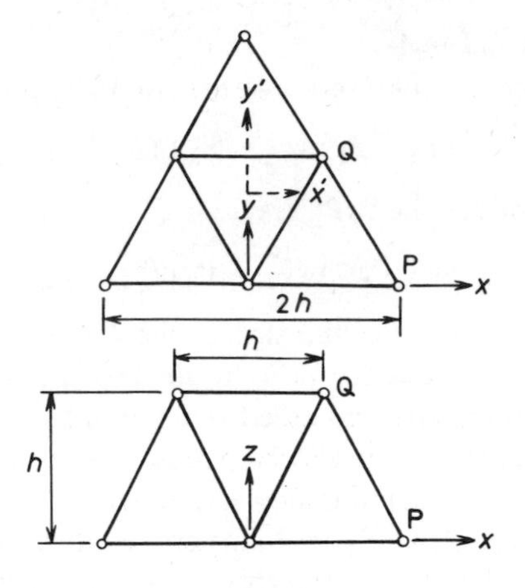

Figure 3.6

Rider

Consider the implication of changing the origin of axes which for the purpose of illustration will be sited at the centroid in plan, and denoted by the new axes $x'yz$.

The coordinates of P are now:

$$\left(h, -\frac{h\sqrt{3}}{3}, 0\right)$$

The coordinates of Q are now:

$$\left(\frac{h}{2}, \frac{h\sqrt{3}}{6}, h\right)$$

whence the direction ratios are:

$$\left(-\frac{h}{2} : -\frac{h\sqrt{3}}{2} : h\right)$$

and the true length of PQ is:

$$\text{(TL)}\quad \text{PQ} = \sqrt{\frac{h^2}{4} + \frac{3h^2}{4} + \frac{4h^2}{4}} = h\sqrt{2}$$

as previously. The siting of axes clearly does not affect the identity of geometrical features referred to and in practice the siting of reference axes depends on considerations of convenience or in some cases by prior commitment.

3.2.5 Directed lines and angles

We observe that if the direction ratios of the line P_1P_2 are:

$$(x_2 - x_1):(y_2 - y_1):(z_2 - z_1)$$

the direction ratios of the line P_2P_1 are:

$$(x_1 - x_2):(y_1 - y_2):(z_1 - z_2)$$

If x_2 is greater than x_1 then the sign of the x direction ratio in the first case is positive, whereas the x direction ratio in the second case is negative, and so the order in which the coordinates are called is important.

Whilst a length cannot, in an absolute sense, be negative, a sign may be allotted (as in the case of a vector) to indicate direction.

If the direct line path from P_1 to P_2 is positive, then the direct line path from P_2 to P_1 is negative and the round trip from P_1 to P_2 and back to P_1 is vectorially zero: and the same considerations are true of angles.

3.2.6 Division of a line segment in a given ratio

The general object point $P(x, y, z)$ lies along the longest diagonal of a rectangular prism and it is clear that the coordinates of the mid-point of the line OP are

$$\left(\frac{o+x}{2}\right), \left(\frac{o+y}{2}\right), \left(\frac{o+z}{2}\right) = \left(\frac{x}{2}, \frac{y}{2}, \frac{z}{2}\right)$$

This is because the mid-point divides the line in the ratio 1:1 and hence the length of the half segment is one half the total length of line. Since the origin is at zero the coordinates of the mid-point are one half the total extent.

The coordinates of a point which divides the line joining two given points in a given ratio, may be deduced as follows.

We see from Fig. 3.7, that if the point R divides the line PQ internally in the ratio of λ/μ the x coordinate of R is: $x = (x_1 + \lambda \cos \phi)$:

$$\cos \phi = \frac{x_2 - x_1}{\lambda + \mu}$$

$$x = x_1 + \lambda\left(\frac{x_2 - x_1}{\lambda + \mu}\right)$$

$$x = \frac{\lambda x_2 + \mu x_1}{\lambda + \mu}$$

and the coordinate of P in terms of y and z may be obtained by a similar route.

The coordinate of a point which divides a given line internally in the ratio λ/μ are:

$$x = \frac{\lambda x_2 + \mu x_1}{\lambda + \mu}, \quad y = \frac{\lambda y_2 + \mu y_1}{\lambda + \mu}, \quad z = \frac{\lambda z_2 + \mu z_1}{\lambda + \mu}$$

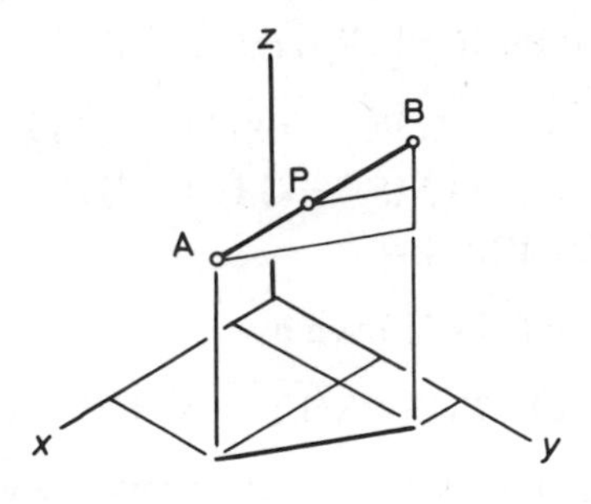

Figure 3.7

and similarly the coordinates of a point which divides a given line externally in the ratio λ/μ are:

$$x = \frac{\lambda x_2 - \mu x_1}{\lambda - \mu}, \quad y = \frac{\lambda y_2 - \mu y_1}{\lambda - \mu}, \quad z = \frac{\lambda z_2 - \mu z_1}{\lambda - \mu}$$

Example

The line of a drive shaft is to pass through a clearance hole in the inclined face of a machine frame. If the centre line of the shaft is defined by its end points the coordinates of which are $P_1(6, 2, 2)$ and $P_2(-6, -2, -1)$ and it is known that, in the absence of a hole, the shaft would intersect the face of the machine frame at a distance of 3 units from P_1 along the line of shaft, determine the coordinates of the hole centre, with respect to the axes of reference within the machine frame.

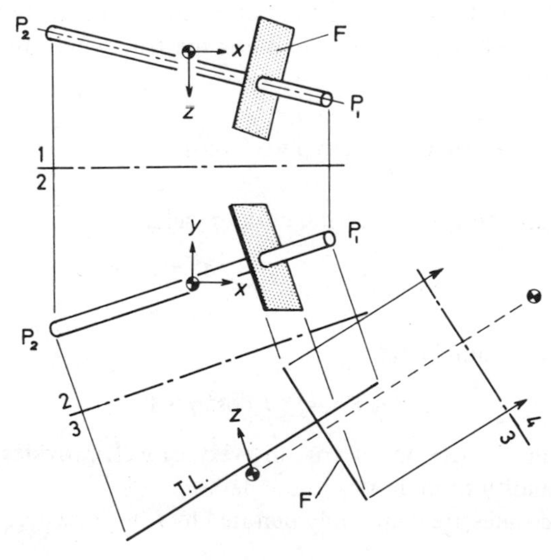

Figure 3.8

The direction ratios of the line P_1P_2 are:

$$(x_2 - x_1):(y_2 - y_1):(z_2 - z_1)$$
$$(-6 -6):(-2 -2):(-1 -2)$$
$$-12:-4:-3$$

The true length of the line P_1P_2 is given by:

$$\text{(TL)} \quad P_1P_2 = \sqrt{12^2 + 4^2 + 3^2} = 13$$

Since the machine face is 3 units from P_1 measured along the centre line of P_1P_2 the machine face divides the drive shaft in the ratio 3:10. Whence

$$\lambda = 3, \quad \mu = 10, \quad \lambda + \mu = 13$$

The coordinates of the point of penetration are

$$\frac{\lambda x_2 + \mu x_1}{\lambda + \mu}, \frac{\lambda y_2 + \mu y_1}{\lambda + \mu}, \frac{\lambda z_2 + \mu z}{\lambda + \mu}$$

$$\frac{42}{13}, \frac{14}{13}, \frac{17}{13}$$

Hence the coordinates of the point of penetration are (3.231, 1.077, 1.308).

3.2.7 Direction cosines

If the true angle between the line OP and the axis ox is α, the true angle between the line OP and the axis oy is β, the true angle between the line OP and the axis oz is γ, then $\cos\alpha$, $\cos\beta$, $\cos\gamma$, are the direction cosines of the line OP.

If $(a: b: c)$ are the direction ratios and e is the true length of the line OP, as defined in Fig. 3.9.

$$\cos\alpha = \frac{a}{e}, \quad \cos\beta = \frac{b}{e}, \quad \cos\gamma = \frac{c}{e}$$

Squaring and adding the three direction cosines yields

$$\cos^2\alpha + \cos^2\beta + \cos^2\gamma = \frac{a^2 + b^2 + c^2}{e^2}$$

But $e^2 = a^2 + b^2 + c^2$, and hence

$$\cos^2\alpha + \cos^2\beta + \cos^2\gamma = 1$$

This equation is an identity and is true in every case. It provides a ready means of checking the validity of direction cosine data.

The direction cosines are commonly denoted by l, m, n, where

$$l = \cos\alpha, \; m = \cos\beta, \; n = \cos\gamma.$$

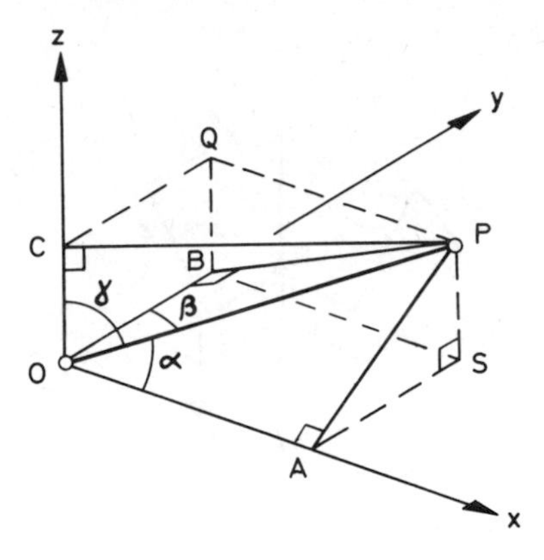

Figure 3.9

and hence

$$l^2 + m^2 + n^2 = 1$$

as previously.

The direction angles α, β, γ, are defined as the angles between the positive direction of the axes *ox*, *oy* and *oz* and the space line to which they refer.

Since $(a:b:c)$ are here direction ratios, these may be positive or negative according to the origin used.

If $a = \pm(x_2 - x_1), b = \pm(y_2 - y_1), c = \pm(z_2 - z_1)$ then

$$\cos\alpha = \frac{a}{e} = \pm\frac{x_2 - x_1}{e}$$

$$\cos\beta = \frac{b}{e} = \pm\frac{y_2 - y_1}{e}$$

$$\cos\gamma = \frac{c}{e} = \pm\frac{z_2 - z_1}{e}$$

When a cosine is negative the angle referred to is the supplement of its positive and

$$-\cos\alpha = \cos(180 - \alpha), -\cos\beta = \cos(180 - \beta), -\cos\gamma = \cos(180 - \gamma)$$

Example

The roof planes of an L shaped bungalow intersect in a line P_1P_2, as shown in Fig. 3.10. If the coordinates of P_1 are $(-15, 0, 9)$ and the coordinates of P_2 are $(-8, 7, 0)$, determine the direction angles of the line of intersection P_1P_2.

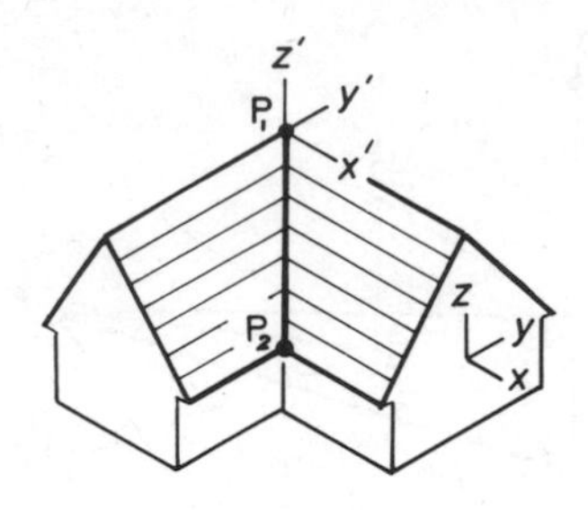

Figure 3.10

Solution

When P_1 is taken as the local origin the direction angles α, β, γ, are the angles made by the line P_1P_2 and the positive directions of the axes ox', oy', oz'.

	x	y	z	
P_2	−8	−7	0	Coordinates
P_1	−15	0	9	(in feet)
D.R.	7	−7	−9	

The true length of P_1P_2 is the square root of the sum of the squares of the direction ratios:

$$\text{(TL)}\quad P_1P_2 = \sqrt{49 + 49 + 81} = \sqrt{179} = 13.379 \text{ ft.}$$

The direction cosines are:

$$\cos\alpha = \frac{7}{13.379}, \quad -\cos\beta = \frac{7}{13.379}, \quad -\cos\gamma = \frac{9}{13.379}$$

hence

$$\alpha = 58.05°, \; \beta = 121.55°, \; \gamma = 132.27°$$

3.2.8 True angle between two intersecting straight lines

The true angle between two intersecting straight lines is commonly defined as the acute angle θ which lies in their common plane.

If, as in Fig. 3.11, the (x, y, z) coordinates of a point on each line, plus the coordinates of their point of intersection is given, the true angle θ may be calculated directly, using the cosine rule.

Let two straight lines L_{12} and L_{13} intersect at the common point 1 as shown in Fig. 3.11. Then the true angle θ_1 lies in the plane containing the three points

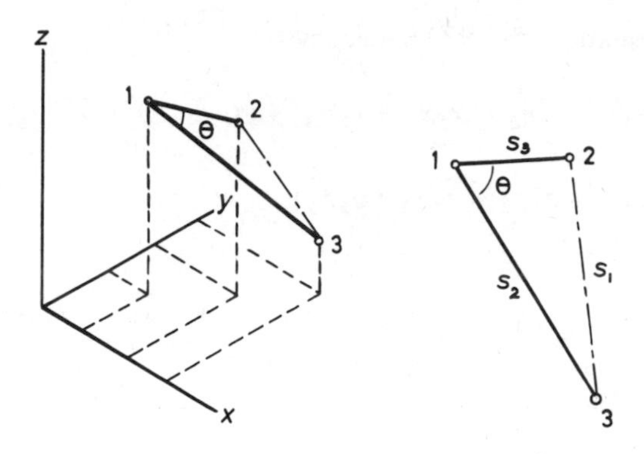

Figure 3.11

1, 2, 3. Let the true lengths L_{12}, L_{13}, L_{23} be represented by S_3, S_2, S_1, in accord with the usual plane geometry notation.

Let the length of the line P_1P_2 equal S_3 and the length of the line P_1P_3 equal S_2 and the length of the line P_2P_3 equal S_1. Let the direction cosines of line P_1P_2 equal l_3, m_3, n_3, and the direction cosines of the line P_1P_3 equal l_2, m_2, n_2, and the direction consines of the line P_2P_3 equal l_1, m_1, n_1.

Applying the cosine rule we write the angle θ_1 in terms of the lengths S_1, S_2, S_3.

The cosine rule yields

$$\cos\theta_1 = \frac{S_2^2 + S_3^2 - S_1^2}{2S_2S_3}$$

If the coordinates of the line S_1 are:

$$(x_3 - x_2), (y_3 - y_2), (z_3 - z_2),$$

and the coordinates of the line S_2 are:

$$(x_3 - x_1), (y_3 - y_1), (z_3 - z_1)$$

and the coordinates of the line S_3 are:

$$(x_2 - x_1), (y_2 - y_1), (z_2 - z_1)$$

It follows that:

$$\cos\theta_1 = \frac{1}{2S_2S_3}\,[(x_3 - x_1)^2 + (y_3 - y_1)^2 + (z_3 - z_1)^2 + (x_2 - x_1)^2$$
$$+ (y_2 - y_1)^2 + (z_2 - z_1)^2 - (x_3 - x_2)^2 - (y_3 - y_2)^2 - (z_3 - z_2)^2]$$

Expanding this equation, the R.H.S. becomes:

$$\frac{1}{2S_2S_3}[2(x_1^2 - x_1x_3 - x_1x_2 + x_2x_3) + 2(y_1^2 - y_1y_3 - y_1y_2 + y_2y_3)$$
$$+ 2(z_1^2 - z_1z_3 - z_1z_2 + z_2z_3)]$$

and subsequently

$$\frac{(x_2 - x_1)}{S_3}\frac{(x_3 - x_1)}{S_2} + \frac{(y_2 - y_1)}{S_3}\frac{(y_3 - y_1)}{S_2} + \frac{(z_2 - z_1)}{S_3}\frac{(z_3 - z_1)}{S_2}$$

Now

$$\frac{(x_2 - x_1)}{S_3} = l_3, \quad \frac{(y_2 - y_1)}{S_3} = m_3, \quad \frac{(z_2 - z_1)}{S_3} = n_3,$$

and

$$\frac{(x_3 - x_1)}{S_2} = l_2, \quad \frac{(y_3 - y_1)}{S_2} = m_2, \quad \frac{(z_3 - z_1)}{S_2} = n_2,$$

hence

$$\cos\theta_1 = l_2l_3 + m_2m_3 + n_2n_3$$

If we redefine our two lines in terms of previous notation: Line L_1 has direction ratios $(a_1 : b_1 : c_1)$, line L_2 has direction ratios $(a_2 : b_2 : c_2)$ and the true angle θ between the two lines is given by the equation:

$$\cos\theta = \frac{a_1a_2 + b_1b_2 + c_1c_2}{\sqrt{a_1^2 + b_1^2 + c_1^2} \cdot \sqrt{a_2^2 + b_2^2 + c_2^2}}$$

N.B. When the two lines are perpendicular $\theta = 90°$, $\cos\theta = 0$ and the above expression becomes

$$(a_1a_2 + b_1b_2 + c_1c_2) = \cos\alpha_1 \cos\alpha_2 + \cos\beta_1 \cos\beta_2 + \cos\gamma_1 \cos\gamma_2 = 0$$

and either of these expressions serve to denote a normal.

3.2.9. True angle between two non-intersecting skew lines

We define the true angle between two skew lines as the apparent angle between their traces when each line is simultaneously viewed true length.

The traces of two skew lines are shown in Fig. 3.12, from which it is clear that neither line is seen true length, and hence the true angle between them is not immediately apparent.

In order to view the two lines' true length it is necessary to implement the standard descriptive method, as shown in Fig. 3.12, in which the two true lengths and the true angle appear in plane 55. The further projection (at an arbitrary angle) into plane 66, is included to show that if the two lines are viewed in a direction which is normal to their common perpendicular, then the two skew lines are seen to be parallel, except when the direction of projection is such that one or other of the two lines appears as a point view (as occurs in plane 44).

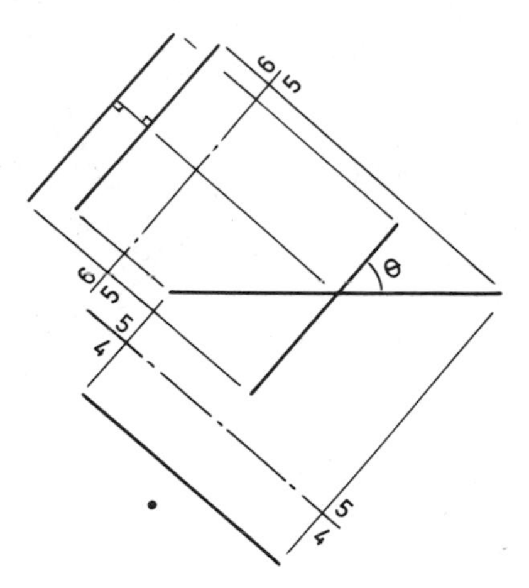

Figure 3.12

As demonstrated above the true angle between two skew lines is the angle formed by the lines in a direction normal to their common perpendicular.

Whilst those well acquainted with the descriptive procedure, Fig. 3.12, will have no difficulty in visualising the lay of skew lines in space many students find the visualisation of skew lines quite difficult and many never master the art. We therefore consider an alternative proof to that already given.

The sum of the indirect line (Component) projections of the one line onto the other is equal to the direct projection of their resultant: and with this fact to hand we may proceed as follows.

The perpendicular projection of a direct line OQ onto the line OP is equal to the sum of the projections of the indirect line O, E, H, Q.

The perpendicular projection of OQ on OP equals OQ cos θ, where θ is the true angle between the lines OP and OQ.

The perpendicular projection of OE = OO_1 = OE cos α_1. The length of the line EH equals the length of the line OF. If O_1F_1 is parallel to the axis OY and FF_1 is parallel to OO_1; OF = O_1F_1.

The perpendicular projection of O_1F_1 = O_1O_2 = EH cos β_1. The length of the line HQ equals the length of the line OG. If O_2G_1 is parallel to the axis OZ and GG_1 is parallel to OO_2; OG = O_2G_1.

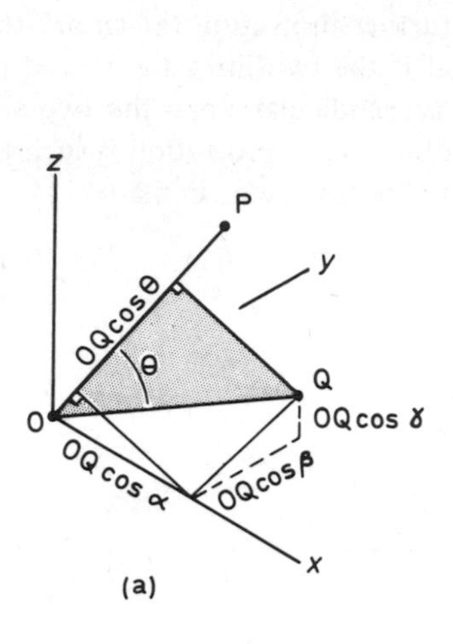

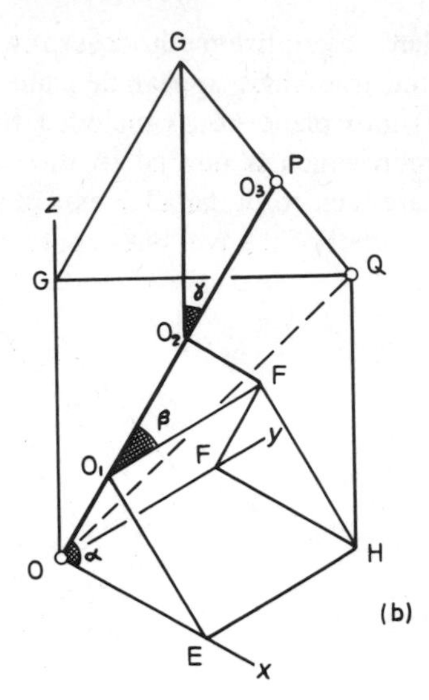

Figure 3.13

The perpendicular projection of $O_2G_1 = O_2P = HQ \cos \gamma_1$. The length OP equals $OQ \cos \theta$ and hence:

$$OQ \cos \theta = OE \cos \alpha_1 + EH \cos \beta_1 + HQ \cos \gamma$$

But

$$OE = OQ \cos \alpha_2$$
$$EH = OF = OQ \cos \beta_2$$
$$HQ = OG = OQ \cos \gamma_2$$

and hence

$$\cos \theta = \cos \alpha_1 \cos \alpha_2 + \cos \beta_1 \cos \beta_2 + \cos \gamma_1 \cos \gamma_2$$

If the direction ratios of the two lines OP and OQ are $(a_1 : b_1 : c_1)$ $(a_2 : b_2 : c_2)$, the true angle θ between the two lines may be conveniently expressed in the form:

$$\cos \theta = \frac{a_1a_2 + b_1b_2 + c_1c_2}{\sqrt{a_1^2 + b_1^2 + c_1^2} \cdot \sqrt{a_2^2 + b_2^2 + c_2^2}}$$

This equation is fundamental to many problems encountered in three dimensional geometry.

Example
A universal coupling suitable for transmitting rotary motion between two included shafts is being considered for the application in Fig. 3.14. If the maximum obliquity of the coupling is not to exceed $\pm 25^\circ$ is this particular coupling suitable for connecting the two shafts, AO, OB, for which data is given in the table

Coordinates	x	y	z	
O	0	0	0	dimensions
A	0	−1.2	−0.5	in metres
B	0.5	1.2	0	

Solution
Shaft AO is in the *yoz* plane, shaft BO is in the *xoy* plane, the true length of both AO and BO is 1.3 metres and hence the direction cosines are:

$$\text{(DC) AO} \left(0: -\frac{12}{13} : -\frac{5}{13}\right)$$

$$\text{(DC) BO} \left(\frac{5}{13} : \frac{12}{13} : 0\right)$$

The true angle θ formed by AOB is given by the equation:

$$\cos\theta = \cos\alpha_1 \cos\alpha_2 + \cos\beta_1 \cos\beta_2 + \cos\gamma_1 \cos\gamma_2$$

$$= \left(0 \cdot \frac{5}{13}\right) + \left(-\frac{12}{13} \cdot \frac{12}{13}\right) + \left(-\frac{5}{13} \cdot 0\right)$$

$$= -\frac{144}{169} = -0.8520$$

$$\theta = -148.334^\circ$$

$$= (180^\circ - 148.334^\circ) = 31.566^\circ$$

The maximum allowable angle (at which this particular coupling may work) is $\pm 25^\circ$ and hence the coupling is not suitable for this application.

3.2.10 The shortest distance from a point to a straight line

If the coordinates of a line AB and a point P are known, the problem of finding the shortest distance between the point and the line may be resolved by observing that the three points AB and P serve to define a plane.

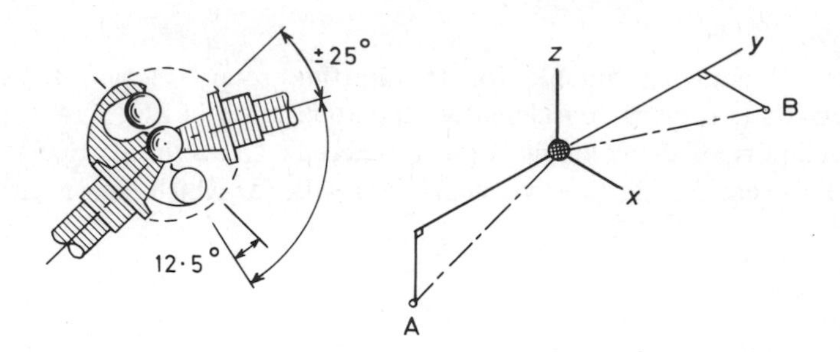

Figure 3.14

If the true angle between the line AP and the line AB is θ, and PN is the perpendicular from P to AB as shown in Fig. 3.15, then the minimum distance d equals PN

$$PN = d = AP \sin\theta$$
$$= AP\sqrt{1 - \cos^2\theta}$$

where the true angle θ is given by the equation:

$$\cos\theta = \cos\alpha_1 \cos\alpha_2 + \cos\beta_1 \cos\beta_2 + \cos\gamma_1 \cos\gamma_2$$

as previously defined and AP is a true length to be obtained from the given co-ordinate dimensions.

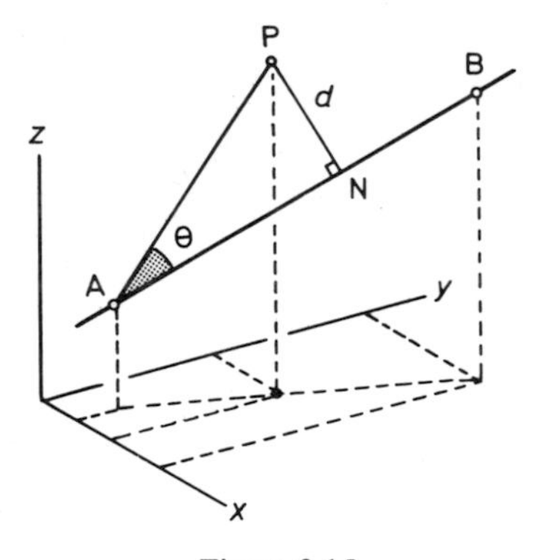

Figure 3.15

Example
If the coordinates of the centre line of a particular control cable AB and a pulley centre P are as tabulated, calculate the minimum distance from the point P to the cable centre line AB, and hence specify the root diameter of a pulley wheel with centre at P, suitable for running a cable AB of 1 cm diameter away from the main cable run. Such layouts are common to aircraft "trim" controls.

	x	y	z	
A	100	60	60	coordinates
B	40	80	30	(in cm)
P	80	50	40	

Solution

The direction ratios of the line AB are:

$$(40 - 100) = -60 : (80 - 60) = 20 : (30 - 60) = -30$$

The true length of the line AB is:

$$\text{(TL)} \quad \text{AB} = \sqrt{60^2 + 20^2 + 30^2} = \sqrt{4900} = 70 \text{ cm}$$

The direction ratios of the line AP are:

$$(80 - 100) = -20 : (50 - 60) = -10 : (40 - 60) = -20$$

The true length of the line AP is:

$$\text{(TL)} \quad \text{AP} = \sqrt{20^2 + 10^2 + 20^2} = \sqrt{900} = 30 \text{ cm}$$

The direction cosines of the line AP are:

$$-\frac{2}{3} : -\frac{1}{3} : -\frac{2}{3}$$

and since

$$\cos\theta = ad + be + cf$$

$$\cos\theta = \left(-\frac{6}{7} \cdot -\frac{2}{3}\right) + \left(\frac{2}{7} \cdot -\frac{1}{3}\right) + \left(-\frac{3}{7} \cdot -\frac{2}{3}\right)$$

$$= \frac{12}{21} - \frac{2}{21} + \frac{6}{21} = \frac{16}{21}$$

whence

$$\cos^2\theta = \frac{256}{441}$$

and since the minimum distance d is given by:

$$d = \text{AP}\sqrt{1 - \cos^2\theta}$$

$$d = 30\sqrt{1 - \frac{256}{441}} = 30\sqrt{\frac{185}{441}}$$

$$d = 19.431 \text{ cm}$$

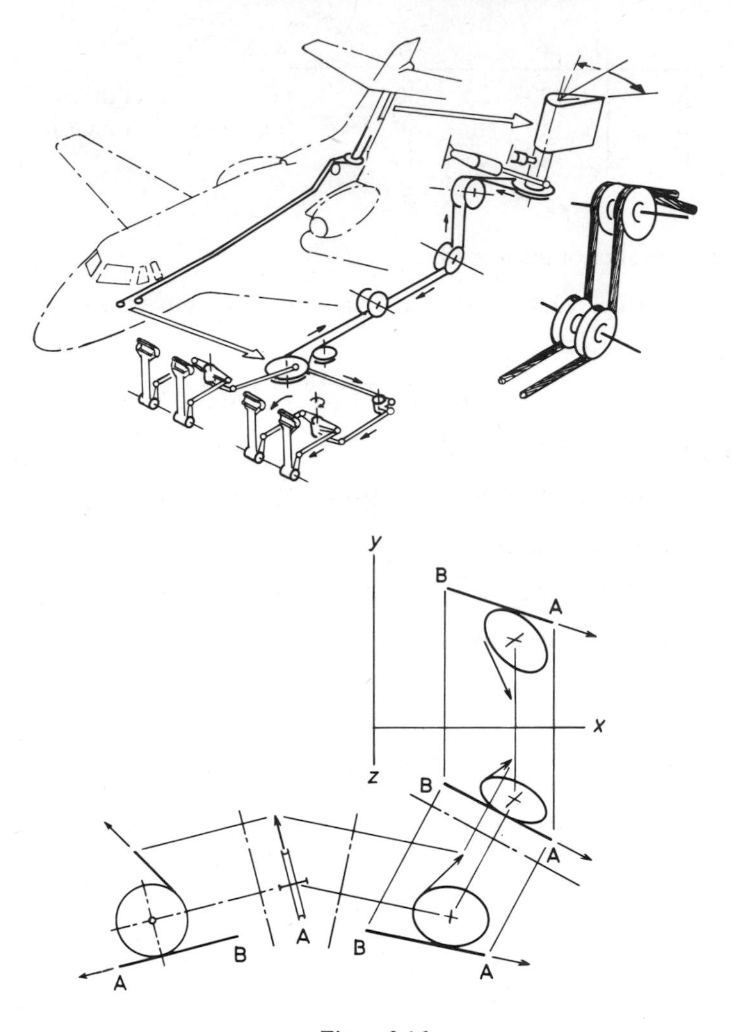

Figure 3.16

This result is confirmed by the orthographic representation, Fig. 3.16, and the pulley root diameter for a 1 cm diameter cable is 18.931 cm.

Example

A branch pipe is to be connected from a point P(7, 5, 5) to a main pipe by means of a 90° tee piece. If A(2, 4, 6) and B(1, 3, 5) are two reference points on the main pipe centre line, determine the minimum centre to centre-line length of the branch pipe and hence the point at which it joins the main pipe as measured from A.

Solution

	x	y	z
A	2	4	6
B	1	3	5
P	7	5	5

$$\text{(D.R.)} \quad \text{AP}(7-2) = 5 : (5-4) = 1 : (5-6) = -1$$

$$\text{(TL)} \quad \text{AP} = \sqrt{5^2 + 1^2 + 1^2} = \sqrt{27}$$

$$\text{(D.R.)} \quad \text{AB}(1-2) = -1 : (3-4) = -1 : (5-6) = -1$$

$$\text{(TL)} \quad \text{AB} = \sqrt{1^2 + 1^2 + 1^2} = \sqrt{3}$$

If the angle formed by PAB is θ

$$\cos\theta = \cos\alpha_1 \cos\alpha_2 + \cos\beta_1 \cos\beta_2 + \cos\gamma_1 \cos\gamma_2$$

$$= \frac{5}{\sqrt{27}} \cdot -\frac{1}{\sqrt{3}} + \frac{1}{\sqrt{27}} \cdot -\frac{1}{\sqrt{3}} + \frac{1}{\sqrt{27}} \cdot \frac{1}{\sqrt{3}}$$

$$= \frac{-5-1+1}{\sqrt{81}} = -\frac{5}{9}$$

whence

$$\theta = 123.75^\circ$$

The length of the perpendicular PN is given by:

$$\text{PN} = \text{AP}\sqrt{1 - \cos^2\theta}$$

$$= \frac{\sqrt{27}\sqrt{56}}{\sqrt{81}} = \frac{\sqrt{56}}{\sqrt{3}} = 4.321 \text{ m}$$

The distance of the foot of the perpendicular from A is:

$$\text{AN} = \text{AP} \cos\theta$$

$$= -\sqrt{27} \cdot \frac{5}{9} = -\frac{5}{\sqrt{3}} = -2.886 \text{ m}$$

The minus sign indicates that a 90° connector may only be used to join these points on the far side of the reference point A as shown in Fig. 3.17, and not between A and B as may have been envisaged.

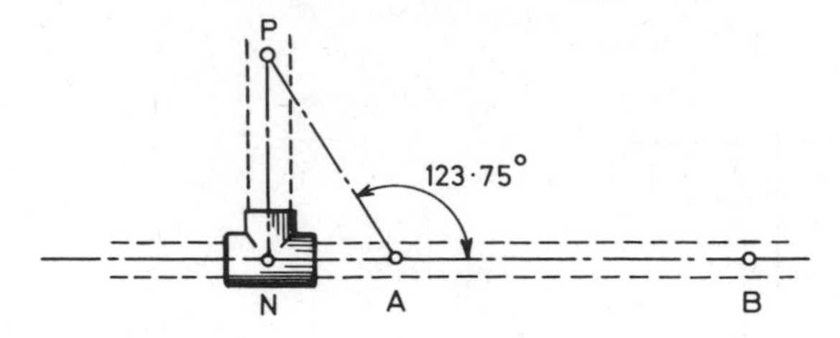

Figure 3.17

3.3 THE DEFINITION OF A PLANE IN 3-SPACE

A plane in 3-space may be defined in several ways and the most common forms are given below.

3.3.1 The equation to a plane in general form

The general equation to a plane is of the form:

$$Ax + By + Cz + D = 0$$

This equation may assume numerous truncated forms, representative of special cases. For instance, the condition $C = 0$, means there is no term in z and the equation $Ax + By + D = 0$ represents all planes which are perpendicular to the reference plane xoy. Similar the condition $B = 0$ represents all planes which are perpendicular to the reference plane xoz and the condition $A = 0$ represents all planes which are perpendicular to the reference plane yoz. The slope and position of these special case planes depends on the choice of constants but their perpendicularity is generally as shown in Fig. 3.18.

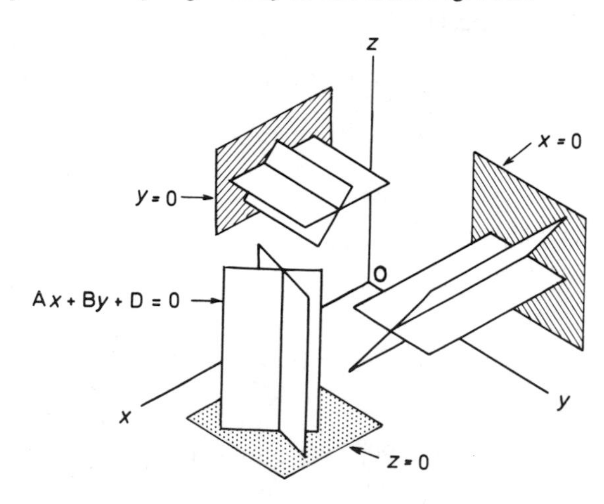

Figure 3.18

The three equations corresponding to C = 0, B = 0, A = 0, define all the planes which are perpendicular to the reference planes *xoy*, *xoz*, *yoz* respectively.

$$Ax + By + D = 0 \perp xoy$$
$$Ax + Cz + D = 0 \perp xoz$$
$$By + Cz + D = 0 \perp yoz$$

The coefficients A, B, C, D, of a particular plane may be readily derived as in the following example.

Example

Determine the coefficients and hence the general equation to a cant rib in an aircraft wing defined by the four points P, Q, R, S, Fig. 3.19, the coordinates of P, Q, R, S, being as given.

	x	y	z
P	0	10	−1
Q	1	5	−6
R	6	10	−6
S	5	15	−1

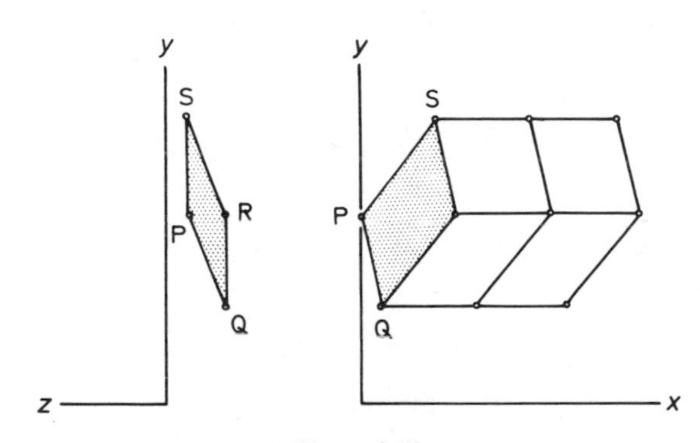

Figure 3.19

Solution

The general equation to a plane is:

$$Ax + By + Cz + D = 0$$

and three independent equations may be written in terms of the coordinates of P, Q and S which are given

$$(P) \quad 0 + 10B - C + D = 0$$
$$(Q) \quad A + 5B - 6C + D = 0$$
$$(S) \quad 5A + 15B - C + D = 0$$

We observe from equations (P) and (S) that A = −B and hence the data is reduced to two equations in three unknowns.

$$4B - 6C + D = 0$$
$$10B - C + D = 0$$

or by dividing through by D and putting B/D = U and C/D = V.

$$4U - 6V = -1$$
$$10U - V = -1$$

here we have two equations in two unknowns and the solution is routine.

$$U = \frac{-5}{56},\quad V = \frac{+6}{56}$$

and this result checks on substitution

$$\frac{A}{D} = \frac{5}{56},\quad \frac{B}{D} = \frac{-5}{56},\quad \frac{C}{D} = \frac{6}{56}$$

and since the general equation to the plane is of the form:

$$Ax + By + Cz + D = 0$$

or

$$\frac{Ax}{D} + \frac{By}{D} + \frac{Cz}{D} + 1 = 0$$

$$\frac{5x}{56} - \frac{5y}{56} + \frac{6z}{56} + 1 = 0$$

or

$$5x - 5y + 6z + 56 = 0$$

This is the equation to the plane P, Q, R, S, in general form.

(N.B. The determinant method of solving simultaneous equations is discussed later in the text).

3.3.2 The equation to a plane in intercept form

A plane may be defined in terms of its (x, y, z) intercepts (a, b, c), its perpendicular p and its direction cosines α, β, γ.

With reference to Fig. 3.20, we obtain two independent expressions for the perpendicular p from the definition of a direction cosine:

$$x = p \cos \alpha,\ y = p \cos \beta,\ z = p \cos \gamma$$

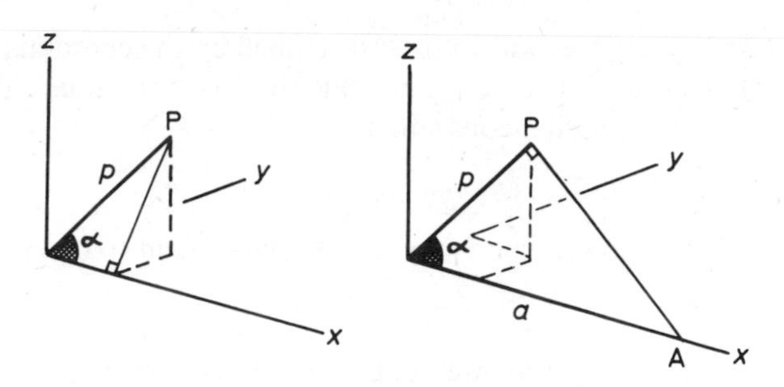

Figure 3.20

and since any line in a plane is normal to the plane's perpendicular p

$$p = a \cos \alpha,\ p = b \cos \beta,\ p = c \cos \gamma$$

Eliminating p from these two sets of equations

$$\cos^2 \alpha = \frac{x}{a},\ \cos^2 \beta = \frac{y}{b},\ \cos^2 \gamma = \frac{z}{c}$$

but since $\cos^2 \alpha + \cos^2 \beta + \cos^2 \gamma = 1$, the intercept equation to a plane is:

$$\frac{x}{a} + \frac{y}{b} + \frac{z}{c} = 1$$

3.3.3 Relationship between the coefficients of the general and intercept forms

The relationship between the coefficients in the general equation and the corresponding intercept form, may be deduced by directly comparing the coefficients.

If the general equation to a plane is:

$$\mathrm{A}x + \mathrm{B}y + \mathrm{C}z + \mathrm{D} = 0$$

$$\frac{\mathrm{A}x}{\mathrm{D}} + \frac{\mathrm{B}y}{\mathrm{D}} + \frac{\mathrm{C}z}{\mathrm{D}} = -1$$

$$\frac{\mathrm{A}x}{\mathrm{D}} - \frac{\mathrm{B}y}{\mathrm{D}} - \frac{\mathrm{C}z}{\mathrm{D}} = +1$$

and hence since the intercept equation is of the form:

$$\frac{x}{a} + \frac{y}{b} + \frac{z}{c} = 1$$

where

$$a = \frac{-\mathrm{D}}{\mathrm{A}},\ b = \frac{-\mathrm{D}}{\mathrm{B}},\ c = \frac{-\mathrm{D}}{\mathrm{C}}$$

We note that since three known points defined by the coordinates (x_r, y_r, z_r) suffice to define a plane, it is always possible to deduce the four unknowns from the three appropriate simultaneous equations.

Example

The general equation to the cant rib, Fig. 3.19, was found to be:

$$5x - 5y + 6z + 56 = 0$$

and to convert this equation to intercept form we divide through by the constant D and reduce the equation as follows:

$$\frac{5x}{56} - \frac{5y}{56} + \frac{6z}{56} = -1$$

$$-\frac{5x}{56} + \frac{5y}{56} - \frac{6z}{56} = +1$$

and hence

$$-\frac{x}{11.2} + \frac{y}{11.2} - \frac{z}{9.33} = 1$$

This is the equation to the plane PQRS in intercept form.

We again observe that the relationship between the coefficients in the general and intercept forms is as follows:

$$a = \frac{-D}{A}, b = \frac{-D}{B}, c = \frac{-D}{C}$$

3.3.4 The equation to a plane in normal form

Let $P(x_1 y_1 z_1)$ be any point in a plane and let the direction ratios of the normal to the plane be $(l: m: n)$.

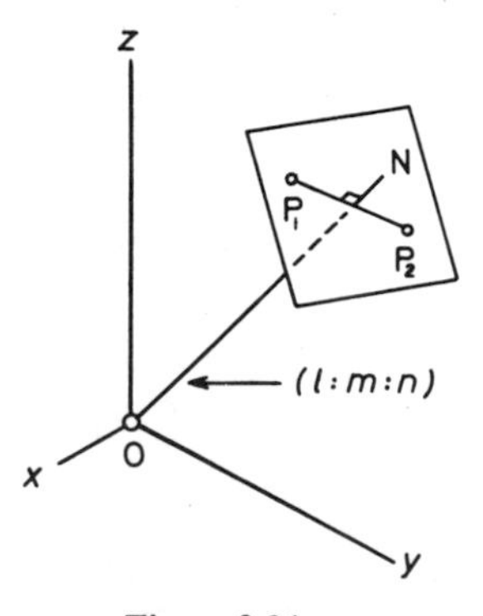

Figure 3.21

If the line P_1P_2 lies in the plane and NN is
The direction ratios of P_1P_2 are:

$$(x_2 - x_1):(y_2 - y_1$$

The direction ratios of NN are:

$$l:m:n$$

The angle between two lines is given by the general equation:

$$\cos\theta = \frac{a_1a_2 + b_1b_2 + c_1c_2}{\sqrt{a_1^2 + b_1^2 + c_1^2}\cdot\sqrt{a_2^2 + b_2^2 + c_2^2}}$$

and when $\theta = 90^9$, $\cos\theta = 0$, and

$$a_1a_2 + b_1b_2 + c_1c_2 = 0$$

whence the equation to a plane in normal form is:

$$l(x - x_1) + m(y - y_1) + n(z - z_1) = 0$$

3.3.5 The tilt of a plane in terms of the direction ratios of its normal

All normals to a plane are parallel to each other and the tilt of a plane flat surface is the same throughout.

Let the direction ratios $(a_1 : b_1 : c_1)$ and $(a_2 : b_2 : c_2)$ define two straight lines AA, BB, and hence a plane, and the direction ratios $(e: f: g)$ define its normal. (N.B. $(e: f: g)$ are synonymous with $(l: m: n)$ in the previous article).

The angle between the line AA and the normal NN, equals the angle between the line BB and the normal NN and the angle between any line in the plane and the normal NN is 90°.

The angle between any two lines is given by the equation:

$$\cos\theta = \frac{ae + bf + cg}{\sqrt{a^2 + b^2 + c^2}\cdot\sqrt{e^2 + f^2 + g^2}}$$

and hence, since $\cos 90 = 0$, the equation to a normal is of the form:

$$ae + bf + cg = 0$$

where $(a: b: c)$ and $(e: f: g)$ are the direction ratios of any line in the plane and its normal.

Since any line in a plane is at right angles to the plane's normal, the two lines AA and BB may be paired with the normal NN as follows:

$$a_1e + b_1f + c_1g = 0$$

$$a_2e + b_2f + c_2g = 0$$

and if we write:

$$\frac{e}{g} = p, \frac{f}{g} = q$$

the direction ratios: $(e\colon f\colon g)$ equal the direction ratios $(p\colon q\colon 1)$, and hence the two equations above may be written:

$$a_1p + b_1q + c_1 = 0$$

$$a_2p + b_2q + c_2 = 0$$

to which the simultaneous solution is:

$$p = \frac{b_1c_2 - c_1b_2}{a_1b_2 - b_1a_2}, \quad q = \frac{a_1c_2 - a_2c_1}{a_2b_1 - a_1b_2} = \frac{c_1a_2 - a_1c_2}{a_1b_2 - b_1a_2}$$

and hence the direction cosines of the normal NN are:

$$(b_1c_2 - c_1b_2)\colon (c_1a_2 - a_1c_2)\colon (a_1b_2 - b_1a_2)$$

The following rule based on the determinant method of solution should be noted. See Chapter 4,

If

$$\begin{vmatrix} a_1 & b_1 & c_1 \\ a_2 & b_2 & c_2 \end{vmatrix}$$

then

$$\begin{matrix} b_1 & & c_1 \\ & \times & \\ b_2 & & c_2 \end{matrix} = (b_1c_2 - c_1b_2)$$

$$\begin{matrix} a_1 & & c_1 \\ & \times & \\ a_2 & & c_2 \end{matrix} = -(a_1c_2 - c_1a_2) = (c_1a_2 - a_1c_2)$$

$$\begin{matrix} a_1 & & b_1 \\ & \times & \\ a_2 & & b_2 \end{matrix} = (a_1b_2 - b_1a_2)$$

3.3.6 The equation to a plane in perpendicular form

The equation to a plane at a perpendicular distance p from the origin may be deduced as follows:

The equation to a plane may be written in the form:

$$lx + my + nz = lx_1 + my_1 + nz_1$$

which on dividing through by $\sqrt{l^2 + m^2 + n^2}$ yields:

$$x \cos \alpha + y \cos \beta + z \cos \gamma = \frac{lx_1 + my_1 + nz_1}{\sqrt{l^2 + m^2 + n^2}}$$

Now if the point $P(x_1, y_1, z_1)$ is the foot of the perpendicular $(x_1, y_1, z_1) = (l, m, n)$. Hence

$$x \cos \alpha + y \cos \beta + z \cos \gamma = \sqrt{l^2 + m^2 + n^2} = p$$

where p is the perpendicular distance required.

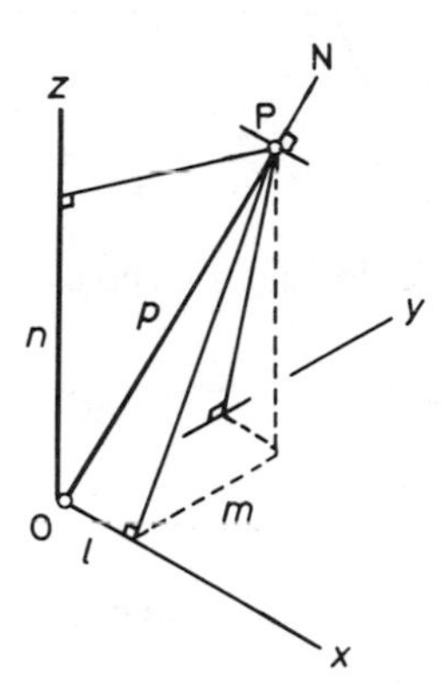

Figure 3.22

An alternative formulation for the perpendicular distance of a plane from the origin may be written in terms of the x, y, z, intercepts a, b, c, as follows.

Since the line OP, Fig. 3.22, is perpendicular to the plane it is perpendicular to the lines AP, BP, CP, and hence the angles APO, BPO, CPO are right angles, and since $OA = a$, $OB = b$, $OC = c$, the direction cosines are:

$$\cos \alpha = \frac{p}{a} = l$$

$$\cos \beta = \frac{p}{b} = m$$

$$\cos \gamma = \frac{p}{c} = n$$

The sum of the squares of the direction cosines is equal to 1, whence:

$$l^2 + m^2 + n^2 = 1$$

and

$$p^2 \left\{\frac{1}{a^2} + \frac{1}{b^2} + \frac{1}{c^2}\right\} = 1$$

or

$$p^2 = \frac{1}{\dfrac{1}{a^2} + \dfrac{1}{b^2} + \dfrac{1}{c^2}}$$

whence

$$p = \sqrt{\frac{1}{\dfrac{1}{a^2} + \dfrac{1}{b^2} + \dfrac{1}{c^2}}}$$

The length of the perpendicular may also be expressed in terms of its direction cosines:

The equation to the plane is

$$\frac{x}{a} + \frac{y}{b} + \frac{z}{c} = 1$$

and multiplying through by p yields the equation:

$$x\,\frac{p}{a} + y\,\frac{p}{b} + z\,\frac{p}{c} = p$$

but

$$\frac{p}{a} = l = \cos\alpha,\ \frac{p}{b} = m = \cos\beta,\ \frac{p}{c} = n = \cos\gamma$$

whence

$$x\cos\alpha + y\cos\beta + z\cos\gamma = p$$

or

$$xl + ym + zn = p$$

whichever form is preferred.

The direction cosines l, m, n, may be written in the form:

$$l = \frac{p}{a} = \frac{x}{p},\ \text{whence } x = \frac{p^2}{a}$$

$$m = \frac{p}{b} = \frac{y}{p},\ \text{whence } y = \frac{p^2}{b}$$

$$n = \frac{p}{c} = \frac{z}{p}, \text{ whence } z = \frac{p^2}{c}$$

$$x \cos \alpha + y \cos \beta + z \cos \gamma = p \text{ (positive)}$$

$$x \cos \alpha - y \cos \beta + z \cos \gamma = p \text{ (positive)}$$

$$-x \cos \alpha - y \cos \beta + z \cos \gamma = p \text{ (negative)}$$

$$x \cos \alpha + y \cos \beta - z \cos \gamma = p \text{ (positive)}$$

Example

The normal to a flat reflective mirror surface passes through zero and the point P(2.3076, 3.0769, 9.2308). Determine the direction cosines of the normal OP and hence deduce the equation to the mirror plane in normal form.

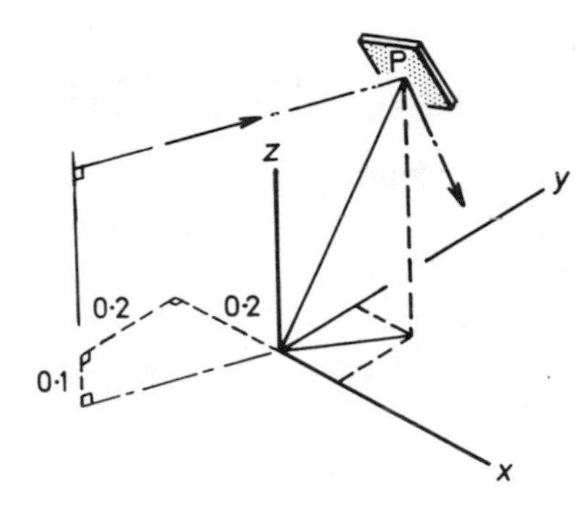

Figure 3.23

Solution

$$\text{(TL)} \quad \text{OP} = \sqrt{2.3076^2 + 3.0769^2 + 9.2308^2} = 10$$

Hence the direction cosines are:

$$\cos \alpha_1 = \frac{2.3076}{10}, \cos \beta_1 = \frac{3.0769}{10}, \cos \gamma_1 = \frac{9.2308}{10}$$

$$\cos \alpha_1 = 0.23076, \cos \beta_1 = 0.30769, \cos \gamma_1 = 0.92308$$

The consistency of the direction cosines should be checked by squaring and adding:

Thus

$$\cos^2 \alpha_1 + \cos^2 \beta_1 + \cos^2 \gamma_1 = 1$$

whence

$$0.23076^2 + 0.30769^2 + 0.92308^2 = 1 \text{ exactly.}$$

The equation to a plane in terms of its direction cosines and normal is of the form:

$$x \cos \alpha + y \cos \beta + z \cos \gamma = p$$

whence the equation to the mirror plane is:

$$0.23076x + 0.30769y + 0.92308z = 10$$

Example

If in the example above a ray from a point source S approaches the flat mirror surface in a direction defined by the direction ratios (0.2 : 0.2 : 0.1), determine the true angle of the incident ray and the direction ratios of the reflected ray. The direction cosines of the mirror surface are (0.2307 : 0.30769 : 0.92308).

The true length of the line corresponding to the given direction ratios is:

$$\text{(TL)} \quad \text{PS} = \sqrt{0.2^2 + 0.2^2 + 0.1^2} = 0.3$$

hence direction cosines of incident ray are

$$(\cos \alpha_2 : \cos \beta_2 : \cos \gamma_2) = \left(\frac{0.2}{0.3} : \frac{0.2}{0.3} : \frac{0.1}{0.3}\right) = \left(\frac{2}{3} : \frac{2}{3} : \frac{1}{3}\right)$$

The true angle between incident ray and flat mirror surface is given by the equation:

$$\cos \theta = \cos \alpha_1 \cos \alpha_2 + \cos \beta_1 \cos \beta_2 + \cos \gamma_1 \cos \gamma_2$$

whence

$$\cos \theta = 0.6667 \text{ and } \theta = 48.19°$$

3.3.7 The perpendicular distance of a given point to a given plane

In the preceding derivation the position of a point P was given relative to the origin (0, 0, 0). We now consider the problem in which the position of a second point P_2 is taken relative to the plane.

Let the x, y, z intercepts of a plane equal (a, b, c) and the coordinates of the point P_1 equal (x_1, y_1, z_1).

Denote the position of a second point P_2 (not in the plane π_1) by the coordinates x_2, y_2, z_2, and let a second plane parallel to the first contain the point P_2.

Since planes π_1 and π_2 are parallel any perpendicular distance drawn between the two planes has the same direction ratios, and in terms of the perpendicular the same direction cosines. Let l, m, n, be the direction cosines.

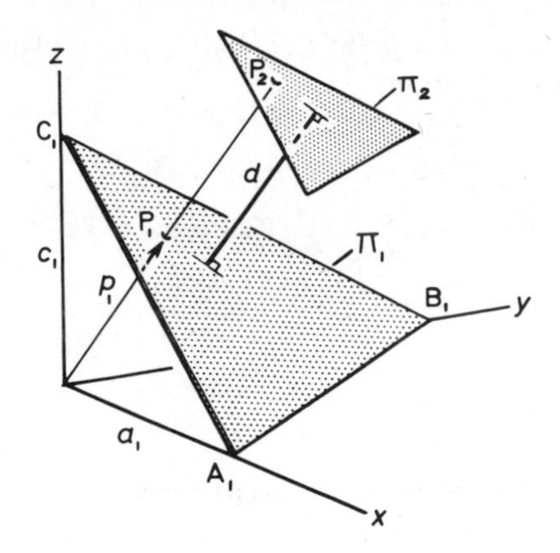

Figure 3.24

As previously shown the equations to planes π_1 and π_2 may be written in the form:

$$x_1 l + y_1 m + z_1 n = p_1$$

$$x_2 l + y_2 m + z_2 n = p_2$$

and if the perpendicular distance between the two planes is d:

$$d = (p_2 - p_1) = (x_2 l + y_2 m + z_2 n - p_1)$$

and since as previously shown:

$$\frac{p_1}{a_1} = \frac{p_2}{a_2} = l, \quad \frac{p_1}{b_1} = \frac{p_2}{b_2} = m, \quad \frac{p_1}{c_1} = \frac{p_2}{c_2} = n$$

$$d = p_1 \left(\frac{x_2}{a_1} + \frac{y_2}{b_1} + \frac{z_2}{c_1} - 1 \right)$$

We also know that:

$$p_1^2 = \frac{1}{\dfrac{1}{a_1^2} + \dfrac{1}{b_1^2} + \dfrac{1}{c_1^2}}$$

and hence the distance d from a point (x_2, y_2, z_2) to a plane with intercepts (a_1, b_1, c_1) is given by the equation:

$$d = \frac{\dfrac{x_2}{a_1} + \dfrac{y_2}{b_1} + \dfrac{z_2}{c_1} - 1}{\sqrt{\dfrac{1}{a_1^2} + \dfrac{1}{b_1^2} + \dfrac{1}{c_1^2}}} = \left(\frac{x_2}{a_1} + \frac{y_2}{b_1} + \frac{z_2}{c_1} - 1\right) p$$

3.3.8 The angle between two planes in intercept form

If two planes are defined by the equations:

$$\frac{x}{a_1} + \frac{y}{b_1} + \frac{z}{c_1} = 1, \frac{x}{a_2} + \frac{y}{b_2} + \frac{z}{c_2} = 1$$

Then as previously shown, the perpendicular distance from each plane to the origin is given by:

$$p_1^2 = \frac{1}{\dfrac{1}{a_1^2} + \dfrac{1}{b_1^2} + \dfrac{1}{c_1^2}}$$

$$p_2^2 = \frac{1}{\dfrac{1}{a_2^2} + \dfrac{1}{b_2^2} + \dfrac{1}{c_2^2}} \quad \text{respectively.}$$

and the direction cosines are

$$l_1 = \frac{p_1}{a_1}, \quad m_1 = \frac{p_1}{b_1}, \quad n_1 = \frac{p_1}{c_1}$$

$$l_2 = \frac{p_2}{a_2}, \quad m_2 = \frac{p_2}{b_2}, \quad n_2 = \frac{p_2}{c_2}$$

The angle between two planes is given by:

$$\cos\theta = l_1 l_2 + m_1 m_2 + n_1 n_2$$

and hence the required angle is:

$$\cos\theta = \frac{p_1}{a_1} \cdot \frac{p_2}{a_2} + \frac{p_1}{b_1} \cdot \frac{p_2}{b_2} + \frac{p_1}{c_1} \cdot \frac{p_2}{c_2}$$

$$= p_1 p_2 \left(\frac{1}{a_1 a_2} + \frac{1}{b_1 b_2} + \frac{1}{c_1 c_2}\right)$$

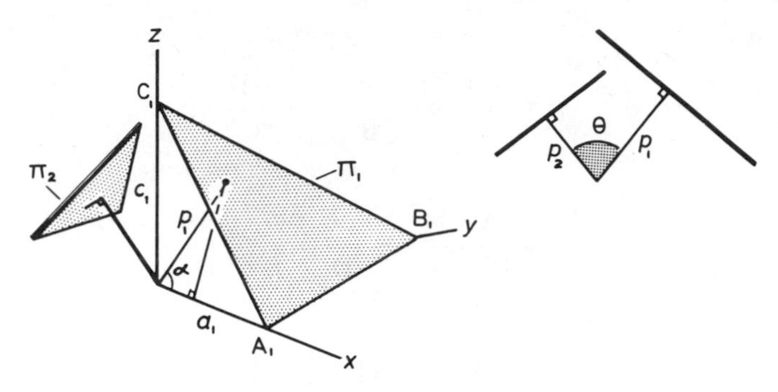

Figure 3.25

3.3.9 The reduction of the equation to a plane from one form to another

(1) General form to normal form. Since the cosine of an angle cannot be more nor less than ±1, the coefficients of a plane in normal form cannot be greater nor less than ±1

The equation $0.5x + 2y + z + 0.5 = 0$ is clearly in general form but may be reduced to normal form by dividing through by 5.

The equation $0.1x + 0.4y + 0.2z + 0.1 = 0$ is the equivalent normal form.

(2) Normal form to intercept form. The equation $0.1x + 0.4y + 0.2z + 0.1 = 0$ may be transformed to intercept form by recalling that:

$$a = \frac{-D}{A}, b = \frac{-D}{B}, c = \frac{-D}{C}, d = \frac{-D}{D} = -1$$

Whence the intercept form is obtained by changing signs, dividing through by 0.1, and inverting all coefficients. Thus the normal form, $-0.1x - 0.4y - 0.2z = 0.1$ becomes $-x - 4y - 2z = 1$, which on inverting coefficients becomes:

$$-\frac{x}{1} - \frac{y}{0.25} - \frac{z}{0.5} = 1$$

and this is the intercept form required.

(3) Interrelationships of the coefficients of a plane. By comparing the coefficients of the general form, with the intercept form, with the perpendicular normal form, we may establish the following relationships:

Given that:

$$Ax + By + Cz = -D \text{ (general form)}$$

$$\frac{x}{a} + \frac{y}{b} + \frac{z}{c} = 1 \qquad \text{(intercept form)}$$

$$lx + my + nz = p \qquad \text{(perpendicular form)}$$

we see that:

$$a = \frac{-D}{A}, b = \frac{-D}{B}, c = \frac{-D}{C}$$

and

$$\frac{l}{p} = \frac{-A}{D}, \frac{m}{p} = \frac{-B}{D}, \frac{n}{p} = \frac{-C}{D}$$

whence

$$\frac{l}{p} = \frac{1}{a}, \frac{m}{p} = \frac{1}{b}, \frac{n}{p} = \frac{1}{c}$$

or

$$a = \frac{p}{l}, \ b = \frac{p}{m}, \ c = \frac{p}{n}$$

Now since

$$l^2 + m^2 + n^2 = 1$$

and

$$l = \frac{-A}{D} p, m = \frac{-B}{D} p, n = \frac{-C}{D} p$$

$$\frac{A^2}{D^2} + \frac{B^2}{D^2} + \frac{C^2}{D^2} = \frac{l}{p^2}$$

and

$$p = \frac{\pm D}{\sqrt{A^2 + B^2 + C^2}}$$

and since

$$A = \frac{-D}{a}, B = \frac{-D}{b}, C = \frac{-D}{c}$$

$$p = \frac{1}{\sqrt{\frac{1}{a^2} + \frac{1}{b^2} + \frac{1}{c^2}}}$$

These relationships are useful for resolving plane data presented in varying forms.

3.3.10 The test that a point or line shall be in a plane

If P_1 is a point known to be in a plane, then a second point P_2 lies in that plane, if the line P_1P_2 lies in the plane. If the line P_1P_2 lies in the plane, the angle between it and the normal to the plane must be 90°.

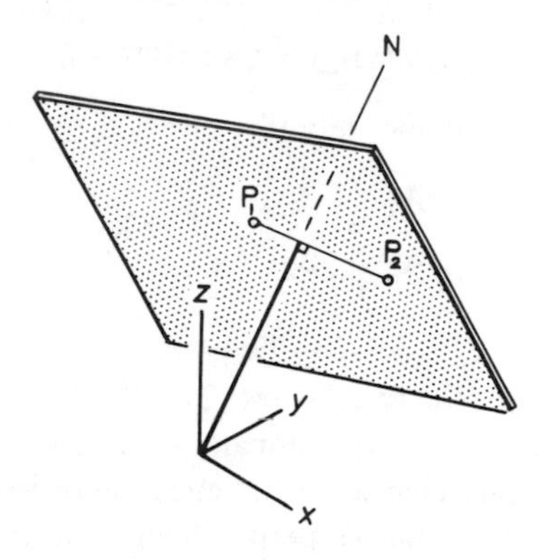

Figure 3.26

If the direction ratios of the line P_1P_2 are:

$$(x_2 - x_1):(y_2 - y_1):(z_2 - z_1)$$

and the direction ratios of the normal are:

$$l:m:n$$

Then since the angle between the two lines must be a right angle,

$$a_1a_2 + b_1b_2 + c_1c_2 = 0$$

whence

$$l(x_2 - x_1) + m(y_2 - y_1) + n(z_2 - z_1) = 0$$

This is the condition to be met by every point that lies in a plane defined in normal form.

(N.B. When $\cos\theta = 0$, either direction ratios or direction cosines may be used).

The condition that two points P_1 and P_2, i.e. a line, shall lie in a plane may thus be stated as follows:

$$l(x_2 - x_1) + m(y_2 - y_1) + n(z_2 - z_1) = 0$$

and similarly for three points P_1, P_2 and P_3

$$lx_1 + my_1 + nz_1 = lx_2 + my_2 + nz_2 = lx_3 + my_3 + nz_3$$

where l, m, n, are the direction ratios or cosines of the normal to the plane and (x_r, y_r, z_r) are the coordinates of any point P_r in the plane.

3.3.11 Test for parallel planes

Two planes are parallel if and only if the coefficients of x, y, and z are equal or proportional. Thus if the equations of two planes are:

$$A_1x + B_1y + C_1z + D_1 = 0$$

$$A_2x + B_2y + C_2z + D_2 = 0$$

The two planes are parallel if and only if:

$$\frac{A_1}{A_2} = \frac{B_1}{B_2} = \frac{C_1}{C_2} = K$$

Example

We may ascertain whether the two planes: $2x + 8y - 5z - 6 = 0$ and $4x + 16y - 10z - 18 = 0$ are parallel or not by comparing the ratios of like coefficients. The ratios $2/4 = 8/16 = 5/10$ are identical and hence the two planes are parallel.

When the equations are in normal perpendicular form:

$$x \cos \alpha_1 + y \cos \beta_1 + z \cos \gamma_1 - p_1 = 0$$

$$x \cos \alpha_2 + y \cos \beta_2 + z \cos \gamma_2 - p_2 = 0$$

they are parallel if the following conditions are true:

$$\cos \alpha_1 = \cos \alpha_2, \cos \beta_1 = \cos \beta_2, \cos \gamma_1 = \cos \gamma_2$$

$$-\cos \alpha_1 = \cos \alpha_2, -\cos \beta_1 = \cos \beta_2, -\cos \gamma_1 = \cos \gamma_2$$

3.3.12 Test for perpendicular planes

Two planes are perpendicular if and only if the sum of the product of each pair of coefficients is equal to zero. If $\cos \alpha_1 \cos \alpha_2 + \cos \beta_1 \cos \beta_2 + \cos \gamma_1 \cos \gamma_2 = 0$ the planes are perpendicular.

If the two planes are defined by the equations:

$$A_1 x + B_1 y + C_1 z + D_1 = 0$$

$$A_2 x + B_2 y + C_2 z + D_2 = 0$$

The coefficients of x, y and z are the direction ratios of the normals to the planes.

That is, the planes are perpendicular if:

$$A_1 A_2 + B_1 B_2 + C_1 C_2 = 0$$

3.3.13 The plane parallel to a given plane through a given point

If $Ax + By + Cz + D = 0$ is the given plane and (x_1, y_1, z_1) the given point, then the direction ratios of the first and second planes are equal.

If we denote planes 1 and 2 by the equations:

$$Ax + By + Cz + D = 0$$

and

$$Ax_1 + By_1 + Cz_1 + D = 0$$

and eliminate D we obtain the necessary condition

$$A(x - x_1) + B(y - y_1) + C(z - z_1) = 0$$

3.3.14 The line of intersection between two given planes

If the equations

$$A_1x + B_1y + C_1z + D_1 = 0$$

$$A_2x + B_2y + C_2z + D_2 = 0$$

define two planes and the direction ratios of the line of intersection are $\lambda: \mu: \nu$, then the normals to planes 1 and 2 are perpendicular to the line of intersection and hence,

$$A_1\lambda + B_1\mu + C_1\nu = 0$$

$$A_2\lambda + B_2\mu + C_2\nu = 0$$

Solving in terms of λ/μ and λ/ν yields the required result.

$$\lambda: \mu: \nu = (B_1C_2 - B_2C_1): (C_1A_2 - C_2A_1): (A_1B_2 - A_2B_1)$$

3.3.15 The plane through the intersection of two given planes

If the equations of two given planes π_1, π_2, are written in the general form:

$$A_1x + B_1y + C_1z + D_1 = 0 \quad (\pi_1)$$

$$A_2x + B_2y + C_2z + D_2 = 0 \quad (\pi_2)$$

Then the equation to a third plane which passes through the common line of intersection, is of the form:

$$A_1x + B_1y + C_1z + D_1 + k(A_2x + B_2y + C_2z + D_2) = 0$$

where k is the constant which allows the plane π_3 to be directed through any chosen point not on the line of intersection.

The angle between the planes may, of course, be obtained using the standard expression as previously.

If A_1, B_1, C_1 are given; the choice of the coefficients A_2, B_2, C_2, is arbitrary but subject to the above condition.

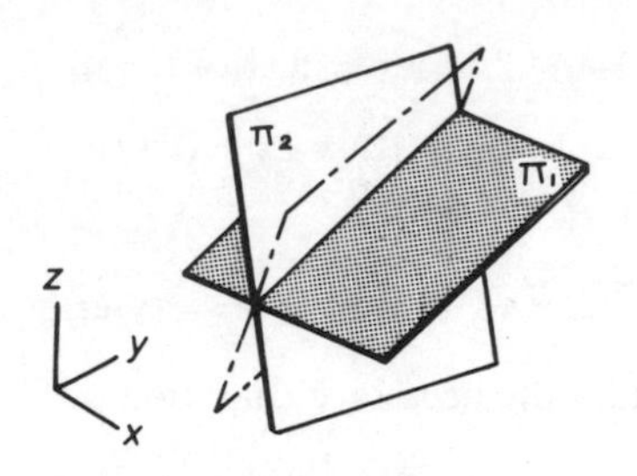

Figure 3.27

If, for example, $A_1 = 4, B_1 = 2, C_1 = 3$ we may choose any values for A_2, B_2, C_2, such that:

$$4A_2 + 2B_2 + 3C_2 = 0$$

noting that 12 is the least common multiple we may write:

$$\frac{4A_2}{12} + \frac{2B_2}{12} + \frac{3C_2}{12} = 0$$

and $A_2 = 1, B_2 = 4, C_2 = -4$, is clearly a tenable condition and the plane $x + 4y - 4z + D_2 = 0$ is perpendicular to the plane $4x + 2y + 3z + D_1 = 0$.
A descriptive illustration of this statement is given in Fig. 3.28.

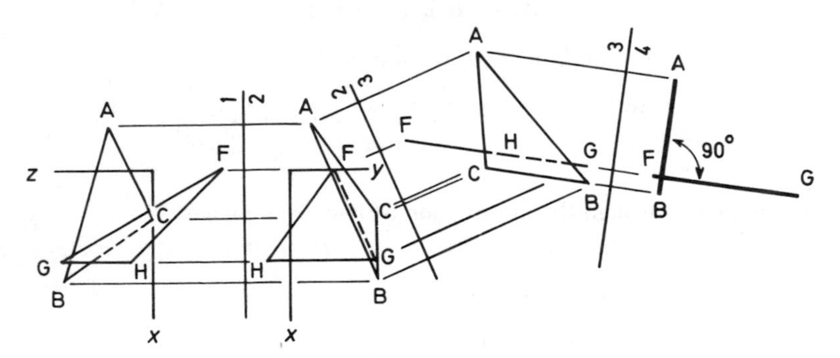

Figure 3.28

Example
Given that the plane π_1 passes through the origin and is perpendicular to the plane π_2, the equation to which is: $3x + 4y + 12z = 143$, determine the equation to their line of intersection.

Solution
The condition to be met by two perpendicular planes is:

$$A_1A_2 + B_1B_2 + C_1C_2 = 0$$

In this instance the coefficients of the plane π_2 are given as $A_2 = 3$, $B_2 = 4$, $C_2 = 12$, but the coefficients of the plane π_1 may be arbitrarily chosen to satisfy the perpendicular condition, $A_1A_2 + B_1B_2 + C_1C_2 = 0$.

Thus, $A_1A_2 = 2$, $B_1B_2 = -1$, $C_1C_2 = -1$, is clearly a tenable condition, and the coefficients of all planes perpendicular to the plane π_2 are:

$$A_2 = \frac{2}{3},\ B_2 = -\frac{1}{4},\ C_2 = -\frac{1}{12}$$

whence the equation to all planes perpendicular to the plane π_2 is:

$$\frac{2}{3}x - \frac{y}{4} - \frac{z}{12} + D = 0$$

and since the particular plane π_1 is to pass through the origin, $D = 0$

$$8x - 3y - z = 0$$

is the equation to the plane π_1.

The line of intersection between the planes π_1 and π_2 is the only line which contains points common to both these planes, and hence the equations to planes π_1 and π_2 must have a common solution.

That is:

$$3x + 4y + 12z - 143 = 8x - 3y - z$$

and hence

$$-5x + 7y + 13z = 143$$

Points on the line of intersection may be obtained by assuming any desired value for x, y or z and solving for the other two coordinates simultaneously.

If we choose $z = 11$, then the equation to the plane of intersection reduces to:

$$-5x + 7y = 0$$

whence

$$y = \frac{5}{7}x$$

and since

$$8x - 3y - z = 0$$

$$8x - 3y \quad = 11$$

$$8x - \frac{3.5}{7}x \quad = 11$$

whence

$$x = \frac{77}{41},\ y = \frac{55}{41},\ z = 11$$

is a point on the line of intersection produced by the planes π_1 and π_2.

The values of (x, y, z) obtained above, must also satisfy the equation for π_2 and substitution of these values confirms that they are true.

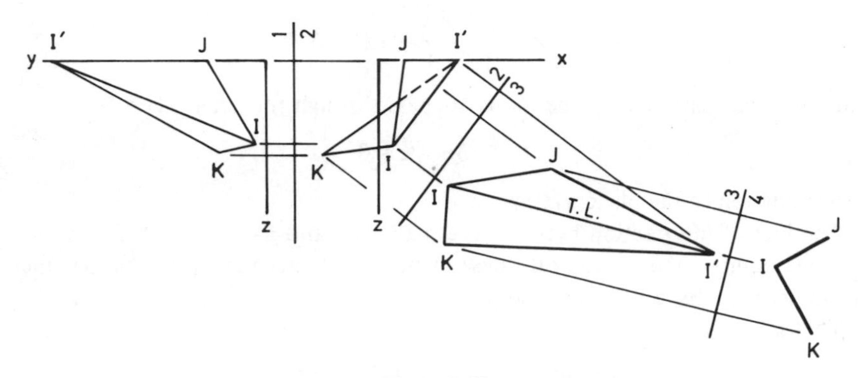

Figure 3.29

In the event of it being necessary to express the plane of intersection in normal form, then this transformation may be swiftly effected as follows:

We observe that:

$$-\frac{5x}{143} + \frac{7y}{143} + \frac{13z}{143} = 1$$

and multiply throughout by the unknown perpendicular p, we obtain:

$$-\frac{5px}{143} + \frac{7py}{143} + \frac{13pz}{143} = p$$

which is of the form:

$$x \cos \alpha + y \cos \beta + z \cos \gamma = p$$

Now since:

$$\cos^2 \alpha + \cos^2 \beta + \cos^2 \gamma = 1$$

$$\left(\frac{5p}{143}\right)^2 + \left(\frac{7p}{143}\right)^2 + \left(\frac{13p}{143}\right)^2 = 1$$

whence

$$\left(\frac{25 + 49 + 169}{20449}\right) p^2 = 1$$

and

$$p = 9.17345$$

$$\cos \alpha = \frac{-5\,(9.17345)}{143} = -0.3163$$

$$\cos \beta = \frac{7\,(9.17345)}{143} = 0.4491$$

$$\cos \alpha = \frac{13\,(9.17345)}{143} = 0.8340$$

The equation to the plane of intersection in normal form is:

$$-0.3163x + 0.4491y + 0.8340z = 9.1735$$

and any particular point on the line of intersection may be obtained as previously demonstrated.

3.4 ORTHOGRAPHIC TRACE EQUATIONS

The need to establish mathematical relationships between objects in 3-space and their associated orthographic traces in 2-space is a fundamental problem of direct relevance and bearing on computational graphics and manufacture.

3.4.1 Orthographic trace equations of a line in 3-space

Any straight line in 3-space may be defined in terms of its orthographic traces and two mutually perpendicular projections suffice to define the spatial position and orientation of any given line completely.

If the equation to the straight line trace in plane xoy is known to be:

$$y = mx + b$$

and the equation to the straight line trace in plane xoz is known to be:

$$z = nx + c$$

then the spatial position and orientation of the real line is known completely.

It is clear from Fig. 3.30, that the third trace in plane zoy is dependent upon the position and orientation of the two given traces in mutually perpendicular

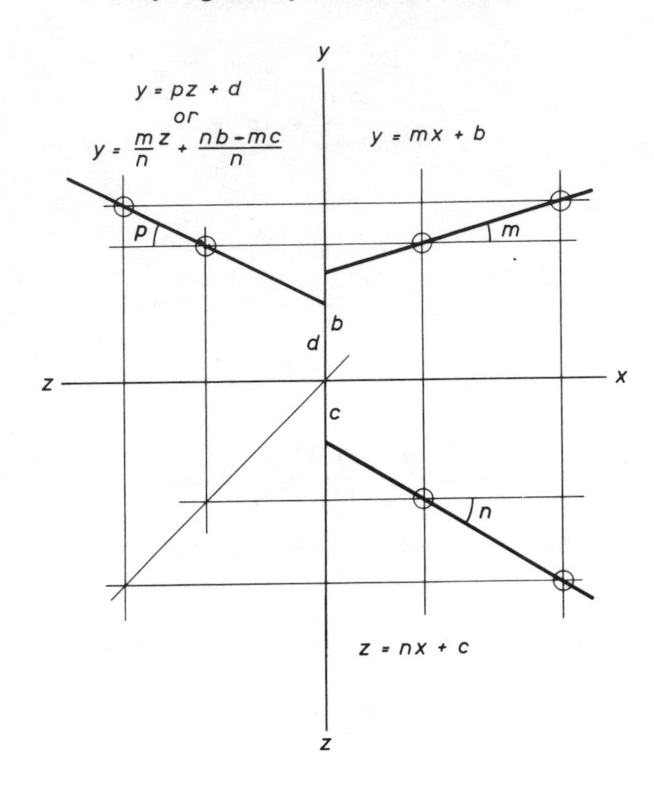

Figure 3.30

planes and it is immaterial which two of the three mutually perpendicular planes are used to delineate the object. The third trace is always conditional on the other two.

The equation to the trace in the plane *zoy* may be obtained directly from Fig. 3.30.

As seen from the figure, the extreme points of the trace in the plane *zoy* are located by the coordinates:

y	x
b	c
$mx + b$	$nx + c$

The slope of the line is thus $m/n = p$, say, and the intercept with the *oy* axis is

$$\frac{nb - mc}{n} = d, \text{ say.}$$

The equation to the trace in the plane *zoy* is therefore:

$$y = pz + d$$

and the three mutually perpendicular traces are given by the three straight line equations:

$$y = mx + b \quad \text{plane } xoy$$

$$z = nx + c \quad \text{plane } xoz$$

$$y = pz + d \quad \text{plane } zoy$$

Bearing in mind that m, n and p are here the tangents of the three respective traces, we may write the direction ratios and hence the direction cosines of the real line as follows:

For a unit increase in x, y increases by an amount $(mx) = m$ and z increases by $(nx) = n$.

Hence the direction ratios are:

$$1 : m : n$$

The true length of the space line is:

$$\sqrt{1 + m^2 + n^2}$$

and hence the direction cosines of the space line are:

$$\frac{1}{\sqrt{1 + m^2 + n^2}} : \frac{m}{\sqrt{1 + m^2 + n^2}} : \frac{n}{\sqrt{1 + m^2 + n^2}}$$

Example

Two mutually perpendicular views of a line AB, which lies skew to the planes of projection are as indicated by its traces A_0B_0, A_1B_1, Fig. 3.31. Calculate the true length and direction cosines of the line. Check that the sum of the squares of the direction cosines equals one and express the three angles in terms of degrees.

Solution

The coordinates of A are (4, 2, 7) and of B (8, 9, 3).

The direction ratios are:

$$(8 - 4) = 4, (9 - 2) = 7, (3 - 7) = -4$$

The true length AB is given by the square root of the sum of the square of the direction ratios.

$$\text{(TL)} \quad AB = \sqrt{4^2 + 7^2 + 4^2} = \sqrt{81} = 9$$

The direction cosines are:

$$\frac{4}{9} : \frac{7}{9} : \frac{-4}{9}$$

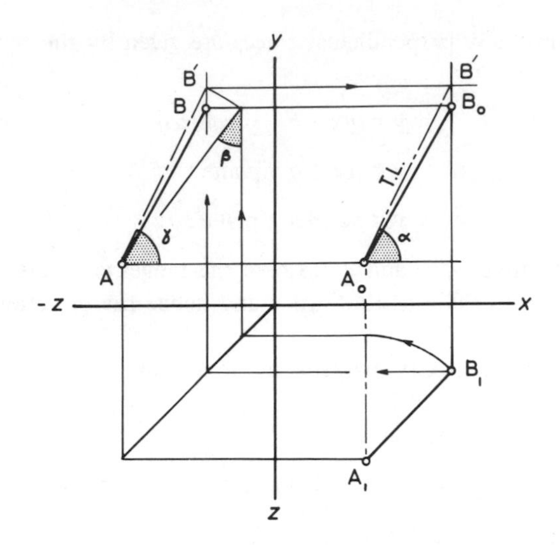

Figure 3.31

and this result may be checked by squaring and adding.

$$\frac{16 + 49 + 16}{81} = 1 \quad \text{correct.}$$

$$\text{arc cos } \frac{4}{9} = 63.612°$$

$$\text{arc cos } \frac{7}{9} = 38.942°$$

$$\text{arc cos } \frac{-4}{9} = (180 - 63.612) = 116.388°$$

A practical instance in which this type of calculation is necessary, is afforded by the skew flap hinge fitting shown in Fig. 3.32. This fitting is based on an actual piece from an in-service aircraft and it is clear that the direction of the skew shaft axis AB must be precisely set and accurately machined in relation to the through bore CD.

In this particular instance the front view of the component is such that the axis CD lies normal to an orthographic plane and hence the axis C_1D_1 appears as a dot view, and the corresponding trace in plan appears at the true length line C_0D_0.

Instances do occur, however, in which two or more axes are skew to the orthographic planes of projection and the need to calculate the true angle between two skew lines arises.

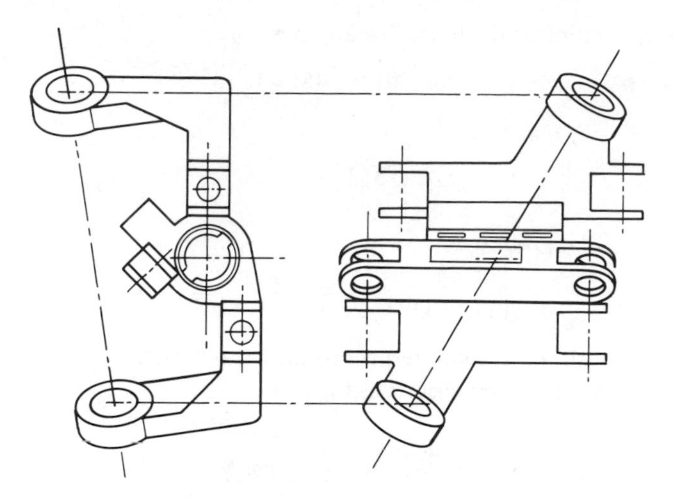

Figure 3.32

Example

Suppose the spatial orientation of an aircraft undercarriage leg is as defined by the two orthographic projections, Fig. 3.33, all data being as given in the figure. The local axes ox_1 oy_1 oz_1 are right handed axes, with positive y directed aft and positive z directed upwards, such that ox is the port wing axis. The problem is to determine the direction of the hinge axis about which the undercarriage folds.

As may be seen from Fig. 3.33, oz is the common axis, which relates the two orthographic views, and we therefore base the calculation of direction ratios on a unit length in the oz direction. As shown in the figure the unit length is negative in this instance.

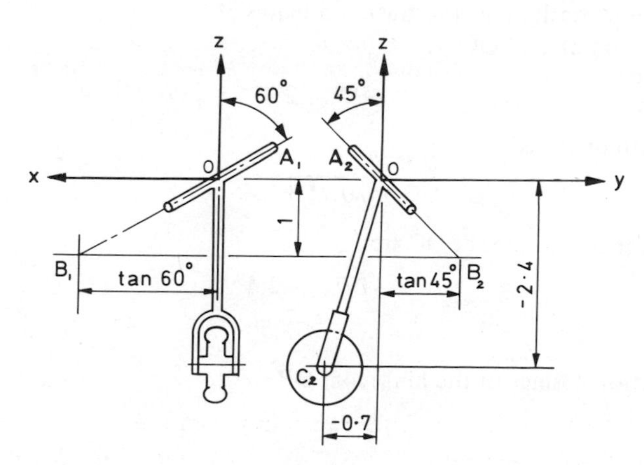

Figure 3.33

The direction ratios of the hinge line are thus:

$$\tan 60 : \tan 45 : -1$$

that is

$$\sqrt{3} : 1 : -1$$

The true length OP is given by

$$\text{(TL)} \quad OP = \sqrt{3+1+1} = \sqrt{5}$$

The inclination of the hinge, with respect to the *ox, oy, oz*, axes is now expressed in terms of its three direction cosines as follows:

$$\cos\alpha = \frac{a}{e} : \cos\beta = \frac{b}{e} : \cos\gamma = \frac{c}{e}$$

whence

$$\cos\alpha = \frac{\sqrt{3}}{\sqrt{5}} : \cos\beta = \frac{1}{\sqrt{5}} : \cos\gamma = -\frac{1}{\sqrt{5}}$$

and the true angles are:

$$\alpha = 39.232^\circ, \beta = 63.435^\circ, \gamma = -63.435^\circ$$

The angles calculated above are the direction angles of the hinge axis with respect to the *oxyz* axes. The true angle between the hinge axis and undercarriage leg axis may be computed as follows.

Since OC is zeroed at the origin, the direction ratios of OC (the undercarriage leg axis) are numerically equal to the coordinates of C.

The direction ratios of OC are:

$$0 : -0.7 : -2.4$$

The true length of OC is:

$$\text{(TL)} \quad OC = \sqrt{0.7^2 + 2.4^2} = 2.5$$

Hence the direction cosines of OC are:

$$0 : \frac{-0.7}{2.5} : \frac{-2.4}{2.5}$$

But the direction cosines of the hinge axis OP are:

$$\frac{\sqrt{3}}{\sqrt{5}} : \frac{1}{\sqrt{5}} : \frac{-1}{\sqrt{5}}$$

The true angle between the hinge axis and the leg axis is given by:

$$\cos\theta = ad + be + cf$$

where $a : b : c$, $d : e : f$, are the two sets of direction cosines respectively.

$$\cos\theta = \left(0\ \frac{\sqrt{3}}{\sqrt{5}}\right) + \left(\frac{-0.7}{2.5} \cdot \frac{1}{\sqrt{5}}\right) + \left(\frac{-2.4}{2.5} \cdot \frac{-1}{\sqrt{5}}\right)$$

$$= 0.3041$$

whence

$$\theta = 72.29°$$

3.4.2 Orthographic trace equations in terms of direction cosines

We know from plane geometry that the equations to a line through two trace points which lie in an orthographic plane are of the form:

$$m = \frac{y - y_1}{x - x_1},\quad n = \frac{z - z_1}{x - x_1},\quad p = \frac{y - y_1}{z - z_1}$$

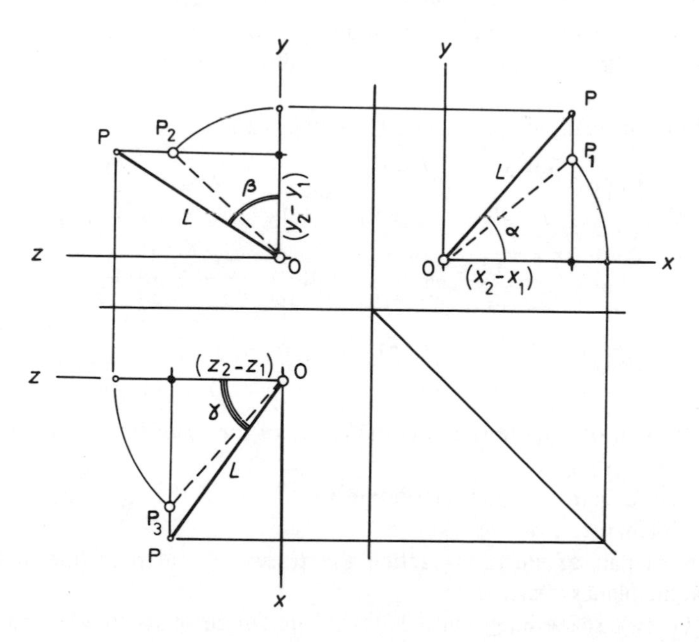

Figure 3.34

If the true length of the line $P_1P_2 = L$, then the direction cosines are:

$$\frac{x_2 - x_1}{L} = \cos\alpha = l, \quad \frac{y_2 - y_1}{L} = \cos\beta = m, \quad \frac{z_2 - z_1}{L} = \cos\gamma = n$$

and hence the equations to the line P_1P_2 may be expressed in either one of the following forms:

$$\frac{x - x_1}{\cos\alpha} = \frac{y - y_1}{\cos\beta} = \frac{z - z_1}{\cos\gamma}$$

or

$$\frac{x - x_1}{l} = \frac{y - y_1}{m} = \frac{z - z_1}{n}$$

We observe, however, that the two three part equations above are combinational. That is, there are three different ways of pairing two from three and due to the nature of the terms involved, any single pairing defines the line completely.

If we split the equation:

$$\frac{x - x_1}{x_2 - x_1} = \frac{y - y_1}{y_2 - y_1} = \frac{z - z_1}{z_2 - z_1}$$

into its three separate pairings we obtain the three trace equations to the space line: and we observe once again that if any pair of trace equations are independently chosen, the third trace equation is conditional on the choice of the other two.

The three mutually perpendicular trace pairings are:

$$(1)\quad \frac{x - x_1}{x_2 - x_1} = \frac{y - y_1}{y_2 - y_1} \quad \text{and} \quad \frac{y - y_1}{y_2 - y_1} = \frac{z - z_1}{z_2 - z_1}$$

$$(2)\quad \frac{x - x_1}{x_2 - x_1} = \frac{y - y_1}{y_2 - y_1} \quad \text{and} \quad \frac{x - x_1}{x_2 - x_1} = \frac{z - z_1}{z_2 - z_1}$$

$$(3)\quad \frac{y - y_1}{y_2 - y_1} = \frac{z - z_1}{z_2 - z_1} \quad \text{and} \quad \frac{x - x_1}{x_2 - x_1} = \frac{z - z_1}{z_2 - z_1}$$

The first pair of equations define the traces of the space line in the two orthographic planes xoy, yoz.

The second pair of equations define the traces of the space line in the two orthographic planes xoy, xoz.

The third pair of equations define the traces of the space line in the two orthographic planes yoz, xoz.

If P_1P_2 is a space line denoted by L, the length of its three mutually perpendicular traces may be denoted by L_{xy}, L_{yz}, L_{xz}.

The three separate pairs of trace equations also define three separate planes, which contain the real line L and each of its three traces L_{xy}, L_{yz}, L_{xz}, respectively.

The first pair of trace equations define a plane π_1 which contains the real line L and the trace L_{xy} and is therefore perpendicular to the plane *xoy*.

The second pair of trace equations define a plane π_2 which contains the real line L and the trace L_{yz} and is therefore perpendicular to the plane *yoz*.

The third pair of trace equations define a plane π_3 which contains the real line L and the trace L_{xz} and is therefore perpendicular to the plane *xoz*.

These three planes are clearly shown in Fig. 3.35.

Conversely, the line L is the intersection of the three planes, defined by the three pairs of lines L L_1, L L_2, L L_3, and since it is known that the planes containing the line pairs: L and L_1, L and L_2, L and L_3, are perpendicular to the orthographic planes *xoy, yoz, xoz*, respectively. The equations

$$\frac{x - x_1}{x_2 - x_1} = \frac{y - y_1}{y_2 - y_1} \text{ and } \frac{y - y_1}{y_2 - y_1} = \frac{z - z_1}{z_2 - z_1}$$

suffice to define the intersects of the three planes 1, 2, 3

$$y - y_1 = \frac{y_2 - y_1}{x_2 - x_1}\left(x - x_1\right)$$

$$y - y_1 = m(x - x_1) \text{ and } y - y_1 = n(z - z_1)$$

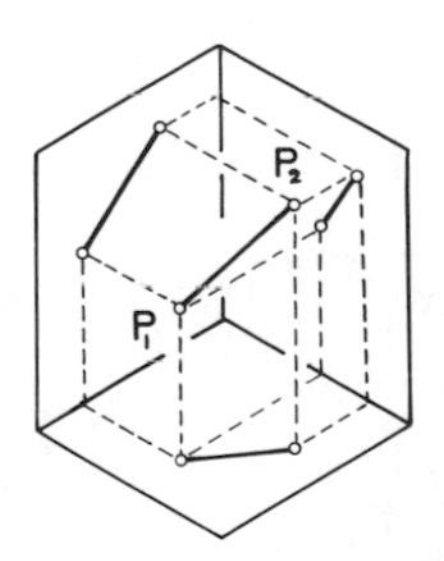

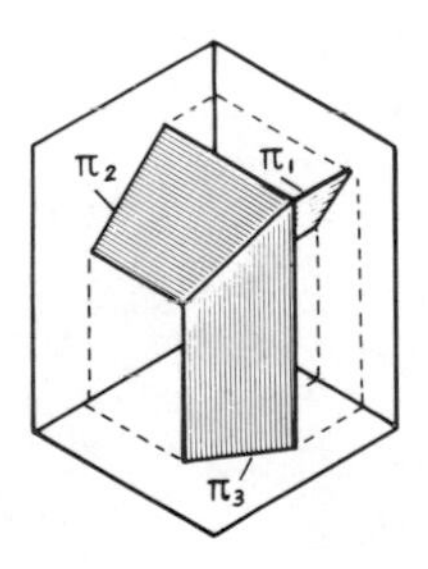

Figure 3.35

In the absence of auxiliary projection planes the orthographic method is to divide the paper or visual display screen into four quadrants. The three mutually perpendicular traces may then be drawn in three of the quadrant spaces. The arrangement of views depends on the direction of viewing and the relationship between adjacent views depends on whether the first or the third angle method of projection is used.

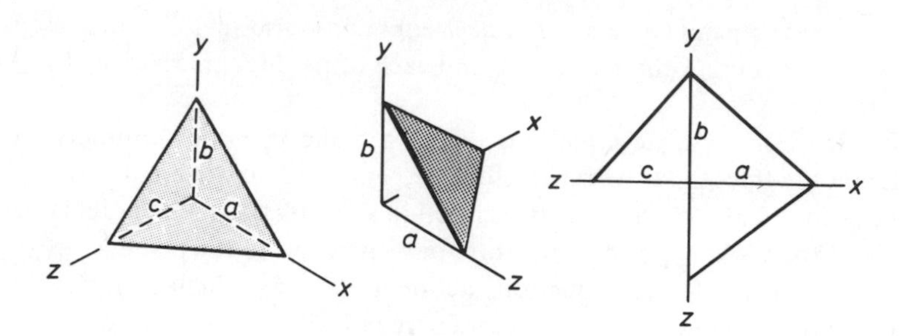

Figure 3.36

3.4.3 Orthographic trace equations to a plane given in intercept form

If a plane ABC, Fig. 3.36, intercepts a set of mutually perpendicular axes at the points, A, B, C, where OA = a, OB = b, OC = c, then the equation to the plane may be written in intercept form as follows:

$$\frac{x}{a} + \frac{y}{b} + \frac{z}{c} = 1$$

whence

$$z = c\left(1 - \frac{x}{a} - \frac{y}{b}\right)$$

and the lines of intersection (traces) formed by the plane ABC and the reference planes xoz, yox, zoy, are:

$$x = a\left(1 - \frac{z}{c}\right)$$

$$y = b\left(1 - \frac{x}{c}\right)$$

$$z = c\left(1 - \frac{y}{b}\right)$$

(N.B. These three equations refer to the traces on the planes xoz, yox, zoy, and do not represent other points on these planes).

In order to represent points within the bound formed by the traces we use the intercept equation and set the appropriate coordinate to zero.

Thus we map all points within the triangle AOC by setting $y = 0$ in the intercept equation to the plane:

$$\frac{x}{a} + \frac{y}{b} + \frac{z}{c} = 1$$

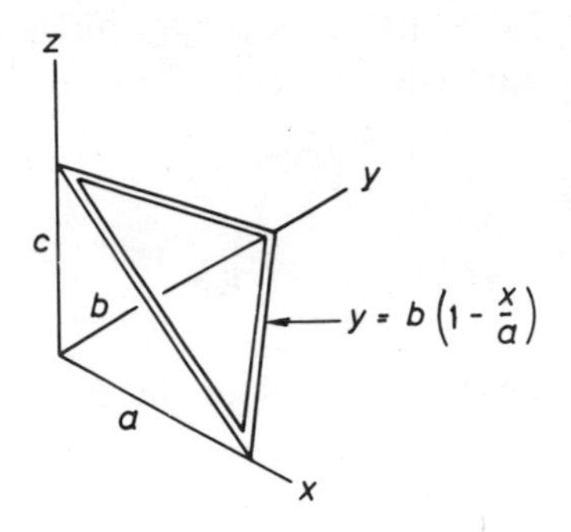

Figure 3.37

whence we obtain

$$\frac{x}{a} + \frac{z}{c} = 1 \quad \text{(plane } xoz\text{)}$$

$$\frac{x}{a} + \frac{y}{b} = 1 \quad \text{(plane } yox\text{)}$$

$$\frac{y}{b} + \frac{z}{c} = 1 \quad \text{(plane } zoy\text{)}$$

The space within the volume OABC may be partitioned by similar route by setting the appropriate coordinate to the plane position required.

3.4.4 Orthographic trace equations to a small element of surface

There are many problems in which one is required to express the area/surface relationship between a small element on a doubly curved surface and its projection onto an orthographic plane.

If

$$\text{OA} = a, \text{OB} = b, \text{OC} = c$$

$$\text{AB} = \sqrt{a^2 + b^2}, \text{AC} = \sqrt{a^2 + c^2}, \text{BC} = \sqrt{b^2 + c^2}$$

if angle OAC = α and angle OBC = β and the angle ACB = γ. (These angles not to be confused with the usual direction angles). We have from the cosine rule

$$\cos \gamma = \frac{c}{\sqrt{a^2 + c^2}} \cdot \frac{c}{\sqrt{b^2 + c^2}} = \sin \alpha \sin \beta$$

whence

$$\sin \gamma = \sqrt{1 - \sin^2 \alpha \sin^2 \beta}$$

If the element of surface PQRS has sides PQ, PR parallel to the lines AC and BC respectively the element PQRS projects to the rectangular element $dy \times dx$ in the xoy plane.

Whence

$$PQ = \frac{dx}{\cos\alpha},\quad PR = \frac{dy}{\cos\beta}$$

The area of the element PQRS is equal to the sum of the areas of the two triangles PQR, QRS.

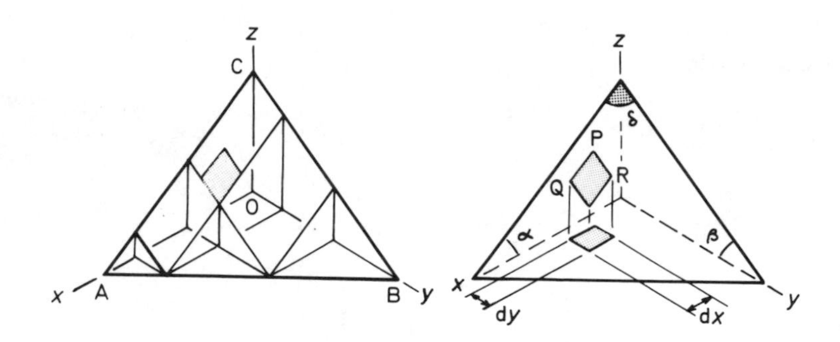

Figure 3.38

Whence

$$\text{Area of surface element} = PQ \,.\, PR \,.\, \sin\gamma$$

$$= \frac{dx}{\cos\alpha} \cdot \frac{dy}{\cos\beta} \cdot \sin\gamma$$

but

$$\sin\gamma = \sqrt{1 - \sin^2\alpha\sin^2\beta}$$

whence

$$\text{Area of surface element} = \frac{dx}{\cos\alpha} \cdot \frac{dy}{\cos\beta} \cdot \sqrt{1 - \sin^2\alpha\sin^2\beta}$$

Hence

$$\frac{\text{Area of surface element PQRS}}{\text{Area of projected area } dx\,.\,dy} = \frac{\sqrt{1} - \sin\alpha\sin\beta}{\cos\alpha\,.\,\cos\beta} = \mathrm{T}$$

Case when axes *xoy* are oblique: If the angle *xoy* is equal to ω, the length AB = $\sqrt{a^2 + b^2 - 2ab \cos \omega}$ and by the cosine rule, as previously:

$$\cos \gamma = \sin \alpha \,.\, \sin \beta + \cos \alpha \,.\, \cos \beta \,.\, \cos \omega$$

$$\sin \gamma = \sqrt{1 - (\sin \alpha \sin \beta + \cos \alpha \cos \beta \cos \omega)^2}$$

$$\text{Area of surface element} = \frac{dx}{\cos \alpha} \cdot \frac{dy}{\cos \beta} \cdot \sin \gamma$$

$$\text{Area of projected element} = dx \cdot dy \cdot \sin \omega$$

$$\frac{\text{Area of surface element}}{\text{Area of projected element}} = \frac{\sqrt{1 - (\sin \alpha \sin \beta + \cos \alpha \cos \beta \cos \omega)^2} \,.\, \sin \omega}{\cos \alpha \cos \beta}$$

3.4.5 Polar coordinate definition of an auxiliary plane

The line OP, Fig. 3.39, may be defined in terms of its radial distance r and its angular inclination as given by the two angles θ and ϕ, Fig. 3.39. Or alternatively, the line OP may be expressed in terms of its radial distance and two direction cosines.

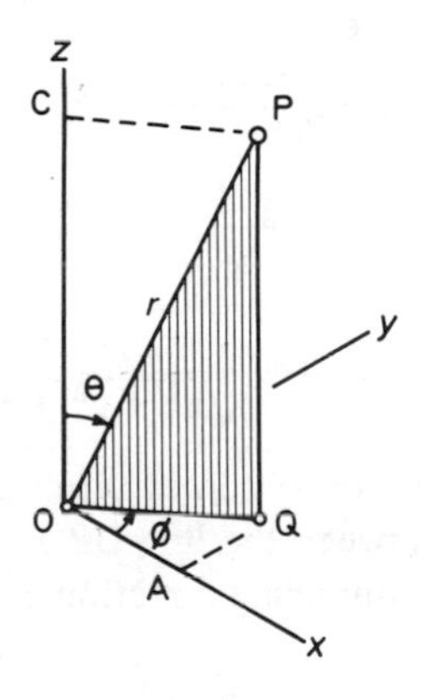

Figure 3.39

(i) Polar definition of a point-position in terms of r, and two projected angles.

Let the distance OP = r, the angle AOQ = ϕ and the angle COP – θ. The angle QAO and the angle PCO are right angles.

$$\text{OC} = r \cos \theta$$

$$\text{OS} = \text{CP} = r \sin \theta$$

$$\text{OA} = \text{OS} = \cos \phi = x$$

Hence

$$x = r \cos\phi \sin\theta$$

$$AS = OS \sin\phi = OB$$

but

$$OS = CP = r \sin\theta$$

Hence

$$OB = r \sin\phi \sin\theta = y$$

and

$$OC = r \cos\theta = z$$

To summarise:

$$x = r \cos\phi \sin\theta$$

$$y = r \sin\phi \sin\theta$$

$$z = r \cos\theta$$

In terms of rectangular coordinates

$$r = \sqrt{x^2 + y^2 + z^2}$$

$$\tan\phi = \frac{y}{x}$$

$$\cos\theta = \frac{z}{\sqrt{x^2 + y^2 + z^2}}$$

(ii) Polar definition of a point-position in terms of r and its direction ratios.

If α, β, γ, are the angles between the line OP and the ox, oy, oz axes respectively, and l, m, n, are the corresponding direction cosines

$$\cos\alpha = l = \frac{x}{r}$$

$$\cos\beta = m = \frac{y}{r}$$

$$\cos\gamma = n = \frac{z}{r}$$

whence

$$\frac{x}{r} = \cos\phi \sin\theta$$

$$\frac{y}{r} = \sin\phi \sin\theta$$

$$\frac{z}{r} = \cos\theta$$

Hence

$$\cos\alpha = \cos\phi \sin\theta$$

$$\cos\beta = \sin\phi \sin\theta$$

$$\cos\gamma = \cos\theta$$

(iii) The auxiliary plane defined in terms of r, ϕ and θ. The orthographic representation of the space point, line and plane defined by r, ϕ and θ, Fig. 3.39 is shown in Fig. 3.40, from which it is clear that the angle ϕ defines an auxiliary view and hence trace of the line OP.

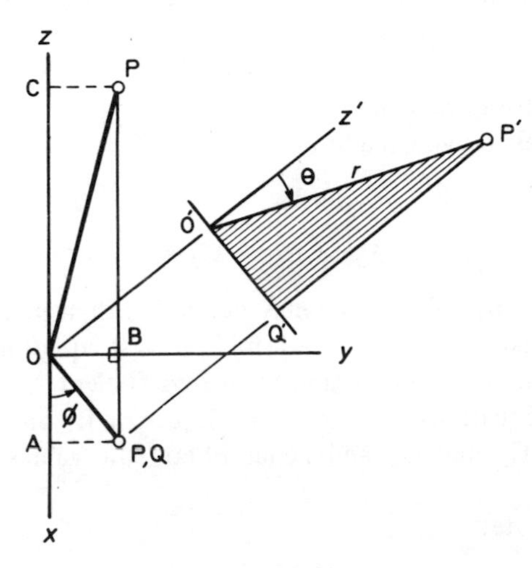

Figure 3.40

4 Determinants and matrices

A mathematician, like a painter or a poet, is a maker of patterns. If his patterns are more permanent than theirs, it is because they are made with ideas.

G. H. Hardy

4.1 DETERMINANTS

At several stages in the preceding discussion it has been necessary to solve linear simultaneous equations, the long-hand solution of which may be systematised and the work and chances of error reduced by the use of the determinant method.

4.1.1 Determinants of order two

Given the equations to two straight lines

$$A_1x + B_1y + C_1 = 0$$

$$A_2x + B_2y + C_2 = 0$$

for which $(A_1B_2 - B_1A_2) \neq 0$, we may determine their point of intersection as follows. The long-hand method is to multiply the first equation by the coefficient A_2 and multiply the second equation by the coefficient A_1: and on subtracting the two resulting equations, determine the values of B_1 and B_2 in terms of the known constants C_1 and C_2 and hence obtain the values of A_1 and A_2 by substitution.

Thus we may write:

$$B_2(A_1x + B_1y + C_1) - B_1(A_2x + B_2y + C_2) = 0$$

$$B_2A_1x - B_1A_2x + B_2B_1y - B_1B_2y + B_2C_1 - B_1C_2 = 0$$

whence

$$x = \frac{B_1C_2 - B_2C_1}{A_1B_2 - A_2B_1}$$

and

$$y = -\frac{A_1C_2 - A_2C_1}{A_1B_2 - A_2B_1}$$

from which it follows that:

$$\frac{x}{B_1C_2 - B_2C_1} = \frac{-y}{A_1C_2 - A_2C_1} = \frac{1}{A_1B_2 - A_2B_1}$$

or

$$\frac{x}{\begin{vmatrix} B_1 & C_1 \\ B_2 & C_2 \end{vmatrix}} = \frac{-y}{\begin{vmatrix} A_1 & C_1 \\ A_2 & C_2 \end{vmatrix}} = \frac{1}{\begin{vmatrix} A_1 & B_1 \\ A_2 & B_2 \end{vmatrix}}$$

Any set of coefficients (contained within parallel bars) which represent the difference between two or more products, is called a determinant. The expressions

$$B_1C_2 - B_2C_1 = \begin{vmatrix} B_1 & C_1 \\ B_2 & C_2 \end{vmatrix} \text{ and } b_1c_2 - b_2c_1 = \begin{vmatrix} b_1 & c_1 \\ b_2 & c_2 \end{vmatrix}$$

are determinants

It has been tacitly assumed that the two simultaneous equations above have a common solution, which in an infinite 2-space is true of all but parallel lines, subject to the further condition that $A_1B_2 - A_2B_1 \neq 0$.

We have seen that:

$$x = \frac{b_1c_2 - b_2c_1}{a_1b_2 - a_2b_1} \qquad y = -\frac{a_1c_2 - a_2c_1}{a_1b_2 - a_2b_1}$$

and it is clear that three possible forms arise

(i) Lines intersect $\quad a_1b_2 - a_2b_1 \neq 0$
(ii) Lines parallel $\quad (a_1b_2 - a_2b_1) = 0, (a_1c_2 - a_2c_1) \neq 0$
(iii) Lines coincident $\quad (a_1b_2 - a_2b_1) = 0, (a_1c_2 - a_2c_1) = 0$

Example

The constant coefficients for two straight lines are as listed in the accompanying table. Use the determinant method to determine whether the two lines are

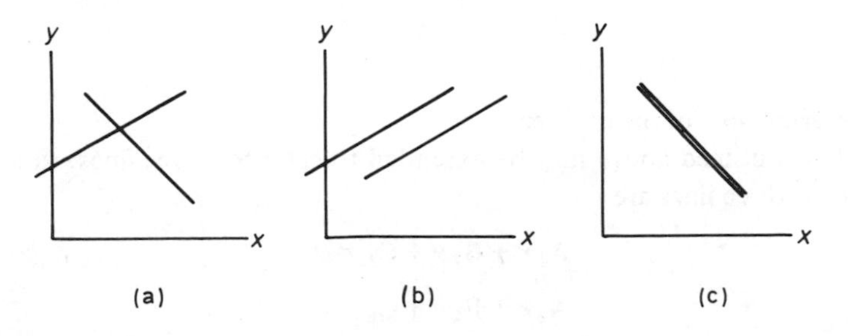

Figure 4.1

(a) coincident, (b) parallel, (c) have a common point of intersection. Also write the equation to each line and illustrate by a diagram.

	a	b	c
Line 1	–2	2	1
Line 2	4	–4	6

Solution

$$(a_1 b_2 - a_2 b_1) = (2.4) - (4.2) = 0$$

Hence lines do not intersect.

$$(a_1 c_2 - a_2 c_1) = (-2.6) - (4.1) = -16$$

Hence lines are parallel.

The equation t line 1 is $-2x + 2y + 1 = 0$.
The equation to line 2 is $4x - 4y + 6 = 0$.
The two lines appear as illustrated in fig. 4.2.

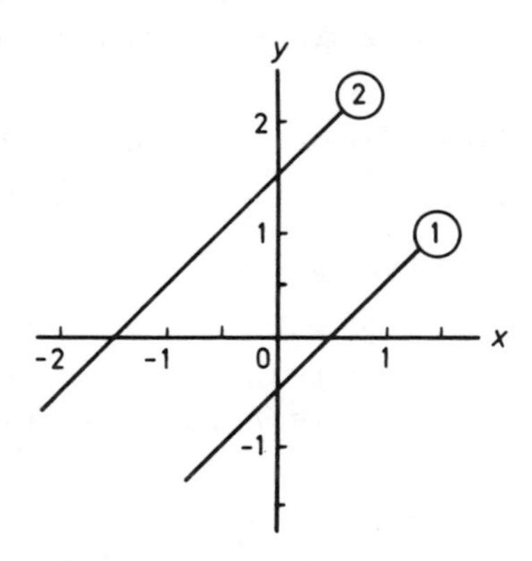

Figure 4.2

4.1.2 Determinants of order three

The method outlined above may be extended to cater for three lines: thus if the equations to three lines are

$$A_1 x + B_1 y + C_1 = 0$$

$$A_2 x + B_2 y + C_2 = 0$$

$$A_3 x + B_3 y + C_3 = 0$$

We begin by writing the solution to the second and third equations in the determinant form

$$\frac{x}{\begin{vmatrix} B_2 & C_2 \\ B_3 & C_3 \end{vmatrix}} = \frac{-y}{\begin{vmatrix} A_2 & C_2 \\ A_3 & C_3 \end{vmatrix}} = \frac{1}{\begin{vmatrix} A_2 & B_2 \\ A_3 & B_3 \end{vmatrix}}$$

and subject to the condition that the minor determinant

$$\begin{matrix} A_2 & B_2 \\ A_3 & B_3 \end{matrix} \neq 0$$

see article 4.1.3, go on to assume that the solution to the second and third equations also satisfies the first equation.

For this assumption to be true we must have:

$$A_1 \begin{vmatrix} B_2 & C_2 \\ B_3 & C_3 \end{vmatrix} = -B_1 \begin{vmatrix} A_2 & C_2 \\ A_3 & C_3 \end{vmatrix} = C_1 \begin{vmatrix} A_2 & B_2 \\ A_3 & B_3 \end{vmatrix} = 0$$

That is:

$$A_1(B_2C_3 - B_3C_2) = -B_1(A_2C_3 - A_3C_2) = C_1(A_2B_3 - A_3B_2) = 0$$

The expression

$$(A_1B_2C_3 - A_1B_3C_2) = -(B_1A_2C_3 - B_1A_3C_2) = (C_1A_2B_3 - C_1A_3B_2) = 0$$

is written:

$$\begin{vmatrix} A_1 & B_1 & C_1 \\ A_2 & B_2 & C_2 \\ A_3 & B_3 & C_3 \end{vmatrix} = \Delta$$

where Δ is called the value of the determinant.

If $\Delta = 0$ the three lines intersect at a common point. If $\Delta_{12} = 0$, but $\Delta_{13} \neq 0$, lines 1 and 2 intersect but lines 1 and 3 do not.

(N.B. Readers who have not previously studied the application of determinants are advised to read the appropriate texts on the subject but the following rules are included to provide continuity of reading.

4.1.3 Rules for evaluating determinants

The minor of the element a_r is obtained by sticking out the appropriate column and row.

Thus the minor of the element a_1 is:

$$\begin{vmatrix} a_1 & b_1 & c_1 \\ a_2 & b_2 & c_2 \\ a_3 & b_3 & c_3 \end{vmatrix} = \begin{vmatrix} b_2 & c_2 \\ b_3 & c_3 \end{vmatrix}$$

The determinant on the right-hand side is the minor of the element a_1. The minor of the element b_2 is:

$$\begin{vmatrix} a_1 & b_1 & c_1 \\ a_2 & b_2 & c_2 \\ a_3 & b_3 & c_3 \end{vmatrix} = \begin{vmatrix} a_1 & c_1 \\ a_3 & c_3 \end{vmatrix}$$

The determinant on the right-hand side is the minor of the element b_2, and a similar procedure yields the minors for all other elements.

N.B. The reader should check the truth of the elements and minors used in the previous section, repeated below for convenience.

$$a_1 \begin{vmatrix} b_2 & c_2 \\ b_3 & c_3 \end{vmatrix} = -b_1 \begin{vmatrix} a_1 & c_1 \\ a_3 & c_3 \end{vmatrix} = c_1 \begin{vmatrix} a_2 & b_2 \\ a_3 & b_3 \end{vmatrix} = 0$$

The determinant format may also be used to express the condition that two given points lie on a straight line and the equation to this line may be obtained by evaluating the determinant.

Suppose that a line is known to pass through two given points $P_1(1, 5)$, $P_2(5, 13)$, then the equation to this line may be concisely expressed as a determinant.

$$\begin{vmatrix} x & y & 1 \\ 1 & 5 & 1 \\ 5 & 13 & 1 \end{vmatrix} = 0$$

The solution is

$$x[(5.1) - (1.13)] - y[(1.1) - (1.5)] + [(1.13) - (5.5)] = 0$$

whence $2x - y + 3 = 0$ is the equation to the line.

Having introduced the determinant method as an easy, error-free approach to the solution of simultaneous equations, it is now clear that the equations into which the numerical substitutions are to be made are far from easy to remember.

For example if three planes are defined by the equations

$$A_1x + B_1y + C_1z + D_1 = 0$$

$$A_2x + B_2y + C_2z + D_2 = 0$$

$$A_3x + B_3y + C_3z + D_3 = 0$$

Then

$$x = \frac{D_1B_2C_3 + D_2B_3C_1 + D_3B_1C_2 - D_3B_2C_1 - D_2B_1C_3 - D_1B_3C_2}{A_1B_2C_3 + A_2B_3C_1 + A_3B_1C_2 - A_3B_2C_1 - A_2B_1C_3 - A_1B_3C_2}$$

and two similar expressions are required for y and z.

N.B. It should be noted, *en passant*, that if A_r is written for D_r in the numerator of the equation above, both numerator and denominator are identical and this suggests that the expressions for y and z may be readily obtained by writing the denominator as written, and then rewriting the numerator on making the following substitution.

when x is required write (D_1 for A_1), (D_2 for A_2), (D_3 for A_3)

when y is required write (D_1 for B_1), (D_2 for B_2), (D_3 for B_3)

when z is required write (D_1 for C_1), (D_2 for C_2), (D_3 for C_3)

A useful rule for expanding a determinant of order three is to write the determinant with its first two columns repeated. Adding the products of the three diagonal elements sloping down from right to left and subtracting the products of the three diagonal elements sloping down from left to right.

Thus if we write the determinant of order three as given and repeat columns one and two as shown:

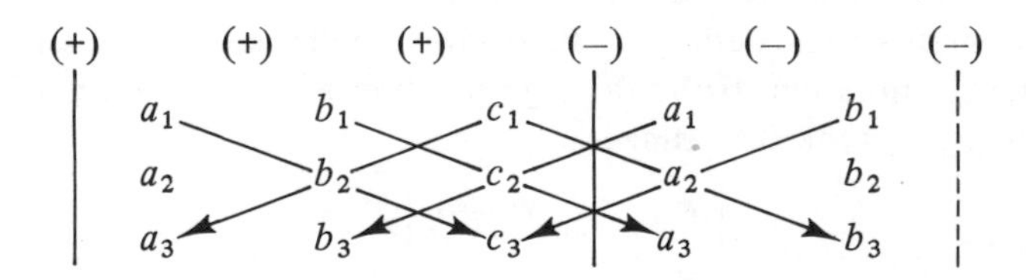

we obtain:

$$a_1b_2c_3 + b_1c_2a_3 + c_1a_2b_3 - c_1b_2a_3 - a_1c_2b_3 - b_1a_2c_3$$

(N.B. The order of terms is different but in all other respects this expression agrees with the master denominator term, already given.

Solution of simultaneous equations–Cramer's rule

If a set of homogeneous simultaneous equations

$$a_1x + b_1y + c_1z = 0$$

$$a_2x + b_2y + c_2z = 0$$

$$a_3x + b_3y + c_3z = 0$$

have a solution other than $x = y = z = 0$ then the determinant Δ must equal zero.

The solution to a set of non-homogeneous simultaneous equations

$$a_1x + b_1y + c_1z = k_1$$

$$a_2x + b_2y + c_2z = k_2$$

$$a_3x + b_3y + c_3z = k_3$$

may best be obtained by Cramer's rule which states that

$$\frac{x}{\Delta x} = \frac{y}{\Delta y} = \frac{z}{\Delta z} = \frac{1}{\Delta}$$

where Δ_x is obtained by substituting k_1, k_2, k_3, in place of the coefficients a_1, a_2, a_3, and similarly for Δ_y, Δ_z, whence

$$\frac{x}{\begin{vmatrix} k_1 & b_1 & c_1 \\ k_2 & b_2 & c_2 \\ k_3 & b_3 & c_3 \end{vmatrix}} = \frac{y}{\begin{vmatrix} a_1 & k_1 & c_1 \\ a_2 & k_2 & c_2 \\ a_3 & k_3 & c_3 \end{vmatrix}} = \frac{z}{\begin{vmatrix} a_1 & b_1 & k_1 \\ a_2 & b_2 & k_2 \\ a_3 & b_3 & k_3 \end{vmatrix}} = \frac{1}{\begin{vmatrix} a_1 & b_1 & c_1 \\ a_2 & b_2 & c_2 \\ a_3 & b_3 & c_3 \end{vmatrix}}$$

and it is clear that the method may be extended to cater for any number of linear simultaneous equations.

4.1.4 Equation to a plane in determinant form

The equation of a plane defined by three given points can be concisely expressed in determinant notation. If the three given points are $P_1(x_1, y_1, z_1)$, $P_2(x_2, y_2, z_2)$, $P_3(x_3, y_3, z_3)$ the determinant is

$$\begin{vmatrix} x & y & z & 1 \\ x_1 & y_1 & z_1 & 1 \\ x_2 & y_2 & z_2 & 1 \\ x_3 & y_3 & z_3 & 1 \end{vmatrix} = 0$$

Example

The coordinates of three points in a plane are: (a, o, o), (o, b, o), (o, o, c), and we know in advance that the equation is that of a plane in intercept form.

$$\begin{vmatrix} x & y & z & 1 \\ a & o & o & 1 \\ o & b & o & 1 \\ o & o & c & 1 \end{vmatrix} = 0$$

We may evaluate this determinant by expanding by minors.

Solution

We strike out the row and column in which the given element occurs, and write the equation in x, y, z and c with alternating signs (+, –, +, –) as illustrated below.

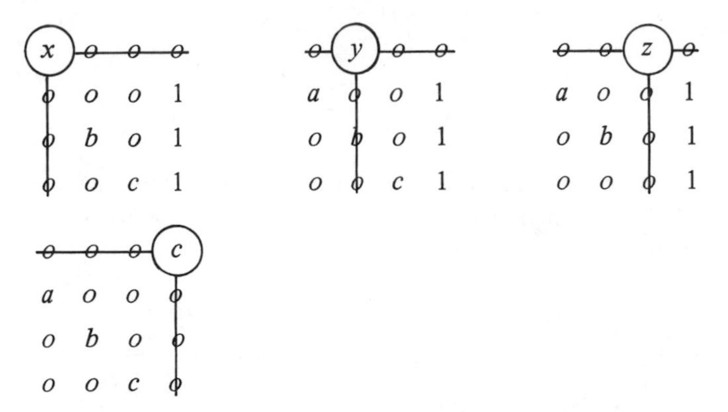

whence

$$x\begin{vmatrix} o & o & 1 \\ b & o & 1 \\ o & c & 1 \end{vmatrix} - y\begin{vmatrix} a & o & 1 \\ o & o & 1 \\ o & c & 1 \end{vmatrix} + z\begin{vmatrix} a & o & 1 \\ o & b & 1 \\ o & o & 1 \end{vmatrix} - \begin{vmatrix} a & o & o \\ o & b & o \\ o & o & c \end{vmatrix} = 0$$

$$x\left(o\begin{vmatrix} o & 1 \\ c & 1 \end{vmatrix} - o\begin{vmatrix} b & 1 \\ o & 1 \end{vmatrix} + 1\begin{vmatrix} b & o \\ o & c \end{vmatrix}\right) - y\left(a\begin{vmatrix} o & 1 \\ c & 1 \end{vmatrix} - o\begin{vmatrix} o & 1 \\ o & 1 \end{vmatrix} + 1\begin{vmatrix} o & o \\ o & c \end{vmatrix}\right)$$

$$+z\left(\begin{vmatrix} b & 1 \\ o & 1 \end{vmatrix} - o\begin{vmatrix} o & 1 \\ o & 1 \end{vmatrix} + 1\begin{vmatrix} o & b \\ o & o \end{vmatrix}\right) - \left(a\begin{vmatrix} b & o \\ o & c \end{vmatrix} - o\begin{vmatrix} o & o \\ o & c \end{vmatrix} + o\begin{vmatrix} o & b \\ o & o \end{vmatrix}\right) = 0$$

from which

$$bcx + acy + abz - abc = 0$$

and

$$\frac{x}{a} + \frac{y}{b} + \frac{z}{c} = 1$$

as predicted.

Example

A simple cubic lattice of slender bars may be stiffened by the inclusion of six bars arranged in the form of a tetrahedron and it is clear that if the axes are sited as shown in Fig. 4.3, then the internal plane formed by the lines P_1P_2, P_2P_3,

P_3P_1 is represented by the intercept equation. If the cubic lattice is of side a, the intercept equation to the plane containing the points P_1, P_2, P_3 is of the form:

$$\frac{x}{a} + \frac{y}{a} + \frac{z}{a} = 1$$

and

$$x + y + z = a \text{ in this case.}$$

If however the axes are sited at O, Fig. 4.3(b) and if the coordinates of O, P, Q are as tabulated, the general equation to the plane OPQ may be obtained using the determinant method as follows:

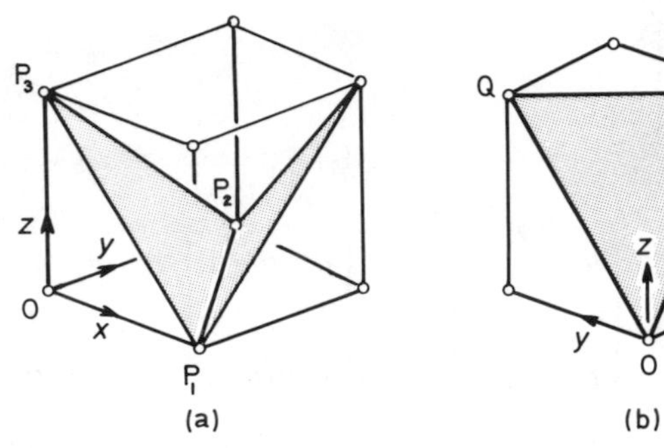

Figure 4.3

Solution

$$\begin{array}{c} \\ O \\ P \\ Q \end{array} \begin{vmatrix} x & y & z & 1 \\ 0 & 0 & 0 & 1 \\ 2 & 0 & 4 & 1 \\ 0 & 3 & 4 & 1 \end{vmatrix} = 0$$

Expanding as previously explained:

$$x \begin{vmatrix} 0 & 0 & 1 \\ 0 & 4 & 1 \\ 3 & 4 & 1 \end{vmatrix} - y \begin{vmatrix} 0 & 0 & 1 \\ 2 & 4 & 1 \\ 0 & 4 & 1 \end{vmatrix} + z \begin{vmatrix} 0 & 0 & 1 \\ 2 & 0 & 1 \\ 0 & 3 & 1 \end{vmatrix} - \begin{vmatrix} 0 & 0 & 0 \\ 2 & 0 & 4 \\ 0 & 3 & 4 \end{vmatrix} = 0$$

$$x\left(0\begin{vmatrix} 4 & 1 \\ 4 & 1 \end{vmatrix} - 0\begin{vmatrix} 0 & 1 \\ 3 & 1 \end{vmatrix} + 1\begin{vmatrix} 0 & 4 \\ 3 & 4 \end{vmatrix}\right) - y\left(0\begin{vmatrix} 4 & 1 \\ 4 & 1 \end{vmatrix} - 0\begin{vmatrix} 2 & 1 \\ 0 & 1 \end{vmatrix} + 1\begin{vmatrix} 2 & 4 \\ 0 & 4 \end{vmatrix}\right)$$

$$+z\left(0\begin{vmatrix} 0 & 1 \\ 3 & 1 \end{vmatrix} - 0\begin{vmatrix} 2 & 1 \\ 0 & 1 \end{vmatrix} + 1\begin{vmatrix} 2 & 0 \\ 0 & 3 \end{vmatrix}\right) - \left(0\begin{vmatrix} 0 & 4 \\ 3 & 4 \end{vmatrix} - 0\begin{vmatrix} 2 & 4 \\ 0 & 4 \end{vmatrix} + 0\begin{vmatrix} 2 & 0 \\ 0 & 3 \end{vmatrix}\right) = 0$$

whence

$$-12x - 8y + 6x = 0$$

(N.B. The plane passes through the origin and hence the constant term is zero).

4.1.5 Determinants of order four

In order to deal with lines and planes in 3-space we need to consider determinants of order four, and it will here suffice to note how minors are used to effect the expansion.

If

$$\Delta = \begin{vmatrix} a_1 & b_1 & c_1 & d_1 \\ a_2 & b_2 & c_2 & d_2 \\ a_3 & b_3 & c_3 & d_3 \\ a_4 & b_4 & c_4 & d_4 \end{vmatrix}$$

We may extract the minors of the elements a_1, b_1, c_1, d_1, using the method previously explained and hence express the fourth order determinant as a determinant of order three. Extracting minors we obtain:

$$\Delta = a_1 \begin{vmatrix} b_2c_2d_2 \\ b_3c_3d_3 \\ b_4c_4d_4 \end{vmatrix} - b_1 \begin{vmatrix} a_2c_2d_2 \\ a_3c_3d_3 \\ a_4c_4d_4 \end{vmatrix} + c_1 \begin{vmatrix} a_2b_2d_2 \\ a_3b_3d_3 \\ a_4b_4d_4 \end{vmatrix} - d_1 \begin{vmatrix} a_2b_2c_2 \\ a_3b_3c_3 \\ a_4b_4c_4 \end{vmatrix}$$

(N.B. The alternating sequence of signs (+, –, +, –) should be noted).

4.1.6 Minors and cofactors

The expansion of determinants of order three and four yields an alternating series of signs. For instance, the expansion of the determinant of order three is achieved by re-expressing the determinant in terms of its minors.

If we consider the determinant

$$\Delta \equiv \begin{vmatrix} a_1 & b_1 & c_1 \\ a_2 & b_2 & c_2 \\ a_3 & b_3 & c_3 \end{vmatrix}$$

We express the minor of any term as explained in article 4.1.3. For instance the minor of the element b_2 is

$$\begin{vmatrix} a_1 & c_1 \\ a_3 & c_3 \end{vmatrix}$$

and the full expansion of Δ takes the form a_1 (minor of a_1) – a_2 (minor of a_2) + a_3 (minor of a_3) whence

$$a_1b_2c_3 - a_1b_3c_3 + a_2b_3c_1 - a_2b_1c_3 + a_3b_1c_2 - a_3b_2c_1$$

and from this expansion the alternation of signs is apparent.

Now in order to simplify computation we introduce the notion of a signed minor, called a cofactor. The cofactor of an element in the ith row and jth column is the minor of that element times $(-1)^{i+j}$.

For instance the element b_2 is in the second row of the second column of the determinant Δ, $i = 2, j = 2$, and the signed minor of b_2 is

$$(-1)^{2+2}\begin{vmatrix} a_1 & c_1 \\ a_3 & c_3 \end{vmatrix} = +\begin{vmatrix} a_1 & c_1 \\ a_3 & c_3 \end{vmatrix}$$

Similarly the element c_2 is in the third row of the second column whence $i = 3$, $j = 2$ and the signed minor of c_2 is

$$(-1)^{3+2}\begin{vmatrix} a_1 & b_1 \\ a_3 & b_3 \end{vmatrix} = -\begin{vmatrix} a_1 & b_1 \\ a_3 & b_3 \end{vmatrix}$$

and the same procedure yields the signed cofactor of all other elements. The following table in which the elements b_2 and c_2 are circled is a useful aid to memory.

$$\begin{matrix} + & - & + & - \\ - & \textcircled{+} & \textcircled{-} & + \\ + & - & + & - \\ - & + & - & + \end{matrix}$$

4.2 MATRICES

In our treatment of determinants we encountered many square arrays in which the number of rows and columns of coefficients were equal. We enclosed these arrays within two vertical straight lines. In matrix algebra we encounter both square and rectangular arrays and distinguish between determinants and matrices by enclosing the latter between square brackets.

4.2.1 Matrix formulation

Matrix notation provides a convenient way of writing linear equations and is the method widely used to effect coordinate transformations.

The two equations

$$x' = ax + by$$
$$y' = dx + ey \tag{4.1}$$

represent a transformation from an initial condition denoted by the coordinates (x, y), to a transformed-or final condition denoted by the coordinates (x', y').

The two equations are in the present instance conveniently written in matrix form as follows

$$\begin{bmatrix} x' \\ y' \end{bmatrix} = \begin{bmatrix} x \\ y \end{bmatrix} \begin{bmatrix} a & b \\ d & e \end{bmatrix}$$

and we observe that the constants (ab, de) are grouped in the same formation in which they appear in the ordinary formulation, as they are in a determinant.

The format used above is for this reason well-suited to the representation of equations written down the page in columns. It is, however, more generally the case that (x, y, z) coordinate data is more conveniently written across the page in rows, rather than columns, and it is very important to note that the column formulation

$$\begin{bmatrix} x' \\ y' \end{bmatrix} = \begin{bmatrix} x \\ y \end{bmatrix} \begin{bmatrix} a & b \\ d & e \end{bmatrix} \tag{4.2}$$

does not have the same interpretation as the row formulation

$$[x'y'] = [xy] \begin{bmatrix} a & b \\ d & e \end{bmatrix} \tag{4.3}$$

The meaning of the column matrix (4.2) is

$$x' = (ax + by) \quad y' = (dx + ey)$$

The meaning of the row matrix (4.3) is

$$x' = (ax + dy) \quad y' = (bx + ey)$$

Since we are mainly concerned with the manipulation of coordinate data, we adopt the latter method of writing coordinates in rows and readily extend this format to cater for the three-dimensional case.

When operating in three dimensions the 3×3 format for a single point is

$$[x'y'z'] = [xyz] \begin{bmatrix} a & b & c \\ d & e & f \\ g & h & i \end{bmatrix} \tag{4.4}$$

and we may readily increase our data file by listing all subsequent points in columns. If $(x_1y_1z_1), (x_2y_2z_2)$ are two separate points then

$$\begin{bmatrix} x'_1y'_1z'_1 \\ x'_2y'_2z'_2 \end{bmatrix} = \begin{bmatrix} x_1y_1z_1 \\ x_2y_2z_2 \end{bmatrix} \begin{bmatrix} a & b & c \\ d & e & f \\ g & h & i \end{bmatrix} \tag{4.5}$$

records this information.

4.2.2 Pre- and post-multiplication

If A and B are two matrices we obtain the pre- and post-multiplication products by observing the following rule.

The rule to be observed when evaluating a matrix product is to multiply row by column, term by term.

If we consider a pair of 2 × 2 matrices we see that the product BA is obtained by pre-multiplying A by B and the product AB is obtained by post-multiplying A by B.

$$\text{BA} = \begin{bmatrix} b_{11}b_{12} \\ b_{21}b_{22} \end{bmatrix} \begin{bmatrix} a_{11}a_{12} \\ a_{21}a_{22} \end{bmatrix} = \begin{bmatrix} b_{11}a_{11} + b_{12}a_{21} & b_{11}a_{12} + b_{12}a_{22} \\ b_{21}a_{11} + b_{22}a_{21} & b_{21}a_{12} + b_{22}a_{22} \end{bmatrix}$$

and

$$\text{AB} = \begin{bmatrix} a_{11}a_{12} \\ a_{21}a_{22} \end{bmatrix} \begin{bmatrix} b_{11}b_{12} \\ b_{21}b_{22} \end{bmatrix} = \begin{bmatrix} a_{11}b_{11} + a_{12}b_{21} & a_{11}b_{12} + a_{12}b_{22} \\ a_{21}b_{11} + a_{22}b_{21} & a_{21}b_{12} + a_{22}b_{22} \end{bmatrix}$$

We observe that BA is not in general equal to AB.

Example

Use the rules of pre- and post-multiplication to show that if

$$\text{A} = \begin{bmatrix} 1 & 2 \\ 3 & 4 \end{bmatrix} \text{ and B} = \begin{bmatrix} 5 & 6 \\ 7 & 8 \end{bmatrix}$$

the product [AB] is different from the product [BA].

Solution

The product [AB] is obtained using the rule:

1st row of A × 1st col. of B	1st row of A × 2nd col. of B
2nd row of A × 1st col. of B	2nd row of A × 2nd col. of B

Using the numerical values above. The first row of A is (1, 2) the first column of B is (5, 7)'. Hence the top left-hand element of the product matrix [AB] is: $(1 \times 5) + (2 \times 7) = 19$ and the full complement of elements is written:

$$\left[\begin{array}{c|c} (1 \times 5) + (2 \times 7) & (1 \times 6) + (2 \times 8) \\ \hline (3 \times 5) + (4 \times 7) & (3 \times 6) + (4 \times 8) \end{array}\right] = \begin{bmatrix} 19 & 22 \\ 43 & 50 \end{bmatrix}$$

whence

$$[AB] = \begin{bmatrix} 19 & 22 \\ 43 & 50 \end{bmatrix}$$

The product [BA] is obtained in a similar way using the rule:

1st row of B × 1st col. of A	1st row of B × 2nd col. of A
2nd row of B × 1st col. of A	2nd row of B × 2nd col. of A

and the full complement of elements is written:

$$\left[\begin{array}{c|c} (5 \times 1) + (6 \times 3) & (5 \times 2) + (6 \times 4) \\ \hline (7 \times 1) + (8 \times 3) & (7 \times 2) + (8 \times 4) \end{array}\right] = \begin{bmatrix} 23 & 34 \\ 31 & 46 \end{bmatrix}$$

whence

$$[BA] = \begin{bmatrix} 23 & 34 \\ 31 & 46 \end{bmatrix}$$

The matrix product [AB] is not therefore the same as the matrix product [BA].

4.2.3 The null matrix

Particular cases may arise in which all elements in the products [AB] and/or [BA] are zeros. The product of the two matrices below is a case in point.

$$\begin{bmatrix} 2 & 4 \\ 4 & 8 \end{bmatrix} \begin{bmatrix} 4 & -4 \\ -2 & 2 \end{bmatrix} = \begin{bmatrix} 0 & 0 \\ 0 & 0 \end{bmatrix}$$

4.2.4 The unit identity matrix

If we insert $a = e = 1, b = d = 0$, into the transformation matrix equation 4.3, we obtain

$$[x', y'] = [x, y] \begin{bmatrix} 1 & 0 \\ 0 & 1 \end{bmatrix}$$

and it is immediately apparent that $x' = x$, $y' = y$. That is, the point $P(x', y')$ coincides with the point $P(x, y)$ and no visible change in the position or orientation of the point P has taken place. The matrix I, written

$$I = \begin{bmatrix} 1 & 0 \\ 0 & 1 \end{bmatrix}$$

is called the identity operator and might be used to represent a complete rotation of a point or line through 360°.

4.2.5 The transpose of a matrix

We have seen that columns and rows must be carefully distinguished, but sometimes have reason to transpose columns and rows.

If for example:

$$A = \begin{bmatrix} 1 & 2 \\ 3 & 4 \end{bmatrix} \quad A^T = \begin{bmatrix} 1 & 3 \\ 2 & 4 \end{bmatrix}$$

whence

$$AA^T = \begin{bmatrix} 5 & 11 \\ 11 & 25 \end{bmatrix}$$

If however we consider the matrix

$$A = \begin{bmatrix} \cos\theta & \sin\theta \\ -\sin\theta & \cos\theta \end{bmatrix} \quad A^T = \begin{bmatrix} \cos\theta & -\sin\theta \\ \sin\theta & \cos\theta \end{bmatrix}$$

we see that:

$$AA^T = \begin{bmatrix} \cos\theta\cos\theta + \sin\theta\sin\theta & -\cos\theta\sin\theta + \sin\theta\cos\theta \\ -\sin\theta\cos\theta + \cos\theta\sin\theta & \sin\theta\sin\theta + \cos\theta\cos\theta \end{bmatrix}$$

whence

$$AA^T = \begin{bmatrix} 1 & 0 \\ 0 & 1 \end{bmatrix} = I$$

When $AA^T = I$ we say the matrix is orthogonal.

4.2.6 Matrix inversion

If $AX = B$ is a matrix equation then it is frequently necessary to express X in terms of the matrices A and B.

Given the equation

$$AX = B$$

we pre-multiply both sides by A^{-1}

$$A^{-1}AX = A^{-1}B$$

but since $A^{-1}A = I$ and $IX = X$ it follows that

$$X = A^{-1}B$$

The matrix A^{-1} is called the inverse of the matrix A, and we therefore seek a method of finding the inverse of A. The process is called matrix inversion and we shall adopt the adjoint matrix approach.

4.2.7 The adjoint matrix

The adjoint of a matrix A written adj A is the inverse of the matrix of cofactors of A.

If

$$A = \begin{bmatrix} a_{11} & a_{12} & \cdots & a_{1n} \\ a_{21} & a_{22} & \cdot & \cdot \\ \cdot & \cdot & \cdot & \cdot \\ \cdot & \cdot & \cdot & \cdot \\ a_{n1} & a_{n2} & & a_{nn} \end{bmatrix}$$

$$\text{adj } A = \begin{bmatrix} C_{11} & C_{12} & \cdots & C_{1n} \\ C_{21} & C_{22} & \cdot & \cdot \\ \cdot & \cdot & \cdot & \cdot \\ \cdot & \cdot & \cdot & \cdot \\ C_{n1} & C_{n2} & \cdot & C_{nn} \end{bmatrix}$$

where C_{ij} is the cofactor a_{ij}.

N.B. It is usual to denote the elements of the matrix A using a lower-case letter and their cofactors using a capital letter. The product of A and adj A is the matrix $A \text{ adj } A = |A|I$.

$$A \text{ adj } A = \begin{bmatrix} |A| & 0 & \cdot & \cdot & 0 \\ 0 & |A| & \cdot & \cdot & 0 \\ 0 & 0 & |A| & \cdot & \cdot \\ \cdot & \cdot & \cdot & |A| & \cdot \\ \cdot & \cdot & \cdot & \cdot & |A| \end{bmatrix} = |A|I$$

where $|A|$ is the determinant of A, and I is the unit matrix.

Now since $|A| \neq 0$ the matrix A^{-1} is written

$$A^{-1} = \frac{\text{adj } A}{|A|}$$

Example

Determine the inverse M^{-1} of a matrix M, where

$$M = \begin{bmatrix} 1 & 0 & 0 & 0 \\ 1 & 1 & 1 & 1 \\ 0 & 1 & 0 & 0 \\ 0 & 1 & 2 & 3 \end{bmatrix}$$

Solution by the determinant method

If

$$\begin{bmatrix} 1 & 0 & 0 & 0 \\ 1 & 1 & 1 & 1 \\ 0 & 1 & 0 & 0 \\ 0 & 1 & 2 & 3 \end{bmatrix} \equiv \begin{bmatrix} a_1 & b_1 & c_1 & d_1 \\ a_2 & b_2 & c_2 & d_2 \\ a_3 & b_3 & c_3 & d_3 \\ a_4 & b_4 & c_4 & d_4 \end{bmatrix}$$

and the transpose of the signed cofactor matrix is adj M.

We first obtain the signed cofactor matrix by following the method of 4.1.6. whence we may allocate alternating signs to the matrix M and evaluate the first row of the cofactor matrix as follows.

$$a_1 \quad \begin{vmatrix} 1 & -1 & 1 \\ -1 & 0 & 0 \\ 1 & -2 & 3 \end{vmatrix} = 1 \begin{vmatrix} 0 & 0 \\ -2 & 3 \end{vmatrix} +1 \begin{vmatrix} -1 & 0 \\ 1 & 3 \end{vmatrix} +1 \begin{vmatrix} -1 & 0 \\ 1 & -2 \end{vmatrix} = -1$$

$$b_1 \quad \begin{vmatrix} -1 & -1 & 1 \\ 0 & 0 & 0 \\ 0 & -2 & 3 \end{vmatrix} = 0$$

$$c_1 \quad \begin{vmatrix} -1 & 1 & 1 \\ 0 & -1 & 0 \\ 0 & 1 & 3 \end{vmatrix} = -1 \begin{vmatrix} -1 & 0 \\ 1 & 3 \end{vmatrix} = 3$$

$$d_1 \quad \begin{vmatrix} -1 & 1 & -1 \\ 0 & -1 & 0 \\ 0 & 1 & -2 \end{vmatrix} = -1 \begin{vmatrix} -1 & 0 \\ 1 & -2 \end{vmatrix} = -2$$

The first row of the signed cofactor matrix is therefore $| -1, 0, 3, -2 |$ and the evaluation of other rows follows suit. The full cofactor matrix being as written. we must now obtain $|M|$. $|M|$ is obtained by multiplying corresponding terms in the M and cofactor matrices, before summing them as follows:

$$\underset{(M)}{\begin{matrix} 1 & 0 & 0 & 0 \\ 1 & 1 & 1 & 1 \\ 0 & 1 & 0 & 0 \\ 0 & 1 & 2 & 3 \end{matrix}} \times \underset{(CF)}{\begin{matrix} -1 & 0 & 3 & -2 \\ 0 & 0 & -3 & 2 \\ 0 & -1 & 2 & -1 \\ 0 & 0 & 1 & -1 \end{matrix}} = \begin{array}{cccc|c} -1 & 0 & 0 & 0 & -1 \\ 0 & 0 & -3 & 2 & -1 \\ 0 & -1 & 0 & 0 & -1 \\ 0 & 0 & 2 & -3 & -1 \\ \hline -1 & -1 & -1 & -1 & \end{array}$$

We observe that each row and column sums to a constant $|M|$, where $|M| = -1$, in this case.

Now since

$$M^{-1} = \frac{\text{adj } M}{|M|}$$

and $|M| = -1$

$$M^{-1} = \begin{vmatrix} 1 & 0 & 0 & 0 \\ 0 & 0 & 1 & 0 \\ -3 & 3 & -2 & -1 \\ 2 & -2 & 1 & 1 \end{vmatrix}$$

(N.B. The matrix M^{-1} represents terms in a parametric cubic and is a formulation widely used in curve and surface design and will be stored for use in Chapter 10.

5 Geometrical transformations

I am not a thing – a noun. I seem to be a verb. I am *an evolutionary process.*

R. Buckminster Fuller

5.1 INTRODUCTION

In Chapter 3 we encountered the need to translate axes from a global to a local position and the need to rotate axes is another requirement. Indeed the need to translate and rotate both axes and objects will soon become apparent.

5.2 ROTATION OF AXES

A design problem in which the transposition of axes plays an important rôle is encountered when rigging the wing and tailplane of an aircraft.

In order to facilitate design and manufacture, numerous major assemblies such as the port and starboard wing and fuselage etc., are given their own set of local Cartesian coordinate axes and other major components, such as engine mountings and nacelles, are likewise given their own local sets of axes. The joint between wing and fuselage is obviously a crucial connection which must be set at the correct aerodynamic rigging angle(s), of incidence and dihedral and must be made without appreciable geometrical mismatch and/or physical distortion. In order to achieve this end the exact spatial position orientation and inclination of numerous control points and axes must be set with close precision.

All cardinal dimensions are, for these reasons, referred to a set of general axes which relate to the flight attitude of the complete aircraft.

Two different methods of wing rigging are in common use; (a) the wing chord reference plane system, (b) the wing common percent plane system. The two methods are illustrated in Fig. 5.1, but will not be further considered.

Setting the reference axes in rigged position is in either case accomplished in two mathematical stages; (i) the incidence angle θ is set by means of a pure rotation about the original (initial) axis ox. This rotation creates a new set of axes in which oy' and oz' lie in the original oyz plane. The situation before and after setting the incidence angle θ is shown in Fig. 5.1; (ii) the dihedral angle ϕ is

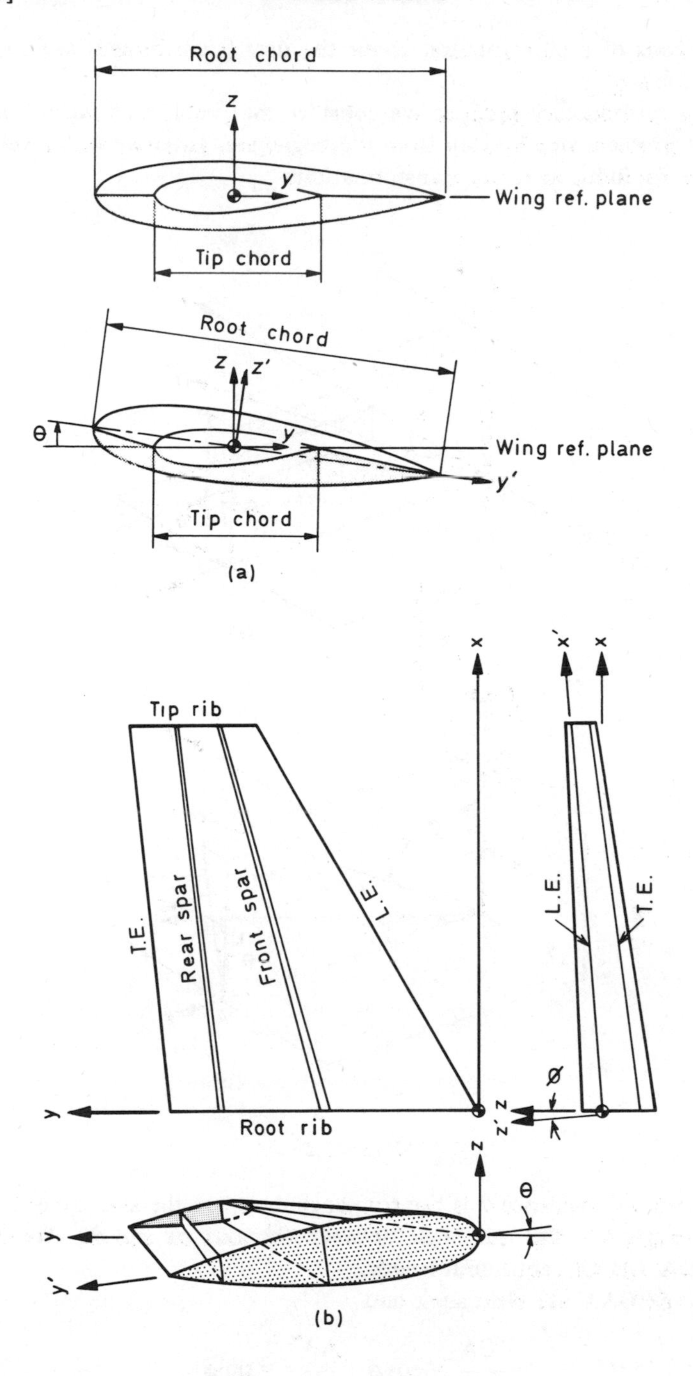

Root chord
z
y
Wing ref. plane
Tip chord
Root chord
z
z'
θ
y
Wing ref. plane
y'
Tip chord
(a)
x
x'
x
Tip rib
T.E.
Rear spar
Front spar
L.E.
L.E.
T.E.
Ø
y
Root rib
z
z'
z
θ
y
y'
(b)

Figure 5.1

set by means of a pure rotation about the new (transformed) axes, this time about the axis oy'.

In this introductory instance we consider the problem *ab initio*, that is we work the problem step by step from the beginning. Later we will develop a set procedure for inducing rotary transformations.

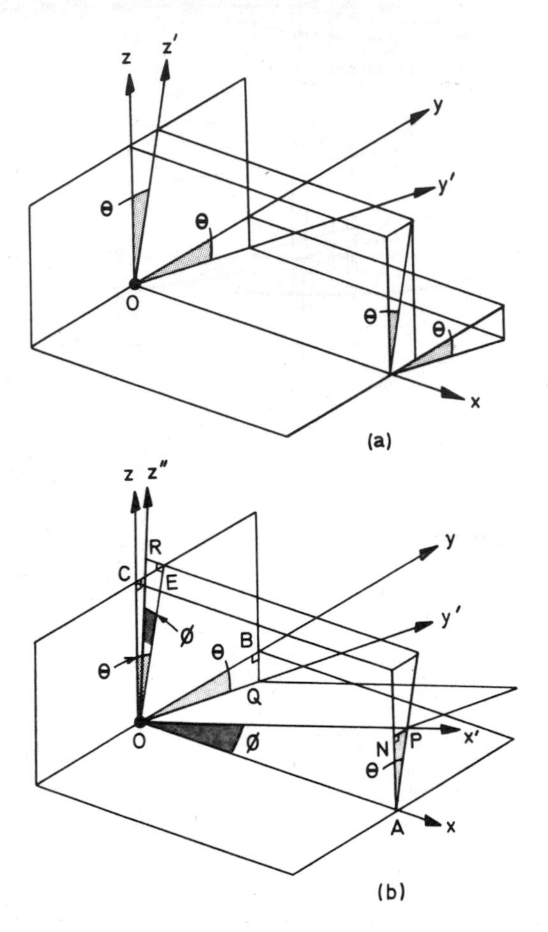

Figure 5.2

The angle of incidence θ is the true angle between the axes oy and oy'. The dihedral angle ϕ is the true angle between the axes ox and ox'. We make the lengths OA, OB, OC, equal unit length.

The angle OAA″ is a right angle and

$$\frac{\mathrm{OA}}{\mathrm{OA''}} = \cos\phi \qquad \frac{\mathrm{AA''}}{\mathrm{OA''}} = \sin\phi$$

and since OA″ is of unit length

$$OA = \cos\phi \quad AA'' = \sin\phi$$

The angle A″AN equals θ and since angle A″NA is a right angle

$$\frac{AN}{AA''} = \cos\theta \quad \frac{A''N}{AA''} = \sin\theta$$

but since AA″ = $\sin\phi$

$$AN = \sin\phi\cos\theta \text{ and } A''N = \sin\phi\sin\theta$$

The (x, y, z) coordinates of A″ are (OA, A″N, AA″). The (x, y, z) coordinates of O are (0, 0, 0).

Since the length of OA = OA″ is of unit length the direction ratios of the line OA″ are also the direction cosines of the line OA″. The (x, y, z) coordinates of the line OA″ are:

$$x = OA = \cos\phi$$

$$y = A''N = \sin\phi\sin\theta$$

$$z = AA'' = \sin\phi\cos\theta$$

We may check this result by squaring and adding the direction cosines on the R.H.S.

$$\cos^2\phi + \sin^2\phi\sin^2\theta + \sin^2\phi\cos^2\theta = \text{RHS}$$

by substituting $(1 - \sin^2\phi)$ for $\cos^2\phi$ and $(1 - \sin^2\theta)$ for $\cos^2\theta$, we obtain

$$1 - \sin^2\phi + \sin^2\phi\sin^2\theta + \sin^2\phi - \sin^2\phi\sin^2\theta = 1$$

The sum of the x, y, z direction cosines squared, should equal unity. This it does and all is as it should be.

$$(\cos\phi, \sin\phi\sin\theta, \sin\phi\cos\theta)$$

is thus the transform to re-orientate a set of x, y, z, unit vectors.

It would, of course, be tedious and time consuming to work each new problem calling for a rotation of axes from scratch and a more general method will be developed in due course.

5.3 TRANSFORMATION OF OBJECT POINTS IN TWO DIMENSIONS

A point $P(x, y)$ may be moved to a new position $P(x', y')$ in various ways, by means of the transformation

$$[x'y'] = [xy]\begin{bmatrix} a & b \\ d & e \end{bmatrix} = [(ax + dy), (bx + ey)]$$

that is

$$x' = (ax + dy) \quad y' = (bx + ey)$$

and it is clear that by choosing different values for the constant terms $(ab, \cdot de)$ we have the means to induce numerous transformations. For example, if $b = d = o$, we may induce the following translations and reflections, by allotting different values to a and e.

(i) if $a = a$, $e = 1$, $x' = ax$, $y' = y$ and the point P is moved parallel to the axis ox.
(ii) if $a = 1$, $e = e$, $x' = x$, $y' = ey$ and the point P is moved parallel to the axis oy.
(iii) if $a = a$, $e = e$, $x' = ax$, $y' = ey$ and the point P is moved at an angle of arc tan e/a to the ox axis.
(iv) if $a = -1$, $e = +1$, $x' = -x$, $y' = y$ and the point P is reflected about the oy axis.
(v) if $a = +1$, $e = -1$, $x' = x$, $y' = -y$ and the point P is reflected about the ox axis. If we take $a = e = 0$ and ascribe values to b and d other transformations are produced.
(vi) if $b = 1$, $d = 1$, $x' = y$, $y' = x$ and the point P is reflected about a line $y = x$, that is about a line at 45° to the axis ox.

These six simple cases are illustrated in Fig. 5.3.

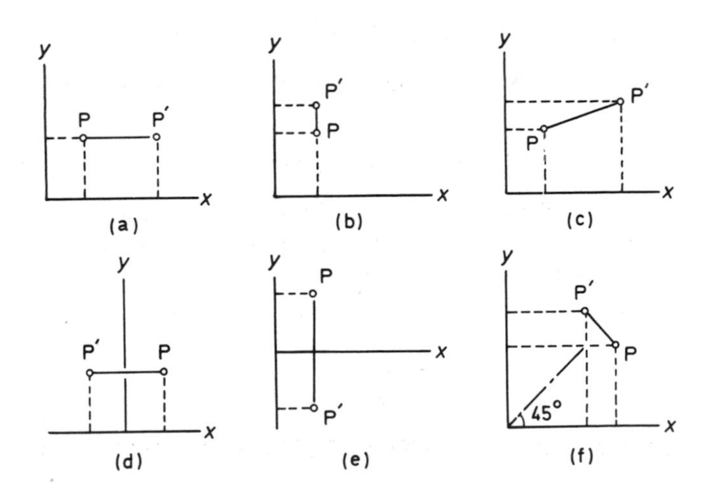

Figure 5.3

5.3.1 Transformation of a straight line

If a straight line L_{12} is defined by two points $P(x_1y_1)$ and $Q(x_2y_2)$. Then

$$P' = [x'_1y'_1] = [x_1y_1]\begin{bmatrix} a & b \\ d & e \end{bmatrix}$$

$$Q' = [x'_2y'_2] = [x_2y_2]\begin{bmatrix} a & b \\ d & e \end{bmatrix}$$

and hence

$$L_{12} = \begin{bmatrix} x_1y_1 \\ x_2y_2 \end{bmatrix}\begin{bmatrix} a & b \\ d & e \end{bmatrix}$$

5.3.2 Scaling matrix

We have seen that if two points $P(x_1y_1)$, $Q(x_2y_2)$ define a straight line then

$$L_{12} = \begin{bmatrix} x_1y_1 \\ x_2y_2 \end{bmatrix} \begin{bmatrix} a & b \\ d & e \end{bmatrix}$$

If P(1, 2) and Q(3, 4) are the two points and we arbitrarily set $a = e = 2$ and $b = d = 0$.

$$L'_{12} = \begin{bmatrix} 1 & 2 \\ 3 & 4 \end{bmatrix} \begin{bmatrix} 2 & 0 \\ 0 & 2 \end{bmatrix} = \begin{bmatrix} 2 & 4 \\ 6 & 8 \end{bmatrix}$$

and we see that all coordinates are scaled by a factor of 2. Matrices of the type

$$\begin{bmatrix} S_x & 0 \\ 0 & S_y \end{bmatrix}$$

always produce a scaling effect, as shown in Fig. 5.4.

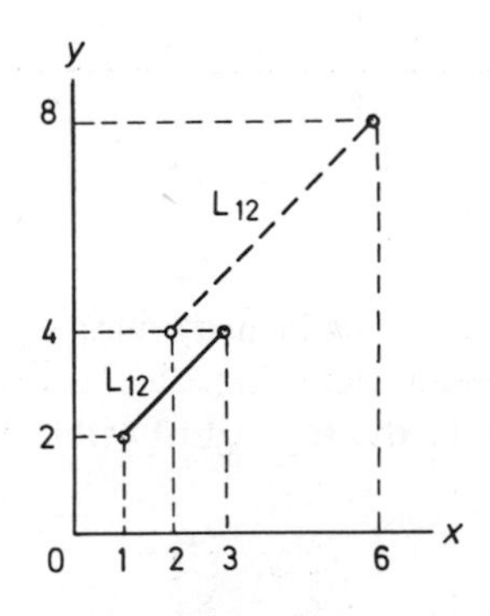

Figure 5.4

5.3.3 Shearing transformations

The effect of applying a transformation in which terms on the leading diagonal are different is to induce unequal scaling. The line L, for which $(x_1y_1) = (1, 2)$ and $(x_2y_2) = (3, 4)$ is transformed by the shearing matrix

$$\begin{bmatrix} 3 & 0 \\ 0 & 1 \end{bmatrix}$$

as follows

$$L' = \begin{bmatrix} 1 & 2 \\ 3 & 4 \end{bmatrix} \begin{bmatrix} 3 & 0 \\ 0 & 1 \end{bmatrix} = \begin{bmatrix} 3 & 2 \\ 9 & 4 \end{bmatrix}$$

The effect of the element 3 in the top left hand corner of the transformation matrix is to multiply the x coordinates (in the left hand column) by 3. The effect of the element 1 in the lower right hand corner of the transformation matrix is to leave the second column of y coordinate elements unchanged. The transformation effected by the matrix

$$\begin{bmatrix} 3 & 0 \\ 0 & 1 \end{bmatrix}$$

is shown in Fig. 5.5, from which we observe that in addition to scaling the x coordinates by three, the initial and transformed lines are no longer parallel. That is the line L has been sheared in the x direction only. (N.B. It is left to the reader to investigate other possible cases).

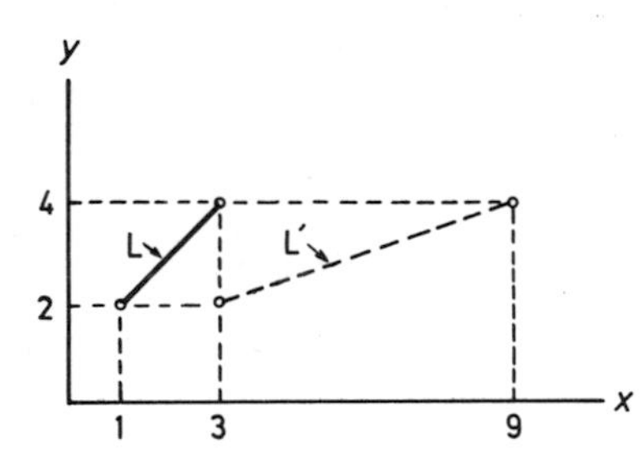

Figure 5.5

5.3.4 Reflection of a line

We have seen that application of the identity matrix produces no change whatsoever, and the next step is to investigate what effect is produced by a transformation matrix with unity elements in the top right/bottom left positions. That is, what effect does the matrix

$$\begin{bmatrix} 0 & 1 \\ 1 & 0 \end{bmatrix}$$

have on the line L(1 2), (3 4).

$$L' = \begin{bmatrix} 1 & 2 \\ 3 & 4 \end{bmatrix} \begin{bmatrix} 0 & 1 \\ 1 & 0 \end{bmatrix} = \begin{bmatrix} 2 & 1 \\ 4 & 3 \end{bmatrix}$$

and similarly the matrix:

$$\begin{bmatrix} 0 & -1 \\ -1 & 0 \end{bmatrix}$$

transforms the line as follows

$$L' = \begin{bmatrix} 1 & 2 \\ 3 & 4 \end{bmatrix} \begin{bmatrix} 0 & -1 \\ -1 & 0 \end{bmatrix} = \begin{bmatrix} -2 & -1 \\ -4 & -3 \end{bmatrix}$$

We see that the effect of each of these transformations is to reflect the original line about axes $x = y$ and $-x = y$ respectively as shown in Fig. 5.6.

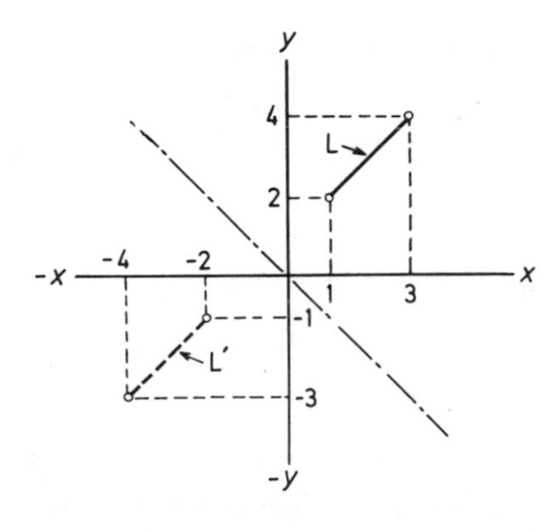

Figure 5.6

5.3.5 Reflection of a plane figure

The above method of specifying a reflection may be extended to deal with the reflection of a plane figure (a triangle for instance) by including information on a third point. If P, Q, R, is a triangle the vertices of which are sited at (2, 1), (2, −1), (5, 0) we may reflect this triangle about the line $x = y$ by applying the transformation

$$\begin{bmatrix} 0 & 1 \\ 1 & 0 \end{bmatrix}$$

Thus we may write:

$$\begin{bmatrix} x'_1 y'_1 \\ x'_2 y'_2 \\ x'_3 y'_3 \end{bmatrix} = \begin{bmatrix} 2 & 1 \\ 2 & -1 \\ 5 & 0 \end{bmatrix} \begin{bmatrix} 0 & 1 \\ 1 & 0 \end{bmatrix}$$

$$= \begin{bmatrix} (2.0) + (1.1) & (2.1) + (1.0) \\ (2.0) - (1.1) & (2.1) - (1.0) \\ (5.0) + (0.1) & (5.1) + (0.0) \end{bmatrix} = \begin{bmatrix} 1 & 2 \\ -1 & 2 \\ 0 & 5 \end{bmatrix}$$

The effect of this transformation is to interchange the initial and final coordinates. (N.B. The x values are replaced by the y values and the y values are replaced by the x values).

The matrix

$$\begin{bmatrix} 0 & -1 \\ -1 & 0 \end{bmatrix}$$

may be used to effect a reflection about the line $-x = y$. Thus we may write:

$$\begin{bmatrix} x'_1 y'_1 \\ x'_2 y'_2 \\ x'_3 y'_3 \end{bmatrix} = \begin{bmatrix} 2 & 1 \\ 2 & -1 \\ 5 & 0 \end{bmatrix} \begin{bmatrix} 0 & -1 \\ -1 & 0 \end{bmatrix}$$

$$= \begin{bmatrix} (2.0) - (1.1) & -(2.1) + (1.0) \\ (2.0) + (1.1) & -(2.1) - (1.0) \\ (5.0) - (0.1) & -(5.1) + (0.0) \end{bmatrix} = \begin{bmatrix} -1 & -2 \\ +1 & -2 \\ 0 & -5 \end{bmatrix}$$

and we have the required reflection about the line $x = -y$. The x values are replaced by the negation of the y values and the y values are replaced by the negation of the x values.

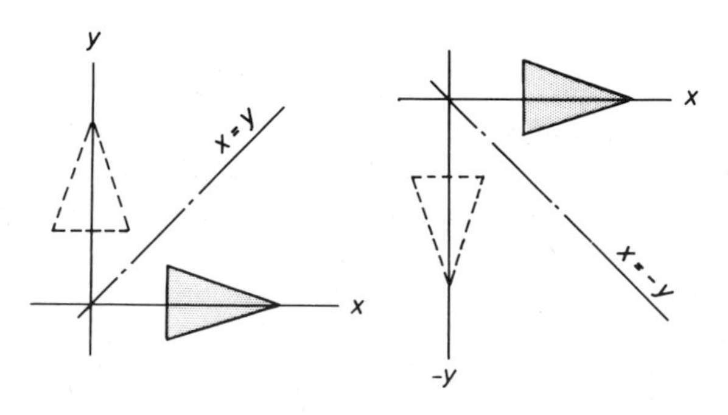

Figure 5.7

5.3.6 Rotation of an object point through an arbitrary angle

We may derive the transformation matrix for an arbitrary angle by considering Fig. 5.8, from which the coordinates of a general point P(x, y) may be written directly.

$$x' = x \cos\theta - y \sin\theta$$

$$y' = x \sin\theta + y \cos\theta$$

When x and y are presented in columns the equivalent matrix expression is

$$\begin{bmatrix} x' \\ y' \end{bmatrix} = \begin{bmatrix} x \\ y \end{bmatrix} \begin{bmatrix} \cos\theta & -\sin\theta \\ \sin\theta & +\cos\theta \end{bmatrix}$$

However, in accord with our agreed convention, article 4.2.1, we write xy coordinates in row formation and care must be taken to decipher the above identity correctly. It is also important not to confuse an object transformation with an axis transformation, see article 5.6.

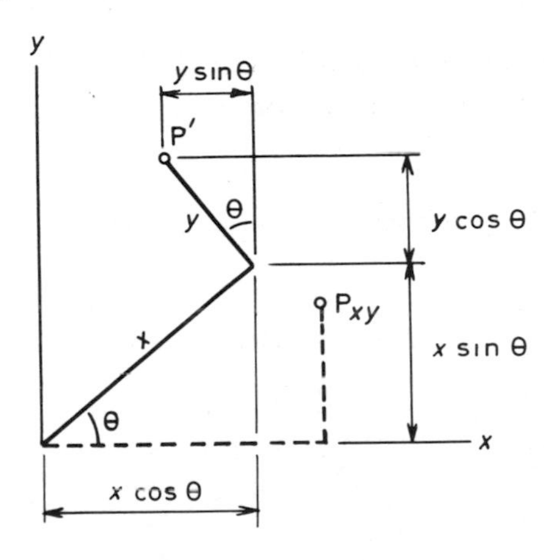

Figure 5.8

Our aim is to write the transformation in the form

$$[x'y'] = [xy] \begin{bmatrix} a & b \\ d & e \end{bmatrix}$$

and by comparing

$$x' = x \cos\theta - y \sin\theta \text{ and } y' = x \sin\theta + y \cos\theta$$

with

$$x' = (ax + dy) \quad y' = (bx + ey)$$

we conclude that

$$a = \cos\theta,\ b = \sin\theta,\ d = -\sin\theta,\ e = \cos\theta$$

hence the general rotation transformation of an object point, written in row formation is

$$[x'y'] = [xy]\begin{bmatrix}\cos\theta & \sin\theta \\ -\sin\theta & \cos\theta\end{bmatrix}$$

when $\theta = 90°, 180°, 270°, 360°$ the matrix

$$\begin{bmatrix}\cos\theta & \sin\theta \\ -\sin\theta & \cos\theta\end{bmatrix}$$

reduces to

$$\begin{bmatrix}0 & 1 \\ -1 & 0\end{bmatrix}_{90°} \begin{bmatrix}-1 & 0 \\ 0 & -1\end{bmatrix}_{180°} \begin{bmatrix}0 & -1 \\ 1 & 0\end{bmatrix}_{270°} \begin{bmatrix}1 & 0 \\ 0 & 1\end{bmatrix}_{360°}$$

N.B. A rotation of 360° returns the object point to its original position. The matrix for a 360° rotation is therefore the unity matrix.

Example
Rotate the point P(5, 0) through a positive angle of 90°.

Solution
The matrix which rotates an object point through a positive angle of 90° is

$$\begin{bmatrix}0 & 1 \\ -1 & 0\end{bmatrix}$$

and

$$[x'y'] = [5 \quad 0]\begin{bmatrix}0 & 1 \\ -1 & 0\end{bmatrix} = [0 \quad 5]$$

The point P′ thus lies at a distance of 5 units along the oy' axis.

5.3.7 Successive transformations

Matrix multiplication is non-commutative and in consequence the order in which two transformations are performed is important.

As may be seen from Fig. 5.9, a translation a followed by a translation b transfers P to P′, as does a translation b followed by a translation a, for this is the basis of the parallelogram law. It is to be observed, however, that a rotation through an angle $\theta°$, followed by a translation c, is not the same as a translation c followed by a rotation $\theta°$. The order in which transformations are performed is therefore important.

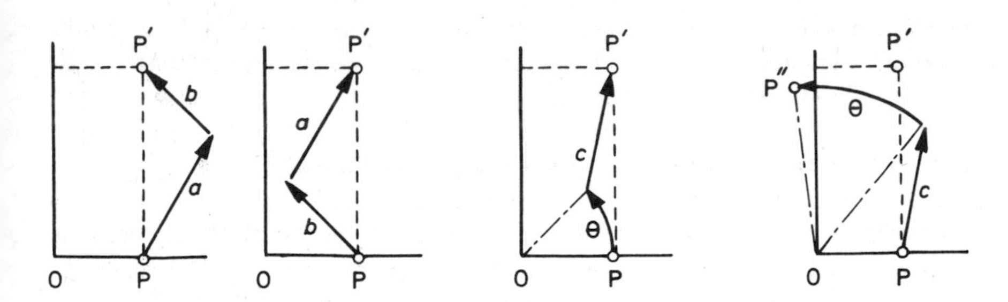

Figure 5.9

5.4 HOMOGENEOUS COORDINATES

In all cases so far considered we have specified the position of a point in 2D space by two coordinate dimensions and a point in 3D space by three coordinate dimensions. We now continued our study by considering the representation of a point in 2D space by three coordinate dimensions and the representation of a point in 3D space by four coordinate dimensions.

It is clearly the case that if $x = X/Z$ and $y = Y/Z$ the two ordinary coordinates (x, y) are related to the three homogeneous coordinates (X, Y, Z). If, for example, $x = \frac{1}{3}$, $y = \frac{4}{3}$, it follows that X = 1, Y = 4, Z = 3, and since $x = \frac{1}{3} = \frac{2}{6}$ and $y = \frac{4}{3} = \frac{8}{6}$, the homogeneous coordinates (X = 1, Y = 4, Z = 3) represent the same point as the homogeneous coordinates (X = 2, Y = 8, Z = 6).

Suppose the homogeneous coordinates of two separate points A and B are (2, 4, 1) and (6, 12, 1) and suppose we add the homogeneous coordinates of A and B together:

	x	y	H	x/H	y/H
A	2	4	1	2	4
B	6	12	1	6	12
C	8	16	2	4	8

the ordinary coordinates of C are $x = 4, y = 8$, and we see that C is the mid-point between A and B.

If we multiply the homogeneous coordinates A by 3, and add the homogeneous coordinates of B, we obtain (2, 4, 1) 3 + (6, 12, 1) = (12, 24, 4) whence

X = 12, Y = 24, Z = 4, are the homogeneous coordinates and $x = 3, y = 6$, the ordinary coordinates of a new point D located at the quarter point between A and B.

We may multiply a set of homogeneous coordinates by any positive or negative number except -1, and apart from this one exception every homogeneous point has a corresponding point in a space of one less dimension. The condition H = 0 has no physical meaning.

5.4.1 Graphical representation of the homogeneous format

The homogeneous format $(x, y, 1)$ is identical to the ordinary three dimensional format (x, y, z) and we may consider the quantity in the third column as an ordinate in a direction mutually perpendicular to the plane oxy, or indeed extend this line of thinking to the more general case of (x, y, h).

The relationship between homogeneous and ordinary coordinates may be readily demonstrated graphically by considering a numerical example.

Suppose the ordinary coordinates of a point are $x = 2, y = 1$, with corresponding homogeneous coordinates X = 2, Y = 1, H = 2. Then the homogeneous coordinates X = 4, Y = 2, Z = 4 represent the same point. The homogeneous equivalent of an ordinary point is thus a straight line and we may illustrate this duality between point and line by allotting different numerical values to H, and plotting the outcome in the form of an isometric (equal scales) projection, as in Fig. 5.10.

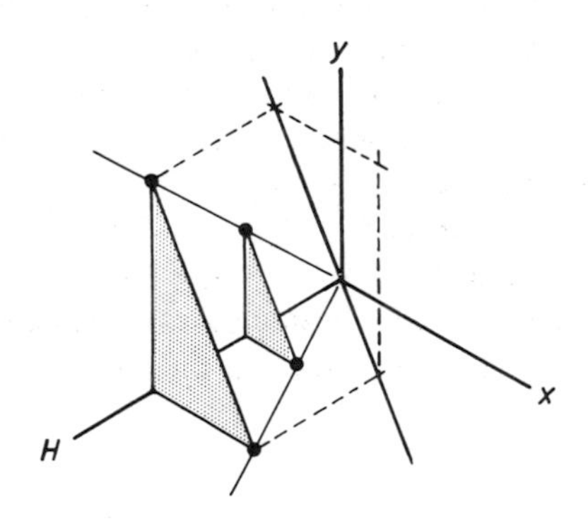

Figure 5.10

5.4.2 Use of homogeneous coordinates

Use of the homogeneous format enables additional elements such as p and q to be included within the matrix and furthermore transforms a 3×2 matrix, which does not have an inverse, into a square 3×3 format which does. The fully populated matrix, for H = 1, may then be written as follows

$$\begin{bmatrix} a & b & p \\ d & e & q \\ u & v & 1 \end{bmatrix}$$

If we set the scale factors a and e equal to 1, we may investigate the effect of the elements p and q as follows

$$[\mathrm{X} \quad \mathrm{Y} \quad \mathrm{Z}] = [x \quad y \quad 1] \begin{bmatrix} 1 & 0 & p \\ 0 & 1 & q \\ 0 & 0 & 1 \end{bmatrix} = [x, y, (px + qy + 1)]$$

whence we conclude that the homogeneous coordinates are

$$\mathrm{X} = x, \mathrm{Y} = y, \mathrm{Z} = (px + qy + 1)$$

and the corresponding ordinary coordinates are

$$x^* = \frac{x}{px + qy + 1}, \quad y^* = \frac{y}{px + qy + 1}$$

We may demonstrate this relationship by allocating set numerical values to p and q and plotting along isometric axes as previously.

If, for the purpose of example, we take $p = 2$, $q = 1$, we have homogeneous coordinates:

$$\mathrm{X} = x, \mathrm{Y} = y, \mathrm{H} = (2x + y + 1)$$

and ordinary coordinates

$$x^* = \frac{x}{2x + y + 1}, \quad y^* = \frac{y}{2x + y + 1}$$

When $\mathrm{H} = 1$

$2x + y = 0$: a straight line located in the oxy plane by the points $x = 0, y = 0$, $x = +1, y = -2, x = -1, y = +2$, etc.

When $\mathrm{H} = 2$

$2x + y = 1$: a straight line located in the oxy plane by the two points $x = 0$, $y = 1$, and $x = 0.5, y = 0$.

When $\mathrm{H} = 3$

$2x + y + 2$: a straight line located in the oxy plane by the two points $x = 0$, $y = 2$, and $x = 1, y = 0$.

When $\mathrm{H} = 5$

$2x + y = 4$: a straight line located in the oxy plane by the two points $x = 0$, $y = 4$, and $x = 2, y = 0$.

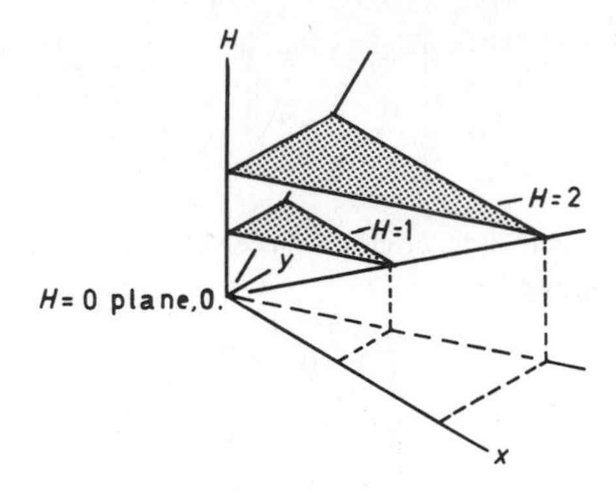

Figure 5.11

An overall scaling factor s is introduced by writing

$$[\mathrm{X} \quad \mathrm{Y} \quad \mathrm{H}] = [x \quad y \quad 1]\left[\begin{array}{cc|c} a & b & p \\ d & e & q \\ \hline u & v & s \end{array}\right]$$

and a local scaling by dividing the diagonal terms by s

$$[\mathrm{X} \quad \mathrm{Y} \quad \mathrm{H}] = [x \quad y \quad 1]\left[\begin{array}{cc|c} \frac{a}{s} & b & p \\ d & \frac{e}{s} & q \\ \hline u & v & 1 \end{array}\right]$$

The function of the four partitions is as follows:

(i) The 3×3 square matrix induces local scaling, shearing, reflection and/or rotation.

(ii) The 2×1 row matrix induces translation.

(iii) The 1×2 column matrix induces projection.

(iv) The 1×1 square matrix induces overall scaling.

If the overall scaling factor s is greater than unity a reduction in scale is effected.

If the overall scaling factor s is less than unity an increase in scale is effected.

5.5 THREE DIMENSIONAL HOMOGENEOUS REPRESENTATION

In order to avail ourselves of full three dimensional versatility, we adopt the use of the homogeneous system throughout.

By analogy with the two dimensional case we write the necessary 4 × 4 transformation as follows:

$$[\mathrm{X}\quad \mathrm{Y}\quad \mathrm{Z}\quad \mathrm{H}] = [x\quad y\quad z\quad 1]\left[\begin{array}{ccc|c} a & b & c & p \\ d & e & f & q \\ g & h & i & r \\ \hline u & v & w & s \end{array}\right]$$

and note the respective functions of the partitioned elements as previously.

5.5.1 Local and overall scaling

Local scaling is achieved by setting all elements, except those on the leading diagonal to zero. The transformation matrix for local scaling is:

$$[x\quad y\quad z\quad 1]\begin{bmatrix} a & 0 & 0 & 0 \\ 0 & e & 0 & 0 \\ 0 & 0 & i & 0 \\ 0 & 0 & 0 & 1 \end{bmatrix} = [ax, ey, iz, 1]$$

Similarly for local and overall scaling we write:

$$[x\quad y\quad z\quad 1]\begin{bmatrix} a & 0 & 0 & 0 \\ 0 & e & 0 & 0 \\ 0 & 0 & i & 0 \\ 0 & 0 & 0 & s \end{bmatrix} = [ax, ey, iz, s] = [x'y'z's]$$

We again note that $s > 1$ produces a reduction, $s < 1$ an increase in scale. We also observe that the above transformation is equivalent to a local scaling of:

$$[x\quad y\quad z\quad 1]\begin{bmatrix} \frac{a}{s} & 0 & 0 & 0 \\ 0 & \frac{e}{s} & 0 & 0 \\ 0 & 0 & \frac{i}{s} & 0 \\ 0 & 0 & 0 & 1 \end{bmatrix} = \left[\frac{a}{s}x, \frac{e}{s}y, \frac{i}{s}z, 1\right] = [x'y'z's]$$

Example

Suppose we consider scaling the unit cube, Fig. 5.12, by amounts $a = 4$, $e = 2$, $i = 4$, $s = 1$. We take the corner points at the positions tabulated in the position matrix. That is, point O at 0001, point A at 1001, point B at 0101, etc.

This is a local transformation which increases the linear dimension in the *ox* and *oz* directions by a factor of four and doubles the linear dimension in the *oy*

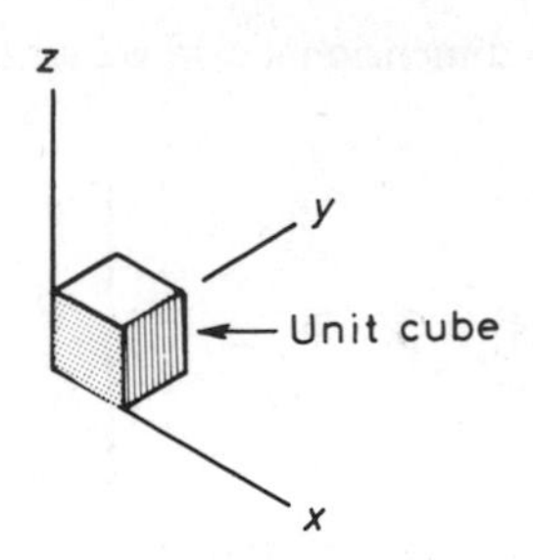

Figure 5.12

direction. We obtain the transformed coordinates O′, A′, B′, C′, etc., by multiplying each row of the position matrix by each column of the scaling matrix term by term. Since there is only one operative element per row the transformation may be written directly:

$$
\begin{matrix} \text{O} \\ \text{A} \\ \text{B} \\ \text{C} \\ \text{D} \\ \text{E} \\ \text{F} \\ \text{G} \end{matrix}
\begin{bmatrix} 0 & 0 & 0 & 1 \\ 1 & 0 & 0 & 1 \\ 0 & 1 & 0 & 1 \\ 1 & 1 & 0 & 1 \\ 0 & 0 & 1 & 1 \\ 1 & 0 & 1 & 1 \\ 0 & 1 & 1 & 1 \\ 1 & 1 & 1 & 1 \end{bmatrix}
\begin{bmatrix} 4 & 0 & 0 & 0 \\ 0 & 2 & 0 & 0 \\ 0 & 0 & 4 & 0 \\ 0 & 0 & 0 & 1 \end{bmatrix}
=
\begin{bmatrix} 0 & 0 & 0 & 1 \\ 4 & 0 & 0 & 1 \\ 0 & 2 & 0 & 1 \\ 4 & 2 & 0 & 1 \\ 0 & 0 & 4 & 1 \\ 4 & 0 & 4 & 1 \\ 0 & 2 & 4 & 1 \\ 4 & 2 & 4 & 1 \end{bmatrix}
\begin{matrix} \text{O}' \\ \text{A}' \\ \text{B}' \\ \text{C}' \\ \text{D}' \\ \text{E}' \\ \text{F}' \\ \text{G}' \end{matrix}
$$

If, for the purpose of example, we now introduce an overall scaling reduction corresponding to $s = 2$ we may transform the coordinates O′, A′, B′, C′, etc., by using the overall scaling matrix

$$
\begin{bmatrix} 1 & 0 & 0 & 0 \\ 0 & 1 & 0 & 0 \\ 0 & 0 & 1 & 0 \\ 0 & 0 & 0 & 2 \end{bmatrix}
$$

from which we obtain

$$\begin{matrix} O' \\ A' \\ B' \\ C' \\ D' \\ E' \\ F' \\ G' \end{matrix}\begin{bmatrix} 0 & 0 & 0 & 1 \\ 4 & 0 & 0 & 1 \\ 0 & 2 & 0 & 1 \\ 4 & 2 & 0 & 1 \\ 0 & 0 & 4 & 1 \\ 4 & 0 & 4 & 1 \\ 0 & 2 & 4 & 1 \\ 4 & 2 & 4 & 1 \end{bmatrix} \begin{bmatrix} 1 & 0 & 0 & 0 \\ 0 & 1 & 0 & 0 \\ 0 & 0 & 1 & 0 \\ 0 & 0 & 0 & 2 \end{bmatrix} = \begin{bmatrix} 0 & 0 & 0 & 2 \\ 4 & 0 & 0 & 2 \\ 0 & 2 & 0 & 2 \\ 4 & 2 & 0 & 2 \\ 0 & 0 & 4 & 2 \\ 4 & 0 & 4 & 2 \\ 0 & 2 & 4 & 2 \\ 4 & 2 & 4 & 2 \end{bmatrix}\begin{matrix} O'' \\ A'' \\ B'' \\ C'' \\ D'' \\ E'' \\ F'' \\ G'' \end{matrix}$$

When the homogeneous coordinates are normalised (that is when divided by 2) they become the ordinary coordinates listed below. Hence the ordinary coordinates are:

$$\begin{bmatrix} 0 & 0 & 0 \\ 2 & 0 & 0 \\ 0 & 1 & 0 \\ 2 & 1 & 0 \\ 0 & 0 & 2 \\ 2 & 0 & 2 \\ 0 & 1 & 2 \\ 2 & 1 & 2 \end{bmatrix}\begin{matrix} O \\ A \\ B \\ C \\ D \\ E \\ F \\ G \end{matrix}$$

The two transformations performed above are illustrated in Fig. 5.13, and we conclude this simple example by observing that the final result illustrated at (b)

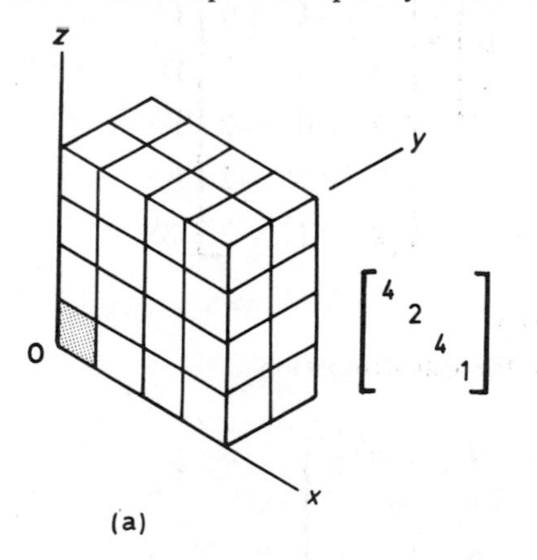

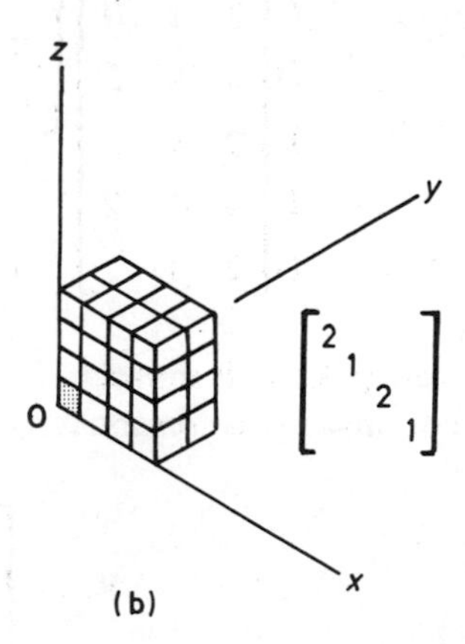

Figure 5.13

could have been obtained directly by using a single transformation in which only local scaling elements, a/s, e/s, i/s, are used.

If we wish to re-position our $2 \times 2 \times 1$ parallelepiped, Fig. 5.13(b), in a rotated orientation, we have only to apply the appropriate rotation matrix.

If the initial position is given by the ordinary coordinates, as tabulated we may induce a rotation of any amount we please.

Example

We illustrate the procedure by considering a pure rotation of +180° about each of the axes ox, oy, oz, in turn.

(i) For object rotation about ox we use the transformation

$$R_{ox} = \begin{bmatrix} 1 & 0 & 0 \\ 0 & \cos\theta_x & \sin\theta_x \\ 0 & -\sin\theta_x & \cos\theta_x \end{bmatrix}$$

which for a rotation of +180° becomes

$$\begin{bmatrix} 1 & 0 & 0 \\ 0 & -1 & 0 \\ 0 & 0 & -1 \end{bmatrix}$$

whence the solution is:

$$\begin{matrix} O \\ A \\ B \\ C \\ D \\ E \\ F \\ G \end{matrix}\begin{bmatrix} 0 & 0 & 0 \\ 2 & 0 & 0 \\ 0 & 1 & 0 \\ 2 & 1 & 0 \\ 0 & 0 & 2 \\ 2 & 0 & 2 \\ 0 & 1 & 2 \\ 2 & 1 & 2 \end{bmatrix} \begin{bmatrix} 1 & 0 & 0 \\ 0 & -1 & 0 \\ 0 & 0 & -1 \end{bmatrix} = \begin{bmatrix} 0 & 0 & 0 \\ 2 & 0 & 0 \\ 0 & -1 & 0 \\ 2 & -1 & 0 \\ 0 & 0 & -2 \\ 2 & 0 & -2 \\ 0 & -1 & -2 \\ 2 & -1 & -2 \end{bmatrix}\begin{matrix} O' \\ A' \\ B' \\ C' \\ D' \\ E' \\ F' \\ G' \end{matrix}$$

The new position is shown in Fig. 5.14(b).

(ii) For object rotation about oy we use the transformation

$$R_{oy} = \begin{bmatrix} \cos\theta_y & 0 & -\sin\theta_y \\ 0 & 1 & 0 \\ \sin\theta_y & 0 & \cos\theta_y \end{bmatrix}$$

which for a rotation of +180° becomes:

$$\begin{bmatrix} -1 & 0 & 0 \\ 0 & 1 & 0 \\ 0 & 0 & -1 \end{bmatrix}$$

whence the solution is:

$$\begin{matrix} O \\ A \\ B \\ C \\ D \\ E \\ F \\ G \end{matrix}\begin{bmatrix} 0 & 0 & 0 \\ 2 & 0 & 0 \\ 0 & 1 & 0 \\ 2 & 1 & 0 \\ 0 & 0 & 2 \\ 2 & 0 & 2 \\ 0 & 1 & 2 \\ 2 & 1 & 2 \end{bmatrix} \begin{bmatrix} -1 & 0 & 0 \\ 0 & 1 & 0 \\ 0 & 0 & -1 \end{bmatrix} = \begin{bmatrix} 0 & 0 & 0 \\ -2 & 0 & 0 \\ 0 & 1 & 0 \\ -2 & 1 & 0 \\ 0 & 0 & -2 \\ -2 & 0 & -2 \\ 0 & 1 & -2 \\ -2 & 1 & -2 \end{bmatrix}\begin{matrix} O' \\ A' \\ B' \\ C' \\ D' \\ E' \\ F' \\ G' \end{matrix}$$

The new position is shown in Fig. 514(c).

(iii) For object rotation about *oz* we use the transformation

$$R_{oz} = \begin{bmatrix} \cos\theta_z & \sin\theta_z & 0 \\ -\sin\theta_z & \cos\theta_z & 0 \\ 0 & 0 & 1 \end{bmatrix}$$

which for a rotation of +180° becomes:

$$\begin{bmatrix} -1 & 0 & 0 \\ 0 & -1 & 0 \\ 0 & 0 & 1 \end{bmatrix}$$

whence the solution is:

$$\begin{matrix} O \\ A \\ B \\ C \\ D \\ E \\ F \\ G \end{matrix}\begin{bmatrix} 0 & 0 & 0 \\ 2 & 0 & 0 \\ 0 & 1 & 0 \\ 2 & 1 & 0 \\ 0 & 0 & 2 \\ 2 & 0 & 2 \\ 0 & 1 & 2 \\ 2 & 1 & 2 \end{bmatrix} \begin{bmatrix} -1 & 0 & 0 \\ 0 & -1 & 0 \\ 0 & 0 & 1 \end{bmatrix} = \begin{bmatrix} 0 & 0 & 0 \\ -2 & 0 & 0 \\ 0 & -1 & 0 \\ -2 & -1 & 0 \\ 0 & 0 & 2 \\ -2 & 0 & 2 \\ 0 & -1 & 2 \\ -2 & -1 & 2 \end{bmatrix}\begin{matrix} O' \\ A' \\ B' \\ C' \\ D' \\ E' \\ F' \\ G' \end{matrix}$$

The new position is shown in Fig. 5.14(d).

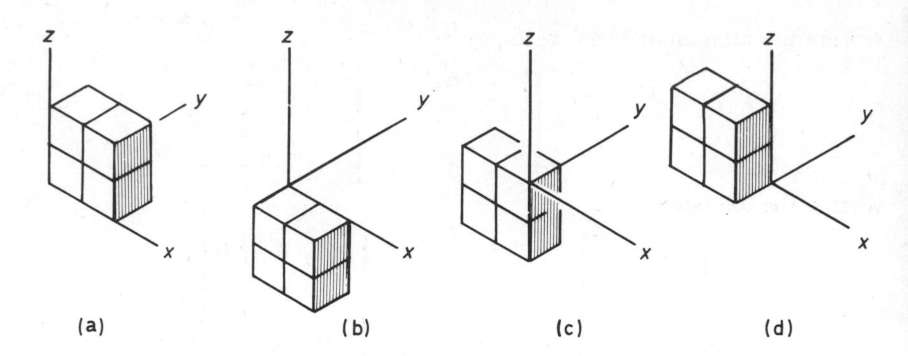

Figure 5.14

Example

The initial coordinates of a lever PQ in a food canning machine are P(0, 1, 0), Q(2, 1, 0). Determine the new coordinates of P and Q following a negative rotation of 45° about the *oy* axis.

For rotation about the *oy* axis we use the transformation matrix

$$R_{oy} = \begin{bmatrix} \cos\theta_y & 0 & -\sin\theta_y \\ 0 & 1 & 0 \\ \sin\theta_y & 0 & \cos\theta_y \end{bmatrix}$$

Solution

Since the direction of θ_y is negative $\sin\theta_y = -1/\sqrt{2}$ $\cos\theta_y = +1/\sqrt{2}$, hence the transformed object coordinates are

$$\begin{bmatrix} P' \\ Q' \end{bmatrix} = \begin{bmatrix} x'_P & z'_P \\ x'_Q & z'_Q \end{bmatrix} = \begin{bmatrix} 0 & 1 & 0 \\ 2 & 1 & 0 \end{bmatrix} \begin{bmatrix} \frac{1}{\sqrt{2}} & 0 & \frac{1}{\sqrt{2}} \\ 0 & 1 & 0 \\ -\frac{1}{\sqrt{2}} & 0 & \frac{1}{\sqrt{2}} \end{bmatrix}$$

The rule of matrix multiplication is: first row A by first column T, term by term, whence

$$\begin{array}{c|c|c} (0.1/\sqrt{2}) + (0.1) + (0.1/\sqrt{2}) & (0.0) + (1.1) + (0.0) & (0.1/\sqrt{2}) + (1.0) + (0.1/\sqrt{2}) \\ \hline (2.1/\sqrt{2}) + (1.0) - (0.1/\sqrt{2}) & (2.0) + (1.1) + (0.0) & (2.1/\sqrt{2}) + (1.0) + (0.1/\sqrt{2}) \end{array}$$

which yields

$$\begin{bmatrix} P' \\ Q' \end{bmatrix} = \begin{bmatrix} 0 & 1 & 0 \\ \sqrt{2} & 1 & \sqrt{2} \end{bmatrix}$$

The transformation is shown in Fig. 5.15(a).

The above transformation could well be used to assess the rotated position coordinates of a wing flap or aileron, where power, clearance considerations are important.

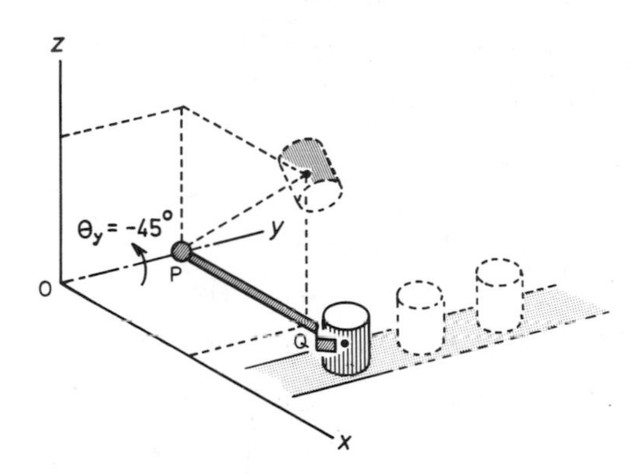

Figure 5.15

5.5.2 Combined rotations

The need to define the outcome of two or more successive rotations about two or more different axis is a problem which arises in the design , study and control of robot manipulators.

The problem is that rotations in three dimensions are non-commutative, which means that a rotation R_{oy} followed by a rotation R_{oz} does not produce the same positional change as the rotation R_{oz} followed by R_{oy}. We demonstrate this fact by means of an example.

Example

Consider the rotation of a robot arm OP, where O = (0, 0, 0), P = (2, 0, 0),

(a) through an angle θ_z about the axis *oz*, followed by a rotation of angle θ_y about the axis *oy*.

(b) through an angle θ_y about the axis *oy* followed by a rotation of angle θ_z about the axis *oz*.

For the purpose of this example we will put

$$\cos\theta_z = \sin\theta_y = \tfrac{4}{5} = 0.8, \cos\theta_y = \sin\theta_z = \tfrac{3}{5} = 0.6$$

Case (a)

For rotation about *oz* followed by rotation about *oy*. We have:

Rotation about *oz*

$$[x \quad y \quad z]\begin{bmatrix} \cos\theta_z & \sin\theta_z & 0 \\ -\sin\theta_z & \cos\theta_z & 0 \\ 0 & 0 & 0 \end{bmatrix} = x' \quad y' \quad z']$$

$$[2 \quad 0 \quad 0]\begin{bmatrix} 0.8 & 0.6 & 0 \\ -0.6 & 0.8 & 0 \\ 0 & 0 & 1 \end{bmatrix} = [1.6, 1.2, 0]$$

followed by rotation about *oy*.

Rotation about *oy*

$$[x' \quad y' \quad z']\begin{bmatrix} \cos\theta_y & 0 & -\sin\theta_y \\ 0 & 1 & 0 \\ \sin\theta_y & 0 & \cos\theta_y \end{bmatrix} = [x'' \quad y'' \quad z'']$$

$$[1.6, \quad 1.2, \quad 0]\begin{bmatrix} 0.6 & 0 & -0.8 \\ 0 & 1 & 0 \\ 0.8 & 0 & 0.6 \end{bmatrix} = [0.96, \quad 1.2, \quad -1.28]$$

The final position of P_a'' is shown in Fig. 5.16(a).

Case (b)

For rotation about *oy* followed by rotation about *oz*. We have:

Rotation about *oy*

$$[x \quad y \quad z]\begin{bmatrix} \cos\theta_y & 0 & -\sin\theta_y \\ 0 & 1 & 0 \\ \sin\theta_y & 0 & \cos\theta_y \end{bmatrix} = [x' \quad y' \quad z']$$

$$[2 \quad 0 \quad 0]\begin{bmatrix} 0.6 & 0 & 0.8 \\ 0 & 1 & 0 \\ 0.8 & 0 & 0.6 \end{bmatrix} = [1.2, \quad 0, \quad -1.6]$$

Rotation about *oz*

$$[x' \quad y' \quad z']\begin{bmatrix} \cos\theta_z & \sin\theta_z & 0 \\ -\sin\theta_z & \cos\theta_z & 0 \\ 0 & 0 & 1 \end{bmatrix}$$

$$[1.2 \quad 0 \quad -1.6]\begin{bmatrix} 0.8 & 0.6 & 0 \\ -0.6 & 0.8 & 0 \\ 0 & 0 & 1 \end{bmatrix} = [0.96, \quad 0.72, \quad -1.6]$$

The final position of P_b is shown in Fig. 5.16(b).

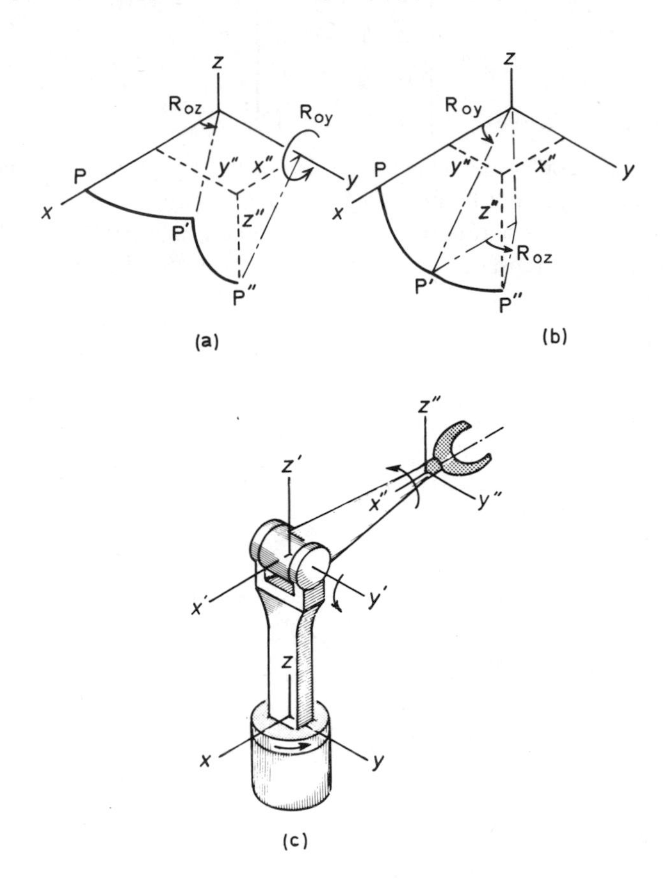

Figure 5.16

5.5.3 Rotation about an arbitrary axis

In all the above examples rotations were made about the origin of axes, but it is clearly a common requirement that rotations about other origins may be required. About the centre of gravity or visual centre, or about a corner point, for example.

In order to open up this facility it is only necessary to include data terms relating to the point about which rotation is required. The method to be used is a three stage operation:

1. we translate the object to the origin;
2. we apply whatever rotation is required;
3. we translate the rotated object back to its original origin.

We have seen that the transformation matrix governing translation is of the form

$$\begin{bmatrix} 1 & 0 & 0 & 0 \\ 0 & 1 & 0 & 0 \\ 0 & 0 & 1 & 0 \\ u & v & w & 1 \end{bmatrix}$$

where u, v, w, are displacements in the directions ox, oy, oz, respectively.

If (u, v, w) are the coordinates about which the rotation is required, $(-u, -v, -w)$ represents the translation of that point to the origin (0, 0, 0) and $(+u, +v, +w)$ represents the translation from the origin back to the original position.

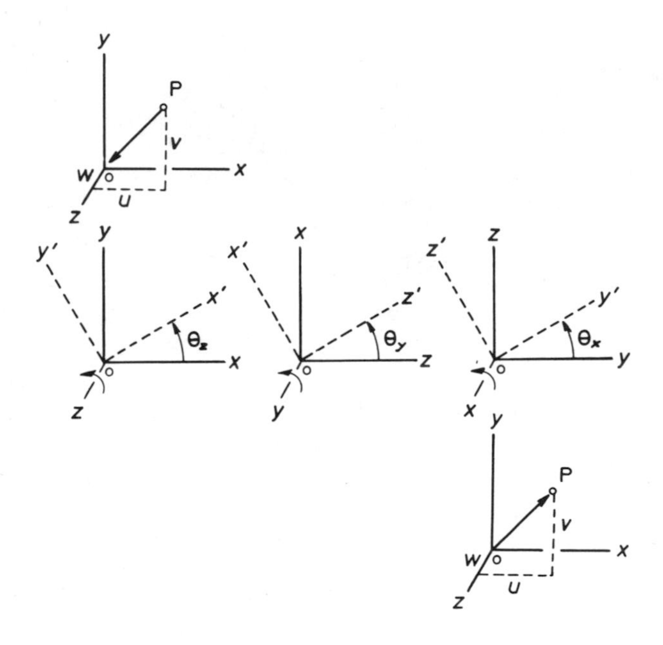

Figure 5.17

Reverting to homogeneous coordinates, to cover all possible contingencies

$$[X \quad Y \quad Z \quad H] = [x \quad y \quad z \quad 1] \begin{bmatrix} 1 & 0 & 0 & 0 \\ 0 & 1 & 0 & 0 \\ 0 & 0 & 1 & 0 \\ -u & -v & -w & 1 \end{bmatrix} [R] \begin{bmatrix} 1 & 0 & 0 & 0 \\ 0 & 1 & 0 & 0 \\ 0 & 0 & 1 & 0 \\ u & v & w & 1 \end{bmatrix}$$

where R is the appropriate rotary transform.

The 4 x 4 homogeneous coordinate matrices for rotation of an object about axes *ox, oy, oz* are as given below:

$$R_{ox} = \begin{bmatrix} 1 & 0 & 0 & 0 \\ 0 & \cos\theta_x & \sin\theta_x & 0 \\ 0 & -\sin\theta_z & \cos\theta_z & 0 \\ 0 & 0 & 0 & 1 \end{bmatrix}$$

$$R_{oy} = \begin{bmatrix} \cos\theta_y & 0 & -\sin\theta_y & 0 \\ 0 & 1 & 0 & 0 \\ \sin\theta_y & 0 & \cos\theta_y & 0 \\ 0 & 0 & 0 & 1 \end{bmatrix}$$

$$R_{oz} = \begin{bmatrix} \cos\theta_y & \sin\theta_y & 0 & 0 \\ -\sin\theta_y & \cos\theta_y & 0 & 0 \\ 0 & 0 & 1 & 0 \\ 0 & 0 & 0 & 1 \end{bmatrix}$$

Suffice to say that in all but simple cases (those in which integer values are used) the manual solution of such problems is lengthy, tedious and prone to error. The concept, however, is simple.

5.5.4 Concatenation

Concatenation is a term used in matrix algebra to represent the product of two separate matrices. The two separately imposed rotations considered in the previous example may be concatenated into a single transformation matrix, as follows.

The two matrices representing a rotation about the axis *oz* followed by a rotation about the axis *oy* are as previously stated.

$$\begin{bmatrix} \cos\theta_z & \sin\theta_z & 0 \\ -\sin\theta_z & \cos\theta_z & 0 \\ 0 & 0 & 1 \end{bmatrix} \begin{bmatrix} \cos\theta_y & 0 & -\sin\theta_y \\ 0 & 1 & 0 \\ \sin\theta_y & 0 & \cos\theta_y \end{bmatrix}$$

and their product or total effect may be obtained by applying the rules of matrix algebra.

$$\left[\begin{array}{cccc|ccc|cccc} \cos\theta_z & \cos\theta_y & 0 & 0 & 0 & \sin\theta_z & 0 & -\cos\theta_z & \sin\theta_y & 0 & 0 \\ -\sin\theta_z & \cos\theta_y & 0 & 0 & 0 & \cos\theta_z & 0 & \sin\theta_z & \sin\theta_y & 0 & 0 \\ 0 & & 0 & \sin\theta_y & 0 & 0 & 0 & 0 & & 0 & \cos\theta_y \end{array}\right]$$

yields

$$\left[\begin{array}{cc|c|cc} \cos\theta_z & \cos\theta_y & \sin\theta_z & -\cos\theta_z & \sin\theta_y \\ -\sin\theta_z & \cos\theta_y & \cos\theta_z & \sin\theta_z & \sin\theta_y \\ \multicolumn{2}{c|}{\sin\theta_y} & 0 & \multicolumn{2}{c}{\cos\theta_y} \end{array}\right]$$

This matrix represents the dual transformation. Rotation through angle θ_z about the axis *oz* followed by rotation through angle θ_y about the axis *oy*.

Similar dual transformations may, of course, be written for other rotation sequences.

Example

We may demonstrate the truth of the above concatenation by substituting numerical values from the previous example. If we substitute $\cos\theta_z = \sin\theta_y = \frac{4}{5} = 0.8$ and $\cos\theta_y = \sin\theta_z = \frac{3}{5} = 0.6$, we should obtain the same final coordinates if the above concatenation is true.

$$[200]\begin{bmatrix} \frac{12}{25} & \frac{15}{25} & -\frac{16}{25} \\ -\frac{9}{25} & \frac{20}{25} & \frac{12}{25} \\ \frac{20}{25} & 0 & \frac{15}{25} \end{bmatrix} = \frac{1}{25}[24, \quad 30, \quad -32]$$

whence the final coordinates, case (a) example, are:

$$\left[\frac{24}{25}, \quad \frac{30}{25}, \quad -\frac{32}{25}\right] = [0.96, \quad 1.20, \quad -1.28]$$

as previously.

5.6 ROTATION OF COORDINATE AXES

Throughout this chapter we have been principally concerned with the transformation of objects within a fixed coordinate system. The axes have been defined and arbitrarily fixed in space. Points, lines and planes have then been moved to new positions and orientations within this fixed frame of reference. The positions of points on the object have been defined in terms of their initial coordinates (x, y, z) and final (transformed) coordinates (x', y', z') or (x'', y'', z''). In all cases these coordinates have been referred to a single fixed frame of reference.

Many applications do, however, arise in which it is desirable to hold the object fixed and transform the initial coordinates to some more convenient orientation. In kinematic and structural analysis it is, for example, commonly necessary to align a set of axes so that one axis lies along the length of a slender member and since there are a number of different member directions in even a simple structure, it is necessary to perform such transformations many times over. In other situations we find it convenient to set up numerous local sets of axes, about the centre of gravity, or about the pivot centres of sub-assemblies, etc. In order to perform this function we clearly need to write transformation matrices that will translate and/or rotate sets of axes as opposed to translating or rotating the objects referred to these axes.

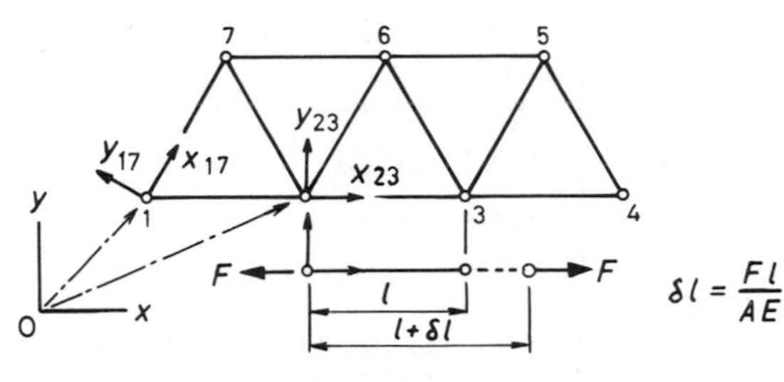

Figure 5.18

The case of a pure translation of axes has already been considered in Chapter 3 from which we recall that if a set of global axes xoy are transformed by a pure translation into a set of local axes $x'oy'$, as shown in Fig. 5.18. Then if the global coordinates of points P and Q are (x_1y_1) and (x_2y_2) respectively, then the local coordinates of P and Q relative to the local axes are (0, 0) and $(x_2 - x_1), (y_2 - y_1)$ respectively. Moving the origin of the global axes to the object point P, in a positive sense, clearly has the same effect on the coordinates of an object point Q, as would an equal movement of the object and the same is true of rotation, as shown in Fig. 5.19. Rotating an object in the negative direction is equivalent to rotating the axes in a positive direction, providing the magnitudes of the two rotations are the same.

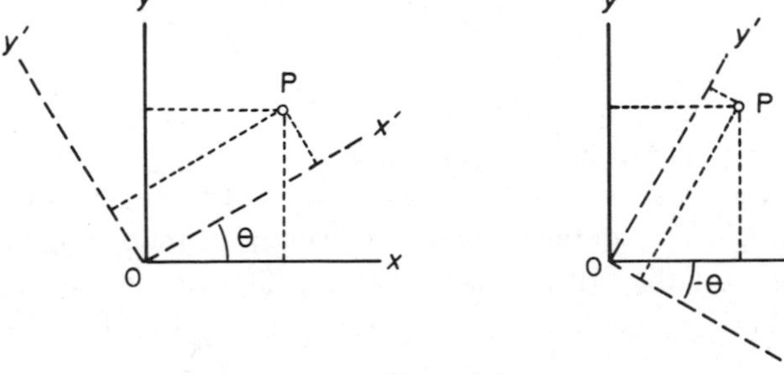

Figure 5.19

As we have seen the transformation

$$R_{oz} = \begin{bmatrix} \cos\theta_z & \sin\theta_z \\ -\sin\theta_z & \cos\theta_z \end{bmatrix}$$

rotates an object through a positive angle θ with respect to a fixed set of axes, and by contrast the transformation

$$R^T_{oz} = \begin{bmatrix} \cos\theta_z & -\sin\theta_z \\ \sin\theta_z & \cos\theta_z \end{bmatrix}$$

rotates a set of axes through a positive angle θ with respect to a fixed object orientation.

If ox transforms to ox', oy to oy', oz to oz' then

	x'	y'	z'
x	$\cos x\,x'$	$\cos y\,x'$	$\cos z\,x'$
y	$\cos x\,y'$	$\cos y\,y'$	$\cos z\,y'$
z	$\cos x\,z'$	$\cos y\,z'$	$\cos z\,z'$

where $\cos x\,x'$ is the direction cosine of the new axis ox' relative to ox and so on. For example, a positive rotation of 30° puts ox' into the first quadrant and oy' into the second quadrant. The x component of ox' is cos 30, so that $\cos xx' = \cos 30 = +\sqrt{3}/2$. The y component of x' is sin 30 but since the y component of x' is $\cos yx'$, $\cos yx' = +\frac{1}{2}$. The x component of oy' is cos 120, so that $\cos xy' = -\frac{1}{2}$, and the y component of oy' is sin 120 and sin 120 equals $\cos yy'$, hence $\cos yy' = \sin 120 = +\sqrt{3}/2$.

5.7 AREA SCALING

As illustrated in Fig. 5.20, a matrix of the type $\left[\begin{smallmatrix} a & b \\ d & e \end{smallmatrix}\right]$ may be used to transform a unit square into a sheared parallelogram and it is easy to verify that the area of the parallelogram is given by

$$A = \left(\frac{de}{2} + \frac{(b+2e)}{2}a - \frac{ab}{2} - \frac{2b+e}{2}\right)d$$

whence area of parallelogram equals $(ae - bd)$. That is, the area of the parallelogram is equal to the value of the transformation matrix evaluated as a determinant.

Moreover, it can be shown that the area of any transformed figure is equal to the area of the original figure times the determinant of its transformation matrix. If A_0 is the area of the original figure and A_T the area for the figure after transformation

$$A_T = A_0\,(ae - bd)$$

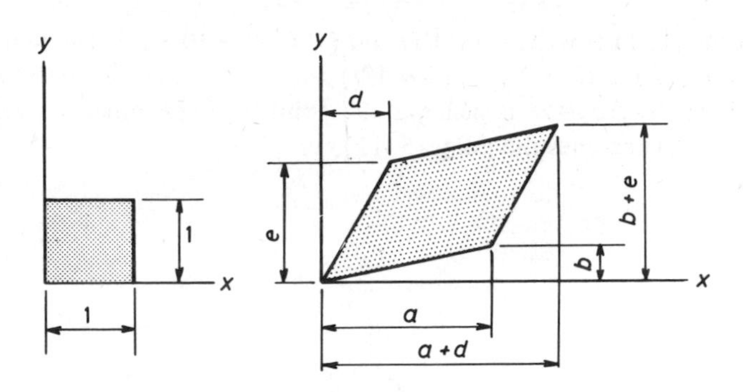

Figure 5.20

Example

The area of a circle is πr^2. Use the method of article 5.7, to show that the area of an ellipse with major axis a and minor axis b is πab.

Solution

The coordinates of a circle are $x_c = r \cos \theta$, $y_c = r \sin \theta$, and by applying the scaling transformation

$$\begin{bmatrix} \frac{a}{r} & 0 \\ 0 & \frac{b}{r} \end{bmatrix}$$

we obtain $x_T = a \cos \theta$, $y_T = b \sin \theta$, the coordinates of an ellipse. Evaluating the scaling transformation as a determinant yields $\Delta = ab/r^2$ and hence the area of the ellipse is $A_c\Delta = \pi r^2 \cdot ab/r^2 = \pi ab$.

5.8 CONFORMAL TRANSFORMATIONS

The technique of conformal mapping enables the shapes of geometrical figures to be greatly changed whilst retaining certain important features. We shall not consider the method in detail but one example and one practical application will serve to introduce the topic.

Suppose the coordinates of a square are (1, 1), (3, 1), (3, 3), (3, 1), with mid-points (2, 1), (3, 2), (2, 3), (1, 2) and centre point (2, 2), see Fig. 5.00. Then we may represent any point by an expression of the form $z = x + yi$, where $i = \sqrt{-1}$. The point (1, 1) is $(1 + i)$, the point (3, 2) is $(3 + 2i)$ and so on. Now suppose we arbitrarily square these later terms and see what happens.

The point (1, 1) is written $(1 + i)$ and $(1 + i)^2 = (0 + 2i)$. The point (3, 2) is written $(3 + 2i)$ and $(3 + 2i)^2 = (5 + 12i)$ and we say that the conformed transform produced by $f(z) = z^2$ is such that the point (1, 1) becomes the point (0, 2). The point (3, 2) becomes the point (5, 12) etc.

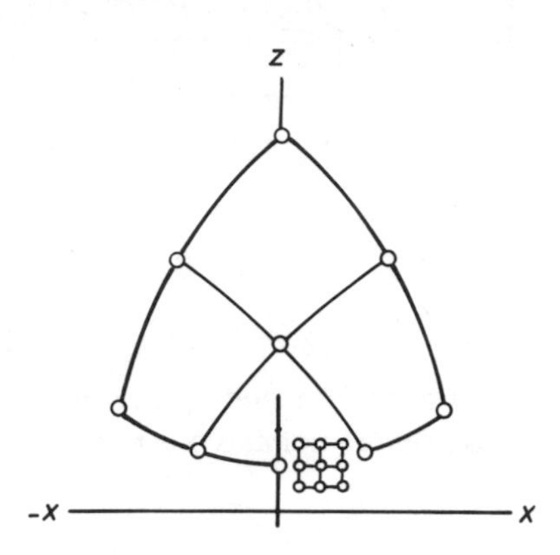

Figure 5.21

The full list of coordinates is given in the table and a plot of the transformation appears as in Fig. 5.21, from which the preservation of all right angles can be clearly şeen.

(1, 3)	(2, 3)	(3, 3)	→	(–8, 6)	(–5, 12)	(0, 18)
(1, 2)	(2, 2)	(3, 2)	→	(–3, 4)	(0, 8)	(5, 12)
(1, 1)	(2, 1)	(2, 3)	→	(0, 2)	(3, 4)	(8, 6)

An important practical illustration of the use of conformed mapping is provided by Joukowski's aerofoil transformation. The hydrodynamic/aerodynamic flow lines representative of a fluid stream passing a circular cylindrical object are broadly as shown in Fig. 5.22, and these may be transformed to an aerofoil shape by means of the function $\xi = z + a^2/z$. When

$$\phi + i\psi = f(x + iy)$$

$$\frac{\partial\phi}{\partial x} = \frac{\partial\psi}{\partial g} \quad \text{and} \quad \frac{\partial\phi}{\partial y} = -\frac{\partial\psi}{\partial x}$$

$$z = x + iy \quad \text{and} \quad \omega = \phi + i\psi$$

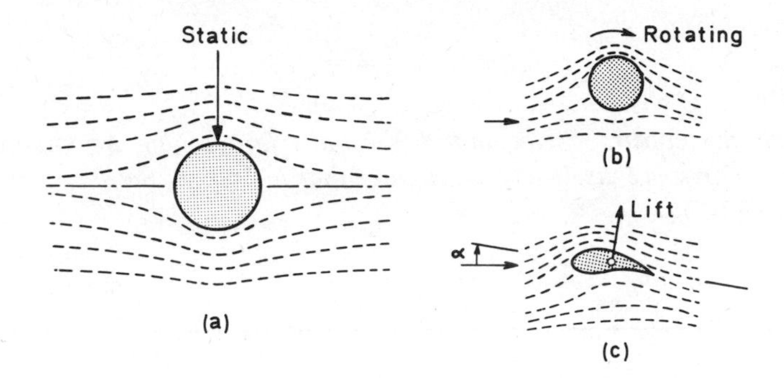
Static
Rotating
(b)
Lift
α
(a)
(c)

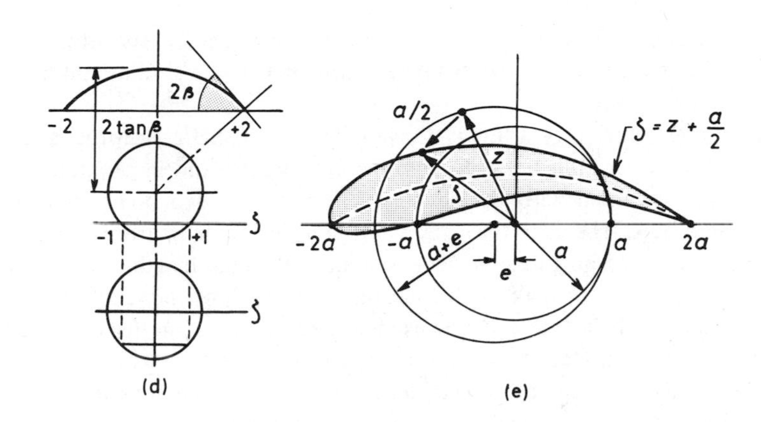
2β
-2
2 tan β
+2
-1
+1
ζ
ζ
(d)
a/2
ζ = z + a/2
z
ζ
-2a
-a
a+e
e
a
a
2a
(e)

Figure 5.22

6 Polyhedral structures

Of all the constraints on nature, the most far-reaching are imposed by space. For space itself has a structure that influences the shape of every existing thing.

Peter S. Stevens

6.1 INTRODUCTION

A geometrical object point is created by every crossing and every termination of one or more lines. Geometrical points are thus said to be sizeless and hence are of dimensionality zero.

The simplest notion of a line is derived from our visual/conceptual awareness of the spatial separation between two distinct points: and in geometry we are often more concerned with the shortest line between two points than with a myriad of others. The fact that we envisage a line to grow from one point to another, leads us to postulate an idealised line which has length but no width or thickness and hence to make a distinction between dimensionality and extent. Extent is simply the length of a line as measured in inches, metres or some other appropriate unit. The fact that lines are characterised by growth along one axis only (be the axis straight or curved) leads us to say that all lines are one dimensional. That is, the dimensionality of a line is always one.

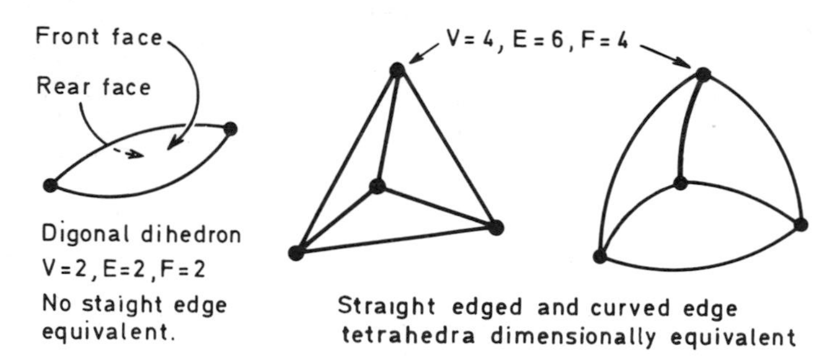

Figure 6.1

By similar reasoning we say that a plane, as also a curved surface, is of dimensionality two: and a cell, that is an enclosure of space is of dimensionality three. Hence, in accordance with this definition the space on the surface of a sphere is two-dimensional whilst the free space within and without the sphere or indeed within and without any form of enclosure is three-dimensional.

The properties of dimensionality are thus independent of dimensional extent— for what is true for the large is true for the small and, what is more, it matters not whether lines and/or edges are straight or curved, nor indeed whether faces are plane or curved, the properties of dimensionality hold regardless. See Fig. 6.1.

6.1.1 Close packed circles in a plane

In order to develop a feel for the way geometrical elements merge together we begin by considering orthogonal and hexagonal close packing of equi-circular pieces, as shown in Fig. 6.2. In the arrangement shown at (a) the circle centres lie on an orthogonal mesh and we see that each circular piece not on a boundary is contained by four mutually perpendicular point contacts which offer no resistance to shearing. A sheared orthogonal grid or lattice is shown at (b) and we observe that as the shear displacement increases, so the original orthogonal lay-up tends to collapse. When adjacent rows have been displaced a distance of half-a-lattice (mesh) spacing, the rows are free to form a more tightly packed nesting as at (c). The close packed hexagonal nesting shown at (c) is clearly a more dense and structurally more stable arrangement than the orthogonal nesting at (a).

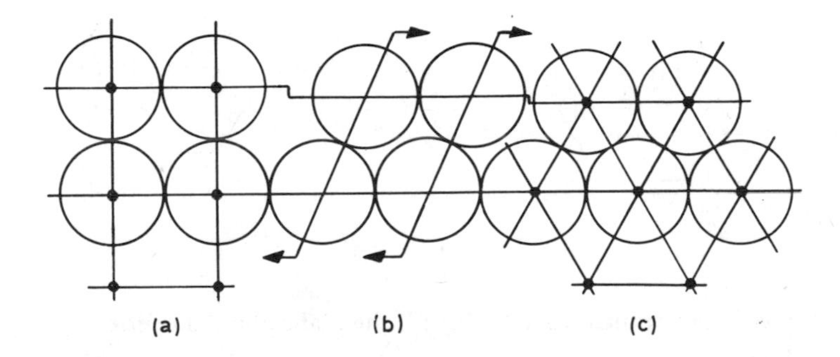

Figure 6.2

A useful measure of solidarity is the packing density as given by the ratio between the area of a unit circle to that of the unit enclosure. The density of an orthogonal pack of circular pieces is the area of a unit circle divided by the area of a unit square. The density of orthogonal nesting is thus 0.7854. The density of a hexagonal pack of circular pieces is the area of a unit circle divided by the area of a unit hexagon. The density of hexagonal nesting is thus 0.9070.

The density of packing obtained from placing circular pieces within a plane closed figure with p equi length sides, called a p-gon; is measured in a similar way. If the side of the p-gon is $2l$, the radius of the inscribed circle is:

$$r = l \cot \frac{\pi}{p}$$

The area of the enclosing p-gon is plr and hence the packing density ρ is area of circle divided by area of p-gon, that is:

$$\rho = \frac{\pi}{plr} = \frac{\pi}{p} \cot \frac{\pi}{p}$$

We find that as p increases ρ increases, but note that p cannot exceed 6. The hexagonal arrangement of circular pieces is thus the most densely packed assembly that may be obtained. (N.B. It is impossible to fill a plane with pentagons, octagons, or indeed with any single higher polygon).

6.1.2 Regular tessellations–mosaics

A polygon with p vertices and p edges is called a p-gon. Every p-gon is divisible into p triangles nested about its centre. A polygon in which all edge lengths and all internal angles are equal is called a regular polygon. If a regular polygon is partitioned into triangles, as shown in Fig. 6.3, the centre angle of each triangle equals $2\pi/p$ and a double base angle equals $\pi/2(1 - 2/p)$.

Figure 6.3

If q, p-gons are brought together to fill the plane about a centre

$$q\pi\left(1 - \frac{2}{p}\right) = 2\pi$$

$$q\left(1 - \frac{2}{p}\right) = 2$$

$$pq - 2q - 2p = 0$$

By adding 4 to both sides, we may factorise the left-hand side. Hence

$$(p - 2)(q - 2) = 4$$

We observe that the minimum polygon is a triangle for which $p = 3$, and the only possible solutions are as shown in the table.

p	q
3	6
4	4
5	fractional
6	3

Equilateral triangles, squares and hexagons are thus the only regular tessellations which fill a plane. The absence of pentagons is to be noted.

Equilateral and square lattices of the type shown in Fig. 6.4(a) and (b) form the basis of many flat plate structural grids, whilst honeycomb nests of the type shown at (c) are widely used as structural sandwich filling.

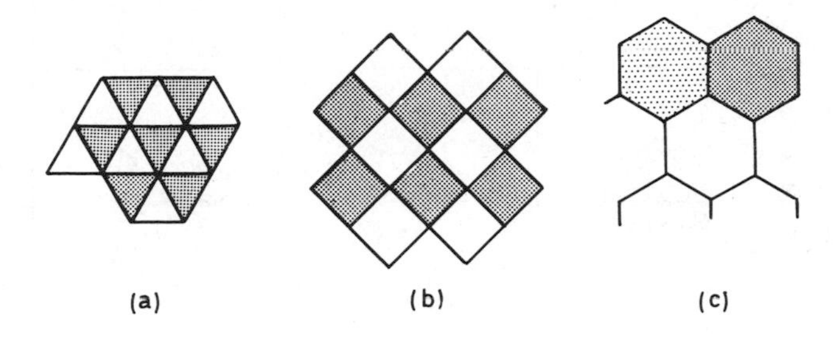

Figure 6.4

6.1.3 Semi-regular tessellations–mosaics

When p-gonal tiles with different numbers of edges are mixed in such a way that the numbers of edges and faces at all nodal junctions is the same, the tessellation or mosaic is said to be of a semi-regular form. There are but eight ways in which equilateral triangles, squares, hexagons, octagons and dodecahedrons may be assembled to fill a plane, but pentagons are excluded.

All semi regular tessellations are characterised by three, four or five way junctions.

There are three, five way, mosaics.

(i) Triangles and hexagons produced by infilling a basic triangular grid.

(ii) Triangles and squares nested to form an interlocking pattern of elliptical decagons.

(iii) Triangles and squares arranged alternately in rows of squares and nested triangles.

There are only two, four way, mosaics

(iv) Triangles, squares and hexagons arranged so that each hexagon is surrounded by an alternation of triangles and squares, to form a pattern of interlocking dodeca polygons.

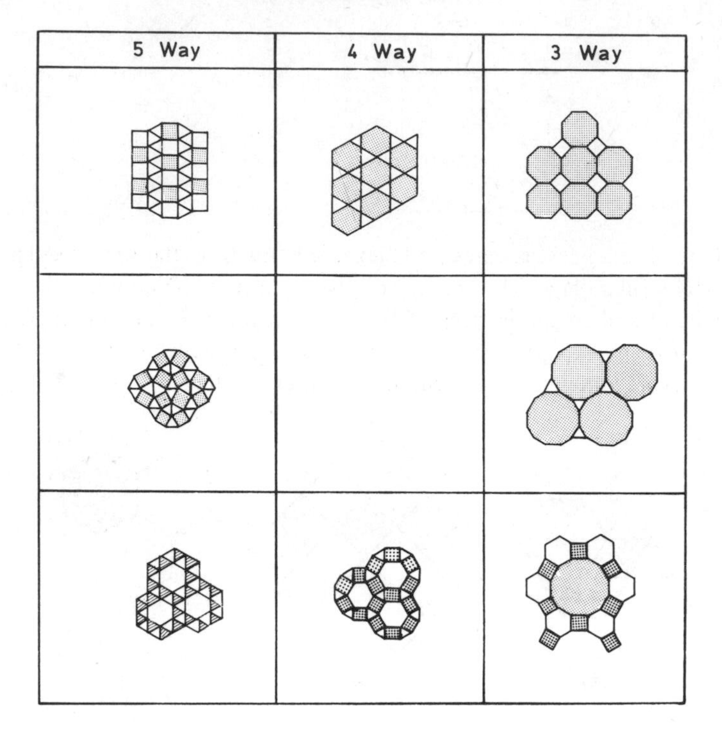

Figure 6.5

(v) Triangles and hexagons a mosaic produced by in-filling the spaces between rows of staggered hexagons, with triangles.

There are three, three way, mosaics.

(vi) Squares and hexagons arranged alternately around twelve sided duogons
(vii) Triangles and duogons arranged in close packed formation.
(viii) Squares and octagons arranged in a close orthogonal pattern.

6.2 DIRICHLET DOMAINS

A Dirichlet domain is a set of points which are closer to a given point than to any other points in the set. In plane geometry, a Dirichlet domain is an area. In three-dimension geometry, a Dirichlet domain is a volume.

The Dirichlet domain of the black dot, Fig. 6.6, is indicated by the dotted line. Any point within the bound defined by the dotted line is clearly closer to the black dot than it is to any one of the eight surrounding clear dots. The area or region within the dotted bound is thus the Dirichlet domain of the black dot.

In certain special cases a dual structure may be superimposed on an existing

structure, for example. The triangular grid, Fig. 6.6 has a hexagonal grid or lattice as its dual and conversely, Fig. 6.6.

Given the original grid, we merely have to bisect each line, produce the bisectors and join them. And when this is done, we see that any point within the bound is closer to its associated centre than it is to any near neighbour. Thus, any point within the hexagonal domain H, is closer to the black dot than to any surrounding centre. Similarly, any point within the triangular domain T is closer to its black dot than to any surrounding centre.

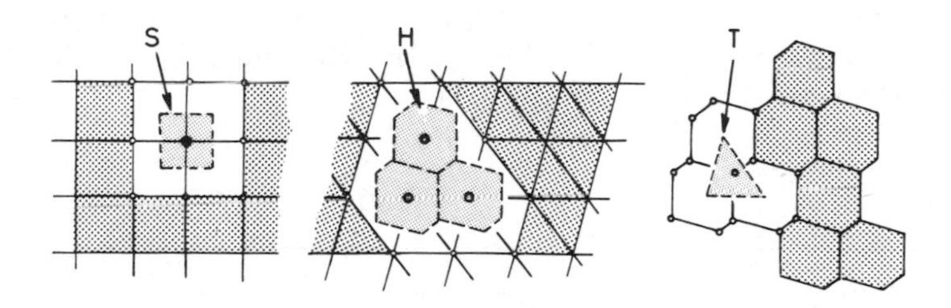

Figure 6.6

The following rules are of importance:

1. Any quadrilateral grid constitutes a Dirichlet domain, providing its vertices are cyclic.
2. All Dirichlet domains are plane fillers but not all plane fillers are Dirichlet domains and we observe that the quadrilateral does not have to be convex.
3. All triangles in which no angle exceeds 90° are Dirichlet domains. Triangles which contain an angle greater than 90° are not Dirichlet domains for clearly the perpendicular bisectors of its sides would not intersect within the triangle.
4. The centres of one domain contitute the vertices of the other.
5. A hexagonal grid may only constitute a Dirichlet domain if and only if its opposite sides are parallel.
6. The number of edges in a planar Dirichlet domain must be either 3, 4 or 6.

6.3 CONVEX POLYHEDRA

We have seen that six equilateral triangles fill a plane around any one point – five, four or three such triangles leave the plane only partially filled. Five equilateral triangles arranged in contact about a point leave a 60° interval unfilled, but if all radial lines are creased and the two free radial edges brought together the centre point is forced out of plane and a five sided pyramid is formed. Four and three equilateral triangles may be treated in a similar fashion to form a four sided and a three sided pyramid respectively. Three squares and three pentagons may also be folded to form a closed three-dimensional apex. It is easy to see that three equilateral (delta) triangles formed to an apex produce a three sided pyramid with an identical equilateral delta base and hence there are three vertices, six edges, and four identical delta faces in all such solid figures. The solid so formed is known as a regular tetrahedron. Similarly it is clear that two square based delta pyramids can be mounted base to base – to form a six vertex, twelve edge, eight faced 'solid' delta figure, known as an octahedron, and a cube consisting of eight vertices, twelve edges, and six faces may be similarly formed. The tetrahedron, octahedron and cube were known to the ancient Egyptians and the icosahedron and dodecahedron (to be discussed in due course) were additionally known to Plato (427–347 BC). The five regular polyhedra, known as the Platonic solids are shown in Fig. 6.7.

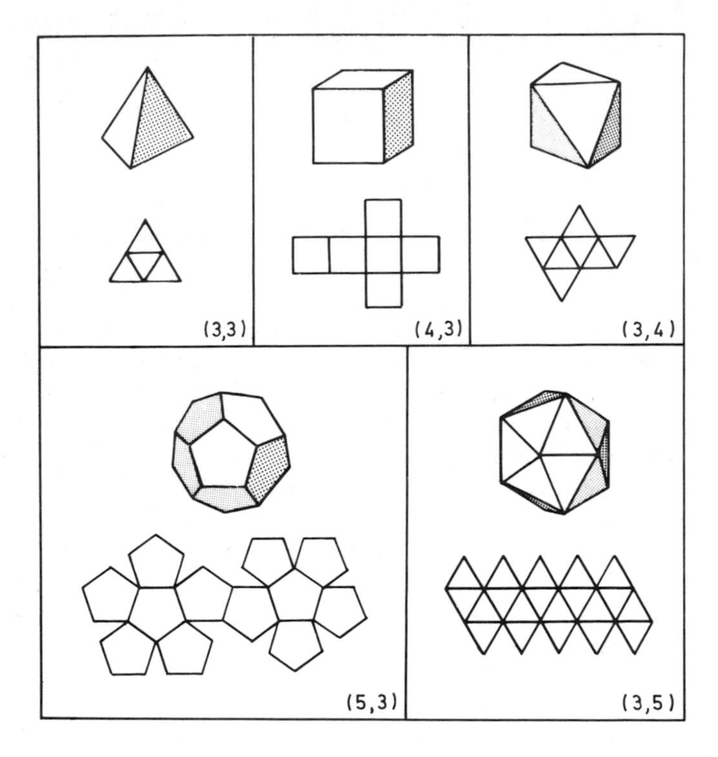

Figure 6.7

A further thirteen semi-regular polyhedra, comprising a mixture of face shapes, were known to Archimedes (287–212 BC) and are shown in Fig. 6.8.

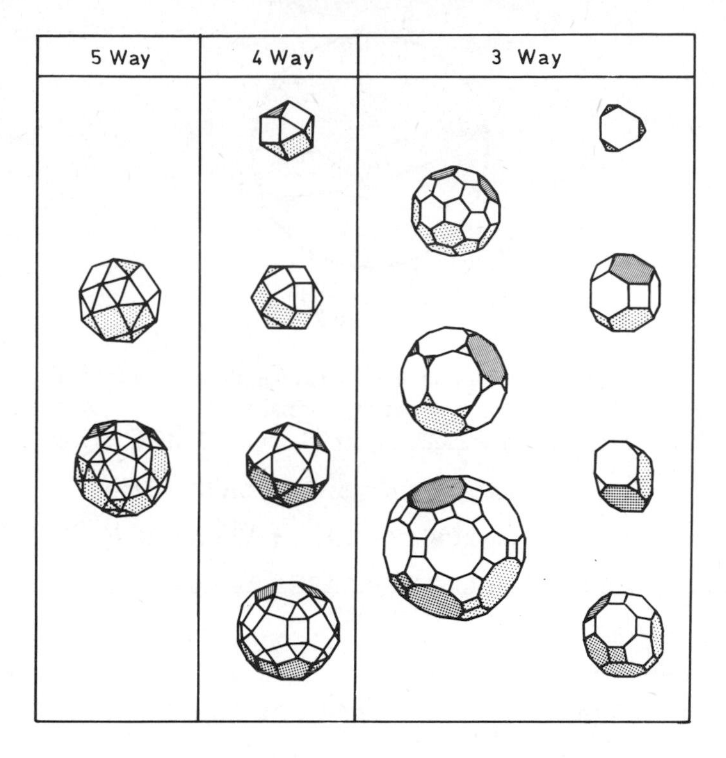

Figure 6.8

6.3.1 The Euler topological equation

It is a matter of observation that the vertices of the regular solids lie on a common sphere, known as the circumsphere and this is also true of the semi regular solids.

If we prove the Euler equation for the case of spherical contours, we automatically obtain proof for the case of straight edges and flat faces. In order to achieve such a proof we need to calculate the surface area of a spherical triangle and this we do as follows.

Two great circles intersect along a diameter and divide the surface of a sphere into four lunes. The surface area contained by a lune is proportional to the angle between its two great circles and since there are 2π radians in a complete circle, the surface area of a lune is in the ratio of $\theta/2\pi$ to the surface area of a complete sphere. Hence

$$\text{Area of lune} = \frac{\theta}{2\pi} \cdot 4\pi r^2 = 2r^2\theta$$

(N.B. Readers unacquainted with the term great circle are referred to article 7.4.1.)

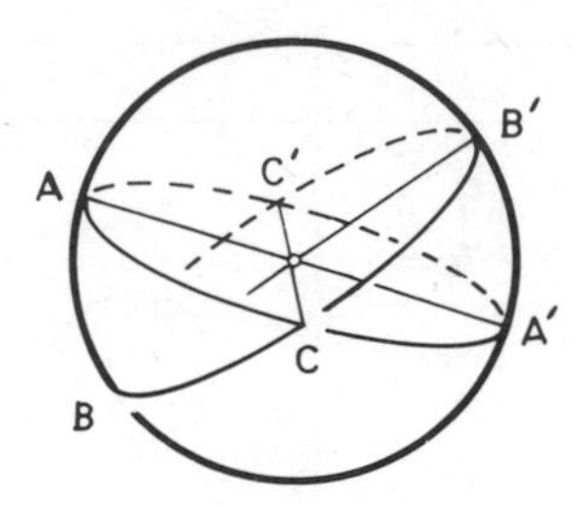

Figure 6.9

If the two great circles produce angles θ_A, θ_B, θ_C, we see from Fig. 6.9, that when BB′, CC′ are viewed in a direction normal to AA′, θ_A = A (the so-called spherical angle) and the plane angles θ_B and θ_C equal B and C respectively.

$$\text{Area of lune ABA'CA} = \Delta\text{ABC} + \Delta\text{A'BC} = 2r^2\,\text{A}$$

$$\text{Area of lune BCB'AB} = \Delta\text{ABC} + \Delta\text{B'CA} = 2r^2\,\text{B}$$

$$\text{Area of lune CAC'BC} = \Delta\text{ABC} + \Delta\text{C'AB} = 2r^2\,\text{C}$$

where Δ means area of

Adding these three equations gives

$$2\Delta\text{ABC} + (\Delta\text{ABC} + \Delta\text{A'BC} + \Delta\text{B'CA} + \Delta\text{C'AB}) = 2r^2\ (\text{A} + \text{B} + \text{C})$$

and since ΔABC = ΔA′BC and (ΔABC + ΔA′BC + ΔB′CA + ΔC′AB) is a half sphere.

$$2\Delta\text{ABC} + 2\pi r^2 = 2r^2\ (\text{A} + \text{B} + \text{C})$$

$$\Delta\text{ABC} = r^2\ (\text{A} + \text{B} + \text{C} - \pi)$$

The quantity in brackets is known as the spherical excess.

The spherical excess of a spherical triangle is thus equal to the sum of its spherical angles minus π, and since a spherical polyhedral face of p edges can be divided into p spherical triangles the area of a polyhedral face with p edges is $r^2\ (s - (p - 2)\pi)$ where s is the sum of the spherical angles (A + B + C + D. . .) and p the number of polyhedral edges per face.

Now if F such polygons partition the surface of a sphere of unit radius, the surface area of the polygons must equal the surface area of the sphere, which is $4\pi r^2 = 4\pi$ in the case of a unit sphere. Thus summing angles, edges and faces

$$4\pi = \Sigma s - \pi\Sigma p + 2\pi \text{F}$$

But Σs is the sum of all the spherical angles of all the faces and is equal to 2π times the numbers of solid angles formed at the centre, that is $\Sigma s = 2\pi$V where V is the sum total of vertices. Σp is simply the sum total of edges. Hence Σp = E. (N.B. It is to be observed that the above derivation is true, regardless of the shape of polygon faces, and providing the number of vertices, edges and faces are correctly counted the equation holds for both the regular and semi-regular solids, regardless of whether these edges and faces are curved or flat.

$$4\pi = 2\pi V - 2\pi E + 2\pi F$$

whence

$$2 = V - E + F \tag{6.1}$$

and this is the Euler topological equation.

6.3.2 The five regular polyhedra

A proof that there can only be five regular convex polyhedra will now be given.

Let p equal the number of sides in each face of a regular polyhedron, q the number of plane angles in each solid angle; then

$$pF = 2E = qV$$

where

$$V - E + F = 2$$

From these two equations we see that:

$$V = \frac{4p}{2(p+q) - pq}, \quad E = \frac{2pq}{2(p+q) - pq}, \quad F = \frac{4q}{2(p+q) - pq}$$

These three expressions must be positive and hence $2(p + q)$ must be greater than pq. Therefore

$$\frac{1}{p} + \frac{1}{q} > \frac{1}{2}$$

and since F cannot be less than 3, $1/q$ cannot be greater than $\frac{1}{3}$ and therefore $1/p$ must be greater than $\frac{1}{6}$ and as n must be an integer not less than 3, q can only equal 3, 4, or 5.

The names given to the regular polyhedra are derived from a count of the number of faces F and these are as given in the table.

p	q	V	E	F	Name
3	3	4	6	4	Tetrahedron
4	3	8	12	6	Hexahedron (cube)
3	4	6	12	8	Octahedron
5	3	20	30	12	Dodecahedron
3	5	12	30	20	Icosahedron

where p is the number of edges per face and q the number of faces per vertex.

6.3.3 Schelegel diagrams

When considering simple polyhedra, the cube for example, we experience no difficulty in counting the number of vertices, edges and faces, but this trite ease of counting is lost when we come to consider more complex figures, and it is here that Schelegel diagrams are useful. The benefit of looking at polyhedra from the most advantageous view-point is apparent from the two different presentations of the cuboctahedron, Fig. 6.10, but in complex cases a Schelegel diagram is much more helpful.

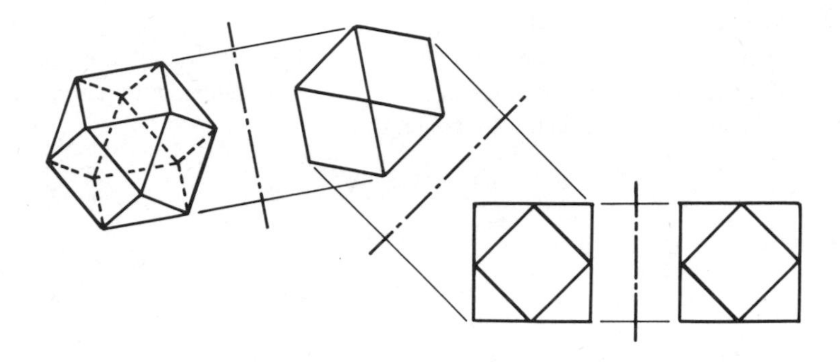

Figure 6.10

If we draw the normal front view of an ordinary cube in orthographic projection, we see but a single square boundary face, as shown in Fig. 6.11. If, however, we draw a full frontal view of an open cubic framework and also take up a perspective view-point, we are able to represent all vertices, all edges and all faces of the cube on a single diagram as shown in Fig. 6.11(b).

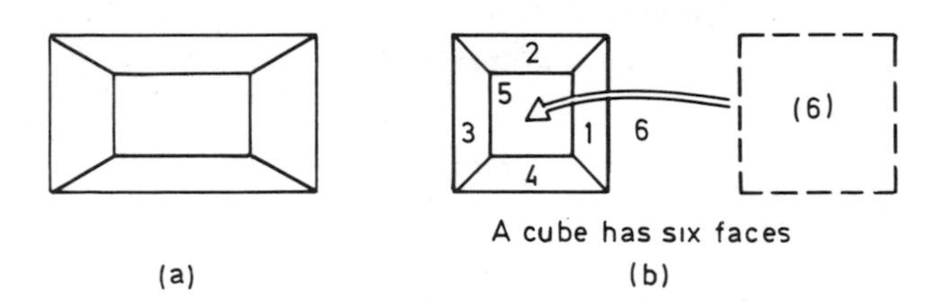

Figure 6.11

From a topological point of view, the shape and size of faces is unimportant and we see that the three-dimensionalities of the Schelegel diagram, Fig. 6.11(b) do in fact accord with the Euler topological equation. We see that there are eight vertices, twelve edges and 'five' faces on the first count and this is one face short.

If, however, we agree to count the space outside the bound (as the missing face) then the Euler equation is satisfied and if we adopt this convention as a general rule it is found that Euler's equation is satisfied regardless of topological complexity. A logical explanation of this rule is to regard Fig. 6.11(b) as the interior of a room in which the frontal wall, or face, is a transparent picture window, whence the frontal face and the bound line, Fig. 6.11(a), are coincident.

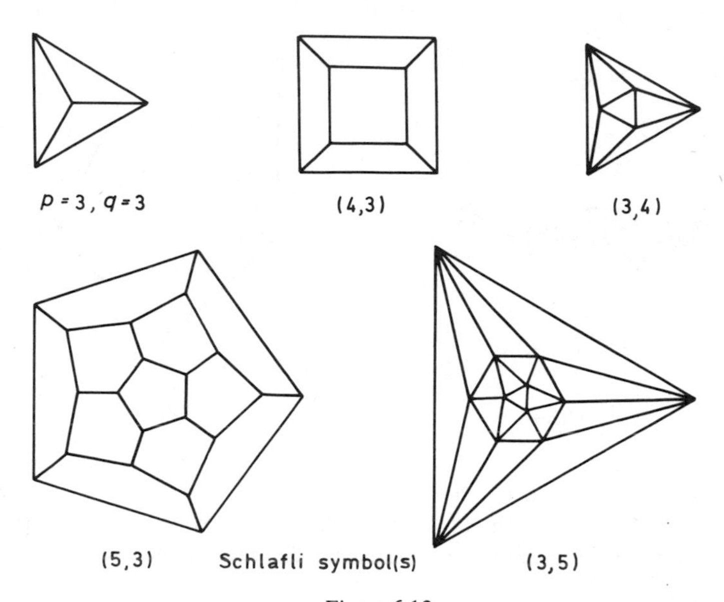

Figure 6.12

6.3.4 The three polyhedral spheres

The vertices of both regular and semi-regular polyhedra lie on a single spherical shell – known as the circumsphere. The centre of polyhedral edges and faces also lie on concentric spheres; these are the inter and in-spheres respectively.

1. The vertices of all regular and semi-regular polyhedra lie on a single circumsphere.
2. The mid-point of every edge of a regular polyhedron lies on a single mid-sphere. The mid-point of edges of a semi-regular polyhedron lies on two or more mid-spheres or interspheres.
3. The mid-point of every face of a regular polyhedron lies on a single inner or in-sphere. The mid-point of every face of a semi-regular polyhedron lie on two or more in-spheres.

We need to know the relationship between radii of the circumsphere, mid-sphere(s) and in-sphere(s) for each polyhedron, each of which is geometrically defined by Fig. 6.13, in which O is the centre of the polyhedron.

If we approach the problem of calculating the circumsphere, mid-sphere and in-sphere radii, using a plane geometry approach, we soon discover the solution is not as obvious as it may seem. All available data is contained in the irregular tetrahedral half wedge, Fig. 6.13, in which AEC is a right-angled triangular element cut from a typical face such that $AE = \frac{l}{2}$ (one half the edge length) and O is the centre of the polyhedron.

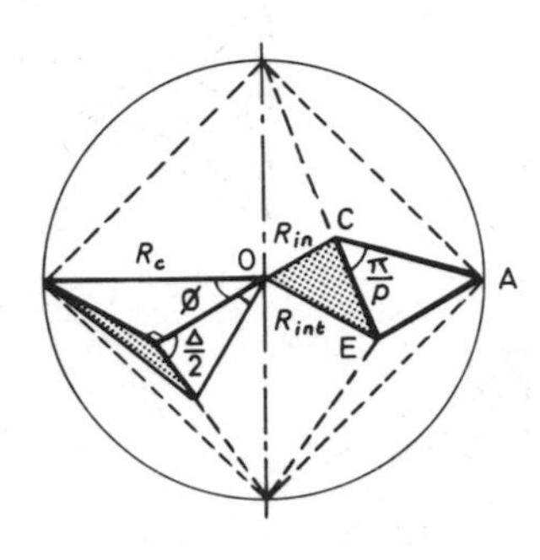

Figure 6.13

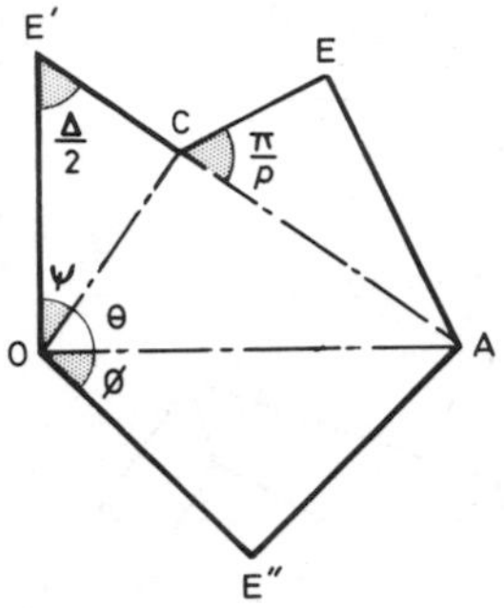

Figure 6.14

It is clear from Fig. 6.13 that AEC, EOC, COA and AEO are right angles, and the geometrical problem of drawing the development, Fig. 6.14 (without further data) will become apparent to anyone who tries to produce it.

It is apparent that

$$\frac{R_{in}}{R_{int}} = \frac{R_{in}}{R_c} \cdot \frac{R_c}{R_{int}} = \cos\psi \tag{6.2}$$

whence

$$\cos\psi = \frac{\cos\theta}{\cos\phi} = \sin\frac{\Delta}{2} \tag{6.3}$$

where $\Delta/2$ is half the dihedral angle.

Now in order to deduce the equations for R_c, R_{int}, R_{in}, in terms of l, p and q, we introduce a dummy function. We know that

$$\sin^2\frac{\pi}{p} + \cos^2\frac{\pi}{p} = \sin^2\frac{\pi}{q} + \cos^2\frac{\pi}{q} = 1$$

and write

$$\sin^2\frac{\pi}{p} - \cos^2\frac{\pi}{q} = \sin^2\frac{\pi}{q} - \cos^2\frac{\pi}{p} = k^2 \text{ say}$$

If we let $\sin \phi = k \operatorname{cosec} \pi/q$ we have a right angled triangle which is geometrically similar to the right angled triangle relating R_c, R_{int}, $\frac{l}{2}$.

By comparing the two similar triangles, Fig. 6.14, we express R_c and R_{int} in terms of k as follows

$$R_c = \frac{1}{2k} \sin \frac{\pi}{q}, \quad R_{int} = \frac{1}{2k} \cos \frac{\pi}{p}$$

whence

$$\frac{R_{int}}{R_c} = \frac{\cos \pi/p}{\sin \pi/q} = \cos \phi \tag{6.4}$$

but since $k = (\sin^2 \pi/q - \cos^2 \pi/p)^{1/2}$

$$R_c = \frac{l}{2} \frac{\sin \pi/q}{(\sin^2 \pi/q - \cos^2 \pi/p)^{1/2}} \tag{6.5}$$

and

$$R_{int} = \frac{l}{2} \frac{\cos \pi/p}{(\sin^2 \pi/q - \cos^2 \pi/p)^{1/2}} \tag{6.6}$$

and from triangle COE, Fig. 6.14, we have

$$R_{in}^2 = R_{int}^2 - \left(\frac{l}{2}\right)^2 \cot^2 \frac{\pi}{p}$$

Substituting for R_{int}^2, rearranging and collecting terms we obtain

$$R_{in} = \frac{l}{2} \frac{\cot \pi/p \cos \pi/q}{(\sin^2 \pi/q - \cos^2 \pi/p)^{1/2}} \tag{6.7}$$

Now from eq 6.5

$$\frac{R_c}{\frac{1}{2}l \sin \pi/q} = (\sin^2 \pi/q - \cos^2 \pi/p)^{1/2} = \frac{R_{in}}{\frac{1}{2}l \cot \pi/p \cos \pi/q}$$

$$R_c = R_{in} \tan \frac{\pi}{p} \cdot \tan \frac{\pi}{q} \tag{6.8}$$

6.3.5 Dihedral angle between polyhedral faces

By substituting $\cot \pi/p = \cos \pi/p / \sin \pi/p$ into equation 6.7 we conclude that

$$R_{in} = \frac{l}{2} \frac{\cos \pi/p}{(\sin^2 \pi/q - \cos^2 \pi/p)^{1/2}} \cdot \frac{\cos \pi/q}{\sin \pi/p}$$

and hence

$$\frac{R_{in}}{R_{int}} = \frac{\cos \pi/q}{\sin \pi/p} = \cos \psi = \sin \frac{\Delta}{2} \tag{6.9}$$

We may now obtain $\cos \theta$ in terms of p and q by combining equations 6.3, 6.4, and 6.9

$$\cos \theta = \cos \phi \cos \psi$$

where

$$\cos \phi = \frac{\cos \pi/p}{\sin \pi/q} \quad \text{and} \cos \psi = \frac{\cos \pi/q}{\sin \pi/p} \tag{6.10}$$

whence

$$\cos \theta = \frac{\cos \pi/p}{\sin \pi/q} \cdot \frac{\cos \pi/q}{\sin \pi/p} = \cot \pi/p \tag{6.11}$$

Given the angles Δ, θ, ϕ, ψ, the direction cosines of all vertices, edges and faces on both regular and semi-regular polyhedra may be found.

6.4 DEFINITION OF DIMENSIONALITY

There are four orders of dimensionality in the work we are about to undertake, and the following defintions are to be understood.

(i) A vertex is a point-position which has neither size nor dimension.
(ii) An edge is a line which may be curved or straight. A line is not an edge unless both its ends terminate in a vertex. To every edge there are always two vertices.
(iii) A face is a closed two-dimensional surface which may be concave, convex, or plane. Every polyhedral face is bounded by edges and has a vertex at every corner. All faces have two sides (an obverse and reverse).
(iv) A cell is any three-dimensional enclosure which divides space into two completely separate regions (all cells have an inside and an outside).
We note the following necessities:
(v) If there are to be no unconnected edges, then every edge must be associated with two vertices.
(vi) Every face has an equal number of vertices and edges.
(vii) In a two-dimensional model every edge divides two faces.
(viii) In a three-dimensional model any number of faces may converge on an edge.
(ix) The number of cells adjacent to an edge is always equal to the number of faces joined to that edge.

(x) If there are no internal partitions, a polyhedron is singly connected, but if, as in a soap froth a number of cells are joined the complex is multi-connected.

6.5 THE VALENCY CONCEPT

Every vertex of a polyhedron is common to a certain number of edges, faces and cells. Every edge is common to a certain number of vertices, faces and cells. Every face is common to a certain number of vertices, edges and cells. Every cell is common to a certain number of vertices, edges and faces.

These interactions give rise to the following combinations:

(i)	The number of vertices per edge is	2
	The number of vertices per face is	p
	The number of vertices per cell is	v
(ii)	The number of edges per vertex is	m
	The number of edges per face is	p
	The number of edges per cell is	e
(iii)	The number of faces per vertex is	q
	The number of faces per edge is	n
	The number of faces per cell is	f
(iv)	The number of cells per vertex is	r
	The number of cells per edge is	n
	The number of cells per face is	2

We note that the number of vertices per edge is always 2.
We note that the number of cells per face is always 2.
We note that the number of vertices per face is always equal to the number of edges per face.
We note that the number of faces per edge is always equal to the number of cells per edge.

For convenience of reference these intereactions are displayed as a matrix table.

		V	E	F	C
→	V	•	2	p	v
→	E	m	•	p	e
→	F	q	n	•	f
→	C	r	n	2	•

If we require the valency symbol for the number of edges per vertex – we locate row E and move into the matrix as far as column V. The required valency symbol is m. If we require the valency symbol for the number of faces per cell – we enter the matrix at row F and move into the matrix as far as column C. The required valency symbol is f.

6.5.1 The valency concept – regular polyhedra

We now demonstrate the meaning of the concept valency as applied to a simple isolated cube.

We know that a cube has V = 8, E = 12, F = 6. We know that a cube satisfies the Euler topological equation

$$V - E + F - C = 0$$

That is, we adopt the convention that a single isolated cube divides space into two separate cells and therefore take C = 2.

A cube is a very simple poly hedron and we may readily see that the number of vertices per face is 4, that is $p = 4$. The number of edges per vertex is 3, hence $m = 3$. The number of edges per face is 4, hence $p = 4$ (the number of vertices per face is equal to the number of edges per face). The number of faces per vertex is 3 hence $q = 3$. The number of faces per edge is 2 hence $n = 2$. The number of cells per vertex is 2 (not one which to the unwary could be a source of error). The number of cells per edge is 2 (and again we must emphasise that a cube is said to divide space into two). The number of cells per face is clearly 2 (as it always is).

We note that in the case of a single cube, as opposed to a complex of cubes, the number of vertices per cell, the number of edges per cell, and the number of faces per cell is the same as V, E and C, and this observation concludes the problem.

The above method is applicable to all regular polyhedra, and their topological equivalents, but the resolution of semi-regular polyhedra is clearly more difficult.

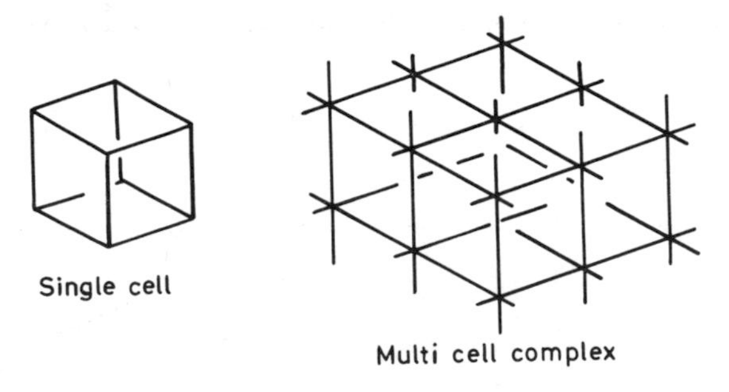

Figure 6.15

6.5.2 The valency concept – Semi-regular polyhedra

The method above must clearly be operated on a summation basis when a different number of edges, faces and cells per vertex are considered.

Suppose, as previously, that there are m edges per vertex, where m is an integer 2, 3, 4. . . etc. We know that there are 2 edges to a bivalent vertex, 3 edges to a trivalent vertex and 4 edges to a tetravalent vertex etc. If we denote the number of bivalent vertices by V_2, the number of trivalent vertices by V_3 and the number of tetravalent vertices by V_4 etc. We may write the total number of edges in terms of the sum of the vertices.

$$2V_2 + 3V_3 + 4V_4 + \ldots = \sum_{m=2}^{\infty} mV_m$$

The total number of vertices is denoted by V and V equals the R.H.S. whence

$$V = \sum_{m=2}^{\infty} mV_m \qquad (6.12)$$

We also know that every edge is common to two vertices and it follows that 2E also equals the R.H.S. whence

$$2E = \sum_{m=2}^{\infty} mV_m \qquad (6.13)$$

By a similar line of argument we may say that if there are f faces per cell, the total number of faces must equal the sum of all the faces, whatever their type.

$$2F = \sum_{f=2}^{\infty} fC_f \qquad (6.14)$$

We note that although $f = 3$ is the minimum for a straight-edged, flat-faced polyhedron, a lens-like dihedron has only two faces, hence f can equal 2.

6.5.3 The valency concept – a statistical approach

There are instances in which it is necessary to account the valency of a number of completely different multi-cell formations and in this event a statistical approach is called for. Living cells, soap froths and metallic crystal structures are instances where a statistical count may be necessary.

The valency equations derived above may be modified to deal with this contingency, simply by writing in terms of an average.

(i) Vertex/edge equations

If we take the equation:

$$V = \sum_{m=2}^{\infty} mV_m$$

we may say that the average number of edges per vertex is $\bar{m}$, where $\bar{m}$ is defined as follows:

$$\bar{m} = \frac{\sum_{m=2}^{\infty} mV_m}{V}$$

and since

$$2E = \sum_{m=2}^{\infty} mV_m$$

$$2E = \bar{m}V \qquad (6.15)$$

(ii) Edge/face equations

With regard to edges and faces, we know the number of edges per face = p, and the number of faces per edge = n, and the sum totals of each of these dimensionalities are equal.

Hence we may write:

$$\sum_{n=2}^{\infty} nE_n = \sum_{p=2}^{\infty} pF_p$$

If we again adopt the concept of an average we may write

$$\bar{n} = \frac{\sum_{n=2}^{\infty} nE_n}{E}$$

and similarly

$$\bar{p} = \frac{\sum_{p=2}^{\infty} pF_p}{F}$$

We note however that these two summations are equal and hence

$$\bar{n}E = \bar{p}F \qquad (6.16)$$

(iii) Edge/cell equations

The same approach may be adopted to obtain the relationship between edges and cell.

$$\sum_{n=2}^{\infty} nE_n = \sum_{e=2}^{\infty} eC_e$$

whence

$$\bar{n} = \frac{\sum_{n=2}^{\infty} nE_n}{E}$$

and

$$\bar{e} = \frac{\sum_{n=2}^{\infty} eC_e}{C}$$

but since these two summations are equal to each other

$$\bar{n}E = \bar{e}C \tag{6.17}$$

(iv) Vertex face equations

It may further be shown that:

$$\sum_{q=2}^{\infty} qV_q = \sum_{p=2}^{\infty} pF_p$$

and by the procedure above:

$$\bar{q} = \frac{\sum_{q=2}^{\infty} qV_q}{V}$$

and

$$\bar{p} = \frac{\sum_{p=2}^{\infty} pF_p}{F}$$

whence

$$\bar{q}V = \bar{p}F \tag{6.18}$$

(v) Vertex/cell equations

Again by a similar route:

$$\sum_{r=2}^{\infty} rV_r = \sum_{v=2}^{\infty} vC_v$$

$$\bar{r} = \frac{\sum_{r=2}^{\infty} rV_r}{V}$$

and

$$\bar{\nu} = \frac{\sum_{\nu=2}^{\infty} \nu C_\nu}{C}$$

whence since the two summations are equal

$$\bar{r}V = \bar{\nu}C \tag{6.19}$$

(vi) Face/cell equations

The basic statistical identity connecting the total number of faces and cell has already been stated in the form:

$$2F = \sum_{f=2}^{\infty} fC_f$$

and it remains only to note that every face divides two cells and hence the average valency $\bar{f}$ is written:

$$\bar{f} = \frac{\sum_{f=2}^{\infty} fC_f}{C}$$

and hence

$$2F = fC \tag{6.20}$$

(vii) Pure valency equations
We have seen that

$$\bar{n}E = \bar{p}F$$

$$\bar{n}E = \bar{e}C$$

$$\bar{q}V = \bar{p}F \cdot$$

hence

$$\bar{q}V = \bar{n}E = \bar{p}F = \bar{e}C \tag{6.21}$$

whence

$$V = \frac{\bar{n}E}{\bar{q}}, \quad F = \frac{\bar{n}E}{\bar{p}}, \quad C = \frac{\bar{n}E}{\bar{e}} \tag{6.22}$$

and we know from Euler's topological equation that

$$V - E + F - C = 0$$

and hence

$$\frac{1}{\bar{q}} - \frac{1}{\bar{n}} + \frac{1}{\bar{p}} - \frac{1}{\bar{e}} = 0 \tag{6.23}$$

We may also combine three other pairs, of valency equations, by virtue of their common factors. Equations 6.15, 6.16 and 6.18 yield $\bar{m}\mathrm{V} = 2\mathrm{E}$ and $\bar{q}\mathrm{V} = \bar{n}\mathrm{E}$ and combine to give

$$\frac{\bar{m}}{2} = \frac{\bar{q}}{\bar{n}} \tag{6.24}$$

Equations 6.16, 6.17 and 6.19 yield $\bar{r}\mathrm{V} = \bar{v}\mathrm{C}$ and $\bar{q}\mathrm{V} = \bar{e}\mathrm{C}$ and combine to give

$$\frac{\bar{q}}{\bar{r}} = \frac{\bar{e}}{\bar{v}} \tag{6.25}$$

Equations 6.16, 6.17, 6.18 and 6.20 yield $2\mathrm{F} = \bar{f}\mathrm{C}$ and $\bar{p}\mathrm{F} = \bar{e}\mathrm{C}$ and combine to give

$$\frac{\bar{f}}{2} = \frac{\bar{e}}{\bar{p}} \tag{6.26}$$

We see from Fig. 6.16, that a vertex with m edges gives rise to m additional vertices and m additional edges.

We see from Fig. 6.16(a) that

$$\mathrm{V} = (1 + m), \mathrm{E} = (2m), \mathrm{F} = (m + 1)$$

and may verify that these quantities satisfy the Euler topological equation:

$$(1 + m) = 2m + (m + 1) = 2$$

If there are m edges and q faces associated with the original vertex and if we recognise that the face m' is not directly associated with the original vertex, we may consider the interaction of valencies between the original vertex and its p associated edges and q associated faces, as follows.

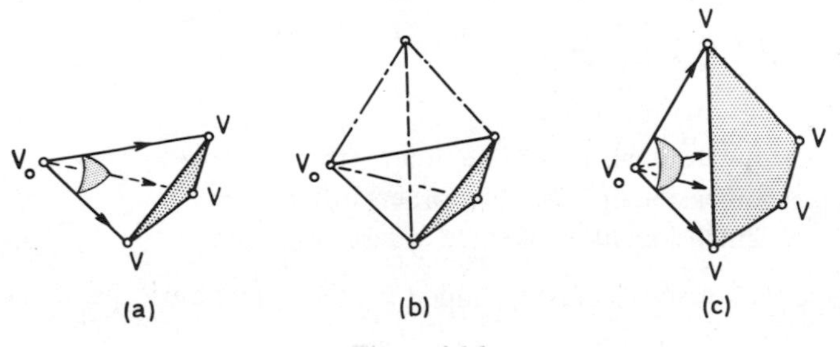

Figure 6.16

We note from Fig. 6.16 that m and q are not necessarily the same and we therefore say that there are m edges and q faces associated with the original vertex. We note, however, that the addition of an edge implies the addition of a face and hence we may say that if there are $(m + 1)$ edges, there are $(q + 1)$ faces, associated with the vertex under consideration. Moreover, if we are to be quite general we may say that there are r cells about the vertex and hence we may write the interaction between the valencies m, q, r, as follows.

We have seen that

$$\mathrm{V} - \mathrm{E} + \mathrm{F} - \mathrm{C} = 0$$

$$(m + 1) - 2m + (m + 1) - 2 = 0$$

and with respect to a single vertex we are free to substitute $(q + 1)$ for $(m + 1)$, and r for the number of cells per vertex.

For the case of a single vertex

$$\mathrm{V} = 1, \mathrm{E} = m, \mathrm{F} = (q + 1), \mathrm{C} = r$$

Hence

$$1 - m + q + 1 - r = 0$$

From which

$$m - q + r = 2 \tag{6.27}$$

We also know that the valencies, v, e, f, are numerically equal to the actual number of vertices, edges and faces in a singly connected polyhedral cell and hence the equation

$$v - e + f = 2 \tag{6.28}$$

is satisfied automatically.

Table 6.1

	V col	V row	E col	E row	F col	F row
1	$\bar{m}\mathrm{V} =$	$2\mathrm{E}$	$\bar{n}\mathrm{E} =$	$\bar{p}\mathrm{F}$	$2\mathrm{F} =$	$\bar{f}\mathrm{C}$
2	$\bar{q}\mathrm{V} =$	$\bar{p}\mathrm{F}$	$\bar{n}\mathrm{E} =$	$\bar{e}\mathrm{C}$		
3	$\bar{r}\mathrm{V} =$	$\bar{v}\mathrm{C}$				

line 1 relates dimensionalities of one order difference
line 2 relates dimensionalities of two orders difference
line 3 relates dimensionalities of three orders difference.

To these six equations must be added the two equations:

$$m - q + r = 2 \tag{6.29}$$

and

$$v - e - f = 2 \qquad (6.30)$$

and this completes the list of all the important statistical interactions between the valencies of polyhedral cells.

The above statistical derivation is essentially that first given by Arthur L. Loeb, but the notation and matrix presentation are different. Readers who have, or who wish to study Loeb's work must take care not to mix the two notations.

The equations we have derived are for average valencies, and the average valencies of a regular polyhedron are in fact the common valencies themselves. The cube is a regular polyhedron for which

$\bar{q} = q =$ the number of faces per vertex $q = 3$

$\bar{n} = n =$ the number of faces per edge $n = 2$

$\bar{p} = p =$ the number of edges per face $p = 4$

$\bar{e} = e =$ the number of edges per cell $e = 12$

The relationship between average valencies and actual valencies is the same and we should have

$$\frac{1}{q} - \frac{1}{n} + \frac{1}{p} - \frac{1}{e} = 0$$

whence

$$\frac{1}{3} - \frac{1}{2} + \frac{1}{4} - \frac{1}{12} = 0$$

and this result is clearly correct. We should also have

$$\frac{1}{\bar{q}} - \frac{1}{\bar{n}} + \frac{1}{\bar{p}} - \frac{1}{\bar{e}} = 0$$

and since $\bar{q} = 3.8/8$, $\bar{n} = 2$, $p = 4.6/6$, $e = 12.1/1$, this result is also correct and the case for a cube is "proved".

Example

Consider a vertex deep inside a simple cube lattice. We know, or can deduce, that there are six edges, twelve faces and eight cells eminating, or surrounding, that vertex, and we may determine the valencies m, q and r, of this vertex, by considering it in isolation. The actual number of edges, faces and cells, connected to the vertex, are the valencies required.

We therefore conclude that:

The number of edges per vertex = 6 = m

The number of faces per vertex = 12 = q

The number of cells per vertex = 8 = r

and we should find that these valencies satisfy the equation

$$m - q + r = 2$$

and on substituting the relevant values of m, q, r

$$6 - 12 + 8 = 2$$

we see that the relationship between valencies is true.

Example

Consider the case of the semi-regular cuboctahedron, with V = 12, E = 24, which yields 8 triangular and 6 square faces, that is, 14 faces in all.

As may be seen from Fig. 6.17,

The number of edges per vertex is $m = 4$.

The number of faces per vertex is $q = 4$

The number of faces per edge is $n = 2$

The number of edges per face is, however, $\bar{p}$ where

$$\bar{p} = \frac{\sum_{p=2}^{\infty} pF_p}{F} = \frac{qV}{F}$$

whence

$$\bar{p} = \frac{(3.8) + (4.6)}{14} = \frac{48}{14} = \frac{24}{7}$$

and this is the correct average value.

The number of edges per cell is $e = 24 = E$ since there is only one cell.

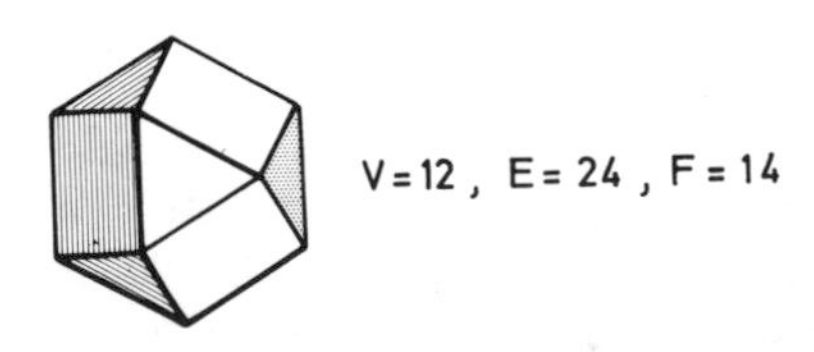

Figure 6.17

6.6 DUALITY OF STRUCTURE

A dual structure is any structure to which there corresponds another structure in which any pair of dimensionalities are transposed, the other dimensionalities being unaltered. Table 6.2 lists the properties of two dual pairs.

Table 6.2

	Polyhedron	*V*	*E*	*F*
1	Pentagonal dodecahedron	20	30	12
	Icosahedron	12	30	20
2	Cuboctahedron	12	24	14
	Rhombic dodecahedron	14	24	12

6.6.1 Interchange of dimensionalities

A glance at the Euler topological equation V – E + F – C = 0, suggests that there are likely to be many structural pairs in which the one is dual to the other. In so far that interchanging V and F produces no new figure the tetrahedron has no dual, but is said to be dual to itself. In the case of a cube, however, interchanging V and F transforms the cube into an entirely different polyhedral figure. A cube has eight vertices, twelve edges, and six faces: an octahedron has six vertices, twelve edges and eight faces and hence the cube and the octahedron are duals.

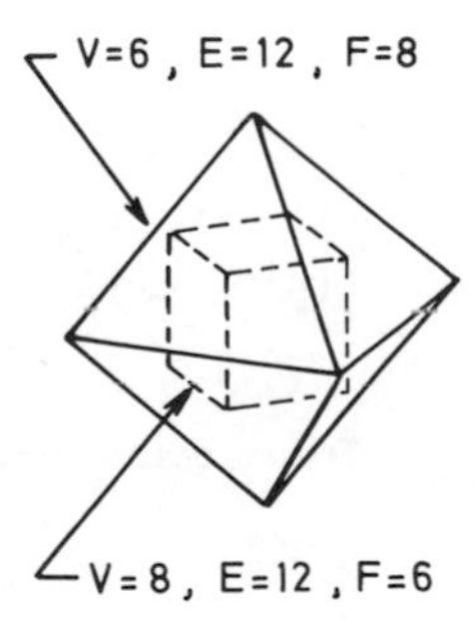

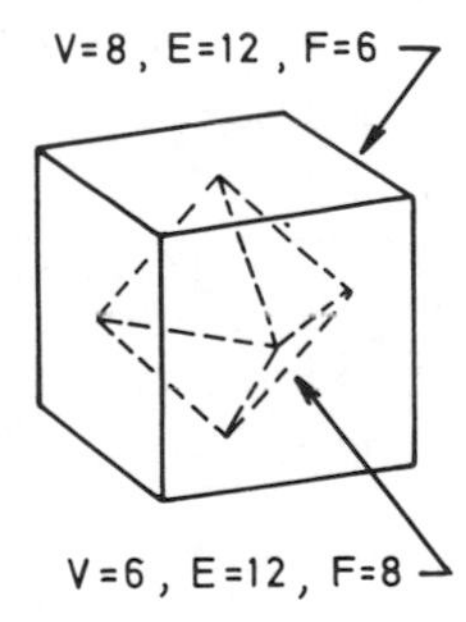

Figure 6.18

6.6.2 Stellation and truncation

A dual pair of structures may be produced (a) by stellation, (b) by truncation, of a given polyhedral figure.

A cube, for example, may be face stellated by point raising the centre of each face, in the manner indicated in Fig. 6.19.

The polyhedron shown at (b) is a stellated cube, which means that six vertices, twenty four edges and twenty four less six, extra faces have been added to the original cube. If the six added vertices are raised to a point where each pair of faces (f, f) lie in a single plane, the original cube edge e is lost and the two separate faces f, f, degenerate to a single rhomb, as shown in Fig. 6.19(c). The degenerate stellation has half the number of faces and twelve less edges than the more general stellation, Fig. 6.19(b). The polyhedron at (c) is the rhombic dodecahedron, for which V = 14, E = 24, F = 12 as shown in Table 6.3.

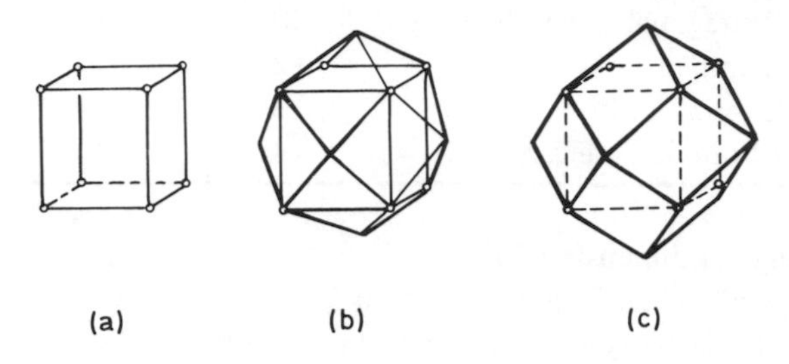

Figure 6.19

Consider, now, the process of vertex truncation as applied to the octahedron (the octahedron being dual to the cube).

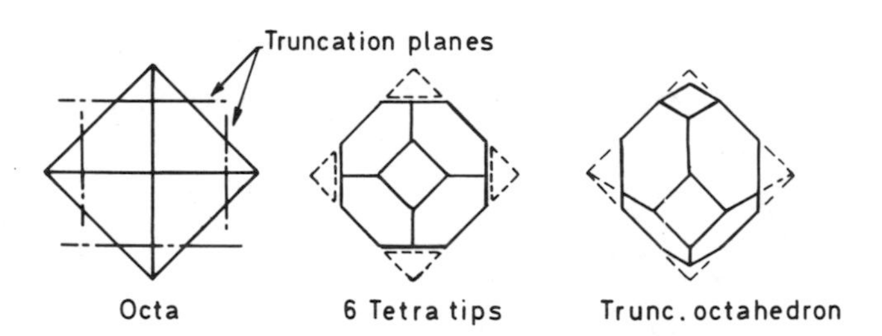

Figure 6.20

We recall that the octahedron has six vertices, eight edges and eight triangular faces, and note that apex truncation produces six new (square) faces and hence twenty four new edges. Each new square face has four vertices and there is one new face to the six original vertices, hence there are twenty four new vertices and no old vertices in the truncated figure. However, twelve old edges are joined by twenty four new edges and hence there are thirty six edges in all. The eight original triangular faces are transformed into hexagonal faces and six square faces are added, hence there are fourteen faces in all.

Table 6.3

	Polyhedron	V	E	F
1	Cube	8	12	6
2	Stellated cube	14	36	24
3	Rhombic dodecahedron	14	24	36
4	Truncated octahedron	24	36	14
	Octahedron	6	12	8

We observe from Table 6.3 the following interactions:

1. the cube is dual to the octahedron by virtue of the direct interchange of vertices for faces – the number of edges being the same.
2. the stellated cube is dual to the rhombic dodecahedron by virtue of the interchange of edges per faces – the number of vertices being the same.
3. the stellated cube is also dual to the truncated octahedron by virtue of the interchange of vertices for faces and faces for vertices – the number of edges being the same.
4. the rhombic dodecahedron and the truncated octahedron are related by a complete interchange of vertices, edges and faces.

Many other duals are listed by Cundy and Rollett in their book 'Mathematical Models'. Suffice to say there are five pairs of regular and thirteen pairs of semi-regular duals.

6.7 SPACE-FILLING POLYHEDRA

The concept of filling a plane with regular and/or semi-regular polyhedra has been explained in article 6.1.3 and it is a natural response to apply this concept in three dimensions. The basic problem is to identify which of the regular and semi-regular polyhedra are space fillers. (N.B. Early work on the equipartitioning of space was done by Lord Kelvin (1824–1907) who was the first to show that coupled tetrakaidecahedra divide space, as in soap froth, most economically. It would appear that space may be divided and completely filled with polyhedra in three possible ways:

1. by certain regular polyhedra, all of the same kind,
2. by certain semi-regular polyhedra, all of the same kind,
3. by certain combinations of regular and semi-regular polyhedra.

There are three important non-isotropic space fillers (or ways in which space may be equi-partitioned) in which identical polyhedra are used. The cube, the truncated octahedron, the rhombic dodecahedron are the regular and semi-regular polyhedra involved.

The truncated octahedron is a close mathematical model of soap froth and represents a minimum surface to volume ratio, which is also a feature of living cells. The rhombic dodecahedron is found in back-to-back formation in the planar hexagonal honeycomb cell of the bee.

There are also a great many ways in which space may be filled (or ways in which space may be equipartitioned) in which more than one type of polyhedra is involved. The most important of these are the cubocahedron and octahedron which together model the closely packed atomic structure of certain crystals and the most densely packed of all configurations – the octet module.

6.7.1 Non-isotropic space fillers, all elements alike

We consider the four most important cases for which all elements are alike:

1. the space-filling cube;
2. the space-filling non-regular octahedron;
3. the space-filling rhombic dodecahedron;
4. the space-filling truncated octahedron.

6.7.2 The space-filling cube lattice

The most obvious way of filling space is by the repeated three-dimensional repetition of cubes. The cube is a regular polyhedron and hence the cubic lattice qualifies as one of the polyhedral space fillers, we are seeking. We observe, however, that the distance between adjacent vertices is not the same in all directions. If the side of the cube is l, its face diagonal is $l\sqrt{2}$ and its body diagonal is $l\sqrt{3}$, as shown in Fig. 6.21, hence the cubic lattice is not isotropic.

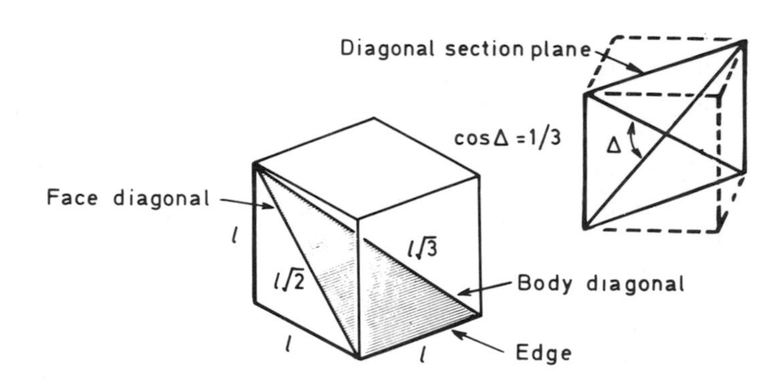

Figure 6.21

6.7.3 The space-filling non-regular octahedron

It is clear that a cube may be truncated in such a way that all eight vertices are replaced by triangular faces. When the truncation is such that the new vertices coincide with the intersection of the face diagonals of the original cube, a central non-regular octahedron is formed. The overall height of the octahedron is clearly equal to the cube edge length. The girth of the octahedron is a square of side equal to $\sqrt{2}$ times the cube edge length. Hence the octahedron is non-regular.

As may be seen from Fig. 6.22, the eight corner tetrahedra are of the same shape, size and volume as the eight tetrahedra which constitute the central octahedron. Moreover, symmetry dictates that space may be filled, or partitioned, by any number of non-regular polyhedra of these proportions.

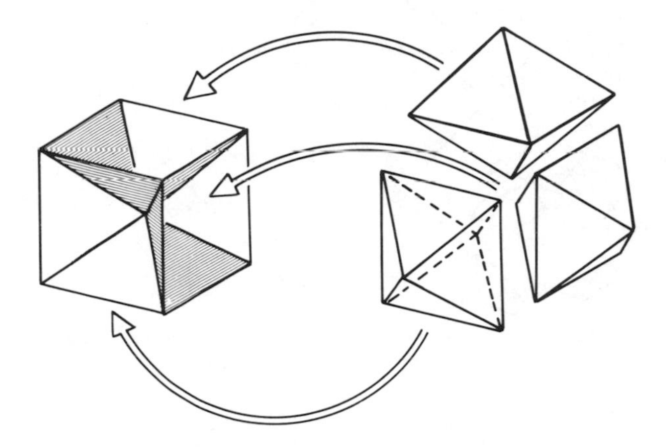

Figure 6.22

6.7.4 The space filling rhombic dodecahedron

The rhombic dodecahedron may also be inscribed in a cube and we again take the cubic lattice as a guide. We take the body centre of the three frontal cubes, Fig. 6.23, and label these OA, OB, OC. We next take the lattice lines 12, 13, 14, as the short diagonals of three rhombic faces and finally add three more rhombic faces which are visible in this orientation.

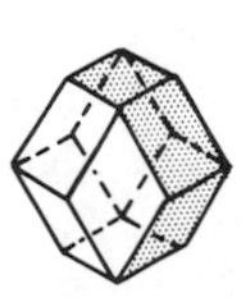

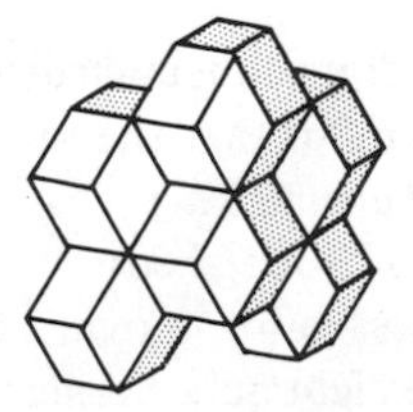

Figure 6.23

6.7.5 The space-filling truncated octahedron

If every edge of a cube is bisected and a square (face) of side $l/\sqrt{2}$ set up on each cube face, as shown in Fig. 6.24, then straight lines drawn between adjacent vertices form six regular hexagonal faces. The solid figure so formed, is a truncated octahedron, which as its name implies may also be derived by truncating a regular octahedron. Any number of truncated octahedra may be stacked to fill a 3-space, see also the space filling structure of soap froth.

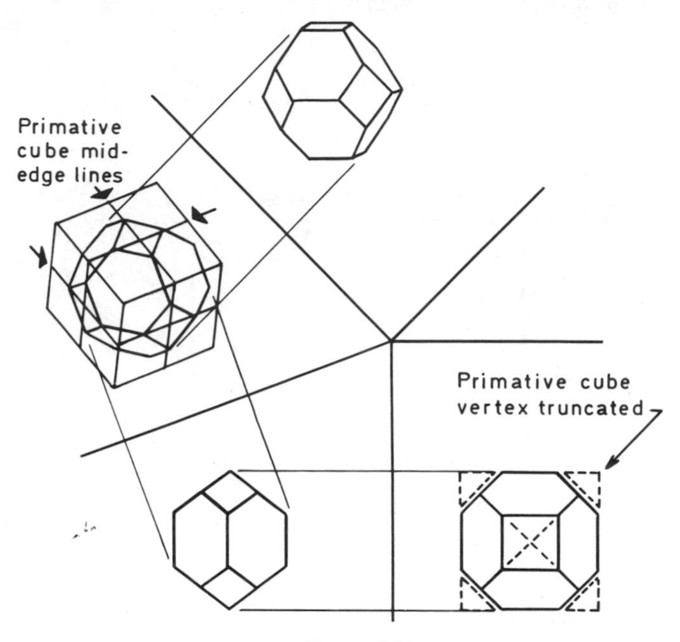

Figure 6.24

6.7.6 Non-isotropic space fillers with mixed elements

There are many ways of filling space or conversely dividing space into non-equipartitions, when an unspecified mixture of polyhedra are used, and numerous possibilities are illustrated by Fig. 6.25. We analyse one example.

6.7.7 The space-filling cuboctahedron and tetrahedron

We see from Fig. 6.26, that the semi-regular cuboctahedron may be inscribed in a cube and observe that when the mid-points of all cube edges are joined in, the resulting figure is a cuboctahedron.

As may be seen from the figure each truncated corner is a non-regular tetrahedron, and since eight cube corner points coincide with each internal point of a cubic lattice there are eight/one eighth tetrahedral spaces left at each corner in the nesting.

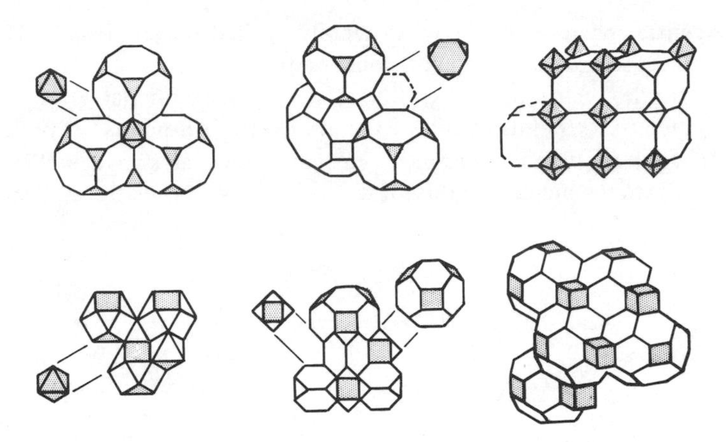

Figure 6.25

As may be readily seen from the figure these eight/one eighth tetrahedra make up a regular tetrahedron of the same edge length as the truncated octa. The gaps left at every corner may thus be filled with a regular octahedron and hence the whole space is filled. Suffice to say, the cuboctahedron forms the basis of many metallic crystals.

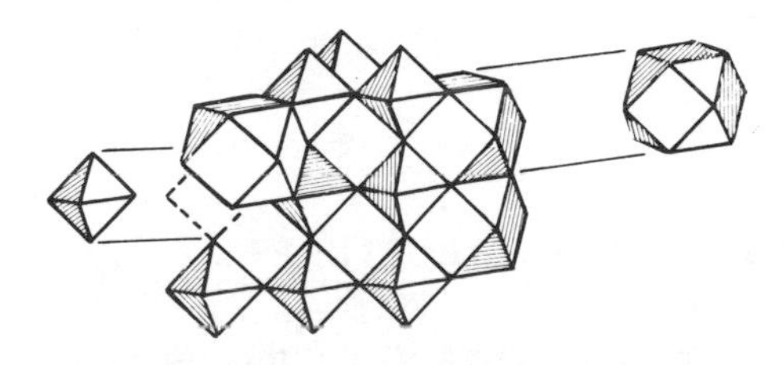

Figure 6.26

6.7.8 The isotropic space-filling octet element

An isotropic space filler is one in which the distance between all nearest vertices is the same, and it might be thought that a stack of regular tetrahedra fulfil this condition; whereas, in point of fact, they do not.

The solution to this particular problem is given, at some length, in Chapter 7, and it suffices here to give only the conclusion.

Space may be isotropically filled by an ordered combination of regular tetrahedra and regular octahedra but not by either of these alone. The number of edges per vertex is 12 and this is the maximum possible number of directions in which true isotropy can be found. Twelve equispheres, close-packed about an equi size centre sphere make the most densely packed cluster possible. The network of centres so produced is the isotropic vector matrix.

The isotropic vector matrix is thus the three-dimensional grid with the greatest symmetry of points about a common centre. All points, or vertices, lie on one of four parallel, equidistant, recurrent planes, as shown in Fig. 6.27. We investigate the practical utility of the isotropic vector matrix in Chapter 7.

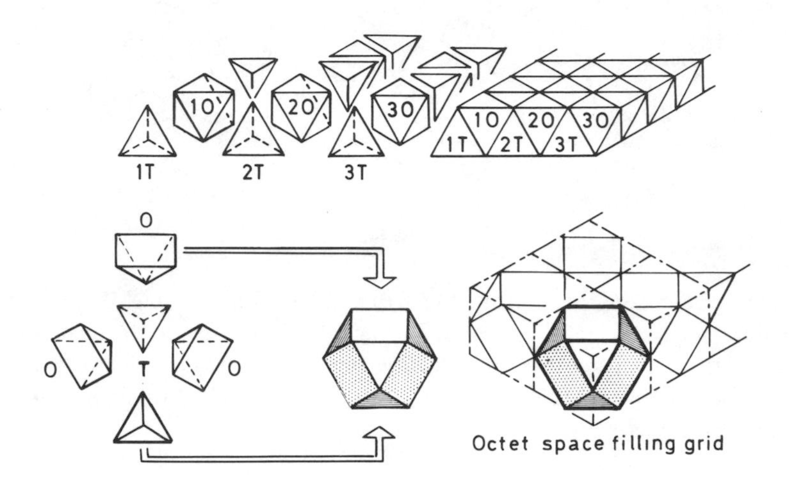

Figure 6.27

6.8 GEOMETRY OF CRYSTAL STRUCTURES

An obvious instance of a space filling system is the formation of metallic crystals, the patterning of which is closely related to the way in which intersecting lines or intersecting planes may be arranged to divide space into congruent repetitive units, called cells. Whilst it may seem unlikely that the atoms in a stable substance should be arranged in a non-triangulated configuration, such as a cube it is nevertheless, sometimes expedient to consider the atoms in crystals to be so arranged.

The simple cubic lattice, in which three mutually perpendicular planes or lines intersect and repeat, is the simplest possible arrangement. As may be seen from Fig. 6.28, there is an atom, or ion, at each corner of a unit cube but each of these eight corner atoms is shared between eight adjacent cubes and since there are a total of eight atoms, each contributing one eighth of its mass to the unit cube, the effective mass is but one atom per unit cube and herein lies one important reason why metals do not crystallise into this, most simple pattern.

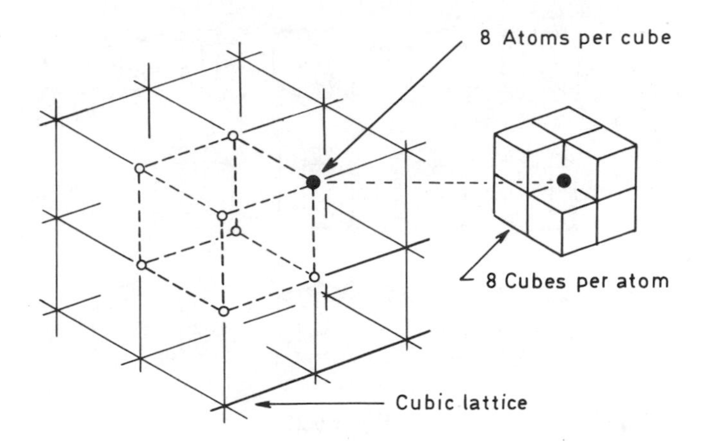

Figure 6.28

Other alternative arrangements based on the intersection of planes readily suggest themselves and there are fourteen – Bravais lattices – in all, but only the seven lattices listed here seem to be of practical value.

Table 6.4

Lattice system	Axes	Respective angles
1 Cubic	$a\ a\ a$	90 90 90
2 Tetragonal	$a\ a\ c$	90 90 90
3 Orthorhombic	$a\ b\ c$	90 90 90
4 Monoclinic	$a\ b\ c$	90 90 $\neq$ 90
5 Triclinic	$a\ b\ c$	$\alpha\ \ \beta\gamma \neq 90$
6 Hexagonal	$a\ a\ a$	90 and 120
7 Rhombohedral	$a\ a\ a$	$\alpha\alpha\alpha \neq 90$

(N.B. It is a matter of some convenience that for the purpose of analysing crystal structures, we may represent atoms as though they were rigid incompressible spheres. The fact that close packed spheres are incapable of transmitting tension is tacitly ignored).

Of the fourteen possible atomic arrangements listed by Bravais, X-ray photographs show that most common metals (zinc and magnesium excepted) are structured in accord with one of three primary space lattice patterns, listed below.

1. Body-centred cubic – b.c.c.
2. Close-packed hexagonal – c.p.h.
3. Face-centred cubic – f.c.c.

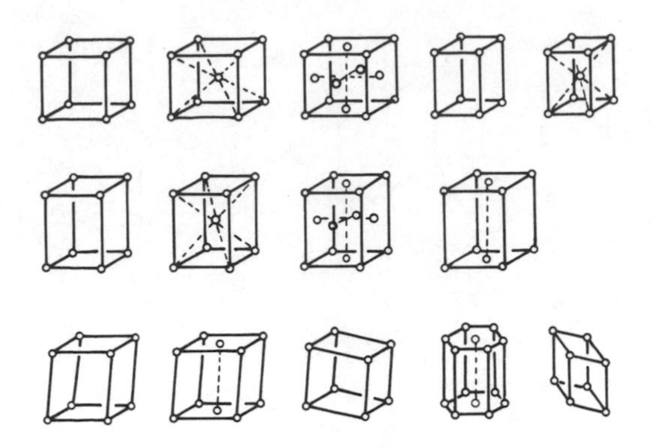

Figure 6.29

6.8.1 Body-centred cubic lattice

The conceptual model of a body centered cubic, b.c.c. comprises eight atoms at the eight corners of a unit cube, with one atom situated within the cube at the point of intersection of the three body diagonals. This impression of the b.c.c. is shown in Fig. 6.30.

An alternative way of modelling the b.c.c. is shown in Fig. 6.30 where the b.c.c. is depicted as a layered cluster of close packed spheres.

As may be seen from Fig. 6.30, the first layer A is arranged in a close packed manner. The second layer B is identical in form but is displaced, relative to the first layer, so that the spherical atoms of the second layer nestle into the hollow cusps formed by the close packed first layer. The third layer is a repeat of the first layer and it too nestles into the second layer.

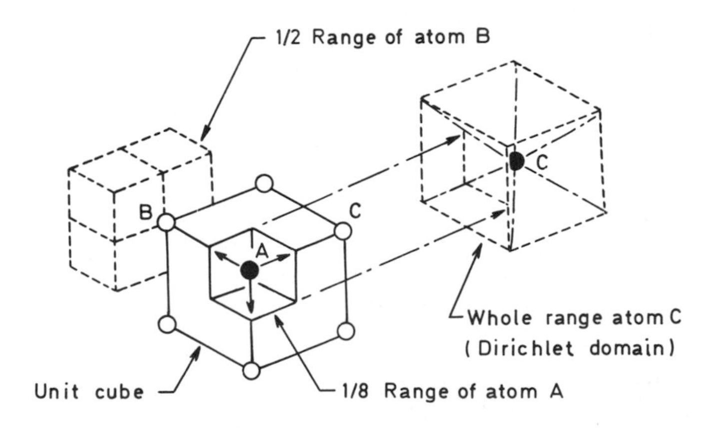

Figure 6.30

Figure 6.30 shows a b.c.c. from which it is clear that nine atoms are present in a unit cube. The centre atom is clearly unique to the unit cube considered but the eight corner atoms are shared with other cubes adjacent. The effective atom mass of a b.c.c. may thus be written as follows.

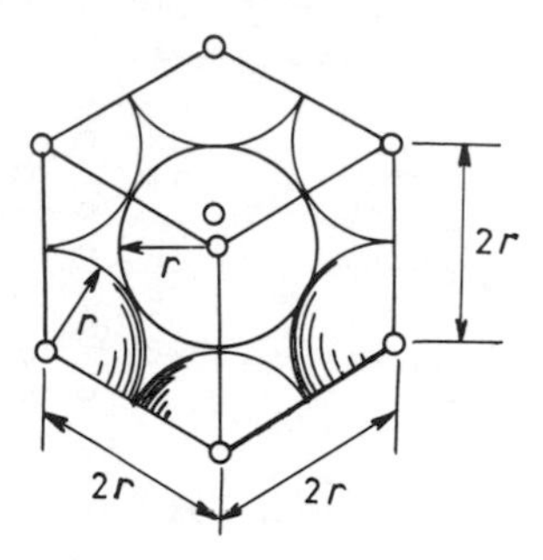

Figure 6.31

The centre atom contributes fully to the mass of the particular unit cube to which it belongs but, since light adjacent cubes meet at each corner, each corner atom contributes but one eighth of its mass.

The mass per unit cube equals:

$$1 + 8.\ \frac{1}{8} = 2 \text{ atoms per unit cube}$$

If every atom in the unit cube is expanded to an effective radius r, the space within the cube is occupied as shown in Fig. 6.31, from which the so-called atom packing factor is derived.

The atomic packing factor is defined as:

$$\text{Packing factor} = \frac{\text{Volume of close packed spheres}}{\text{Volume of unit cube}}$$

$$\text{Volume of unit cube} = 1$$

$$r = \sqrt{3/4} \text{ and } r^3 = 3\sqrt{3}/8$$

hence

$$\text{Volume of close packed spheres} = 2.4\pi r^3/3$$

(N.B. Since there are 2 atoms per unit cube). Hence

$$\text{Packing factor} = \frac{\pi\sqrt{3}}{8} = 0.68017$$

6.8.2 Hexagonal close packed lattice

The first layer of the h.c.p. comprises six atoms arranged in the form of a hexagon, about a single atom centre. The second layer comprises three atoms which nestle into the cusps formed by the first layer, and these are followed by a third layer identical to the first.

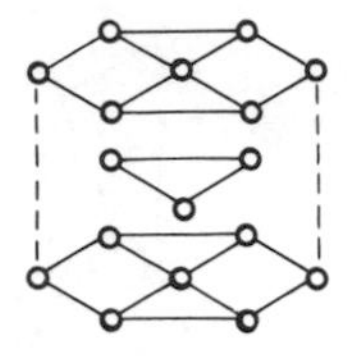

Figure 6.32

As may be seen from Fig. 6.32, there are three internal atoms in the second layer and seven atoms in the first and third layer, beyond which the pattern repeats. There are two face atoms each of which is shared with an adjacent hexagonal unit. There are two times six surrounding atoms each is shared with six adjacent units, as shown in Fig. 6.32. There are three internal atoms in the second layer which are unique to a particular hexagonal cell.

The mass per unit hexagonal cell equals:

$$\frac{2}{2} + \frac{12}{6} + 3 = 6 \text{ atoms per unit cell}$$

In order to calculate the atomic packing factor of an h.c.p. unit we need to calculate the centre to centre distance between eachlayer. We do this with the aid of Fig. 6.33.

$$\text{Volume of hexagonal prism} = 2Ah = 12\Delta h$$

where

$$\text{Area } \Delta = \sqrt{3/4} \text{ and } 2h = \sqrt{8/3}$$

$$\text{Volume of hexagonal prism} = 3\sqrt{2}$$

Since there are six atoms per unit hexagonal cell

$$\text{The total volume of close packed spheres} = 6 . \frac{4\pi r^3}{3} = 8\pi$$

where

$$r = \frac{1}{2} \text{ and } r^3 = \frac{1}{8}$$

Hence

$$\text{Packing factor} = \frac{\pi}{3\sqrt{2}} = 0.74048$$

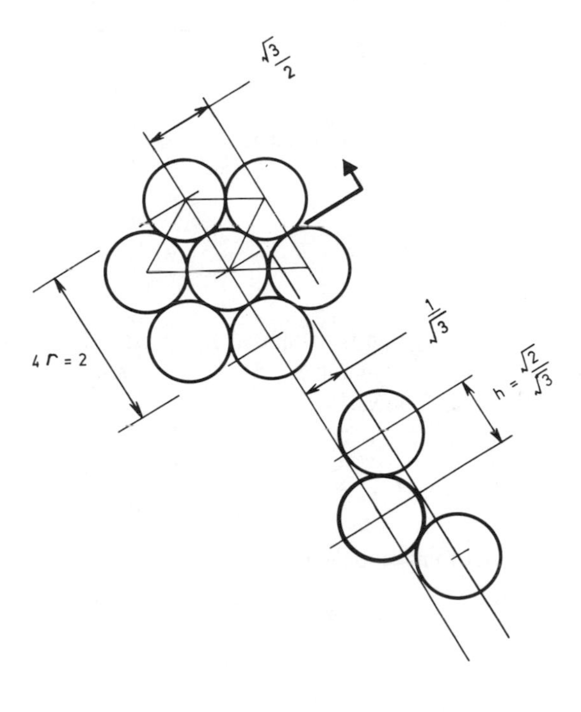

Figure 6.33

6.8.3 Face centered cubic lattice

The conceptual model of a face centered cubic lattice f.c.c. comprises eight corner atoms and six face atoms as shown in Fig. 6.34. The eight corner atoms are clearly shared by eight neighbouring cells and each of the six face atoms is shared by two cells.

The mass of a unit cube may thus be computed as follows:

$$\frac{8}{8} + \frac{6}{2} = 4 \text{ atoms per unit cube}$$

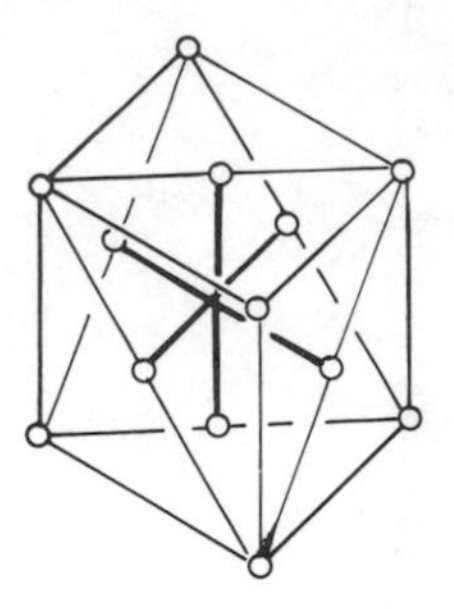

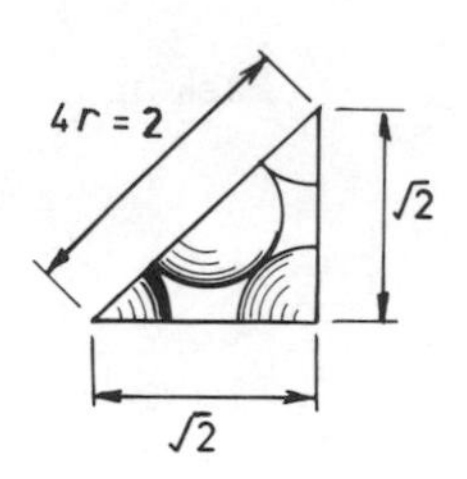

Figure 6.34

If every atom in the unit cube is expanded to an effective radius r the space within the cube is occupied as shown in Fig. 6.34, and the so-called atom packing factor may be calculated as follows.

$$\text{Packing factor} = \frac{\text{Volume of close packed spheres}}{\text{Volume of unit cube}}$$

$$\text{Volume of unit cube} = 1$$

$$\text{Volume of close packed spheres per unit cube} = 4.4\pi r^3/3$$

now

$$r = \frac{\sqrt{2}}{4} \text{ and } r^3 = \frac{2\sqrt{2}}{64}$$

Hence

$$\text{Packing factor} = \frac{\pi\sqrt{2}}{6} = 0.74048$$

In Fig. 6.35, the points 1–6 are expanded to form a close packed hexagonal plane of spheres. This plane of spheres is identical to the hexagonal layer which is characteristic of the close packed hexagonal lattice. Three equi-diameter spheres may thus be laid above and below this hexagonal layer. If the two layers of three sphere are arranged as illustrated in Fig. 6.35(a), the stacking sequence forms a c.p.h. as previously considered and not a face centered packing, as required. If, however, the top layer of spheres is arranged as in Fig. 6.35(b), then a truly face centered packing is produced.

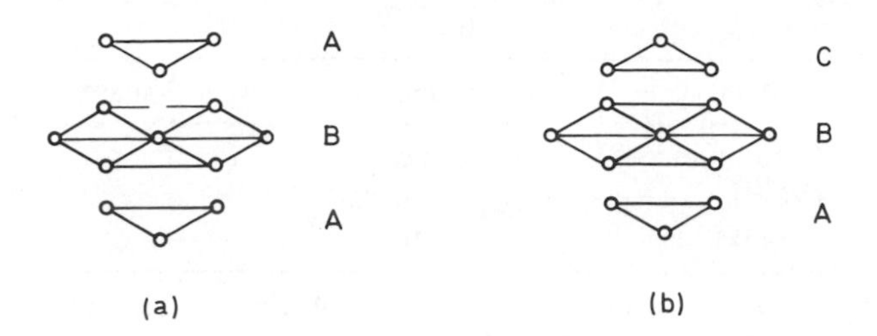

Figure 6.35

6.8.4 Atomic scale of metal lattices

It is sufficient to say that a measure of real lattice pitch is obtained from X-ray diffraction measurements and from these the closest distance of inter-atomic approach is determined. Some representative information is listed in Table 6.5.

Table 6.5

Lattice type	Metal	Atomic dia. Å
b.c.c.	α-Iron	2.48
	Molybdenum	2.72
	Tungsten	2.74
	Vanadium	2.63
h.c.p.	Beryllium	2.25
	Magnesium	3.20
	Titanium	2.93
	Zinc	2.75
f.c.c.	γ-Iron	2.52
	Aluminium	2.86
	Copper	2.55
	Lead	3.49

We must also observe that h.c.p. lattices are characterised by an (axial) aspect ratio: which is 1.57 for Beryllium, 1.62 for Magnesium, 1.60 for Titanium, 1.86 for Zinc. The atomic diameter given for Zinc in Table 6.5 is thus the average inter-atomic distance $(2.65 + 2.90) = 2.75$.

As may be seen from Table 6.6, α phase iron is b.c.c., whereas γ phase iron is f.c.c. and such atomic differences are common.

Table 6.6

Metal	α-phase	β-phase
Chromium	b.c.c.	h.c.p.
Cobalt	h.c.p.	f.c.c.
Nickel	h.c.p.	f.c.c.

Other interesting metallurgical facts are: Antimony and Bismuth have a rhombohedral lattice. α- and β-Tin have a tetrahedral lattice but as previously stated most metals are either b.c.c., f.c.p. or h.c.p.

6.9 DIRICHLET DOMAINS IN POLYHEDRAL CRYSTALS

(1) The simple cubic lattice

If we reconsider the simple cubic lattice (not adopted by metals) we see that each atom commands a cubic space of size equal to the lattice spacing and since every atom in the lattice commands the same amount of space (to a range of half the distance to its nearest neighbour). The dotted cube, Fig. 6.36, is the Dirichlet domain in a simple cubic lattice.

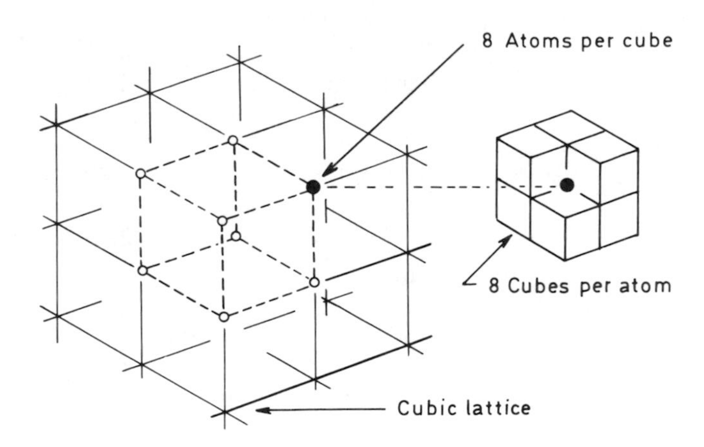

Figure 6.36

(2) The body-centered cubic lattice

If we draw in and bisect the body diagonals joining the central atom (of a b.c.c.) to its eight nearest neighbours, as in Fig. 6.37, we find that the six points of bisection lie on the faces of an octahedron. As shown in Fig. 6.37, each of the six vertices of the octahedron are sited at the centre of the appropriate face of the original cubic lattice. This octahedron is the so-called coordination polyhedron for a b.c.c. lattice.

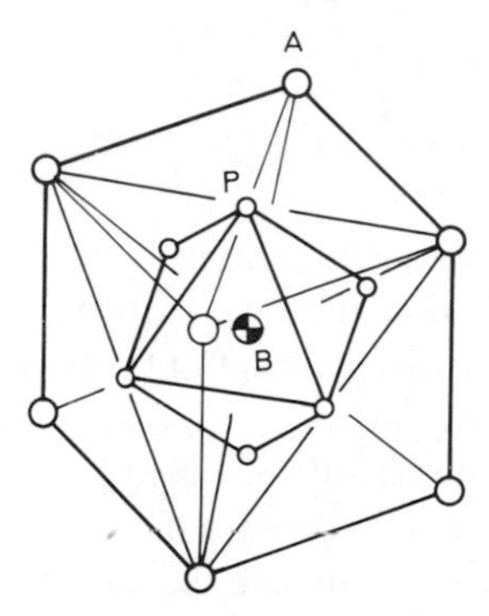

Figure 6.37

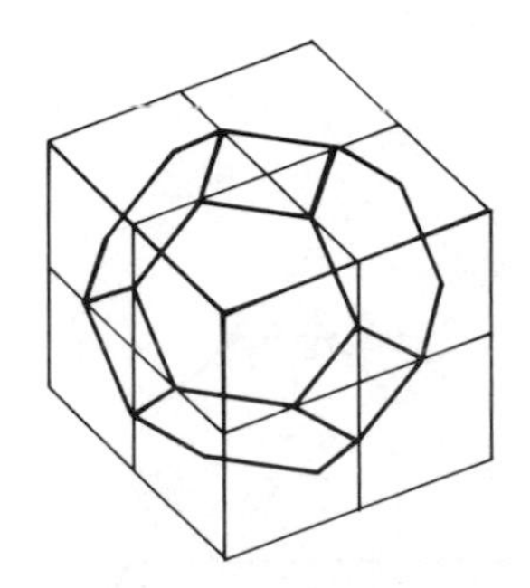

Figure 6.38

The half length of a body diagonal is:

$$\frac{\sqrt{1^2 + 1^2 + 1^2}}{2} = \sqrt{3}/2$$

The bisected half length of a body diagonal is $\sqrt{3}/4$.

The half length of a face diagonal is:

$$\frac{\sqrt{1^2 + 1^2}}{2} = \frac{1}{\sqrt{2}}$$

and the half width of a unit cube is 1/2. Hence the octahedron drawn is not a Dirichlet domain.

If, however, the edges of the octahedron, Fig. 6.37, are bisected a truncated octahedron as shown in Fig. 6.38, is produced. The faces of this truncated octahedron are equidistant between all adjacent atoms and hence the truncated octahedron, Fig. 6.38, is a Dirichlet domain.

(3) The face-centered cubic lattice

Recalling that every point location in a lattice is equivalent to every other, it is easily seen, from Fig. 6.34, that the f.c.c. lattice (as usually pictured) is identically equivalent to the alternative description(s) given.

Thus, the edge centered lattice, Fig. 6.39, is identically interchangeable with the original face centered lattice, Fig. 6.39, and the twelve solid dots, Fig. 6.39(c), correspond to atom positions in Fig. 6.39(b). If the two open circle atoms are extracted from the top and bottom layers of the original lattice, Fig. 6.39(a), we have fourteen equi-spaced atoms as shown in Fig. 6.39(c). By joining adjacent atoms as shown in Fig. 6.39(d), a rhombic dodecahedron is produced.

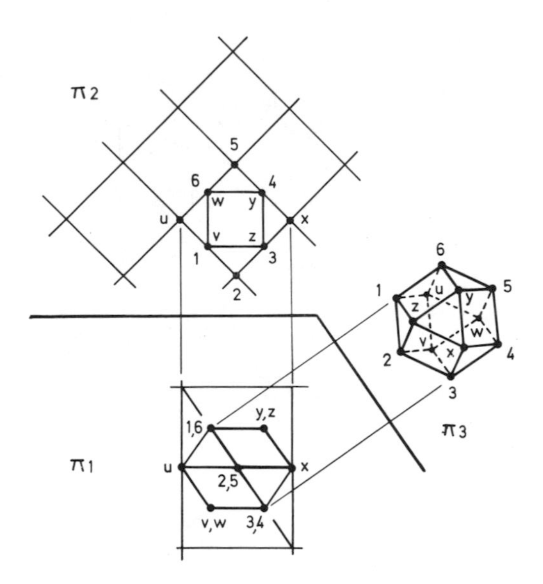

Figure 6.39

An alternative way of viewing a rhombohedral dodecahedron is to regard it as the degenerate stellation of a unit cube.

Any cube inside a cubic lattice is surrounded by six other cubes, so that if each of these six surrounding cubes is disected to leave six square based pyramids

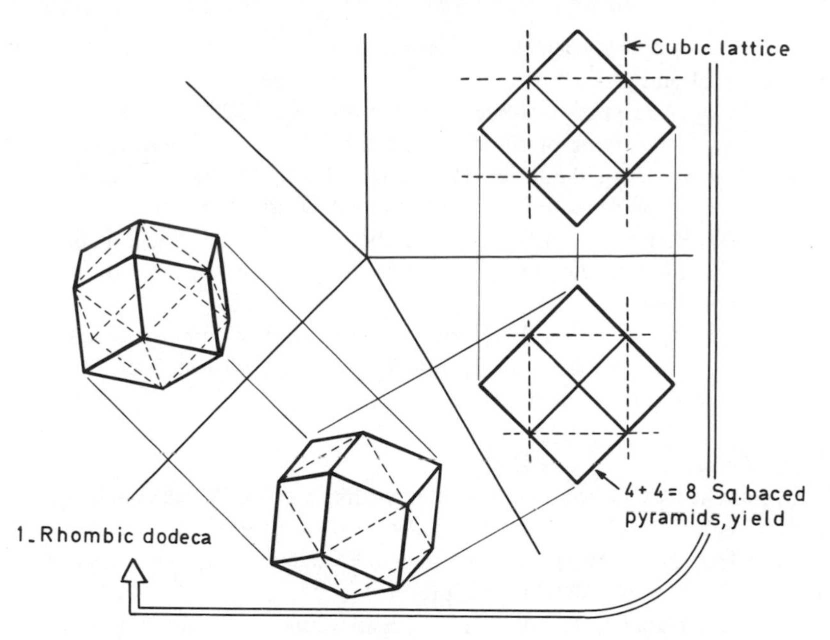

Figure 6.40

with each apex at the body centre of each surrounding cube, the result when added to each face of the original cube is a rhombic dodecahedron.

An isometric view of the face centered cubic lattice, Fig. 6.41 reveals the existence of six atoms which form a hexagonal layer in a diagonal plane. As may be seen from the figure, atoms 1, 2, 3, 4 are common to a particular unit cube, whilst atoms 5 and 6 belong to cubes adjacent.

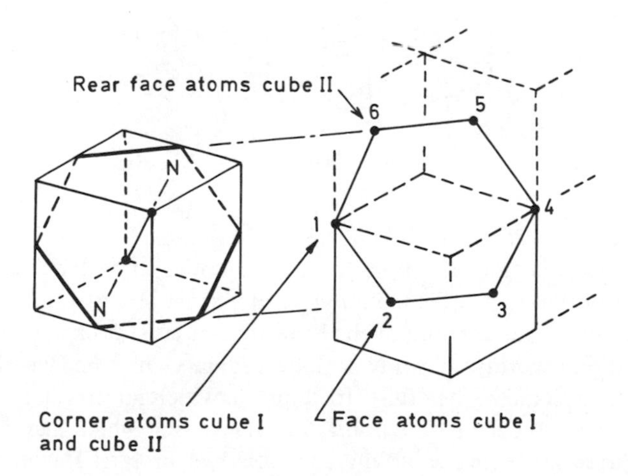

Figure 6.41

The relationship between the hexagonal layer and the cubic lattice is shown in elevation $\pi 1$ and plan $\pi 2$, Fig. 6.41.

Consider again the orthographic presentation, Fig. 6.39.

Mark in the positions of the first and third layers in the hexagonal array. Designate atoms using the letters (U, V, W), (X, Y, Z). Project these points to view $\pi 1$, locating points u and x at opposite ends of the diagonal.

Points (U, V, W) and points (X, Y, Z) lie parallel to the points (16, 25, 34) contained by the hexagonal plane. Project the position of points (u, v, w) and (x, y, z) to view $\pi 2$.

Note that 16, 34, VW, YZ are point views in plane $\pi 1$ and hence appear true length in views A2 and A3. Join all adjacent points, as shown, in all three views. The figure evolved is a cuboctahedron.

(4) The J complex

Every cube has six faces and every cube within a cubic lattice contacts another cube on each of its six faces. If the centre cube is divided into six pyramids, as shown in Fig. 6.42, and each surrounding cube is divided likewise a space filling complex of non-regular octahedra is produced.

Unlike the regular octahedron in which all edges and all face angles are equal, the non-regular octahedron has body diagonals of length $\sqrt{2}$, $\sqrt{2}$, 1 and it is apparent that, when the snub axes of adjacent octahedra are aligned at right angles.

(N.B. The term J (complex) is shorthand for Jackstone).

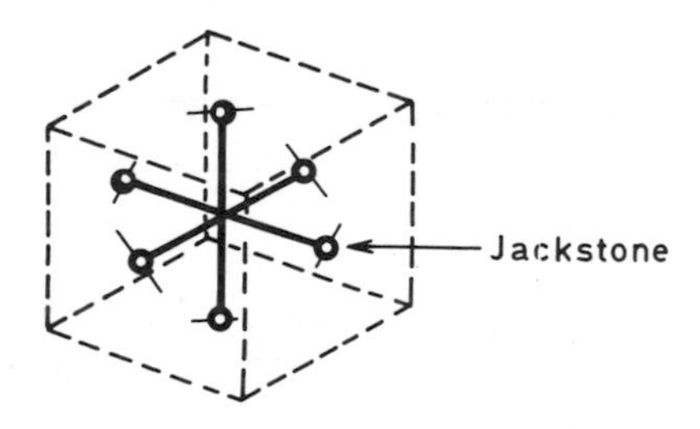

Figure 6.42

(5) Modes of material failure

It is clearly not within our brief to discuss modes of material failure but our conceptual notion of tensile, compressive and shear strain is, in many respects, enlightened by a consideration of the behaviour of close packed coins.

Other topics worthy of study include ordinary mud and crockery glazing, both brittle substances when fully dried out, in which multi-way cracking always occurs in near orthogonal directions. A characteristic which may sometimes be used to retrace the sequence of failure in a crashed aircraft. The lines on a giraffe reflect a very similar pattern and this is another line of interest.

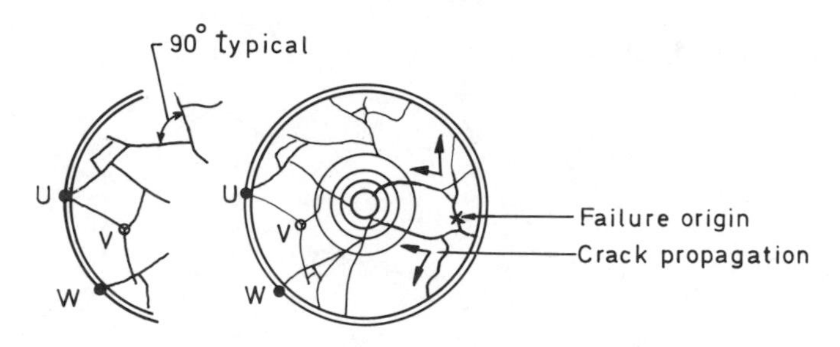

Test bed failure of a turbine engine disc.

Figure 6.43

6.10 SOAP FROTH

Whilst, at first sight, there may not appear to be much structural order in a soap (detergent) froth closer examination reveals many highly structured details. It is a matter of common experience that the free surface of all liquids tends to contract to a minimum size for it is thus that a free weightless bubble of soap froth forms a perfectly spherical enclosure, in which the shrinking action of surface tension is exactly matched by an internal pressure very slightly greater than the ambient pressure outside.

When neighbouring soap films intersect they do so in three way equi-angular junctions. Four way junctions may appear momentarily and may be maintained in the short term by localised impinging pressures. For instance, a soap film supported by three equi-spaced pins could conceivably take up any one of the configurations, Fig. 6.44(a), (b) or (c). However, for a given amount of film the path at (b) is shorter than the path at (a) and the path at (c) is shorter than the path at (b). Hence the configuration at (c) is the one adopted. It is, moreover, an experimental fact that a soap film forms an equi angled three way junction no matter in what positions the pins are placed, subject only to the two exceptions, Fig. 6.44.

The case of four pins, equally or not equally spaced may also be tried to prove the rule. Similarly when five pins are arranged in a pentagonal configuration three triple junctions are formed and when six pins set in a hexagonal formation four triple junctions occur.

With regard to spherical bubble formations, we know that a free floating bubble, blown from a soap/water solution always assumes a minimum surface/ maximum volume configuration. A bubble which descends onto a flat moist surface does not rest on the surface in the manner of a rigid spherical billiard ball, nor does it lie slightly flattened as would a pneumatic tyre. When contact is made between a bubble and a flat wet surface the bubble transforms, almost instantaneously, into a semi spherical dome of increased radius (assuming it does

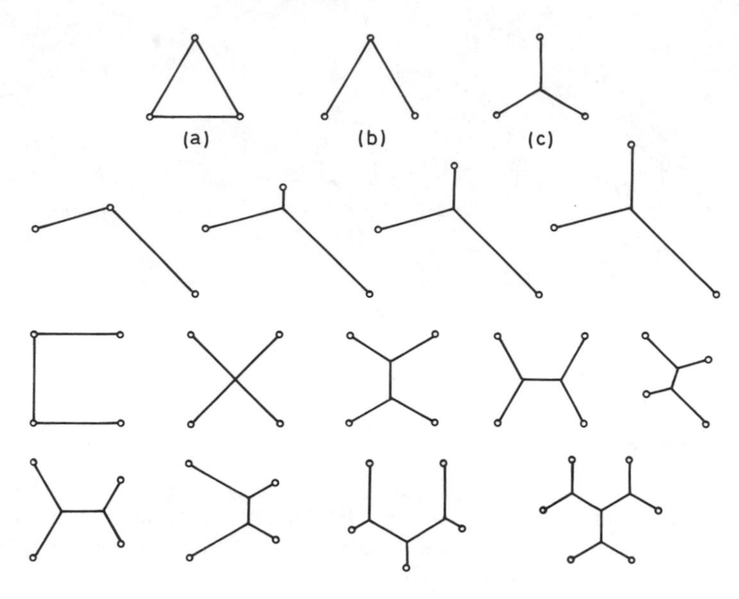

Figure 6.44

not burst). Subject to the condition that no air escapes the change from sphere to half sphere must be accompanied by an increase in size. This is because the volume of a sphere is $4\pi r^3/3$ and that of a half sphere $2\pi r^3/3$ and if these two volumes are equal then

$$4\pi r^3{}_0/3 = 2\pi r^3{}_1/3$$

that is the half sphere has a radius $2^{1/3}$ that of the parent sphere.

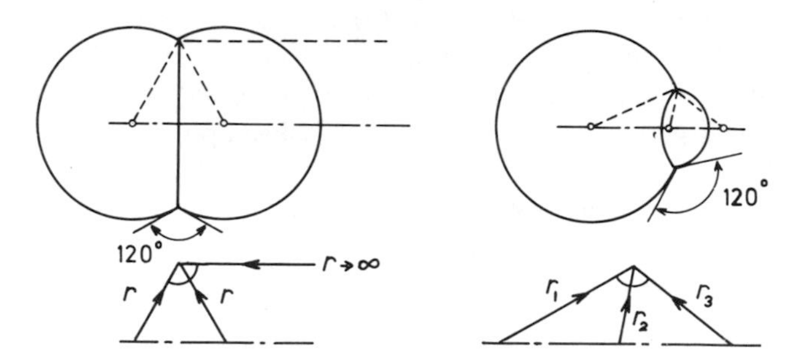

Figure 6.45

Let us next consider the marriage of two bubbles of equal size, for here the outcome is governed by the equalisation of pressure. Since the bubbles are of equal size the pressures in each are equal, and hence the partition between the equi double bubble is plane. The curved bubble surfaces, Fig. 6.45, are of equal radius and symmetrically disposed about the plane partition. The included angle between radii is 60° and we again encounter the equi angle triple junction.

If the two initial bubbles are of different size, the pressure in the smaller is greater than that in the larger and hence the partition which divides the double bubble is curved. The convexity being in the sense of the greater pressure.

If the three radii are r_1, r_2, r_3, where r_3 is the radius of the partition and d the distance between bubble center it is easy to show that

$$d^2 = r^2_1 + r^2_2 - r_1 r_2$$

and

$$\frac{1}{r_1} = \frac{1}{r_2} + \frac{1}{r_3}$$

Moreover, these relationships, first enunciated by Plateau, hold for more complex clusters. Thus the three way junction formed by a triple bubble appears as in a truncated octahedron. The lines of intersection OA, OB, OC, extend of course to the third dimension, and as more and more bubbles cluster a froth complex comprised largely of tetrakaidekahedra is formed. Although any particular cell may be of any compatible polyhedral form, it was discovered by Lord Kelvin that the statistical average cell takes the form of a truncated octahedron. Now a truncated octahedron consists of six square, eight hexagonal faces, thirty six edges and twenty four vertices and when these faces and edges are curved, as in a soap froth, is termed a tetra (four) kai (and) deca (ten) hedron.

(N.B. Readers who wish to pursue this most fascinating subject are referred to the classic work, *On Growth and Form*, by D'Arcy Thompson. We quote but one relevant passage).

> An assembly of rhombic dodecahedron goes far to meet the case. It fills space; its surfaces or interfaces are planes, and therefore surfaces of constant curvature throughout; and they meet together at angles of 120°. Nevertheless, the proof that the rhombic dodecahedron (which we find exemplified in the bee's cell) is a figure of minimum area is not a comprehensive proof; it is limited to certain conditions and practically amounts to no more than this, that of the ordinary space-filling solids with all sides plane and similar, this one has the least surface for its solid content. Lord Kelvin made the remarkable discovery that by means of an assemblage of fourteen sided figures or 'tetrakaidekahedra', space is filled and homogeneously partitioned – into equal, similar and similarly situated cells – with an economy of surface in relation to volume even greater than in an assemblage of rhombic dodecahedra. (*On Growth and Form*, D'Arcy Wentworth Thompson, Cambridge University Press, abridged edition, p. 120).

It is also due to Thompson that we observe in vegetable growth an early indication of organised structure in the formation of quasi-fluid bubble froths. With time and growth these bubbles are drawn, by surface tension forces, into a close and broadly hexagonal packing. The high space-filling efficiency of the straight-sided hexagonal array has, however, a less powerful hold on the growing complex than does the spheroidalising influence of surface tension. The process that causes a fluid soap film to minimise its surface and condense its volume is thus the stronger agent in the formation and eventual coagulation of quasi fluid vegetable tissue. As the process of omnidirectional spheroidalisation proceeds, so the sharp corner junctions of the hexagonal cells becomes less pronounced and when, with passing time, ageing tissue grows more brittle it looses its fluidity and the contractive inward pull of cells exceeds the cohesive strength of sharp junction walls. As the angles become more rounded the cell walls tend to split and each cell tends to a more spherical form. The cell walls become stiffer – pellicles are formed – and the cells that began as a conglomerate of quasi-fluid become as billiard balls in close-packed contact.

(N.B. Our knowledge of pneumatic tensile structures owes much to the study of soap froth and the work of the architect Frei Otto should be read by all who have aspirations in this direction).

7 Synthesis of structure and mechanism

These constraints are independent of specific interactive forces, hence are geometrical in nature.

Arthur L. Loeb

7.1 INTRODUCTION

Triangulation is a fundamental principle employed in many direct load bearing structures and the principles, if not the subtleties, are apparent in all roof trusses and girder bridges of classical design. There is also much current interest in the design and development of industrial robots, many of which have articulate limbs, not unlike those of human beings. Indeed, the replacement of damaged, diseased and worn human joints and limbs (by man-made substitutes), constitutes an important development.

Current practice in structures, robotics and biomechanical engineering owes much to the classical theory of mechanisms.

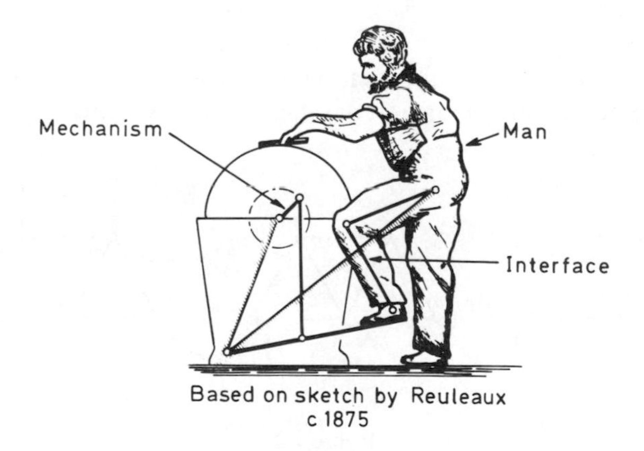

Figure 7.1

7.1.1 Two-dimensional triangulation

A plane triangulated structure is any plane combination of long slender bars, arranged as a triangulated array, designed to receive and transmit any acceptably in-plane loading, induced through its nodes.

The simplest and most perfect two-dimensional load bearing structure is an equilateral triangle, made up of long slender bars designed to carry the input loads in pure tension (when outwardly directed) and in pure compression (when inwardly directed). For all cases in which the intersection of the three (and there must be three) resultant external loads lies within the bound of the triangle the minimum and necessary volume of bar material is the same.

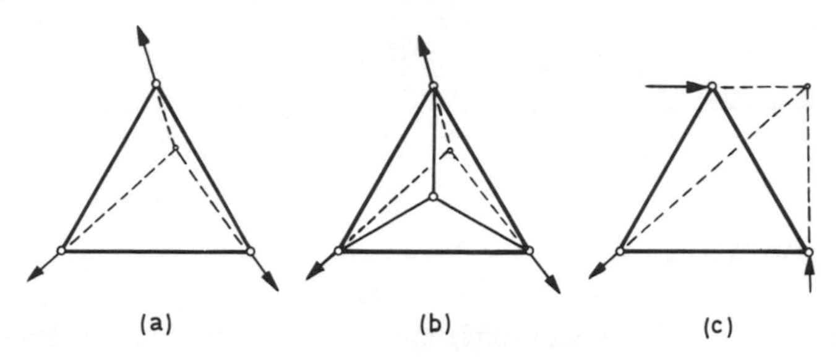

Figure 7.2

The structure, Fig. 7.2(b) for instance, has twice as many members as the structure at (a) and yet the necessary volume, and hence, weight of bars in the two structures is identical, when the same material is used. If, for practical reasons, the triangle cannot be made equilateral, the same weight comparability applies in every case. If, however, the intersection of the three resultant forces (loads) lies outside the bound of the triangle, then two of the bars carry tension, the other compression or vice versa, and the necessary weight of the structure depends on both the shape and the relative direction of loading. For instance, the forces transmitted by the identical frames Figs. 7.3(a), (b) and (c) are clearly different and hence (for the same stress) the necessary weights are different.

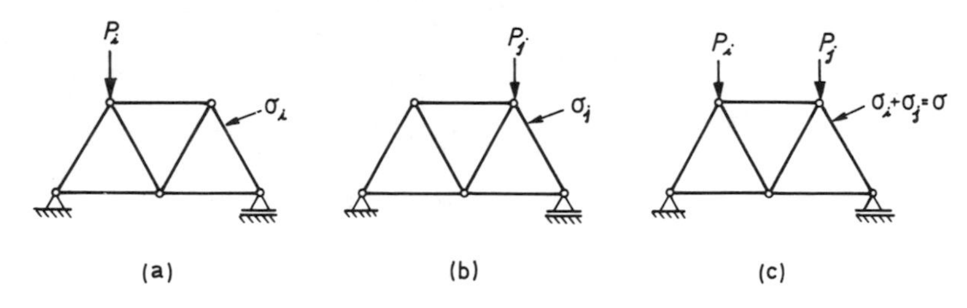

Figure 7.3

7.1.2 Three-dimensional triangulation

If we consider raising the centre node of the six bar frame, Fig. 7.2(b) out of plane, we produce (with the necessary increase in inner bar lengths) a tetrahedron. Suffice to say, that the very shallow plane structure at (a) would be unsuitable for receiving and transmitting loads applied in a direction normal to its plane. The slightly taller tetrahedron at (b) would be capable of accepting modest transverse loads but would inevitably be "overweight" or heavily stressed. In other words, the taller structure is an improvement but in terms of transverse loading is inefficient. The regular tetrahedron at (c) is the perfect 3-space equivalent of the plane triangle and is good for loading in all directions. The taller tetrahedron at (d) is likewise good for vertical loads but less well suited to loads which have an appreciable horizontal component.

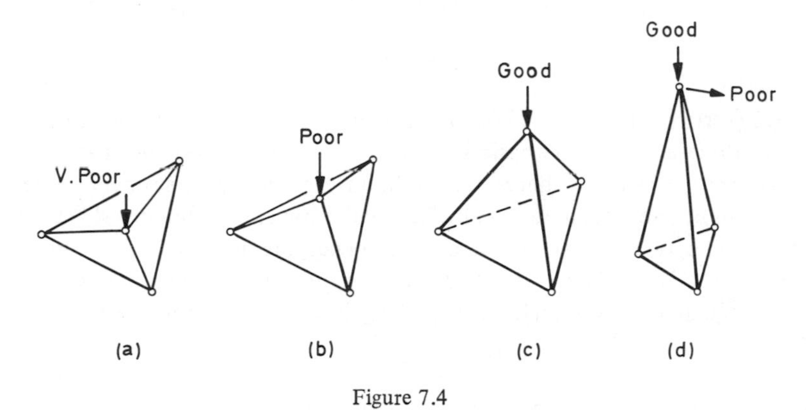

Figure 7.4

7.1.3 Structural topology in two dimensions

It is known from the elementary triangle of forces theorem that three non-parallel coplanar forces in equilibrium pass through a single point. Three material links (members) OA, OB, OC, joined at O, would receive and transmit these loads, but if free to self align at O would not meet the technical requirements of a self stiff structure. A triangulation comprising three links AB, BC, CA, has the advantage that if the direction of load input changes, or is changed, the triangulation retains its shape (though it would of course extend or compress very slightly, under the action of loads applied). Retention of shape is clearly of prime importance and a triangulation such as that shown in Fig. 7.5, is by far the most efficient way of carrying three in plane loads.

If a fourth load is to be injected into the basic triangle, via a new node D, then it is clear that the new node D requires a minimum of two additional links (members) to support it, and these two members act as load paths along which the input through D divides and travels.

The notion of receiving loads and transmitting them through a triangulated network of material members is the concept on which many traditional structures are built.

If we build an extended triangulation of long slender members we see that for each new node two additonal links (members) are required. If the total number

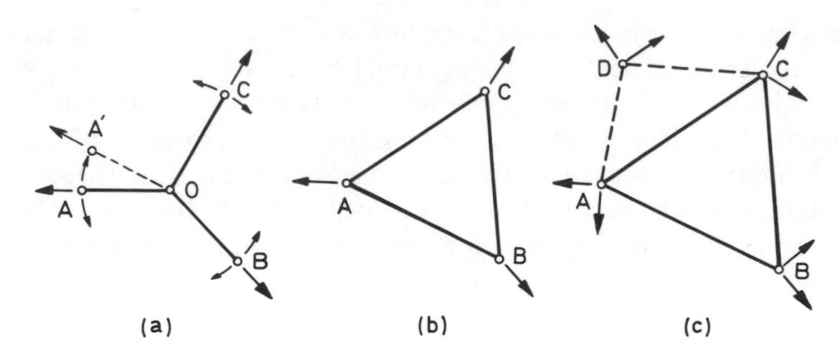

Figure 7.5

of nodes (vertices) is V and the total number of links (members) M, then since there are three nodes in the basic triangle, (V – 3) nodes have been added, and since to every new node there is a minimum of two new members, there are 2(V – 3) additional members, in all. The total number of members M is the sum of the number of members in the basic triangle plus the number of additional members. That is 3 + 2(V – 3). The total number of members in a two dimensional triangulated network with a sufficiency of members is therefore

$$M = 2V - 3 \tag{7.1}$$

The practical utility of any terrestrial structure depends however on it being firmly grounded, and to this end a minimum of three constraints external to the network must be supplied. In the rather hypothetical case of a planar bridge, three external constraints R_1, R_2, R_3, are strictly required to take care of a general loading. However, since there are only three grounding forces, they too constitute a perfect triangulation. Hence equation 7.1 remains true providing the external triangulation is correctly accounted.

A distinction is, however, to be made between freedoms within the network and freedoms between it and its surroundings and the notion of conceptually isolating part of a larger network (structure) by the device of Fig. 7.6(c) is directly relevant to the finite element stress analysis method.

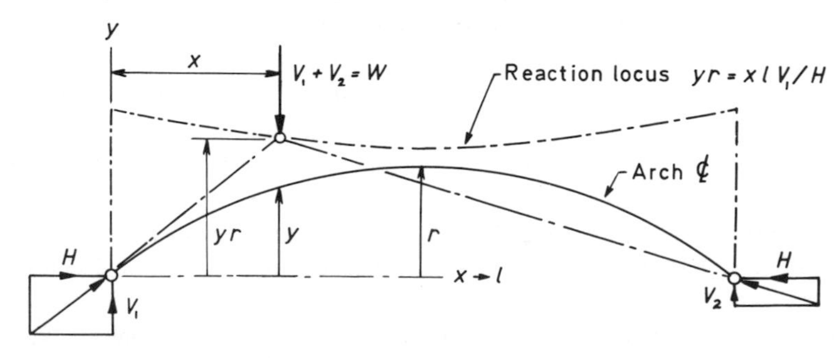

Figure 7.6

7.1.4 Structural topology in three dimensions

The triangulation of four nodes in 3-space clearly produces a tetrahedral configuration in which a minimum of three additional links (members) are required to support every additional node.

Now since there are four vertex nodes in the basic tetrahedron, $V - 4$ additional nodes and $3(V - 4)$ additional members are required to produce a three dimensional triangulated network comprising a total of V nodes and M members, and since there are six members in the original tetrahedron, there are $3(V - 4) + 6$ members in a 3-space triangulation whence

$$M = 3V - 6 \tag{7.2}$$

7.1.5 Statical determinancy

The triangulated structures we have been considering have but a sufficiency of members and are known as statically determinate structures; that is, the forces in the members can be determined by statical theory. Continuing on this assertion, we may argue that in an assembly comprising V vertices and $(V - 1)$ links, a total of $V(V - 1)$ connections may be made between them. Every line (link or member) has two ends and no more than $V(V - 1)/2$ members can be inserted in a statically determinant form.

(N.B. Statical determinancy also implies that members of any length may be freely inserted and hence there is no initial straining).

Now we know that there are $M = 2V - 3$ members in a two dimensional statically determinate structure and $M = 3V - 6$ members in a three dimensional statically determinate structure and since $M = V(V - 1)/2$ in both these cases we conclude that in a two dimensional triangulation

$$V^2 - 5V + 6 = 0$$

whence

$$(V - 2)(V - 3) = 0$$

$V = 2$ corresponds to a line (link or member) with a vertex at each end. $V = 3$ corresponds to three such lines joined into a three vertex triangulation, and these are the only two fundamental options.

Similarly in a three dimensional triangulation

$$V^2 - 7V + 12 = 0$$

whence

$$(V - 3)(V - 4) = 0$$

$V = 3$ corresponds to the plane three vertex triangular element. $V = 4$ corresponds to the four vertex 3-space tetrahedron.

7.1.6 Statical indeterminancy

The degree of statical indeterminancy in a tringulated network is ascertained by accounting the excess of links over and above the statical requirement. The degree of statical indeterminancy I may thus be stated for the 2D case as

$$I = M - 2V + 3 \tag{7.3}$$

and for the 3D case as

$$I = M - 3V + 6 \tag{7.4}$$

whence I is positive for all statically indeterminant forms, and negative when the triangulation is deficient.

(N.B. The failure or removal of a single member in a statically determinate structure promotes its instant collapse, whereas in a statically indeterminate structure up to I members may fail or be removed and hence a so-called fail-safe capability is present).

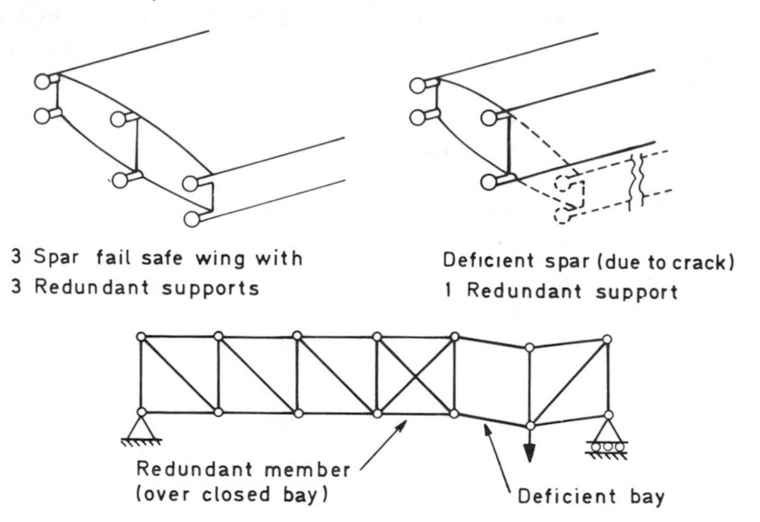

Figure 7.7

It is, however, always important to make sure that no over-closure of one region should occur at the expense of under-closure of another region, see Fig. 7.7, for instance.

The degree of statical indeterminancy in a continuous ring type structure similar to that in Fig. 7.8, may be deduced by converting that structure into a singly connected figure and totting up the number of freedoms suppressed during its reassembly.

(N.B. Any structure in which there is but one unique path between any two points which may be travelled without having to cover any part of the same path

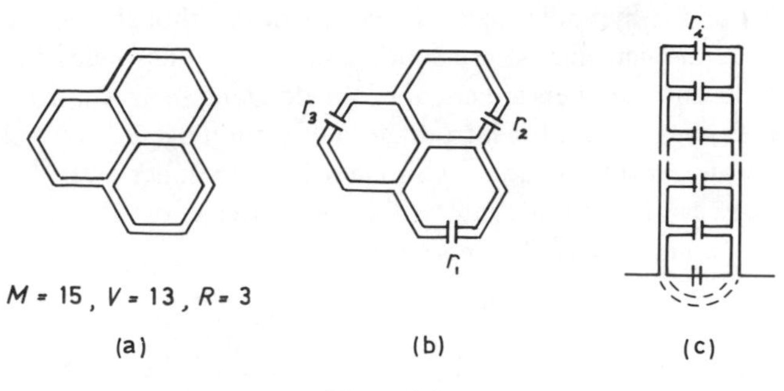

Figure 7.8

twice, is a singly connected figure. A structure in which two options exist is a doubly connected figure and so on).

With reference to Fig. 7.8(a) we see that there are a certain number of vertices V, members M and rings R, and that a singly connected figure is produced by introducing R cuts as at (b). If there are V_0 vertices in the original multi-connected figure, then there are $V_0 + R = V_1$ vertices in the singly connected figure. It is also apparent from Fig. 7.8, that when

$$M = V_0 \qquad R = 1$$
$$M = V_0 + 1 \qquad R = 2$$
$$M = V_0 + 2 \qquad R = 3$$
$$M = V_0 + i \qquad R = i + 1$$

whence

$$M - V_0 = R - 1$$

and

$$R = (M - V + 1)$$

(N.B. Perceptive readers will observe that this equation is similar to the Euler topological equation, except that for reasons soon to become apparent we do not count the outer space cell in this instance).

Now, in order to remake the cut in each ring we need to control an x, y and θ displacement and hence a cut releases three degrees of constraint in a cut member (in 2D) and since there is a cut in each ring the degree of overall indeterminancy in a stiff two-dimensional multi-ring structure is

$$I = 3R = 3(M - V + 1) \tag{7.5}$$

We may similarly determine the degree of statical indeterminancy of a composite ring in which there are both stiff and free nodal junctions. We do this by including a term $r = f$ to represent the number of releases (freedoms introduced).

(N.B. We use the symbol f to represent freedoms, though r and c are often used by structural engineers. The use of r and c sometimes helps to clarify the issue when links and members are cut and/or held along their length).

It is clear, however, that the introduction of a pin hinge in a two dimensional link adds one degree of freedom, whilst cutting a member away from a hinge adds three degrees of freedom, and two degrees of freedom if one of the three cut freedoms is constrained. The expression

$$I = 3(M - V + 1) - \Sigma f$$

where Σf represents the total number of freedoms added (releases made) in an initial structure comprised entirely of stiff zero freedom joints.

Example

Determine the degree of statical indeterminancy in a double bubble fuselage frame (typical of Concorde) (a) for the case when nodes 1 and 2 are stiff and have zero freedom, and (b) for the case when nodes 1 and 2 have rotation.

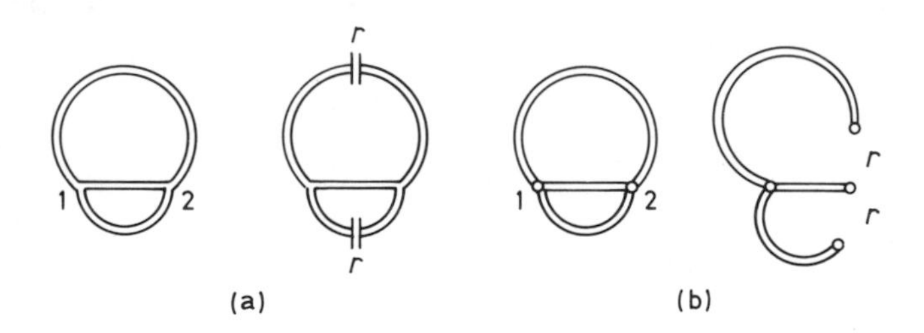

Figure 7.9

Solution

(a) There are two rings, three members and two nodal junctions (vertices), and the degree of statical determinancy is

$$I = 3(M - V + 1) - \Sigma f$$
$$I = 3(3 - 2 + 1) - 0 = 6$$

Hence the degree of statical indeterminancy is 6, and we note that for the case when $\Sigma f = 0$ the degree of statical indeterminancy in a two dimensional stiff jointed frame is equal to three times the number of rings.

(b) There are two rings, three members and two nodal junctions (vertices) each of which has one degree of (rotational) freedom hence

$$I = 3(M - V + 1) - \Sigma f$$

where $\Sigma f = (2.1) = 2$.

Whence

$$I = 2(3 - 2 + 1) - 2 = 4$$

(N.B. We observe that node 1 (or node 2) may be assembled without strain and since the uncoupled node is a three-link hinge there is no rotational constraint on the three members. If the top ring is offered to the straight floor beam, displacements, hence forces, in the x and y directions are required to mate them and similarly for the lower ring. There are thus four degrees of indeterminancy as predicted).

The situation in three dimensions may be resolved by a similar route.

If the number of rings is again accounted by

$$R = M - V + 1$$

then the degree of statical indeterminancy in a three dimensional complex of wholly stiff rings is

$$I = 6R = 6(M - V + 1)$$

and the degree of statical indeterminancy in a three dimensional complex of wholly stiff rings into which g additional freedoms of type f have been introduced is given by

$$I = 6(M - V + 1) - \sum_{g=1}^{g} f \tag{7.6}$$

(e.g. if a single pin hinge is introduced $f = 1$, $g = 1$ hence $\Sigma f = 1$. If two such hinges are introduced $f = 1$, $g = 2$ hence $\Sigma f = 2$).

Example

(a) Determine the degree of statical indeterminancy in the four-member, four-node stiff-jointed space frame, (b) compare result (a) with the number of degrees of freedom in a similar frame when all four nodes are ball and socket joints.

Solution

(a) A four member, four node, stiff jointed space frame contains no additional f freedoms. Hence its degree of statical indeterminancy is equal to the number of rings.

There is only one ring and hence the degree of statical indeterminancy is 6 and this agrees with equation 7.6.

(b) If ball and socket joints are incorporated at all four nodes there are $(4.3) = 12$ additional freedoms. Hence the statical indeterminancy in a four-member, four-vertex, ball jointed space frame is $(6 - 12) = -6$. Freedoms are clearly negative constraints and hence there are $+6$ degree of kinematic freedom.

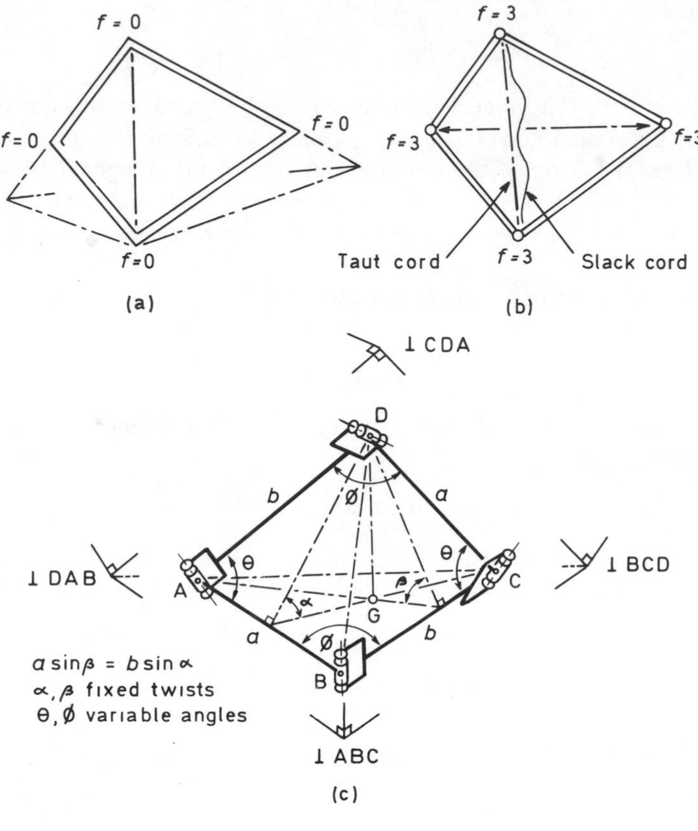

Figure 7.10

(N.B. The reader will find an alternative approach to the assessment of mechanisms in the following article). The mechanisms at (c) is known as Bennetts linkage.

7.1.7 Kinematic freedom and constraint

The converse of statical indeterminancy is kinematic freedom. Any closed assembly of long slender links, for example, which is one link short of meeting the requirements of statical self stiffness, is an articulate mechanism, with one degree of potentially controllable freedom, and mechanisms with two and more degrees of control may also be designed.

The four link undiagonalised assembly, Fig. 7.11, is known as a four bar chain and this is the basic articulation on which countless planar mechanisms are based.

It is easy to see that a single binary link has three degrees of freedom in 2-space, comprising two translational and one rotational freedoms as shown in Fig. 7.11. Two separate links have six degrees of freedom, but the same two links

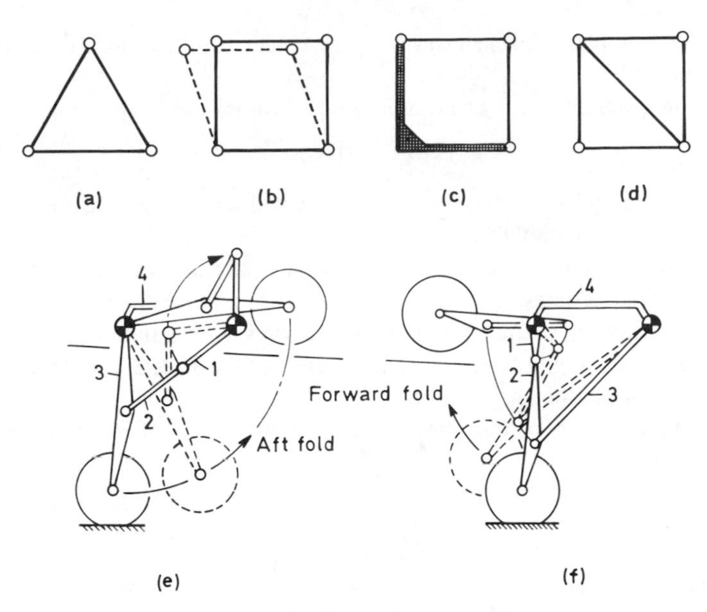

Figure 7.11

when coupled end to end by a single hinge have but four degrees of freedom. A simple pin hinge introduced along the length of a link effectively divides that member and introduces one additional freedom. The two operations of joining two links and splitting a single link, by means of a simple hinge, are clearly complementary.

Three links, indeed any number of links may be coupled end to end but the most important case is that in which the free end of the fourth link is coupled by a hinge to the free end of the first link. Thus producing a four bar chain.

As we have seen, each link has three degrees of freedom and each pin eliminates two degrees of freedom. If there are M binary links and J in-plane hinges, there are 3M – 2J freedoms. A four bar chain which is free to flop around in 2-space is, however, of little practical utility and it is, for this reason, necessary to ground one link. (N.B. Any link may be grounded, the four alternatives being known as inversions). Grounding a link entails removing three degrees of freedom from that link, thereby reducing the number of overall freedoms by that amount. Subtracting 3 from the total of free-freedoms, yields 3M – 2J – 3 and hence the equation defining a plane mechanism with one degree of freedom is

$$3M - 2J - 3 = 1$$

whence

$$3M - 2J - 4 = 0$$

It will further be noted that since $3M - 2J - 4 = 0$, $\frac{3}{2}M - J - 2 = 0$ and it follows that all plane mechanisms with one degree of freedom have an even complement of links.

The more general form of the mobility equation is

$$3(M - 1) - 2J = F \quad (7.7)$$

where F is the number of degrees of freedom and not to be confused with the number of faces F in a polyhedron.

(N.B. The number of degrees of freedom in a mechanism is always positive, usually 1, but sometimes 2, as in a differential. The number of degrees of freedom in a statically determinate structure is always 0. The number of degrees of freedom in a statically indeterminate structure may be 1, 2, 3. . .*n* depending on the degree of indeterminancy built-in).

When more than two links meet at a hinge the count of freedoms is of the form $j = (m - 1)$.

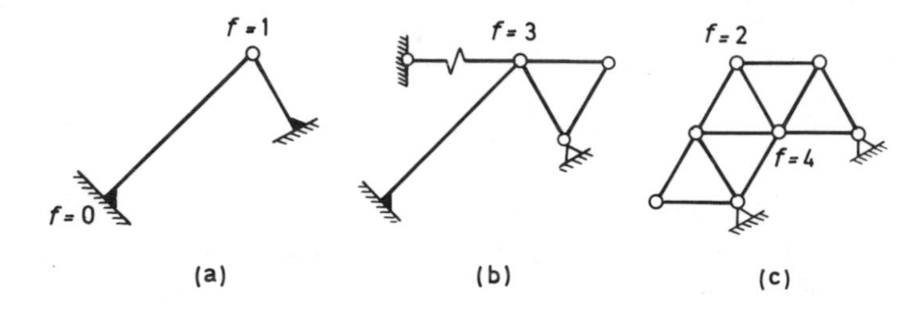

Figure 7.12

There are many practical applications where it is necessary to convert a mechanism into a structure and vice versa. An aircraft undercarriage is a 'mechanism' which needs to be both kinematically articulate and structurally rigid. A symbolic representation of an articulate structure containing a linearly extensible jacking link is shown in Fig. 7.13.

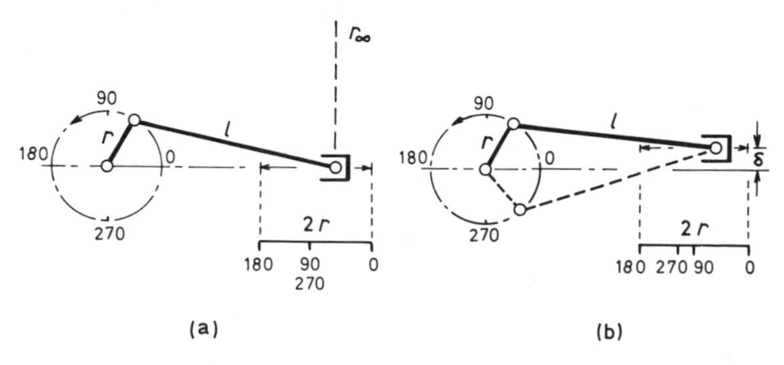

Figure 7.13

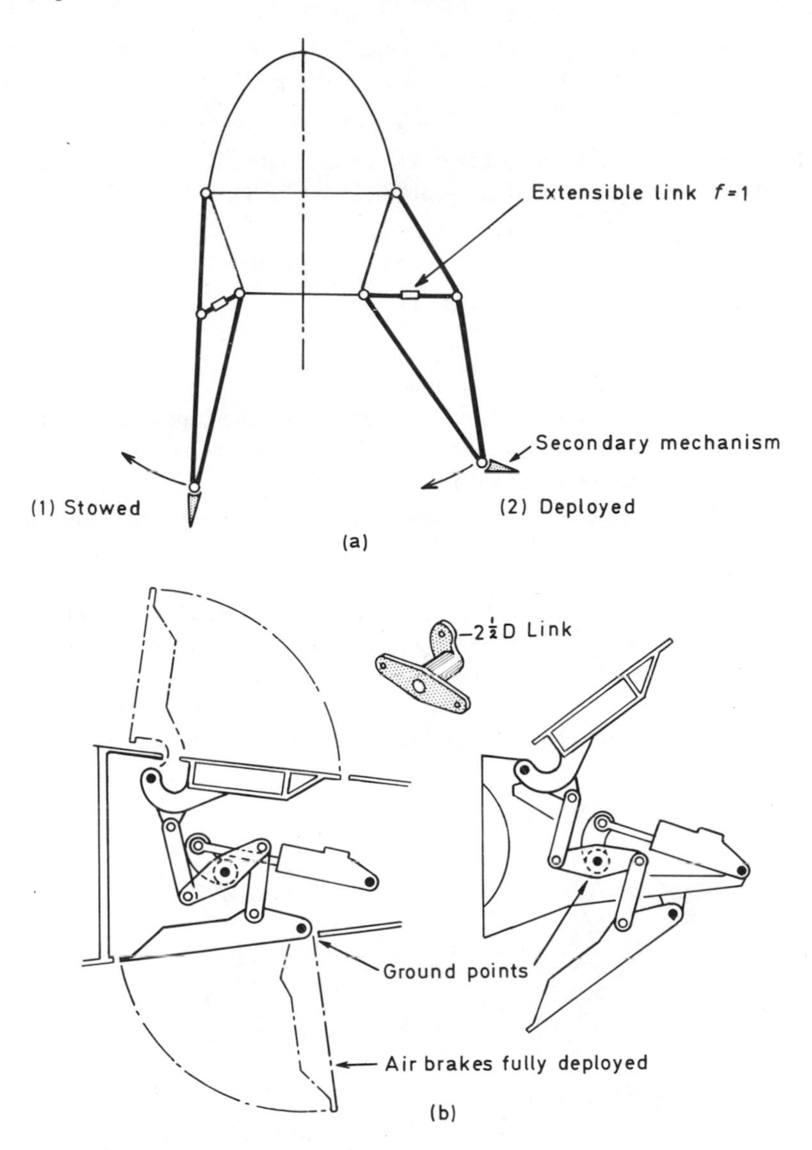

Figure 7.14

7.1.8 Slider crank variation

As seen in the preceding article the freedoms in mechanisms are not necessarily all rotational and it is clear that by increasing the radius arm of one link to infinity, a slider crank mechanism is born. When the line of sliding passes through the first pivot centre the cyclic kinematic motion is symmetric. When the line is offset the motion is asymmetric. A slider crank mechanism is essentially a rotary to linear motion device. It is not generally possible to operate the mechanism as a linear to rotary device.

(N.B. We observe that the count of links and freedoms in a slider crank obeys the mobility equation 7.7, that is $F = 3(4 - 1) - (2 \times 4) = 1$).

7.1.9 Mechanisms with higher order links

We have seen that there is but one four bar chain, the topological content of which cannot be altered and we also recall that in a plane one degree of freedom mechanism there are always an even complement of links.

The possibility of using links with more than two hinge freedoms is intuitively apparent and mechanisms with three hinge, ternary links, and four hinge quarternary links abound.

It is easily seen that there are two possible ways of assembling a six bar chain and four possible ways of assembling an eight bar chain. The two six bar chains, known as the Watt and Stephenson linkages are shown in Fig. 7.15, along with the practical manifestations. The six link Watt mechanism was used on early beam pumping engines. The six line Stephenson mechanism was used on early locomotives.

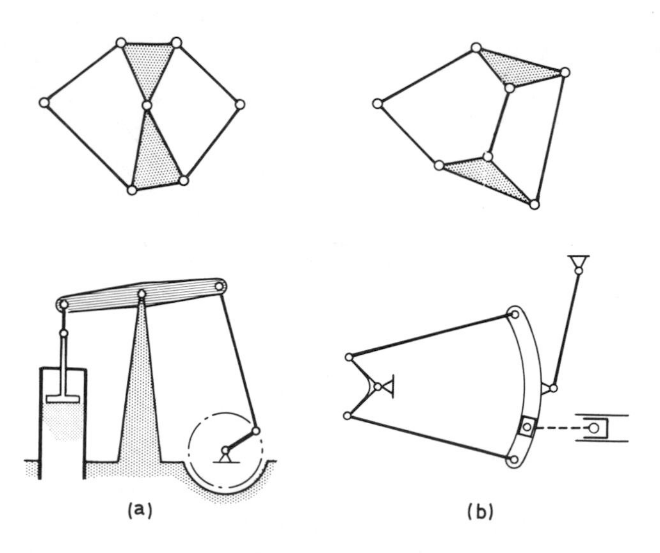

Figure 7.15

7.1.10 Minimum number of binary links

The minimum number of binary links in a plane mechanism is four, and the total number of links in a plane mechanism is therefore

$$M = M_2 + M_3 + M_4 \ldots M_i = \sum_{i=2}^{i} M_i$$

where M_i denotes the number of links with i hinges.

Every binary link connects to two other links, every ternary link connects to three other links and so on.

For mechanisms with two links per hinge

$$2J = 2M_2 + 3M_3 + 4M_4 \ldots iM_i = \sum_{i=2}^{i} iM_i$$

and the condition for one degree of freedom is

$$2J - 3M + 4 = 0$$

hence

$$(2M_2 + 3M_3 + 4M_4 \ldots) - 3(M_2 + M_3 + M_4 \ldots) + 4 = 0$$

and since M_2 cannot be less than 4, the expression

$$M_2 = 4 + \sum_{i=4}^{i} (i - 3) M_i$$

gives the minimum number of binary links.

It is a matter for further study to determine the condition which governs the number of binary links, when more than two members meet at a hinge.

7.1.11 Maximum number of nodes

The number of binary links that may emanate from a link with numerous hinges is equal to the number of hinges. A kinematic chain with one degree of freedom may thus be assembled as in Fig. 7.16, from which it is apparent that there are $(1 + i) + (1 - i) = 2i$ hinges (nodes) in all. Each 'radial' binary link apart from the first and last may be converted to a ternary link as shown in the figure. The total number of hinges J is $(i + 2) + 2(i - 2) = 3i - 2 = J$ and since

$$F = 3(M - 1) - 2J$$

$$F = 3M - 6i + 1$$

and it follows that when $F = 1$, $i = M/2$ is the maximum number of hinges per link in a plane mechanism with one degree of freedom.

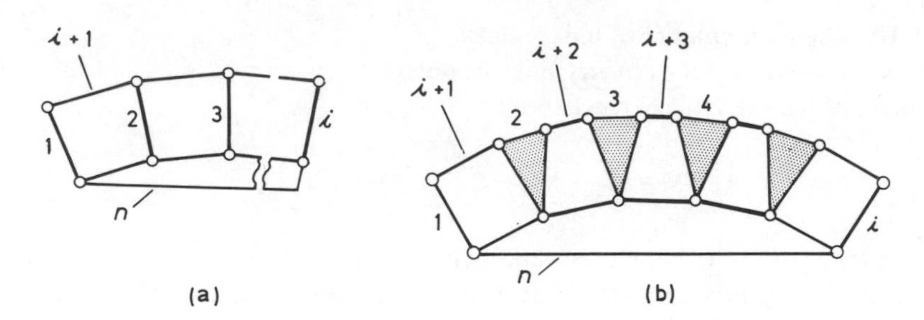

Figure 7.16

7.1.12 Plane mechanisms with compound hinges

As we have seen a plane pin hinge has one degree of rotational freedom, a two axis hinge has two degrees of freedom and a ball type hinge has three degrees of freedom and 3 is the maximum number of freedoms that a 'single' compound hinge can provide. Thus there are $(3 - f)$ freedoms per compound hinge in general. If there are J such hinges

$$F = 3(M - 1) - J(3 - f)$$

(N.B. When $f = 1$ we have plane pin hinges and the above expression reduces to $F = 3(M - 1) - 2J$ as expected).

If we have hinges of different types, as we have in the cam drive, Fig. 7.17. Then

$$F = 3(M - 1) - J_1(3 - f_1) - J_2(3 - f_2)$$

or simply

$$F = 3(M - 1) - 2J_1 - J_2 \tag{7.8}$$

(N.B. when $f = 2$ we have a pin hinge which is free to rotate and to slide along the length of an adjacent, or an open cam like contact. There is no term in f_3 because in a plane mechanism out of plane motions are not accounted and a term of the form $J_3 (3 - f_3)$ where $f_3 = 3$ reduces to zero.)

In the cam mechanism with higher pair open closure, Fig. 7.17, hinges A and B each have one degree of freedom, whilst the open contact at C has two degrees of freedom hence $M = 3, J_1 = 2, J_2 = 1$ and ($f_1 = 1, f_2 = 2$ as always).

$$F = (3.2) - (2.2) - (1.1) = 1$$

and this means that a given input motion produces a unique output motion (conditional to the cams remaining in contact).

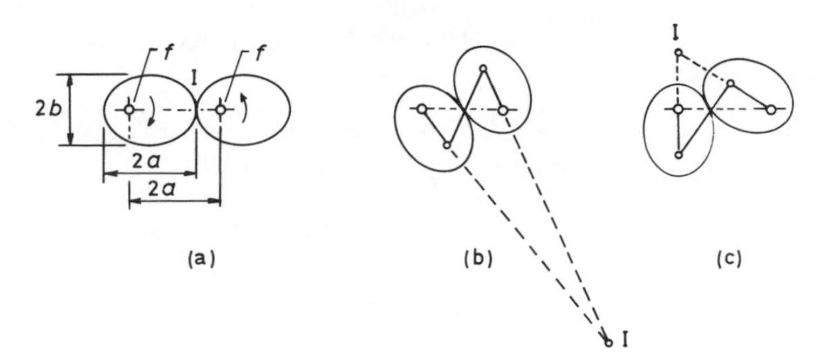

Figure 7.17

7.1.13 Classification of mechanisms

A system of classification used to define mechanism types, which have a mixture of hinges, is outlined in Fig. 7.18. The nomenclature is self explanatory. R denotes a revolving pair (joint/hinge), P is sliding piston, C a cam contact, etc.

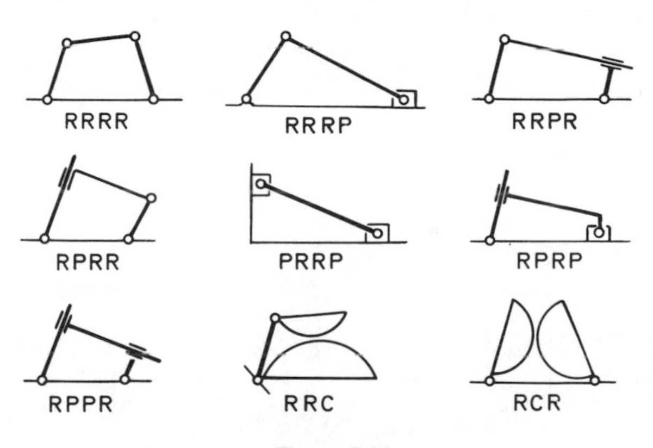

Figure 7.18

7.1.14 Three-space mechanisms with compound hinges

The mobility of a mechanism with compound hinges, free to articulate in 3-space may be deduced from the 2-space case. Each individual link has potentially six degrees of freedom and M uncoupled links have 6M degrees of freedom, and (6M − 6) represents the number of degrees of freedom when one link is grounded. If the M links are coupled by J hinge joints, then 6J freedoms would be eliminated if all J hinges were completely fixed. Now a single pin hinge adds one degree of freedom, a double pin hinge adds two degrees of freedom and a triple (ball and socket) hinge adds three degrees of freedom. If there are J_1 hinges with one degree of freedom, J_2 hinges with two degrees of freedom, J_3 hinges with three degrees of freedom. Then

$$F = 6(M - 1) - 6J + 1J_1 + 2J_2 + 3J_3$$
$$= 6(M - J - 1) + 1J_1 + 2J_2 + 3J_3 \tag{7.9}$$

where $J = J_1 + J_2 + J_3$ and J_1, J_2, J_3 are the number of hinges with 1, 2, 3, degrees of freedom.

The above equation originally due to Gruebler, may be written

$$F = 6(M - 1) - \sum_{J=1}^{J} (6 - f)$$

Example

An interesting 3-space mechanism designed by Stewart, as a space flight training simulator is shown in Fig. 7.19. The design requirement is, or rather was, to produce a mechanism which has from 0-6 controllable degrees of freedom.

Solution

The solution proposed by Stewart is self explanatory, see Fig. 7.19, and the accompanying tables summarise the two extremes. When F = 0 the mechanism becomes a self-stable structure, when F = 6 it becomes a fully articulated 3-space mechanism.

When all hinges are active

Node	1	2	3	Σf
Top	3	3	3	9
Legs	1	1	1	3
Base	2	2	2	6
				$\Sigma 18$

whence

$$F = 6(8 - 1) - (6.9) + 18 = 6$$

when all hinges are locked

Node	1	2	3	Σf
Top	3	3	3	9
Legs	0	0	0	0
Back	1	1	1	3
				$\Sigma 12$

whence

$$F = 6(8 - 1) - (6.9) + 12 = 0$$

(N.B. The reader should note that $J_1 = 3, J_2 = 3, J_3 = 3$ when all hinges are free and $J_1 = 0, J_2 = 0, J_3 = 0$ when all hinges are locked. The above results may then be verified by application of equation 7.9.)

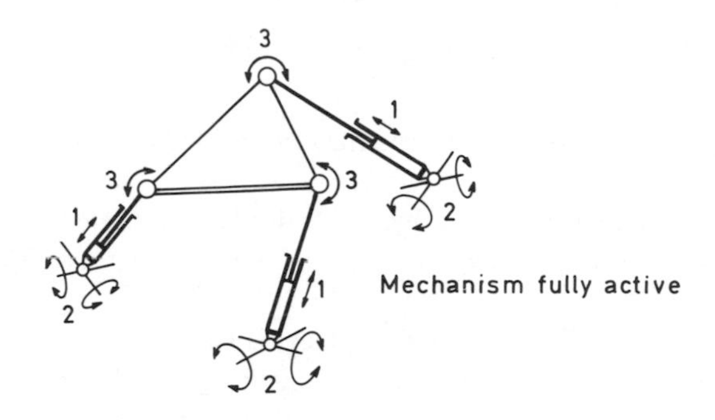

Figure 7.19

7.2 SYNTHESIS OF CONVEX DELTAHEDRA

A convex figure is one in which any line drawn between any two points on the boundary is contained wholly within the figure. A circle is a convex figure, a kidney figure is not. A singly connected figure is one in which there is one and only one arbitrary path between any two points on the figure.

A deltahedron is a closed polyhedral figure in which the surface is wholly convex and made up entirely of delta faces.

A delta face is a triangular face in which all three sides and all three internal angles are equal.

F isolated delta faces have 3F vertices and 3F edges, but when two faces are brought together, a common edge is shared.

In any closed and singly connected figure, there are two faces per edge, and since there must be an integral number of faces, the number of edges must be even. In a complete deltahedron with 3F faces, there are 3F/2 edges in all.

It will be recalled that in a singly connected figure, the maximum number of delta faces per vertex is six. These six delta faces either lie flat in a plane, or lie three and three at a dihedral angle, as shown in Fig. 7.20.

In a wholly convex figure, the maximum number of delta faces per vertex is five and in a closed figure, the minimum number of delta faces per vertex is three. Three, four or five delta faces per vertex are the only tenable choices.

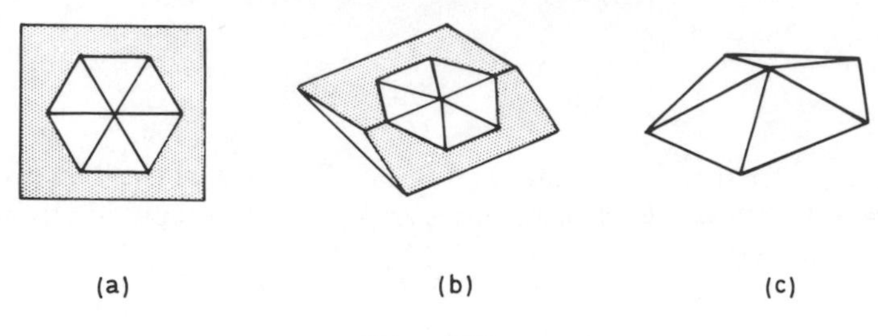

Figure 7.20

Lemma 1 Every deltahedron (which is singly connected) conforms to the Euler Law of structure $V - E + F = 2$.

Lemma 2 Every deltahedron which has a statically determinate structure conforms to the statical law of structural self-stiffness $E = 3V - 6$.

7.2.1 The maximum number of pure delta faces is twenty

If every face is convex with respect to its immediate neighbours, the case of six deltas at a vertex may be excluded and the most complex deltahedron cannot contain more than five edges and five faces per vertex. In a closed figure, every edge is associated with two vertices and hence for the particular case of five edges and five delta faces at every vertex, it follows that there must be $5V/2$ edges in all. That is:

$$2E = 5V$$

From the law of statical self-stiffness:

$$E = 3V - 6$$

whence

$$2E = 6V - 12 = 5V$$

and hence for this particular case $V = 12$. Substituting the value of V into the Euler equation, Lemma 1, yields the corresponding number of edges and faces. The figure with $V = 12$, $E = 30$, $F = 20$ is called an icosahedron and is the deltahedron with the maximum number of faces.

7.2.2 The minimum number of pure delta faces is four

A minimum of three delta faces per vertex is required to make a figure convex and hence there must be at least three edges and three faces per vertex.

For the particular case of three edges and three delta faces per vertex, it follows that there must be $3V/2$ edges in all. That is:

$$2E = 3V$$

From the law of statical self-stiffness:

$$E = 3V - 6$$

whence

$$2E = 6V - 12 = 3V$$

and hence for this particular case, V = 4 and E = 6. Substituting these values into the Euler equation, yields the corresponding number of edges and faces. The figure with V = 4, E = 6, F = 4 is called a tetrahedron and is the deltahedron with the minimum number of faces.

7.2.3 There are only eight pure deltahedra in all

Bearing in mind that in a simply connected figure there can only be 3, 4 or 5 delta faces per vertex and that the total number of faces must be even, the apparent possibilities are: F = 4, 6, 8, 10, 12, 14, 16, 18, 20 (nine apparent possibilities in all).

The cases of F = 4, F = 20 have already been proved and it remains to establish the existence, or non-existence, of the remaining types.

Starting with the tetrahedron it is easily seen that the addition of a single vertex requires three additional edges, and that in a singly connected configuration there are no internal partitions. Hence this deltahedron has V = 5, E = 9, F = 6.

It is similarly easy to visualise a bi-pyramid (an octahedron) in which V = 6, E = 12, F = 8, and the cases for which V = 7, 8, 9 and 10 follow suit, but a deltahedron with V = 11 proves difficult to imagine.

7.2.4 Non-existence of a deltahedron with V = 11

We observe, from Lemma 1 and Lemma 2, that a deltahedron with V = 11, E = 27, F = 18 appears to be a possible structure but will prove that it is not.

Suppose there are x vertices with three edges, y vertices with four edges, z vertices with five edges (for these are the only possibilities). We observe that every edge is shared with two vertices and every face is shared with three vertices. We express these three conditions as follows:

$$x + y + z = 11.1 = 11 \quad \text{vertex condition}$$

$$3x + 4y + 5z = 27.2 = 54 \quad \text{edge condition}$$

$$3x + 4y + 5z = 18.3 = 54 \quad \text{face condition}$$

The equation for edges is, however, the same as the equation for faces and hence there are two equations in three unknowns and an algebraic solution is not strictly possible. We know, however, that x, y and z must all be zero or positive integral values and hence the solution may be obtained by deduction.

Multiplying the first equation by 3 and subtracting from the second equation we obtain:

$$y + 2z = 21$$

Multiplying the first equation by 4 and subtracting from the second equation we obtain:

$$-x + z = 10$$

Multiplying the first equation by 5 and subtracting from the second equation we obtain:

$$2x + y = 1$$

From the first of these (new) equations we conclude that: y cannot be zero and must be odd, otherwise z would be fractional. We note that if y is odd, z is even.

From the second of these (new) equations we conclude that: if z is even, x is either zero or even.

From the third of these (new) equations we conclude that: y cannot be zero, for this would make x fractional. Since y cannot be negative, $y = 1$ and hence $x = 0$.

Substituting this result into the first equation yields $z = 10$.

A deltahedron for which V = 11, E = 27, F = 18, must therefore consist of ten vertices with five edges and five faces plus one vertex with four edges and four faces.

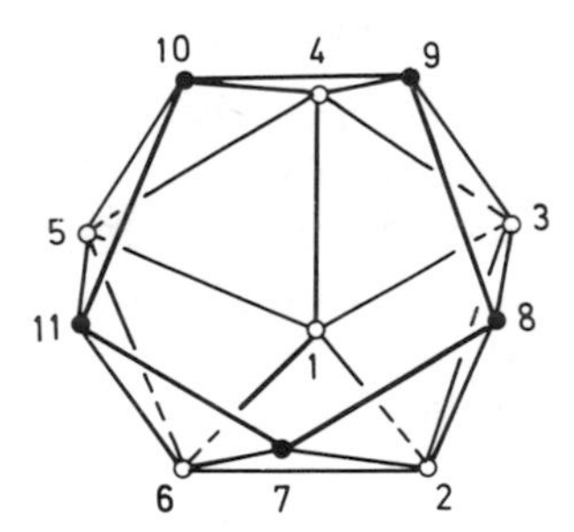

Figure 7.21

Regardless of the order in which the two types of vertex are taken there is only one possible outcome and this does not produce a close convex figure.

There are thus a total of eight singly connected deltahedra and this exhausts the scope of this approach.

The eight convex deltahedra are shown in Fig. 7.22.

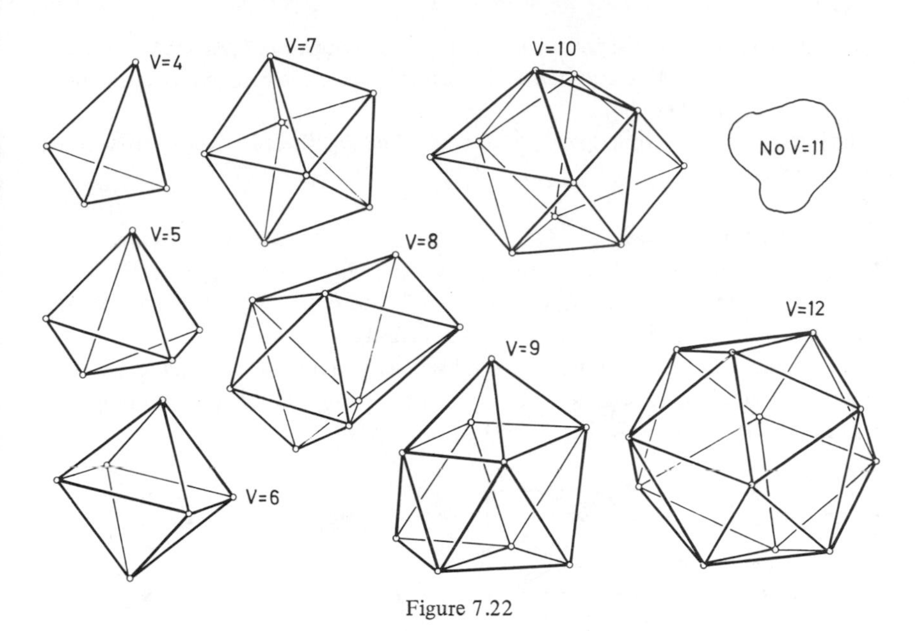

Figure 7.22

7.3 REGULAR OMNIDIRECTIONAL TRIANGULATION

A deltahedron is a closed convex figure which consists of equilateral triangular (delta) faces. All edge lengths are equal and all internal face angles are the same. The question is: how far can the process of pure delta triangulation be continued?

Starting with the simplest convex figure – a tetrahedron – we have V = 4, E = 6, F = 4 and it is a simple and obvious step to initially add further tetrahedral cells as shown in Fig. 7.23.

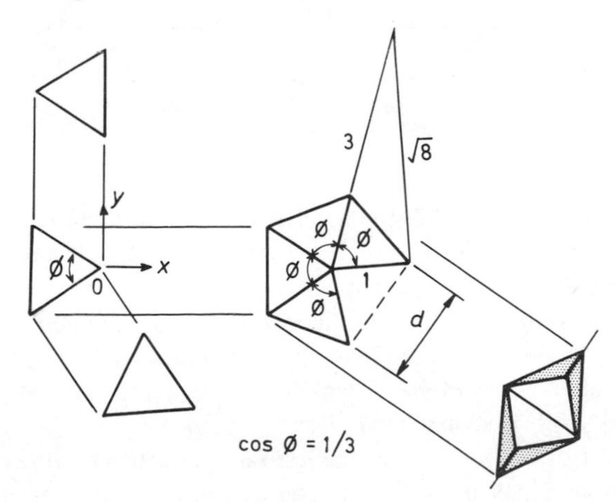

Figure 7.23

But as may be deduced from Fig. 7.23, the distance d is 8.9% greater than the tetrahedral edge length l and hence the tetrahedral space bridged by d cannot be filled by an equi-edged delta.

The conclusion is that space cannot be filled by the piecewise repetition of tetrahedra.

7.3.1 Fuller's octet truss

If in place of the original tetrahedron, we start with a unit edge octahedron (a convex deltahedron which has V = 6, E = 12 and F = 8 delta faces) we discover that the process of adding tetra and octa is infinitely extensible.

An octahedron is most easily visualised as a delta faced, square based bi-pyramid. A single octahedron is shown in Fig. 7.24(a) and a planar complex of eight octahedra is shown in Fig. 7.24(b).

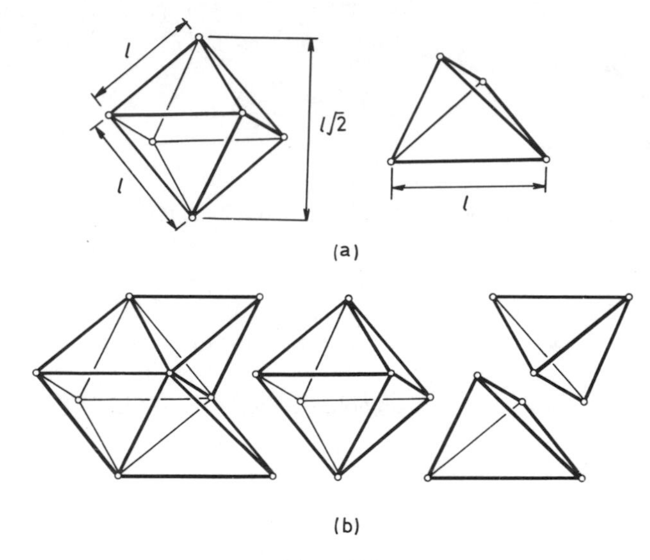

Figure 7.24

As may be seen from Fig. 7.24(b) the distance d is equal to the edge length of the parent unit edged octahedron and the space 0, 1, 2, 3, is a tetrahedral delta (as are all other corresponding spaces).

The manner in which unit edge tetrahedra and octahedra combine to form an omni-planar grid is shown in Fig. 7.24, and our understanding of this unstrained assembly of 'octet' modules has been greatly fostered by the practical and theoretical work of R. Buckminster-Fuller.

When an octahedron is viewed in the less familiar attitude shown in Fig. 7.25, the following points may be observed. The extreme profile projects to a perfect hexagon in plan. The true angle of the sloping face (as seen in side view) is equal

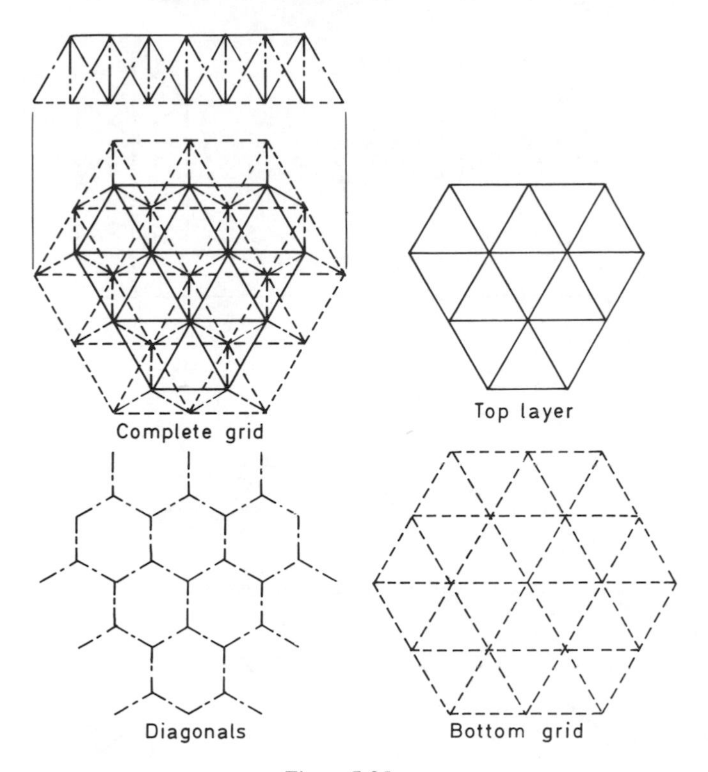

Figure 7.25

to the angle ϕ, and calculation proves that this angle ϕ is the same as the angle ϕ previously assigned to the sloping face of a tetrahedron.

The square cross-section shown in Fig. 7.25, is identified by its vertices 1, 2, 3, 4, and the corresponding positions of these vertices is as marked in side view and plan. It is to be observed that the square cross-section 1, 2, 3, 4, is accompanied by two other mutually perpendicular square cross-sections, identified by the numerals 1, 5, 3, 6, and 2, 6, 4, 5. The presence of these three mutually perpendicular cross-sections is clearly visible in plan.

The planar combination of octahedra and tetrahedra is shown in Fig. 7.26 and various aspects of the octet module are shown in Fig. 7.27 and 7.28.

A notable feature of octet structures is that whilst the entire structure is comprised of equi-length edges, bars or members, the distance between unbridged diagonal vertices is root-two times the unit edge length.

We note that the environment of every internal vertex is identical and consider the arrangement of vertices surrounding any chosen vertex centre. The centre O (previously used to locate our origin of axes) is a typical centre. If we extract the smallest convex 'shell' of vertices from around the centre O, we find that there are **twelve** surrounding vertices which are equi-distant from the centre O. Since

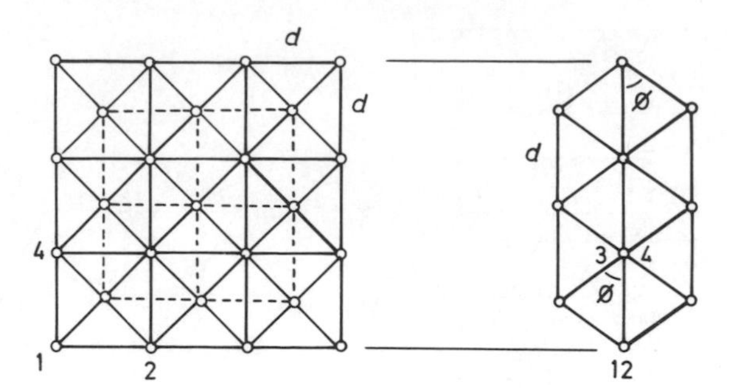

Figure 7.26

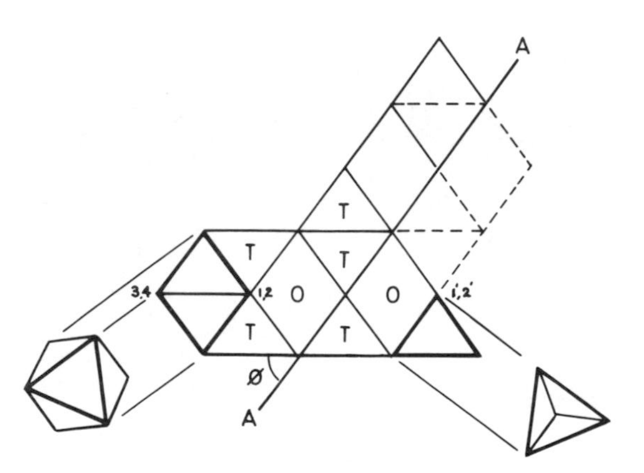

Figure 7.27

the twelve 'shell' vertices lie at a constant distance from the centre O, they lie on the surface of a sphere. These twelve vertices are arranged as in Fig. 7.29.

The convex structure shown in Fig. 7.29, is called a cuboctahedron or, as sometimes described by Fuller, a vector equilibrium.

As may be seen from Fig. 7.29, a cuboctahedron consists of 6 square and 8 delta faces. Each square face is the girth of an octahedron and each triangle is the face of a tetrahedron. Since all half-octa and all tetrahedra home to a common centre, there are 6 half-octa and 8 tetrahedra contained within the bound of the twelve vertex shell.

7.3.2 Close packing of equi-spheres

If we progressively grow a sphere about the thirteen vertex centres of the cub-octahedron, we find that there is exactly the right amount of space (in all directions) to accommodate thirteen identical spheres or conversely, if we start with a single central sphere and progressively pack equi-spheres as closely as

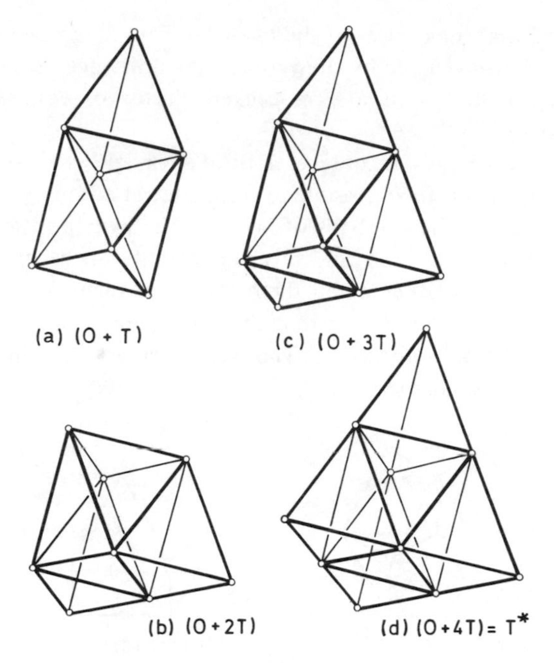

Figure 7.28

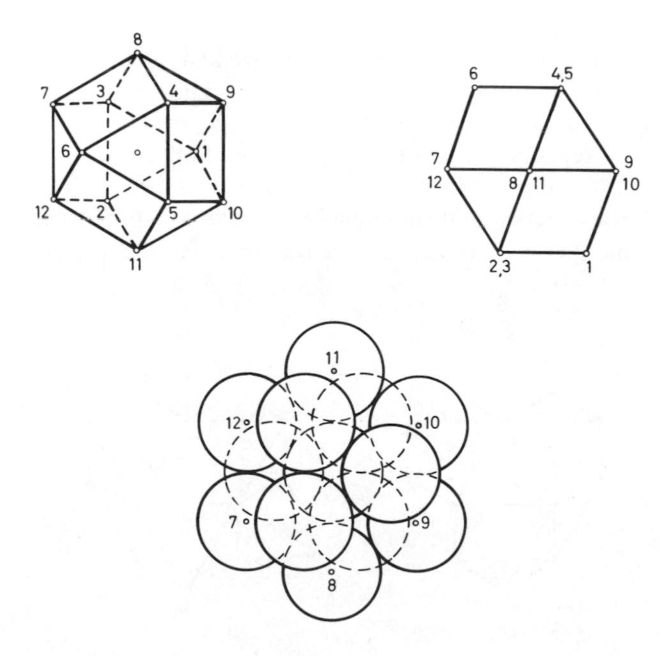

Figure 7.29

possible, we find that the centres of the twelve surrounding spheres lie in exactly the same positions as occupied by the vertices of a unit edge cuboctahedron.

The building sequence for a close-packed cluster of equi-spheres may be pursued as follows:

First, arrange three balls in contact, to form a triangle as in layer 1, Fig. 7.30. In the absence of slip it is then possible to form a tetrahedron by placing a fourth ball in the cusp formed by the balls of layer 1. The axial position of this ball is indicated by a cross in Fig. 7.30, and this ball becomes the centre-ball of layer 2. There are seven balls in layer 2 and these form a hexagonal pattern which fits snugly into layer 1, in such a way that the three balls of layer 1 nest into the cusps formed by the balls in layer 2. The balls of layer 3 nest into the alternate cusps of layer 2, in a similar way.

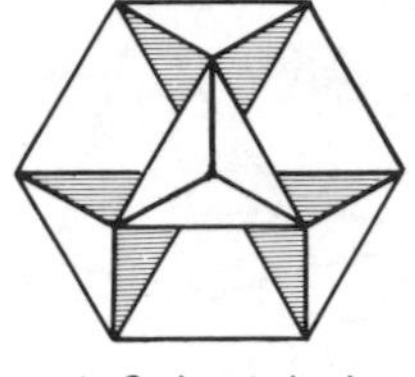

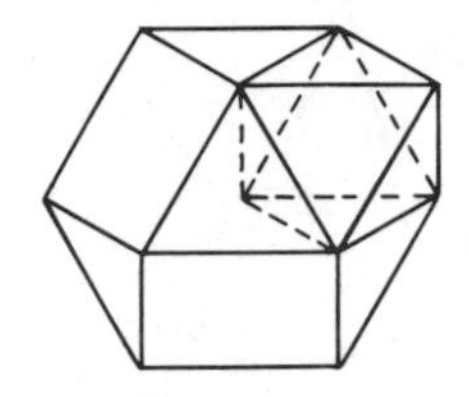

(b) Octahedron inset

Figure 7.30

7.3.3 Relation between octet truss and closely packed spheres

If we take half a cuboctahedron as shown in Fig. 7.31(a), we may add three half-octahedra to each square face and fill the three lower corners (so created) by adding three tetrahedra. We may then proceed to add three further tetrahedra to the corners of the top face and insert a complete octahedron (diagonally) into the central space between them. Finally, we add one tetrahedron above the central octa and thereby complete a three frequency tetrahedron.

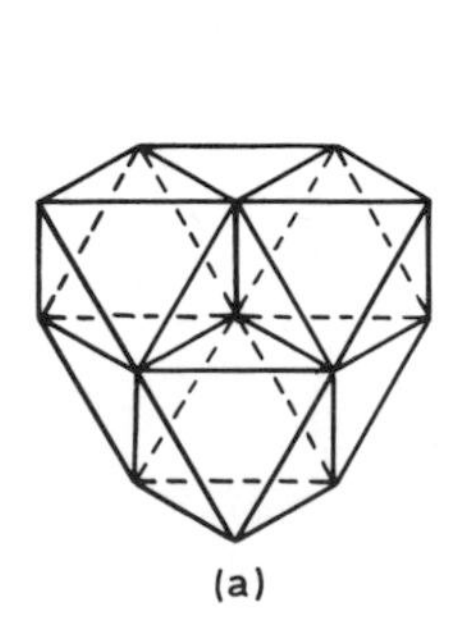

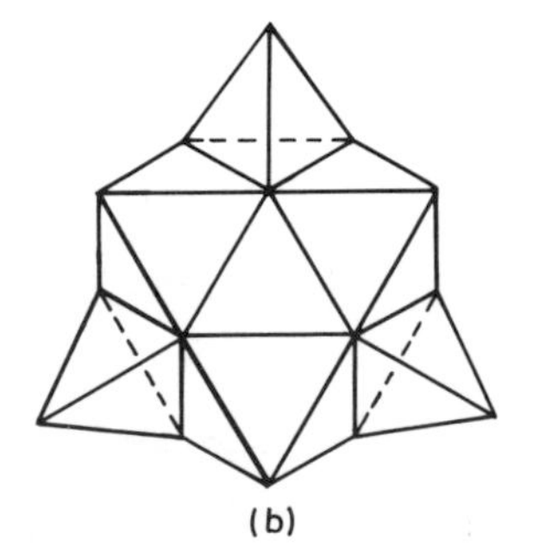

Figure 7.31

7.3.4 The three frequency tetrahedron and octahedron

The two frequency tetrahedron has been derived from, and is consistent with, the all space filling octet grid. The two frequency tetrahedron, Fig. 7.28, is such that a unit tetrahedron lies at each apex and a unit octahedron occupies the centre.

A notable feature of the octet grid is that twelve half-lines radiate from every inner vertex and as may be seen from previous figures, these lines are straight and continuous (though in practice, joints may intervene).

A three frequency tetrahedron may likewise be regarded as a single tetrahedral element, which forms part of a three frequency cuboctahedron.

A three frequency octahedron may be devised via a similar route, and may also be regarded as a single element, suitable for inclusion as part of a three frequency cuboctahedron.

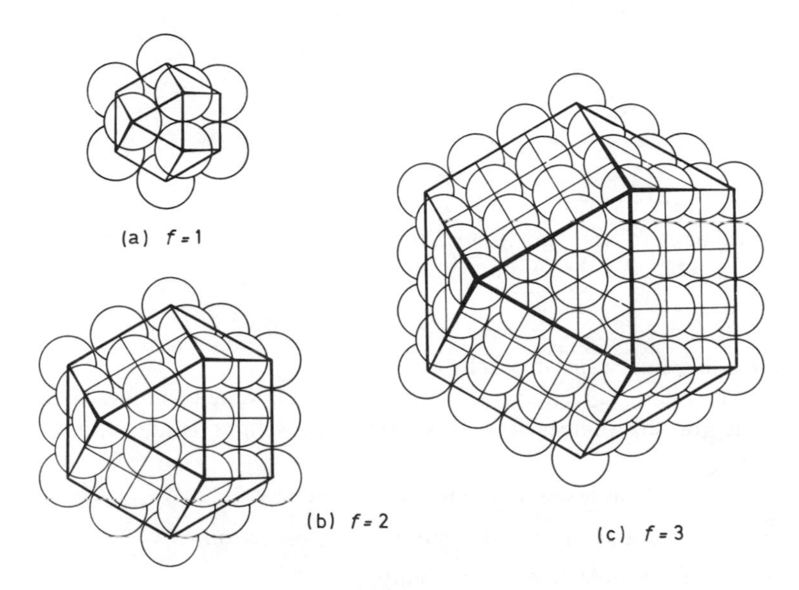

Figure 7.32

7.3.5 Three frequency cuboctahedron

A three frequency cuboctahedron is a cuboctahedron comprised of six, half-three frequency octahedra plus eight three frequency tetrahedra. The fourteen faces of a three frequency cuboctahedron appear divided, as shown in Fig. 7.32.

If a sphere is grown (as previously) from every vertex centre, the space within the three frequency cuboctahedron will fill with equi-diameter unit spheres. As previously shown, there are twelve external spheres in a one frequency cubocta-hedron and it is now proposed to count the number of external vertices and hence spheres contained by a two, three and f frequency cubocta.

7.3.6 The f frequency cuboctahedron

In order to count the number of nodes in the shell of an f frequency cuboctahedron, we start by counting the number of nodes per face. Since the surface shell of a cuboctahedron consists of squares and triangles there are two different arrangements to consider.

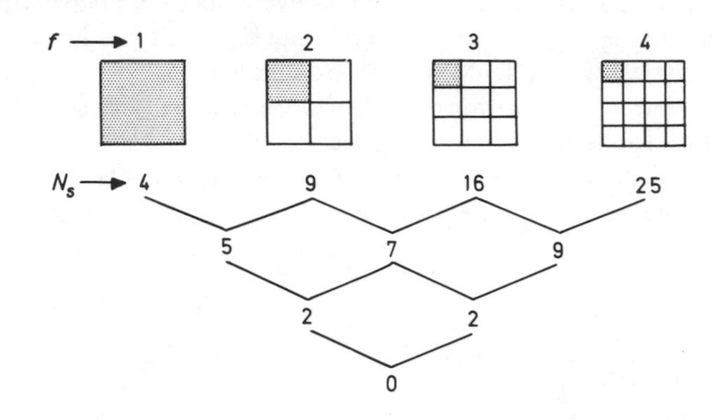

Figure 7.33

As may be seen from the finite difference table, Fig. 7.33, the equation connecting f and N_s is of the form:

$$N_s = af^2 + bf + c$$

where a, b and c are constants to be determined.

Using the known relationships between f and N_s, and noting that for a cuboctahedron the number of faces per vertex $q = 4$ and the number of faces per edge $n = 2$:

A node within the boundary counts 1

A node on an edge but not at a corner counts $\frac{1}{2}$

A node on a corner counts $\frac{1}{4}$

we may write the equations:

$$1 = a + b + c$$

$$4 = 4a + 2b + c$$

$$9 = 9a + 3b + c$$

Subtracting the first equation from the second we obtain:

$$3 = 3a + b$$

Subtracting the second equation from the third we obtain:

$$5 = 5a + b$$

Hence $a = 1, b = 0, c = 0$, and

$$N_s = f^2$$

A corresponding expression may be obtained for a triangular face, as follows: As may be seen from the finite difference table, Fig. 7.34, the equation connecting f and N_Δ is of the form:

$$N_\Delta = af^2 + bf + c$$

where a, b and c are constants to be determined.

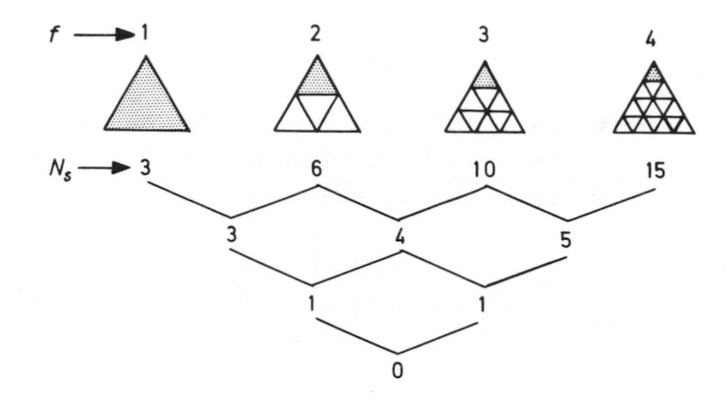

Figure 7.34

Using the known relationship between f and N_Δ and noting that for a cuboctahedron the number of faces per vertex $q = 4$ and the number of faces per edge $n = 2$:

A node within the boundary counts 1

A node on an edge but not at a corner counts $\frac{1}{2}$

A node on a corner counts $\frac{1}{4}$

We may write the equations:

$$0.75 = a + b + c$$

$$2.25 = 4a + 2b + c$$

$$4.75 = 9a + 3b + c$$

Subtracting the first equation from the second we obtain:

$$1.50 = 3a + b$$

Subtracting the second equation from the third we obtain:

$$2.50 = 5a + b$$

Hence $a = \frac{1}{2}$, $b = 0$, $c = \frac{1}{4}$ and

$$N_{\Delta} = \frac{f^2}{2} + \frac{1}{4}$$

In a cuboctahedron there are six square and eight triangular faces and hence the number of surface nodes in an f frequency cuboctahedron equals N where

$$\begin{aligned} N &= 6N_s + 8N_{\Delta} \\ &= 6f^2 + 4f^2 + 2 \\ &= 10f^2 + 2 \end{aligned}$$

Using the expression $N = 10f^2 + 2$ we may calculate the total number of surface nodes in an f-frequency cuboctahedron. When $f = 1$ there are 12 surface nodes, when $f = 2$ there are 42 surface nodes and when $f = 3$ there are 92 surface nodes and so on.

In order to account for the total number of vertices in an f-frequency cuboctahedron, which includes all the associated internal partitions, it is necessary to sum the number of vertices per layer and add one to account for the central vertex.

(N.B. We have used the term nodes to count the number of internal point crossings in the plane face of the polyhedron. We now consider these nodes as members of a vertex complex. The term vertex is therefore reintroduced. The terms node and vertex are commonly used in either sense.)

If we denote the frequencies of each layer by $f_1, f_2, f_3 \ldots f_n$, we merely add the number of vertices per layer as follows:

$$N = 10(f^2_1 + f^2_2 + f^2_3 \ldots + f^2_n) + 2fn + 1$$

When $f_1 = 1$ the total number of vertices is $(10 + 2 + 1) = 13$

When $f_2 = 2$ the total number of vertices is $(10 + 40 + 4 + 1) = 55$

When $f_3 = 3$ the total number of vertices is $(10 + 40 + 90 + 6 + 1) = 147$

Fuller has shown that for any omnitriangulated structure

$$x = 2nf^2 + 2 \tag{7.10}$$

7.3.7 Icosahedral close packing of spheres

At this juncture, it is pertinent to note that the surrounding vertices of an icosahedron may be spheroidalised to form a close-packed cluster, in much the same way as was done for the cuboctahedron. The principal difference between the close-packed cluster derived from the icosahedron and that previously derived for the cuboctahedron is two fold. Firstly, the icosahedron has five edges per vertex and all surface faces are triangular and this accounts for the fact that the surface spheres of an icosa-cluster lie on a smaller circle. As a direct consequence of this arrangement the central void is smaller than that of the cuboctahedron. The surface spheres of the icosahedron are thus of unit diameter but the centre nuclear sphere is smaller.

It is an easy matter to verify that the diameter of this smaller sphere is 0.902 or approximately 8% less than the unit value.

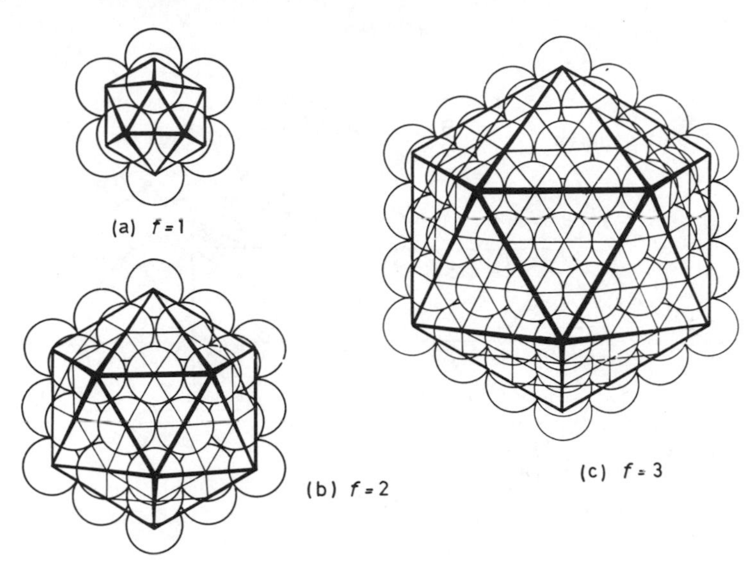

Figure 7.35

The plan and elevation of an icosahedron is shown in Fig. 7.36, and reference should be made to this figure.

Since $A_1 B_1$ equals 0.5 cosec36 and edge **AB** is of unit length and seen true length in elevation,

$$\sin^{-1} \alpha = 0.8506508$$

$$\alpha = 58.282546°$$

$$2\alpha = 116.5650492°$$

hence

$$\theta = 63.4349508°$$

From Figure 7.36(b)

$$\sin \frac{\theta}{2} = \frac{0.5}{r + 0.5}$$

Hence

$$r = 0.45105649$$

and

$$(r + 0.5) = 0.951 = \text{diameter of centre sphere}$$

We note that tan 63.43495° = 2.

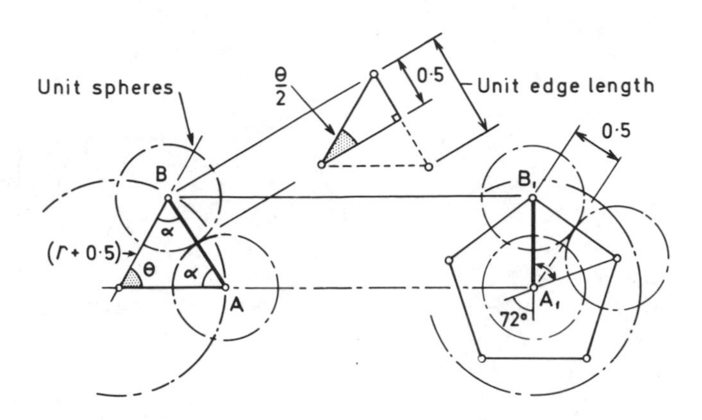

Figure 7.36

7.4 SPHERICAL TRIGONOMETRY

The sphere forms the basis of many dome-like structures and the calculation of its surface and constituent properties is a necessary part of design.

Every point on the surface of a sphere is equidistant from a fixed point called the centre. The straight line that joins any point on the surface to the centre is called the radius. The radius of a given sphere is constant.

Any plane section of a sphere is a circle. All sections of a sphere are circles, all sections that cut the sphere across a full diameter are great circles. All other sections are small circles.

All points that lie at the extremities of a full diameter are technically pole points. The North and South poles of the earth are particular pole points which lie normal to the great circle through the equator. To every pair of non-pole points there corresponds one unique small circle, but when two points are in direct diametral opposition there are an infinite number of orange-slice great circle sections passing through them. These various cross sections are shown in Fig. 7.37.

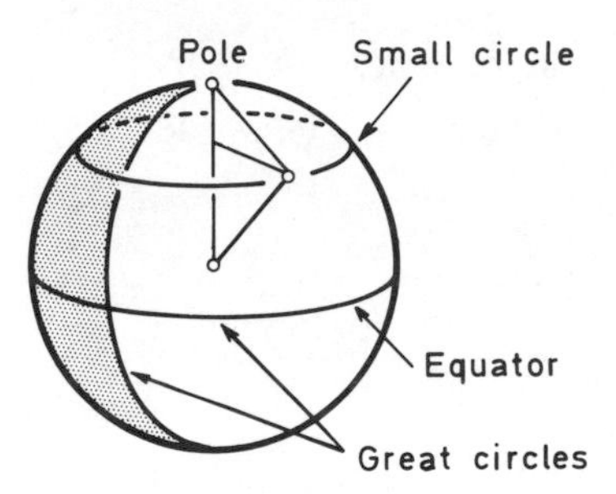

Figure 7.37

(1) The great circle arc drawn from the pole of a great circle to any point on its circumference is a quadrant. The angles shown in Fig. 7.38, are all 90°.

$$\angle \text{POA} = \angle \text{POB} = 90^\circ$$

$$\angle \text{POC} = \angle \text{POD} = 90^\circ$$

(2) The angle subtended at the centre of a sphere, by two great circles is equal to the angle between the two planes in which the two great circles lie. Thus in Fig. 7.38, the angle subtended by the pair of great circle arcs AP and BP is AOB = 90° and in Fig. 7.40 the angle between the two great circles AB and AC is the plane angle BOC = α.

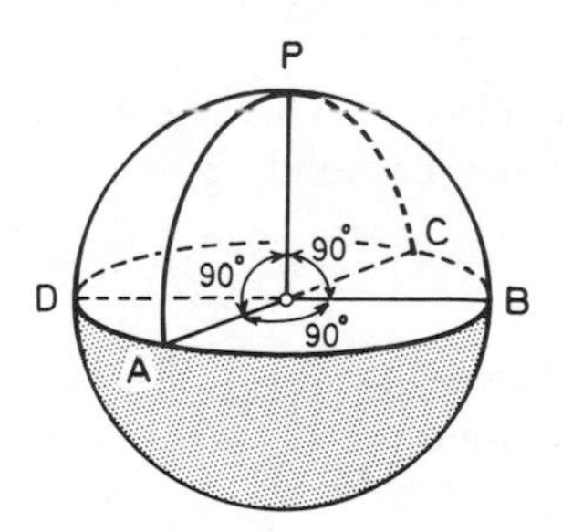

Figure 7.38

(3) The arc subtended by a small circle is related to the arc subtended by a parallel great circle.

$$\frac{\text{arc } bc}{\text{rad } ob} = \frac{\text{arc BC}}{\text{rad OA}}$$

$$\frac{\text{arc } bc}{\text{arc BC}} = \frac{\text{rad } ob}{\text{rad OB}} = \frac{r}{\text{R}}$$

whence

$$\frac{r}{\text{R}} = \sin\theta$$

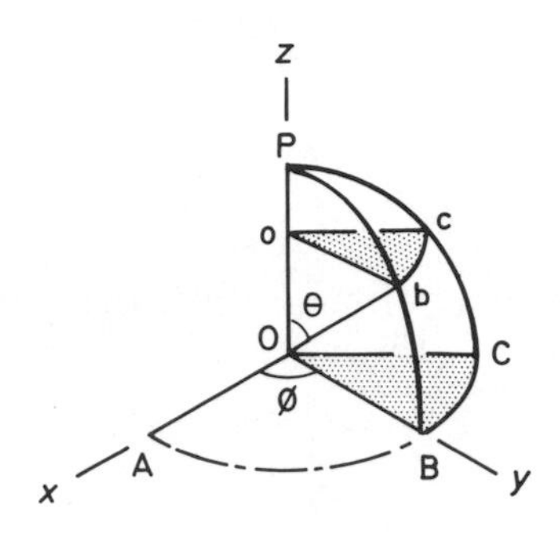

Figure 7.39

7.4.1 Spherical triangles

The smaller figure formed on the surface of a sphere by three intersecting great circles is called a spherical triangle. ABC, Fig. 7.40, is a spherical triangle.

The three planes containing the three great circles form a solid angle at the centre and meet as do the three faces and three edges at the apex of a plane tetrahedron.

(N.B. A demonstration piece to illustrate the relationship between plane, solid and spherical angles may be readily produced from card.

7.4.2 Spherical triangle notation

If the spherical triangle is denoted by capital letters A, B, C, then A, B, C are the three spherical angles contained by the spherical arcs adjacent to each of these angles. (N.B. This is so because the lines AD and AE, Fig. 7.40, are tangent to the sphere at A).

As in plane geometry the sides of spherical triangles are denoted by lower case letters. Thus the side opposite the spherical angle A is a, the side opposite the spherical angle B is b, the side opposite the spherical angle C is c. However, since the great circle arcs BC = a, AC = b, AB = c are proportional to their respective plane angles α, β, γ and since arc length = plane angle × radius. The arc lengths a, b, c, are equal to the plane angles α, β, γ, when the radius is of unit length.

7.4.3 The derivation of spherical trigonometric equations

If ABC is a spherical triangle, the radius OA is normal to the surface and the two lines AD, AE, are at right angles to OA, so that AD and AE are tangential to the surface at A.

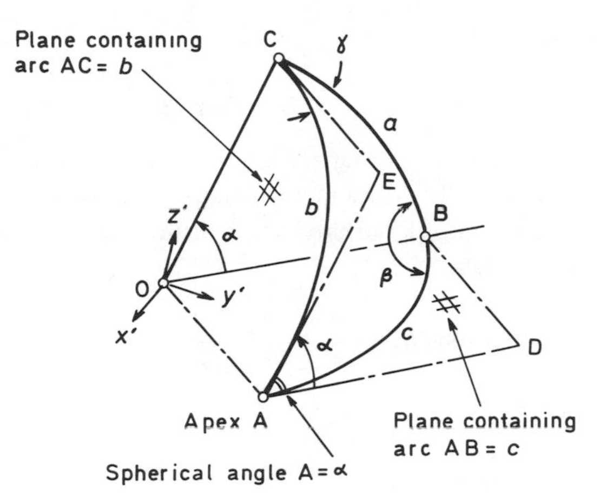

Figure 7.40

Using the cosine rule of plane trigonometry we have a relationship between the plane angle A and the plane angle a.

$$DE^2 = AD^2 + AE^2 - 2.AD.AE.\cos A$$

$$DE^2 = OD^2 + OE^2 - 2.OD.OE.\cos a$$

The first relationship is obtained from triangle ADE. The second relationship follows from triangle ODE.

$$\angle OAD = \angle OAE = 90^\circ$$

and

$$OD^2 = OA^2 + AD^2 \text{ and } OE^2 = OA^2 + AE^2$$

whence we obtain:

$$O = -AD.AE\cos A - OA^2 + OD.OE\cos a$$

by substitution and subtraction.

$$AD.AE.\cos A + OA^2 = OD.OE.\cos a$$

and

$$\frac{AD}{OD} \cdot \frac{AE}{OE} \cdot \cos A + \frac{OA}{OD} \cdot \frac{OA}{OE} = \cos a$$

expressed in terms of sines and cosines we have:

$$\sin c \,.\, \sin b \,.\, \cos A + \cos b \,.\, \cos c = \cos a$$

whence

$$\cos A = \frac{\cos a - \cos b \cos c}{\sin b \sin c}$$

Although it has been assumed that the sides containing A are less than quadrants, it is possible to show that the above relationship is true of any angle.

If true of the angle A the relationship is true of the angles B and C. Hence, by cycling subscripts we obtain:

$$\cos A = \frac{\cos a - \cos b \cos c}{\sin b \sin c}$$

$$\cos B = \frac{\cos b - \cos c \cos a}{\sin c \sin a} \qquad (7.11)$$

$$\cos C = \frac{\cos c - \cos a \cos b}{\sin a \sin b}$$

We may obtain the corresponding expressions for sine, by writing $1 - \sin^2 A = \cos^2 A$.

Whence

$$\sin A = \frac{\sqrt{1 - \cos^2 a - \cos^2 b - \cos^2 c + 2 \cos a \cos b \cos c}}{\sin b \sin c}$$

(N.B. The positive root is taken because $\sin a$, $\sin b$, $\sin c$, are necessarily positive).

It follows that:

$$\frac{\sin A}{\sin a} = \frac{\sqrt{1 - \cos^2 a - \cos^2 b - \cos^2 c + 2 \cos a \cos b \cos c}}{\sin a \sin b \sin c}$$

whence it is clear that:

$$\frac{\sin A}{\sin a} = \frac{\sin B}{\sin b} = \frac{\sin C}{\sin c} \qquad (7.12)$$

The sines of the angles of a spherical triangle are thus proportional to the sines of the opposite sides.

We have seen that:

$$\cos a = \cos b \cos c + \sin b \sin c \cos \mathrm{A}$$
$$\cos c = \cos a \cos b + \sin a \sin b \cos \mathrm{C}$$

and

$$\sin c = \sin a \cdot \frac{\sin \mathrm{C}}{\sin \mathrm{A}}$$

hence by substituting for $\cos c$ and $\sin c$ in the first equation we obtain:

$$\begin{aligned} \cot a \sin b &= \cos b \cos \mathrm{C} + \cot \mathrm{A} \sin \mathrm{C} \\ &= \cot \mathrm{A} \sin \mathrm{C} + \cos b \cos \mathrm{C} \end{aligned}$$

and by cycling angles and sides

$$\begin{aligned} \cot a \sin b &= \cot \mathrm{A} \sin \mathrm{C} + \cos b \cos \mathrm{C} \\ \cot b \sin a &= \cot \mathrm{B} \sin \mathrm{C} + \cos a \cos \mathrm{C} \\ \cot b \sin c &= \cot \mathrm{B} \sin \mathrm{A} + \cos c \cos \mathrm{A} \\ \cot c \sin b &= \cot \mathrm{C} \sin \mathrm{A} + \cos b \cos \mathrm{A} \\ \cot c \sin a &= \cot \mathrm{C} \sin \mathrm{B} + \cos a \cos \mathrm{B} \\ \cot a \sin c &= \cot \mathrm{A} \sin \mathrm{B} + \cos c \cos \mathrm{B} \end{aligned} \tag{7.13}$$

7.4.4 Right-angled spherical triangles

Every spherical triangle has three angles and three sides, plus a radius which is constant.

If, in Fig. 7.41, ABC is a spherical triangle, with C a right angle, and PQR is a plane triangle with R a right angle.

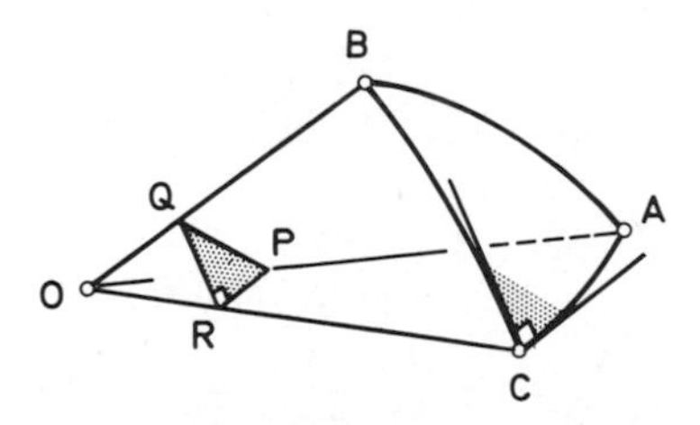

Figure 7.41

$$\mathrm{PQ}^2 = \mathrm{PR}^2 + \mathrm{RQ}^2 = \mathrm{OP}^2 - \mathrm{OR}^2 + \mathrm{OR}^2 - \mathrm{OQ}^2 = \mathrm{OP}^2 - \mathrm{OQ}^2$$

whence

$$\angle \text{PQO is a right angle and}$$

$$\frac{\text{OQ}}{\text{OP}} = \frac{\text{OQ}}{\text{OR}} \cdot \frac{\text{OR}}{\text{OP}}$$

that is:

$$\cos c = \cos a \cos b \tag{7.14}$$

$$\frac{\text{PR}}{\text{OP}} = \frac{\text{PR}}{\text{PQ}} \cdot \frac{\text{PQ}}{\text{OP}}$$

that is:

$$\left.\begin{aligned} \sin b &= \sin \text{B} \sin c \\ \sin a &= \sin \text{A} \sin c \end{aligned}\right\} \tag{7.15}$$

$$\frac{\text{QR}}{\text{OQ}} = \frac{\text{QR}}{\text{PQ}} \cdot \frac{\text{PQ}}{\text{OQ}}$$

that is:

$$\left.\begin{aligned} \tan a &= \cos \text{B} \tan c \\ \tan b &= \cos \text{A} \tan c \end{aligned}\right\} \tag{7.16}$$

$$\frac{\text{PR}}{\text{OR}} = \frac{\text{PR}}{\text{QR}} \cdot \frac{\text{QR}}{\text{OR}}$$

that is:

$$\left.\begin{aligned} \tan b &= \tan \text{B} \sin a \\ \tan a &= \tan \text{A} \sin b \end{aligned}\right\} \tag{7.17}$$

Multiplying the two equations yields

$$\tan \text{A} \cdot \tan \text{B} = \frac{1}{\cos a} \cdot \frac{1}{\cos b} = \frac{1}{\cos c}$$

whence

$$\cos c = \cot \text{A} \cot \text{B} \tag{7.18}$$

We know that:

$$\sin a = \sin \text{A} \sin c$$

and

$$\tan a = \cos \text{B} \tan c$$

cross multiplication yields:

$$\sin a \cdot \cos \text{B} \cdot \tan c = \tan a \cdot \sin \text{A} \cdot \sin c$$

whence

$$\left.\begin{aligned}\cos B &= \sin A \cos b \\ \cos A &= \sin B \cos a\end{aligned}\right\} \qquad (7.19)$$

These six equations comprise ten parts and these suffice to solve every case.

With C a right angle, A, B, a, b, c, are the unknowns, and any two of these may be specified whence we obtain the third.

7.4.5 Napier's rules

The equations of spherical geometry are more difficult to remember than those of plane geometry and Napier's rules of circular parts may be used to recall them.

The five circular parts are $(\pi/2 - A), (\pi/2 - B), (\pi/2 - c), a, b$, and these are arranged round a circle centre as they occur in the triangle.

Any one of the five entries is called the middle part. The two entries on either side are called the adjacent parts. The two remaining entries are called the opposite parts.

Napier's rule states:

sine middle part = product of tangents of adjacent parts

sine middle part = product of cosines of opposite parts

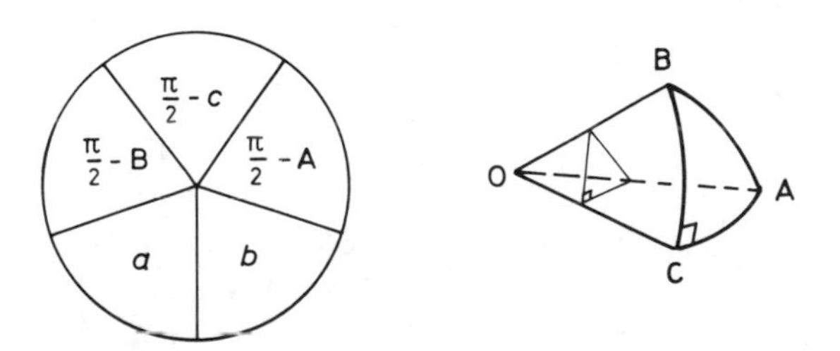

Figure 7.42

If we take $(\pi/2 - c)$ as the middle part $(\pi/2 - A), (\pi/2 - B)$ are the adjacent parts a, b, are the opposite parts.

Applying Napier's rules,

$$\sin\left(\frac{\pi}{2} - c\right) = \tan\left(\frac{\pi}{2} - A\right) \cdot \tan\left(\frac{\pi}{2} - B\right)$$

$$\cos c = \cot A \cdot \cot B$$

and this is equation 7.18

$$\sin\left(\frac{\pi}{2} - c\right) = \cos a \cdot \cos b$$

$$\cos c = \cos a \cdot \cos b$$

and this is equation 7.14

The other equations may be recalled by a similar route.

7.5 GEODESICS AND GEODESIC DOMES

A geodesic is the shortest line path between two points on a surface so that when the surface is spherical all geodesics are great circle arcs and with the sole exception of polar points (points which are diametrically opposed) there is one and only one great circle common to any two arbitrarily chosen points on the sphere.

Since the vertices of all regular polyhedra lie on a circumsphere we may create a set number of great circle arcs by spinning a chosen polyhedra about a succession of polar vertices, as shown in Fig. 7.43 and this is in fact the way Fuller's early domes were designed.

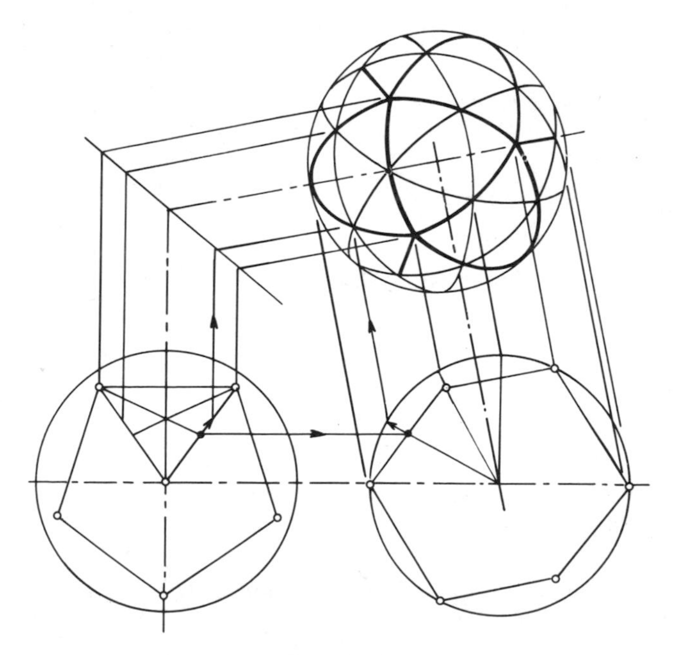

Figure 7.43

7.5.1 Spherical triangulation

The external vertices of a cuboctahedron lie on a sphere but the presence of square faces make the cuboctahedron an unsuitable figure on which to base a dome. The external surface of an icosahedron, however, consists entirely of delta triangular faces and is, in spite of an inherent structural skewness, a far better choice when it comes to designing a dome.

As previously shown the icosahedron is the deltahedron which has the largest number of vertices, edges and faces and is therefore the most nearly spherical of the eight convex deltahedra.

It is perhaps intuitively clear that a far more complex triangulated polyhedral dome may be built, if the stipulation of pure equi-edged, equi-angled faces is relaxed. One method of achieving this end is to employ the sphere-point raising technique.

7.5.2 Sphere-point raising technique

The vertices of all regular and semi-regular polyhedra lie on a circumsphere. The mid-points of all edges lie on an intersphere. The mid-points of all faces lie on an insphere.

The regular (Platonic) polyhedra have a single insphere, but all semi-regular (Archemedian) polyhedra have at least two inspheres.

Bearing in mind that all vertices lie on a circumsphere, it is a natural and obvious step to stellate each face of a parent figure, so that further vertices which lie in the circumsphere are formed. As previously mentioned, the icosahedron serves as a convenient parent figure.

Using the technique of sphere-point-raising, the centre of each flat triangular face is raised to lie on the circumsphere. Each raised sphere-point is joined by three new edges which connect the new vertex to the three original vertices below. A further new edge is then introduced between each adjacent pair of sphere-points.

The method is best illustrated by way of an example, but first recall the following facts:

1. In any true plan the centre point of a triangle lies on the intersection of the median lines, and this is true of the projected intersection of the median lines as seen in any skew view.
2. Since all three vertices of all parent triangular faces lie on a common circumsphere, a line drawn through the point of median intersection, towards the centre of the circumsphere, intersects the plane of each parent triangle at right angles.

7.5.3 Sphere-point-raising technique applied to regular polyhedra

The plan and elevation of a parent icosahedron are shown in Fig. 7.44, and the relevant edge lengths and angles are as given. With an edge length of s, the

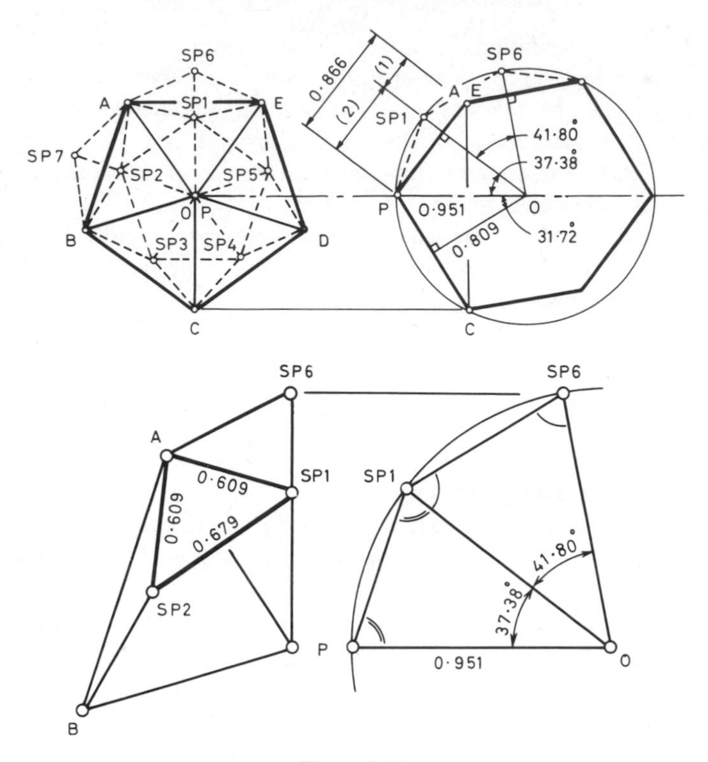

Figure 7.44

pentagon A, B, C, D, E, represents the chords of a small circle, the radius of which is $s/2$ cosec 36°. The length PC_0, in the right hand view is seen true length and is therefore drawn equal to S. The remaining dimensions follow from these observations.

The sphere-point SP, and SP_6 are raised, as already explained, and the chordal distances P, SP_1, S_1, SP_6, calculated from the data given. Noting the axis of symmetry, in the right hand view and noting the planar symmetry in the left hand view, it is clear that P_1, SP_6, is typical of all sphere-point to sphere-point lines.

1. The edge length of every vertex to each adjacent sphere-point = 0.609466s.
2. The edge length of every sphere-point to sphere-point = 0.678616s and these new chords completely envelop the parent edges, only the original icosa vertices remain.

The stellated icosahedron produced by sphere-point raising thus contains but one kind of triangle. This triangle is not, however, an equi-edged, equi-angled delta and hence the correct sequence of assembly is important. The problem of assembly may not at first appear to be difficult and certainly when aided by a model or coloured drawing, the correct sequence of assembly is easy to follow.

A visual difficulty which sometimes confuses the issue, is that two basic figures, pentagons and hexagons, intermingle.

One way of planning the assembly sequence is to plot the new stellated figure as a Schlegel diagram.

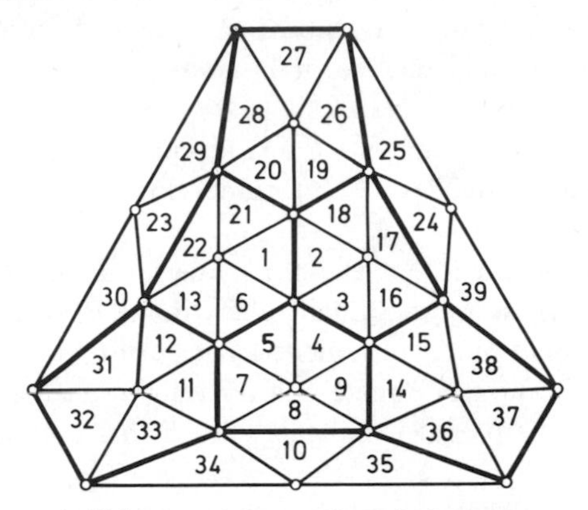

nb. This example numbered clockwise as viewed from inside circumsphere.

(a)

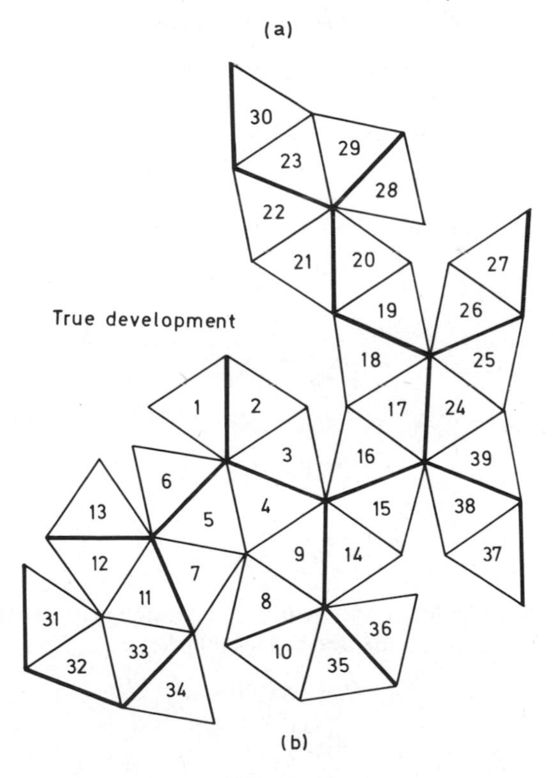

(b)

Figure 7.45

The Schlegel diagram for an icosahedron is shown, by fine lines, in Fig. 7.45. Each parent vertex is shown by a small solid dot and each new sphere-point is identified by a small open circle. Each new edge running between a vertex and an adjacent sphere-point is shown as a full thick line. Each new edge running between adjacent sphere-points is shown as a dashed thick line. But when actually planning an original scheme, it is always helpful to identify different edge lengths using different colours.

The diagram, Fig. 7.45, is drawn for a three-quarter dome (with a six valent vertex on centre) and the reader wishing to develop the figure to its full extent may do so.

The arrangement planned here is a three-quarter dome and has a six valent vertex at its centre. If, however, a half-sphere dome is required then a central five valent vertex offers a more convenient arrangement.

We observe the method of sphere-point raising applied to the icosahedron completely envelops the parent icosahedron, such that only the twelve original icosa vertices remain.

By sphere-point raising the centre of each icosa face we have produced a much more nearly spherical arrangement which is well suited to the size and structural requirements of small to medium size domes.

1. The required apex to sphere point dimension may be conveniently calculated for any desired polyhedron, once the insphere and the circumsphere radii are known. From Fig. 7.47(a).

$$x^2 + y^2 = L^2_{AS} \quad \text{by Pythagoras}$$

$$y(2R_c - y) = x^2 \quad \text{by crossing chords}$$

$$x^2 + y^2 = 2yR_c$$

Hence

$$L^2_{AS} = 2yR_c$$

but

$$y = R_c - R_i$$

and

$$L_{AS} = [2R_c(R_c - R_i)]^{1/2} \tag{7.20}$$

2. The required sphere-point to sphere-point dimension may be calculated using the circumsphere radius and the face-to-face dihedral angle. Thus we may write:

$$L_{SS} = 2R_c \sin(90 - \alpha)$$

$$= 2R_c \cos \alpha \tag{7.21}$$

For an icosahedron of unit edge length

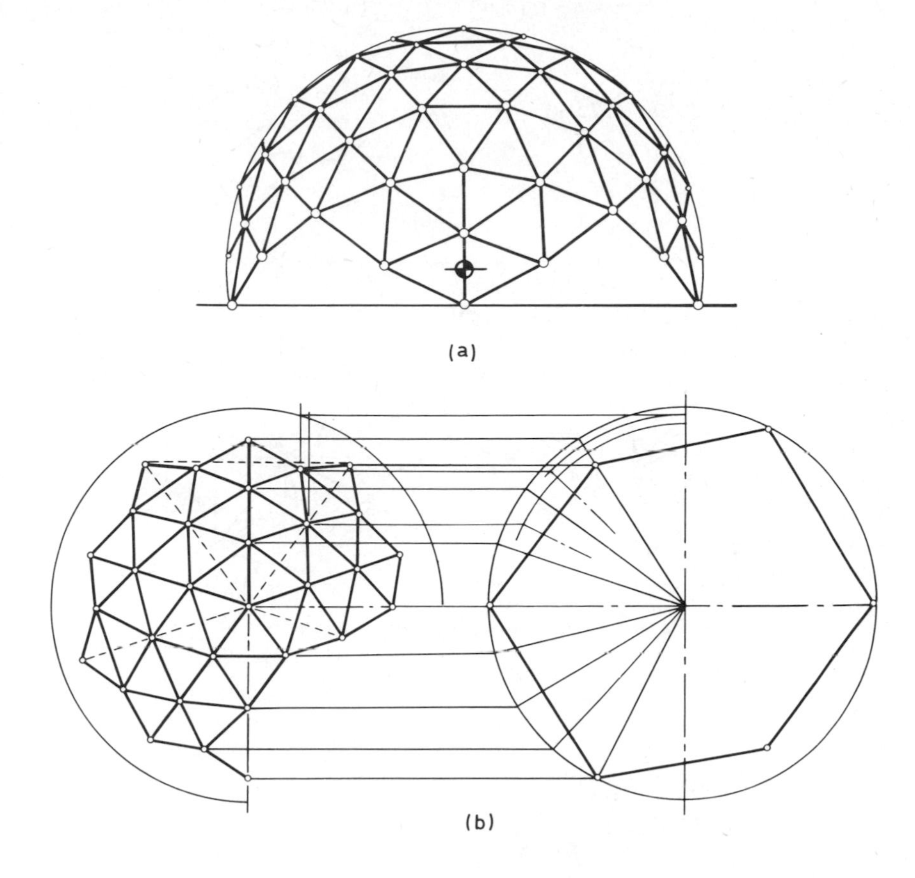

Figure 7.46

$$R_c = 0.9511 \quad R_i = 0.7558 \text{ and } 2\alpha = 138.1833°$$

and hence using the equations above,

$$L_{AS} = 0.60951 \quad L_{SS} = 0.67884$$

It is left as an exercise for the reader to apply the above technique to a regular octahedron.

Example
Calculate the sphere point dimensions for a dodecahedron noting all calculations on the various diagrams, Fig. 7.48.

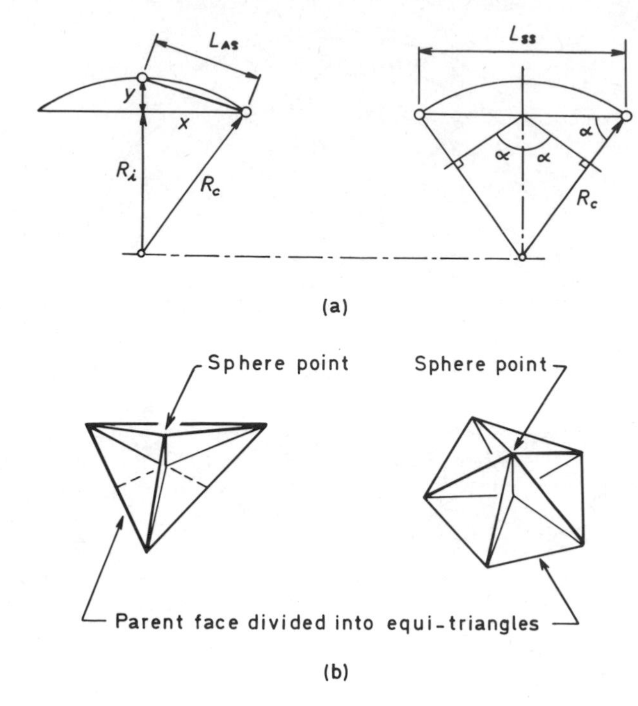

Figure 7.47

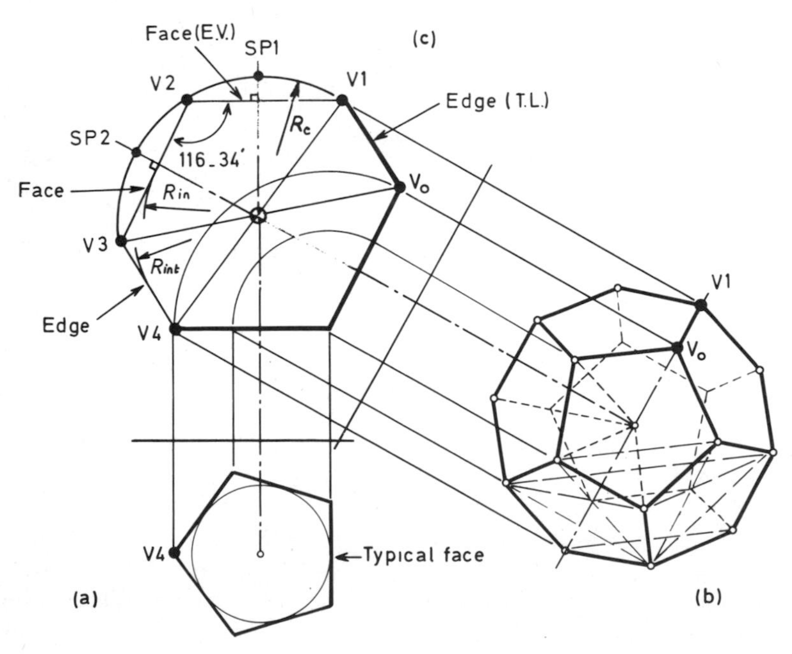

Figure 7.48

Solution
We note only the procedure and sizes here.

(a) We begin by sketching and calculating the various dimensions associated with a regular pentagonal face, diagram (a).
(b) We observe the order in which three consecutive faces blend together, diagram (b).
(c) We use the information from (b) to calculate the insphere, midsphere and circumsphere radii, as illustrated in diagram (c).
(d) We calculate the edge lengths L_{AS}, L_{SS} and use these, together with the original pentagonal edge length, as shown in diagram (d). We do not in this instance join sphere points.
(e) We sketch a typical section as in diagram (e).

7.6 NON-REGULAR TRIANGULATION OF SPHERICAL SHELLS

Whilst it is possible to apply the sphere-point raising technique to any regular or semi-regular polyhedron, the octahedron and the icosahedron are the two polyhedra commonly used. The pentagonal dodecahedron is, however, an alternative.

We know from article 6.2 that the icosahedron constitutes the limit of pure delta triangulation and any attempt to increase the number of facets (triangular faces) necessarily leads to the injection of non-delta faces. Every so-called geodesic dome which contains more than twenty triangular faces in the complete sphere is necessarily built up of non-delta faces and comprises a number of different edge lengths.

For practical reasons we clearly seek to employ as few different edge lengths as possible and from a structural point of view all face shapes should be as near equilateral as topological constraints allow.

There are a number of different ways in which equilateral and isosceles triangles may be divided by patterns of internal triangulation and four methods offering different advantages will shortly be considered.

We know that the vertices of all regular and semi-regular polyhedra lie on a circumsphere and these may be used as scaffolding on which to erect more highly triangulated spherical patterns.

The concept of truncation, stellation and sphere point raising has already been introduced and the concept of frequency (i.e. the division of edges and faces) was introduced in article 6.3.4, and the problem of increasing the density of triangulation (in a spherical surface) will now be introduced.

The three principal methods available to us are based on the division of arcs or chords (widely known as the alternate method), the bisection and further division of faces (the right angle method); the triple overlay procedure (known as the Triacon method).

If A, B, C, are the vertices of a plane triangle lying in the circumsphere, then the arc lengths AB, BC, CA and the chord lengths AB, BC, CA are different. To each arc there is (in general) a different central angle and this angle determines, with the radius of the circumsphere, the absolute difference between arc and chord.

The length of arc equals $\theta^c R_s$ where θ^c is a plane central angle in radians, R_s the circumsphere radius. The length of chord is $2R_s \sin \theta/2$. Or when $R_s = 1$, arc $= \theta^c$, chord $= 2 \sin \theta/2$ = chord factor. We observe that the chord factor (obtained by setting $R_s = 1$) is of wide application since it allows us to calculate the chord length for any radius circumsphere, simply by multiplying chord factor by the radius required.

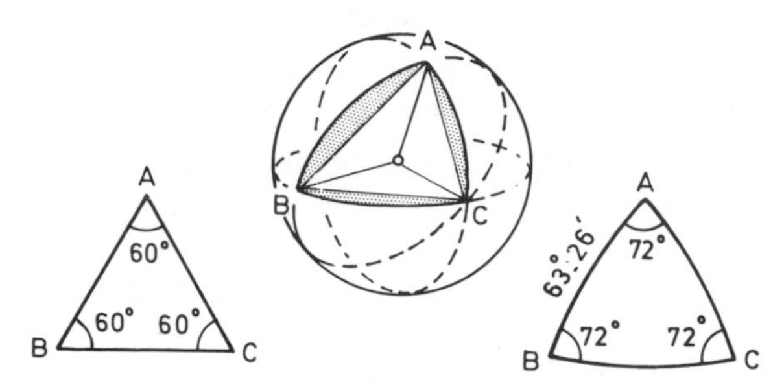

Figure 7.49

7.6.1 Equi-division of edges

If we divide each edge of the parent polyhedron into equal segments and project the divisions onto the circumsphere, as shown in Fig. 7.49(a), we obtain circumsphere chords of different lengths. In the three frequency breakdown, shown in the figure, there are two different lengths, as also (by symmetry) in four frequency division. In a five frequency break-down however there would be three different lengths and so on.

For a circumsphere of unit radius OA = OB =1, AB = $\sqrt{2}$, AP = PQ = QB = $\sqrt{2/3}$ (that is AP = sin $90/f$ in general)

$$PN = AN = (\sqrt{2/3} \cdot 1/\sqrt{2}) = 1/3$$

whence

$$ON = (1 - \tfrac{1}{3}) = \tfrac{2}{3}$$

angle POA = arc tan $\frac{1}{2}$, whence angle POA = 26.5650512°

$$\text{angle POQ} = (90 - 2(26.5650512)) = 36.8698978°$$

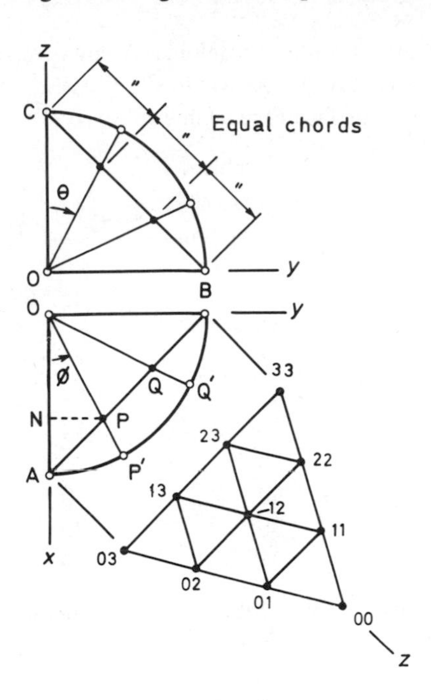

Figure 7.50

and

$$\text{angle AOQ} = (26.5650512 + 36.8698978) = 63.4349490^\circ$$

The points P and Q correspond to the numbered stations (1, 0), (1, 1), (2, 0), (2, 2), (3, 1), (3, 2) in Fig. 7.51(c) and the angular position of the point (2, 1) may be calculated as follows.

The chordal distance (0, 0) to (2, 1) in an equilateral triangle of side $\sqrt{2}$ is equal to $\sqrt{2}/\sqrt{3}$ and if points (0, 0) and (2, 1) lie on a unit circumsphere the central w, Fig. 7.51, equals arc sin $\sqrt{2}/\sqrt{3}$ whence the spherical coordinate θ of the point (2, 1) equals arc sin $\sqrt{2}/\sqrt{3} = 54.7356102^\circ$, when the point (0, 0) is on the axis oz.

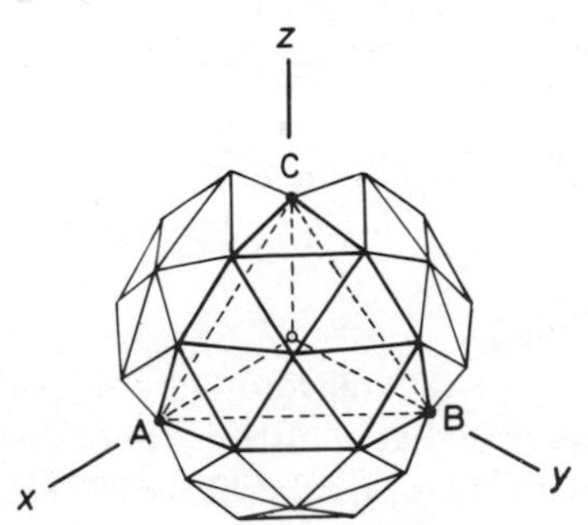

Figure 7.51

If the triangle we are considering belongs to an octahedron with one face lying in the positive quadrant as shown in Fig. 7.52, then the complement of spherical coordinates for this face read as in Table 7.1.

Table 7.1

Pt.	ϕ	θ
0, 0	0	0
0, 1	0	26.5650512
1, 1	90	26.5650512
0, 2	0	63.4349490
1, 2	45	54.7356103
2, 2	90	54.7356103
0, 3	0	90
1, 3	30	90
2, 3	30	90
3, 3	90	90

With four different chord factors (obtained from equations 7.20, 7.21) as listed in Table 7.2, repeating as shown in Fig. 7.52.

Table 7.2

Edge	Length
0, 0/0, 1	0.459506
0, 1/0, 2	0.632439
0, 1/1, 1	0.632439
0, 2/1, 2	0.671421

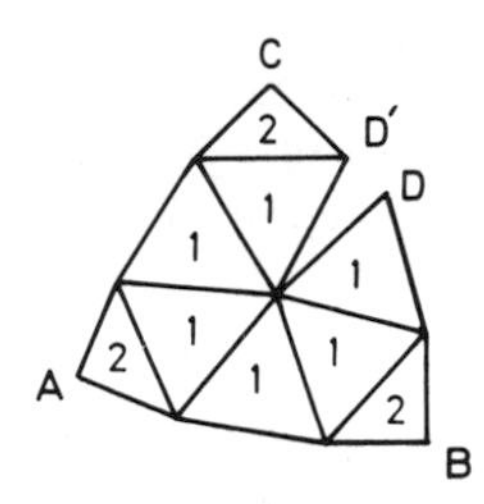

Figure 7.52

7.6.2 Equi-angular divisions

If we divide the edge of a parent octahedron into three equi-angular segments, as shown in Fig. 7.53, we observe that whilst the chord divisions AP, QB are similar in length, the middle chord PQ is slightly shorter. The arc lengths AP′, P′Q′, Q′B are however equal and this, as we shall see, leads to three as opposed to four different chord factors in this type of breakdown.

For a circumsphere of unit radius OA = OB = 1, and since angles ϕ_1, ϕ_2, ϕ_3, are equal, the chords AP′, P′Q′, Q′B, are equal.

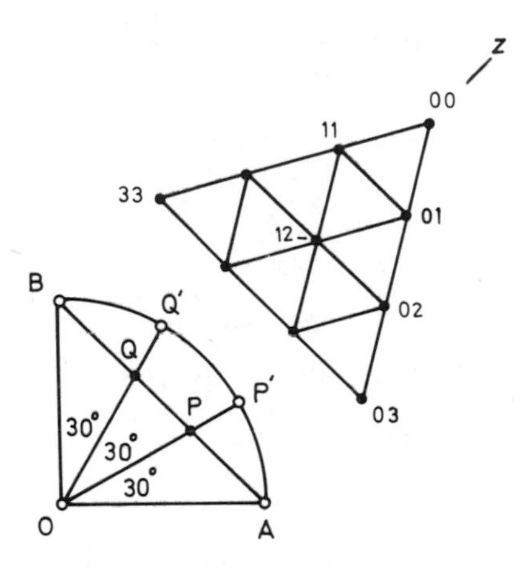

Figure 7.53

For a particular case of a parent octahedron orientated as shown in the figure, the relevant spherical coordinates are as tabulated in Table 7.3.

Table 7.3.

Pt.	ϕ	θ
0, 0	0	0
0, 1	0	30
1, 1	90	30
0, 2	0	60
1, 2	45	54.735613
2, 2	90	60
0, 3	0	90
1, 3	30	90
2, 3	60	90
3, 3	90	90

Given these angles the relevant chord factors are readily calculated, and we note that in a three frequency octahedral breakdown there are three chord factors and three types of triangle as shown in Table 7.4.

Table 7.4

Edge	Length
0, 0/0, 1	0.517638
0, 1/1, 1	0.707107
0, 2/1, 2	0.650115

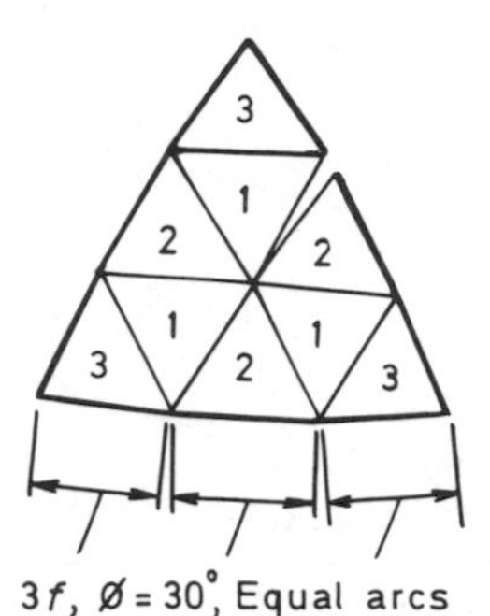

Figure 7.54

As shown in Fig. 7.52, the three frequency triangulation of a spherical octahedral face has a central vertex, which in 7.6.1 is surrounded by six identical isosceles triangles (1) with equal length edges radiating from the central vertex, plus three other isosceles triangles (2) of a different size. There are thus four different edge lengths. In 7.6.2 the central vertex is surrounded by three alternating isosceles triangles (1) and (2), plus three other isosceles triangles (3), but in this case there are but three different edge lengths.

7.6.3 The triacon method, Class II division

The term triacon is used to denote a plan of triangulation in which the centres of all spherical faces are joined to surrounding parent vertices. The method is commonly associated with the icosahedron but is clearly applicable to the octahedron, which is also comprised of equilateral faces.

In Fig. 7.55 the centres of two adjacent spherical icosa faces are joined to near-by vertices. Points ABC, ABD are the vertices of two adjacent faces and the black dots are their two respective sphere points. The original vertices ABCD lie in the circumsphere as do the two sphere points. The two sphere points are linked by a new line which extends from C to D. The original icosa edge AB is removed and a pair of medians drawn through A and B. The outcome is a diamond element which forms part of a 2*f* triacon layout.

The corresponding 4*f* triacon element is shown in Fig. 7.56. As may be seen from Fig. 7.57 a typical median AQ is carried over the original face boundary to the adjacent sphere point R and thence to the far vertex E which belongs to the adjacent face.

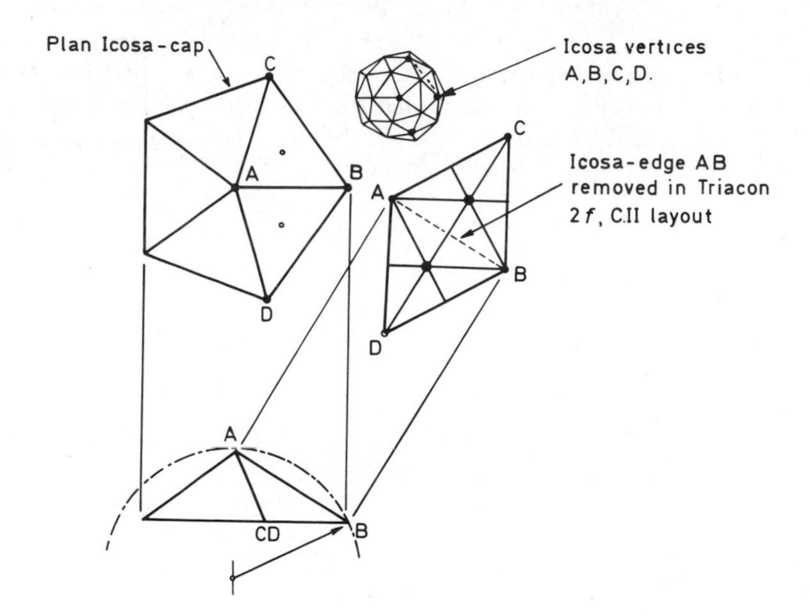

Figure 7.55

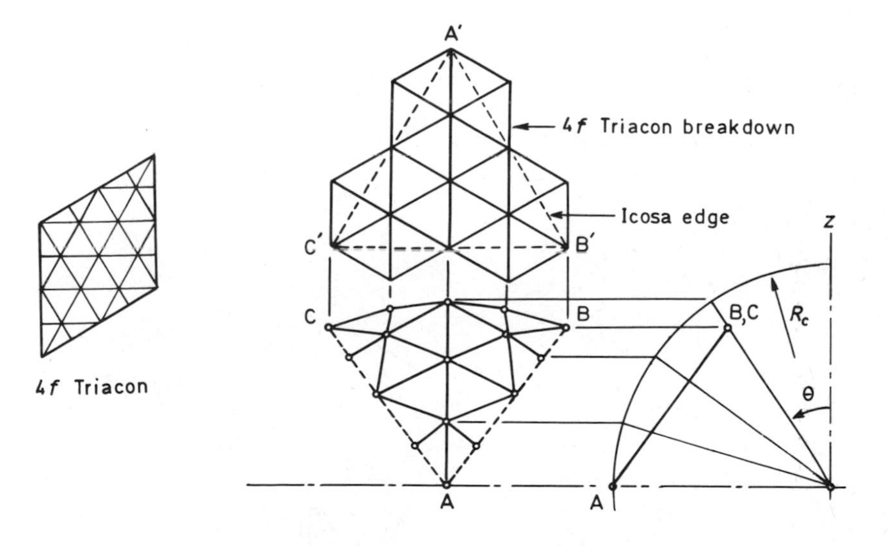

Figure 7.56

If we draw in the medians on two adjacent delta faces, we obtain the triangle BPQ, Fig. 7.57, and by extending the figure to include another delta face, BDE, with medians intersecting at R we see that triangles BQR, DRQ are identical to the triangles BPQ, APQ, and the process is amenable to extension over the whole surface of a complete sphere. See Fig. 7.58 for example.

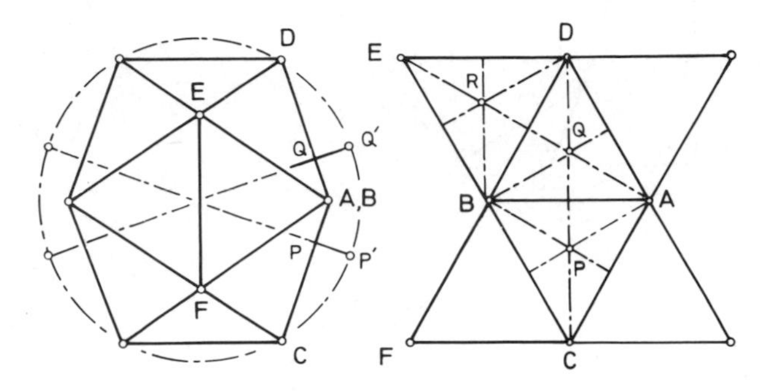

Figure 7.57

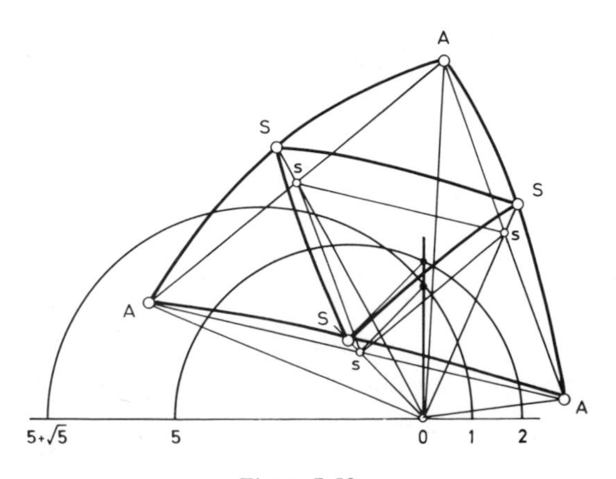

Figure 7.58

7.6.4 Higher frequency breakdowns

Higher frequency breakdowns are obtainable by a similar route and the symmetrices of such breakdowns are worthy of note.

As seen from Fig. 7.58, each developed face of a two frequency spherical icosahedron contains a central equilateral triangle around which the symmetries and asymmetries apply.

(N.B. We observe that when making a demonstration model from stiff card, the graphic symmetry necessarily gives way to a one-pice development of identical shapes.)

7.6.5 Symmetries of structure

The six frequency spherical octahedron, Fig. 7.59, is based on the equi-division of parent edges and contains a central vertex which is surrounded by isosceles triangles (1), to each of which a further isosceles triangle of identical shape is attached. Further triangles (2), (3) and (4) are symmetrically disposed as shown.

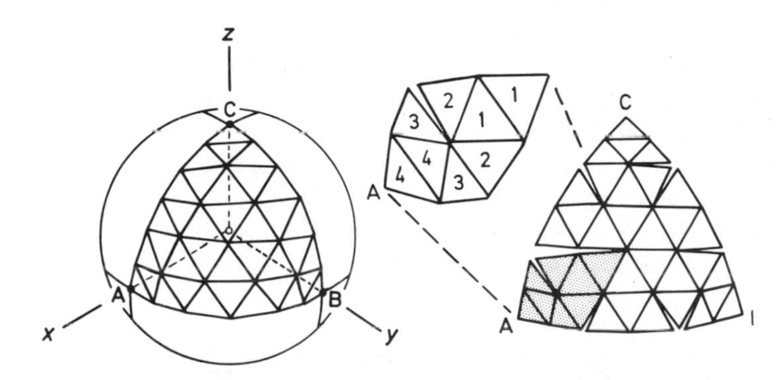

Figure 7.59

We note that triangles (1) and (4) are isosceles, triangles (2) and (3) are not. There are nine different member lengths. Each octa-face contains 36 faces: 12 off (1), 12 off (2), 6 off (3), 6 off (4), but it must be noted that faces (2) and (3) are handed.

A complete sphere therefore contains 96 faces (1), 96 faces (2), 48 faces (3), 48 faces (4), making a total of 288 faces in all.

Typical practical domes are built to heights of a quarter, three eighths, half, five eighths and three quarter diameter and these clearly have a different inventory.

By taking the icosahedron as a parent figure, the number of faces and therefore the roundness of a segmented sphere may be increased, for a given frequency.

If we compare like frequencies, then a $3v$ parent octahedron developed by equal division of edges, yields two types of triangle (1) and (2) as shown in Fig. 7.60.

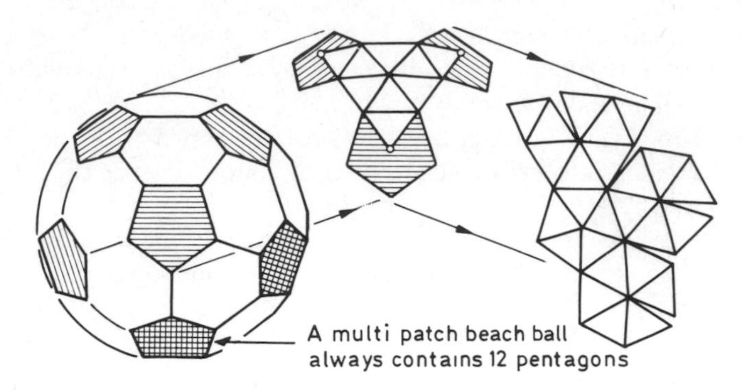

Figure 7.60

There are six off (1) and three off (2), per parent face, and since there are eight parent faces, a total of 48 (1) faces and 24 (2) faces are contained in a complete 3ν octa-sphere. That is, a total of 72 segments.

An icosahedron, however, has twenty faces, each of which produces the same array of triangular segments as does the octahedron. But there are $(20 \times 6) = 120$ (1) faces and $(20 \times 3) = 60$ (2) faces contained by a complete 3ν icosa-phere. That is, a total of 180 segments.

Similarly for the six frequency case. The 6ν octa-sphere contains a total of 288 segments, as compared with a total of 720 segments contained by the 6ν icosa-sphere. There are thus ten times as many segments in a 6ν icosa-sphere as there are in a 3ν octa sphere.

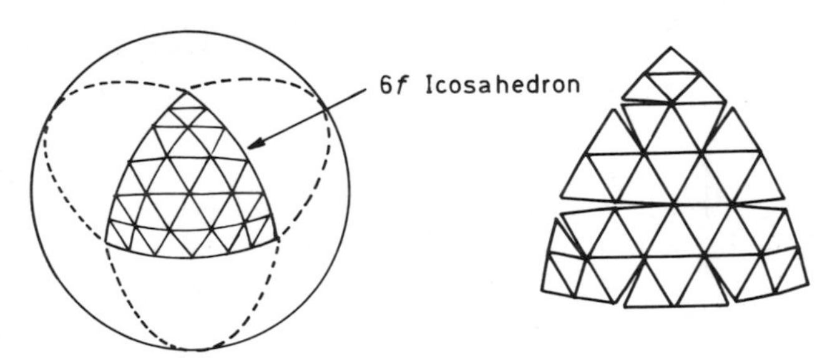

Figure 7.61

Throughout the preceding discussion we have made repeated reference to the regular octahedron, the reason being not that it is the best parent figure on which to base a dome, but because it is with the exception of a cube (which would have served nearly as well) the simplest polyhedron to visualise and deal with.

For reasons of computational convenience we repeatedly positioned two opposite vertices at the north and south pole and this brought the mutually perpendicular vertices into the equatorial plane, and that is not all, we also arranged matters so that the four equatorial vertices were coindicent with the positive and negative *ox* and *oy* axes. Four edges, those lying in the picture plane from equator to pole were thus displayed true length and by virtue of the fact that the equatorial section of a regular octahedron (in this orientation) is a square, the equatorial edges were exactly aligned with two coordinate axes respectively. The plan of the figure in a conventional orthographic representation thus appears in the attitude of a diamond, as shown in Fig. 7.62. We observe that the figure is completely symmetric about the pole axis *oz*. If we rotate the octahedron about its *oz* pole axis, we see that for all but 45° rotational increments the symmetry of the faces (but not the boundary edges) is destroyed. As all faces are the same, we say there are two symmetries about the *oz* axis, and two identical symmetries about the other two axes.

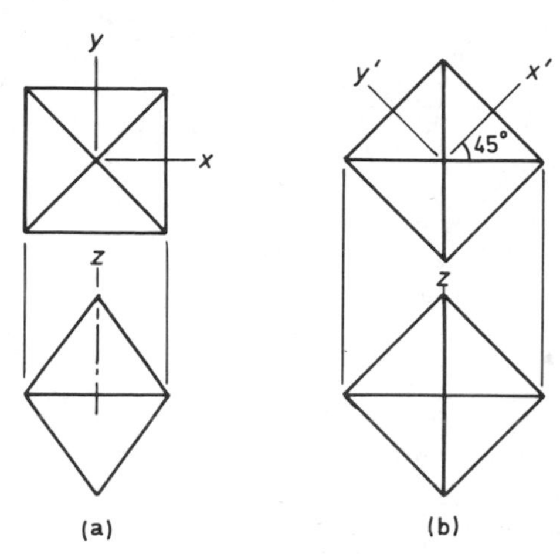

Figure 7.62

Similar symmetrical properties are featured by other polyhedra.

If we rotate an icosahedron about any two diametrically opposite vertices we observe that five different, but indistinguishable, aspects appear. If we rotate the icosahedron about two diametrically opposite mid-edge points we observe two different but topologically indistinguishable positions and if rotated about the mid-points of two diametrically opposite faces, three such identical positions are noted. Since an icosahedron contains 12 vertices, 30 edges and 20 faces and vertices, mid-edges and mid-faces may be selected in pairs, numerous spin-axes may be selected. When icosa vertices are taken there are six different axes of spin and the related equatorial vertices will spin around six different great circle paths, which may be visualised as geodesic lines on the circumsphere. Similarly if diametrically opposite mid-edge points are used a further fifteen great circles will be scribed and a further ten great circles will be similarly generated when diametrically opposite mid-faces are taken. There are thus said to be (6 + 15 + 10) = 31 great circles in the spherical icosahedron. The relative positions of these thirty-one great circles as they appear on the surface of the circumsphere are as shown in Fig. 7.63.

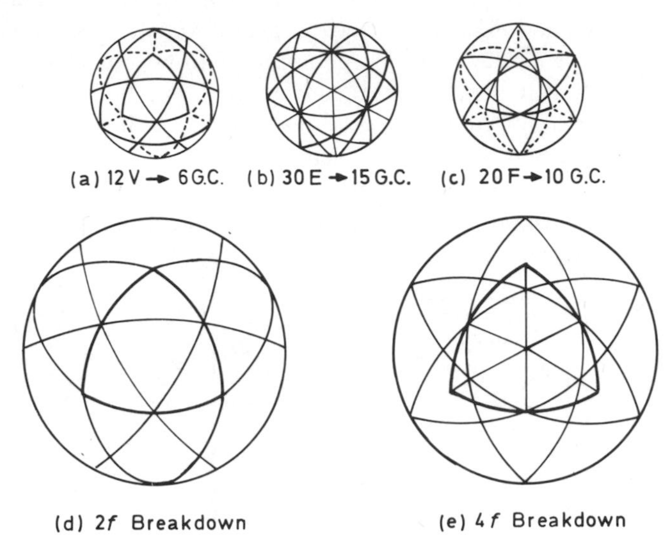

Figure 7.63

As clearly shown by Fig. 7.63(a) three great circles pass through the mid-edge points of each icosa face. Three great circles shown at (b) pass through the mid-point of each icosa face and six great circles pass through the mid-edge point of each icosa face as shown at (c).

If we use both the spherical icosa edges and the great circle arcs of Fig. 7.63(a) we obtain a two frequently spherical icosahedron as shown in Fig. 7.63(d).

By being selective, we may use some icosa edges and some portions of the fifteen great circles, Fig. 7.63(b), to generate the four frequency spherical icosahedron shown in Fig. 7.63(e).

We also observe from Fig. 7.63(b) that ten great circles cross icosa faces at right angles and at first sight, this looks like another promising means of division. (N.B. Although used in the early development of geodesic domes the method leads to a wide variation in face shapes and has been largely superseded. An outline of a related method as applied to a three frequency is, however, given *en passant*.)

7.6.6 The right triangle method

A triangular face may alternatively be divided into any odd or even frequency as demonstrated below. We begin by dividing an equilateral or isosceles triangle into two identical segments, such that the line of division AD, drawn from the

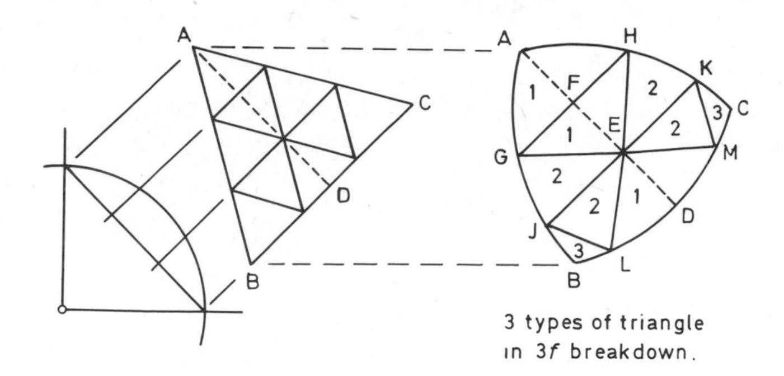

Figure 7.64

appropriate apex bisects the opposite side as shown in Fig. 7.64. The line of division may then be divided into any number of equal parts (in accord with the frequency chosen) and the original triangle partitioned by $(1 - \nu)$ lines drawn through these points and at right angles to the first division, as shown for the case $\nu = 3$ in Fig. 7.64. Two diagonal lines EG, EH, followed by two diagonal lines EL, EM (with LD = DM, equal to GF = FH) yield three identical triangles AGH, EGH, ELM and on joining JL and KM four other identical triangles EGJ, EHK, EJL, EMK are produced, with the two triangles JBL, KCM equal. There are thus three types of triangle in a 3ν breakdown.

7.7 CALCULATION OF CHORD FACTORS

We have seen that in the design and calculation of segmented domes we need to specify the positions of points on the surface and this is best done in terms of spherical coordinates.

We need, for instance, to calculate the lengths of edge members which make up a seemingly complex triangulation, and whilst the techniques of plane geometry are competitive in simple cases, the calculation of a large number of member lengths becomes tedious and prone to error as the number of different member lengths increases.

We know from previous study that the icosahedron is the most complex spherical triangulation that can be assembled from equi-length members or from triangles of equi-length side. Twelve vertices, thirty edges and twenty faces is the most densely populated delta faced sphere we can build. Beyond this, triangulation is possible only at the expense of a variety of different edge lengths.

7.7.1 Chord factors by plane geometry

Before demonstrating the application of spherical coordinates we will first calculate the edge lengths that result from sphere-point raising a parent octahedron, using a plane geometrical technique.

Example

Calculate the edge lengths of a spherical two-frequency octahedron which result from the equi-division of edges and the application of the sphere-point raising technique as illustrated in Fig. 7.65.

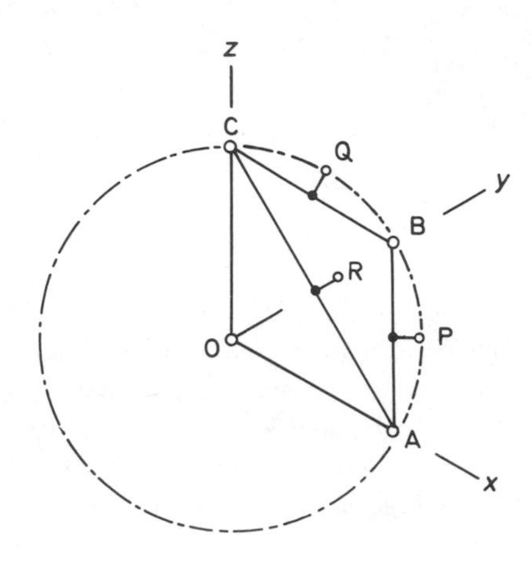

Figure 7.65

Solution

If O is centre and A, B, C, are the vertices of a typical face, with P, Q, R, the mid-points of edges AB, BC, CA respectively and P′, Q′, R′, the corresponding sphere-points, we see that if OA = OB = OC = 1, AB = $\sqrt{2}$ (angle APO = 90°) AP = $\sqrt{2}/2$ and OP = $\sqrt{2}/2$.

Now PP′ = (OP′ – OP) = $(1 - \sqrt{2}/2) = (2 - \sqrt{2}/2)$

and
$$AP' = \sqrt{(AP)^2 + (PP')^2}$$

$$AP' = \sqrt{\left(\frac{\sqrt{2}}{2}\right)^2 + \left(\frac{2-\sqrt{2}}{2}\right)^2} = 0.765367$$

AP = PB = $\sqrt{2}/2$ = PQ and since OP′ = OQ′ = 1, angle POQ = 90°

$$\frac{P'Q'}{PQ} = \frac{OP'}{OP}$$

$$P'Q' = OP' \cdot \frac{PQ}{OP} = 1 \cdot \frac{\sqrt{2}}{2} \cdot \frac{2}{\sqrt{2}} = 1$$

The chord AP′ = 0.765367 is typical of all chords from apex to sphere point. The chord P′Q′ = 1.000 is typical of all lines or edges between sphere points and these are clearly the only two lengths involved.

The plane geometry solution above is to be compared with the spherical coordinate solution, article 7.7.2.

7.7.2 Chord factors by spherical coordinate

We know that the position of any point on a sphere may be expressed in terms of its spherical coordinates.

If (x, y, z) are the Cartesian coordinates of a point on a sphere of radius r.

$$r = \sqrt{x^2 + y^2 + z^2}$$

$$\tan \phi = \frac{y}{x} \quad \text{and} \cos \theta = \frac{z}{\sqrt{x^2 + y^2 + z^2}}$$

and we have proved that:

$$l = \cos \alpha = \frac{x}{r} = \cos \phi \sin \theta$$

$$m = \cos \beta = \frac{y}{r} = \sin \phi \sin \theta$$

$$n = \cos \gamma = \frac{z}{r} = \cos \theta$$

whence

$$x = r \cos \phi \sin \theta$$
$$y = r \sin \phi \sin \theta$$
$$z = r \cos \theta$$

in accord with Fig. 7.66.

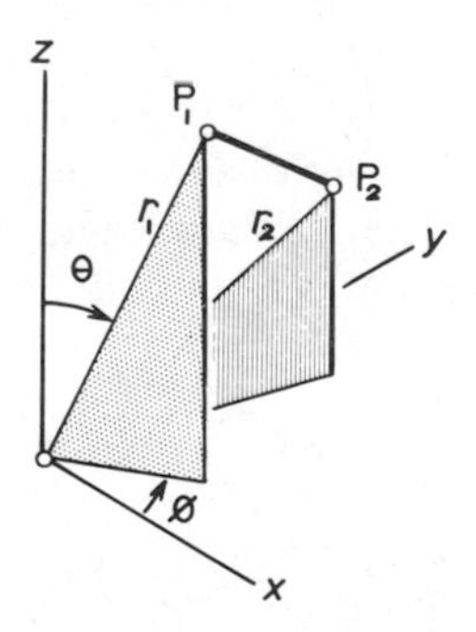

Figure 7.66

We also know that the distance between any two points in 3-space is given by the equation:

$$d = P_1P_2 = \sqrt{(x_2 - x_1)^2 + (y_2 - y_1)^2 + (z_2 - z_1)^2}$$

and hence the chord length (chord factor) between any two points on a surface may be readily calculated in terms of r, ϕ and θ. On making the appropriate substitutions we see that:

$$d = [(r_2 \cos \phi_2 \sin \theta_2 - r_1 \cos \phi_1 \sin_1 \theta)^2 + (r_2 \sin \phi_2 \sin \theta_2 - r_1 \sin \phi_1 \sin \theta_1)^2 + (r_2 \cos \theta_2 - r_1 \cos \theta_1)^2]^{½}$$

and on expanding and collecting terms

$$d = [r_2^2 \sin^2 \theta_2 + r_1^2 \sin \theta_1 + r_2^2 \cos^2 \theta_2 + r_1^2 \cos^2 \theta_1 - 2r_1 r_2 (\cos \phi_1 \cos \phi_2 \sin \theta_1 \sin \theta_2 + \sin \phi_1 \sin \theta_1 \sin \phi_2 \sin \theta_2 + \cos \theta_1 \cos \theta_2)]^{½}$$

$$= [r_1^2 (\sin^2 \theta_1 + \cos^2 \theta_1) + r_2^2 (\sin^2 \theta_2 + \cos^2 \theta_2) - 2r_1 r_2 \cos (\phi_1 - \phi_2) \sin \theta_1 \sin \theta_2 + \cos \theta_1 \cos \theta_2]^{½}$$

whence

$$d = [r_1^2 + r_2^2 - 2r_1 r_2 \cos (\phi_1 - \phi_2) \sin \theta_1 \sin \theta_2 + \cos \theta_1 \cos \theta_2]^{½}$$

The distance d is clearly the chordal length between any two points on any surface when $r_1 = r_2 = r$, for all values of ϕ and θ, the surface is spherical, and we obtain the following simplification:

$$d = r[2 - 2 \cos (\phi_1 - \phi_2) \sin \theta_1 \sin \theta_2 + \cos \theta_1 \cos \theta_2]^{½} \tag{7.22}$$

where $r = 1$

$$d = [2 - 2\{\cos (\phi_1 - \phi_2) \sin \theta_1 \sin \theta_2 + \cos \theta_1 \cos \theta_2\}]^{½}$$

(N.B. When $r = 1$ the distance d is called the chord factor.)

Example

Write down the spherical coordinates of an octahedron centred on zero as shown in Fig. 7.67, and hence calculate the edge length of an octahedron based on a unit circumsphere. Calculate this edge length using both the Cartesian and sperical coordinate methods.

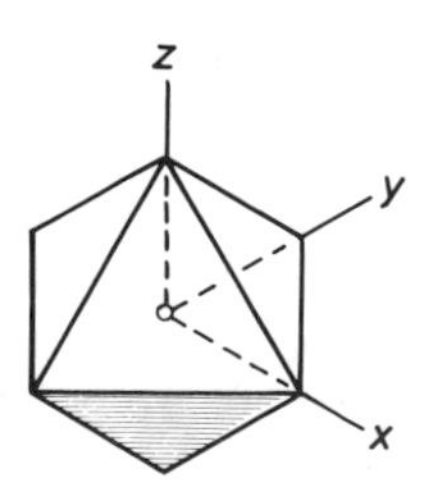

Figure 7.67

Solution
The Cartesian and spherical coordinates of the six vertices are:

	x	y	z	r	ϕ	θ
Point 1	1	0	0	1	0	90
2	0	1	0	1	90	90
3	0	0	1	1	0	0
4	–1	0	0	1	180	90
5	0	–1	0	1	270	90
6	0	0	–1	1	0	180

Since all edge lengths are equal, any convenient edge may be chosen. The edge 1, 3 = d, for instance

(i) In Cartesian coordinates

$$d = \sqrt{(1-0)^2 + (0-0)^2 + (0-1)^2} = \sqrt{2}$$

hence the typical edge length is $\sqrt{2}$.
(ii) In spherical coordinates

$$d = [2 - 2\{\cos(\phi_1 - \phi_2)\sin\theta_1 \sin\theta_2 + \cos\theta_1 \cos\theta_2\}]^{1/2}$$

The point 1 is at $r = 1, \phi = 0, \theta = 90$
The point 3 is at $r = 1, \phi = 0, \theta = 0$
Whence $\phi_1 = 0, \phi_2 = 0, \theta_1 = 90, \theta_2 = 0$ and

$$d = [2 - 2\{1.0 + 0.1\}]^{1/2} = \sqrt{2} \text{ as previously}$$

Example
If A(0, 90), B(90, 90), C(0, 0) are the vertices of an octahedral face in the positive quadrant and P(45, 90), Q(90, 45), R(0, 45), are typical sphere points, as shown in Fig. 7.68. We may calculate the length of the two typical chords. AP = PB = BQ = QC = CR = RA and PQ = QR = RP as follows:

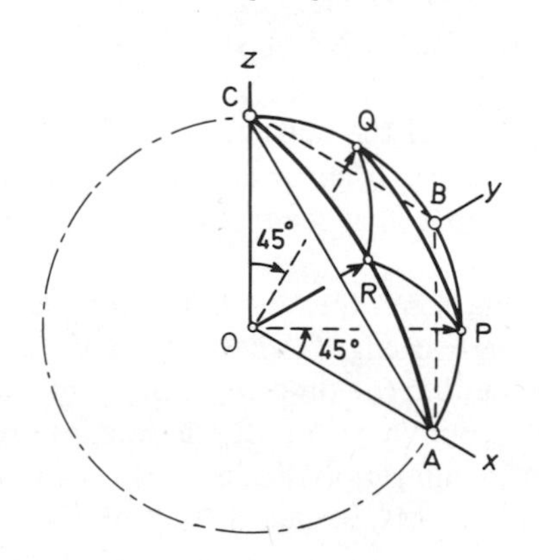

Figure 7.68

Solution

Since ABC . PQR are points on the same circumsphere: the distance between any two point is:

$$d = [2 - 2\{\cos(\phi_1 - \phi_2) \sin\theta_1 \sin\theta_2 + \cos\theta_1 \cos\theta_2\}]^{1/2}$$

In this simple case we need only the sines and cosines of the angles 0, 45°, 90°, and there are only two typical chord lengths to consider. AP is typical of the distance from an original vertex to a sphere point. PQ is typical of the distance from sphere point to sphere point.

$$\begin{aligned} AP &= [2 - 2\{\cos(90 - 90) \sin 0 \sin 45 + \cos 0 \cos 45\}]^{1/2} \\ &= [2 - 2 \cos 45]^{1/2} = 0.765367 \\ PQ &= [2 - 2\{\cos(90 - 45) \sin 45 \sin 90 + \cos 45 \cos 90\}]^{1/2} \\ &= [2 - 2 \cos 45 \sin 45]^{1/2} = 1.000000 \end{aligned}$$

The two chord factors, for a sphere of unit radius based on a parent octahedron are therefore 0.765367 and 1.000000 respectively.

Since there are twelve edges in the parent octahedron there are twelve sphere points, and six original vertices to be joined. There are two types of edge in the sphere pointed figure and these join vertex to sphere point and sphere point to sphere point respectively. We observe that each octa edge is shared by two adjacent faces, but each sphere point 'breaks' each original edge in two. Hence there are 24 edges, or bars, of type AP contained in the sphere. There are eight faces in the parent octahedron, to each of which there are three sphere points, and hence (8.3) = 24 edges, or bars, of type PQ are required to complete the sphere.

7.8 DESIGNATION OF COORDINATES

The geometrical properties of the octahedron make it an easy example to deal with. The plane centre angles of an octahedron are all 90°, and one face (one eighth volume) may be set so that the three vertices are aligned with the *ox, oy, oz* axes respectively.

The octa edge AB, Fig. 7.69, then lies in the *xoy* plane, the octa edge BC lies in the *yoz* plane and the octa edge CA lies in the *zox* plane. Frequence divisions along these edges project onto the three mutually perpendicular reference planes and yield the Cartesian coordinates of all grid line intersections. Since all intersections are produced by integer divisions, numbered outwards from the origin along the three coordinate axes, we see that all nodes can be readily identified.

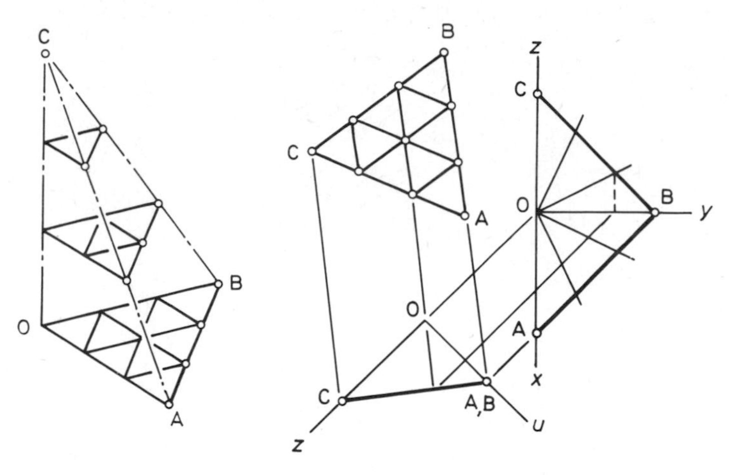

Figure 7.69

We note that the equilateral triangles which divide the octahedral delta face project to right angled triangles in the *xoy*, *yoz*, *zox* planes and hence each point of face sub division is uniquely identified by a set of three integers which correspond to their *x*, *y*, *z* projections.

If the true shape of the delta face is drawn and labelled, as in Fig. 7.69, we see that along one edge *x* is always zero, and the two other edges may similarly be identified with zero *y* and zero *z*. This observation may be put to good use when seeking the identification of a node in a high frequency breakdown. Thus we observe that the node P in the thirty-two frequency breakdown.

In the case of an icosahedron, however, five delta triangles meet at an icosa-apex and it is not possible to position all three vertices along a reference axis. If one vertex of a delta face is aligned with the axis *oz* and another vertex is aligned with the axis *oy*, with edge AB parallel to the $\overline{xoy}$ plane as shown in Fig. 7.70, then the vertex A has coordinates $(\bar{x}_A, \bar{y}_A, \bar{z}_A)$ with respect to the centroidal axes, and the triangle ABC projects to the triangle $\overline{ABO}$ in the $\overline{xoy}$ plane.

If we now set up a new pair of axes *xoy* through the vertices A and B, the shape and size of the projected triangle (but not the absolute *z* coordinate) is maintained. The angle AOB, in Fig. 7.70 is 360/5 = 72° and angle OAB = angle OBA = 54°. The angle CBO is 31.71747° (the arc tan of the golden rectangle). tangle).

Frequency divisions along the edge BC (in elevation) project to plan, as shown for the four frequency breakdown, Fig. 7.71. Lines parallel to sides AB, OA and OB divide the triangle AOB in the usual way and all points of intersection may be identified as previously explained.

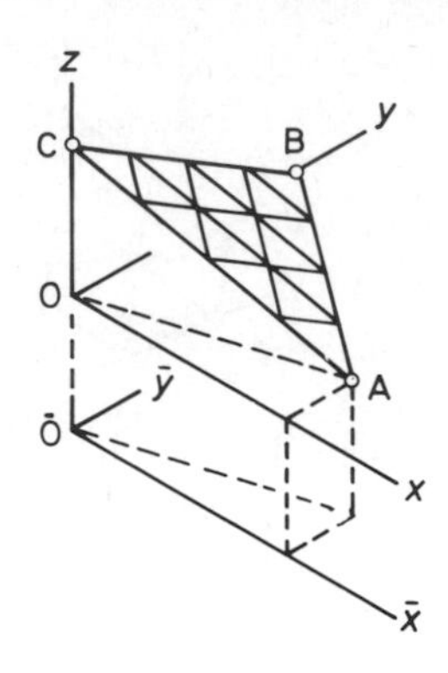

Figure 7.70

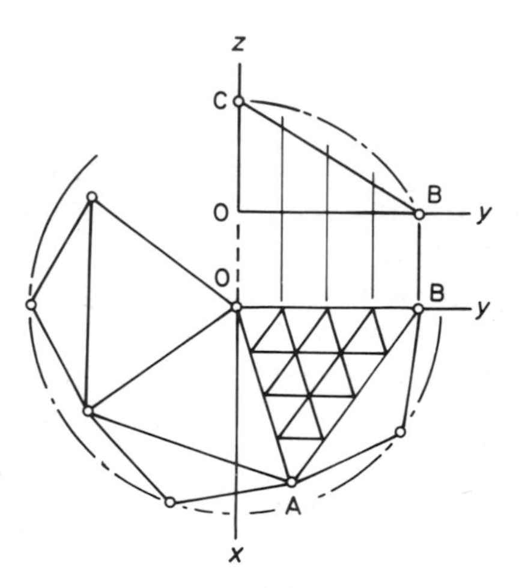

Figure 7.71

The face coordinates of an icosa breakdown may be calculated as follows. If the frequency of sub-division is f, the edge length of the icosa face e, then the projected length of an edge is $e \sin(90 - \tau) = e \cos \tau$ and each segment is e/f $\cos \tau$. The coordinates of the point (1, 1) are thus $(0, e/f \cos \tau, e/f\,(f-1) \sin \tau)$ where $\tau = 31.71747°$ (arc tan $0.61803 = 31.71747°$). The $60°$ apex angle of the delta face projects into the xoy plane to become $60/\cos \tau = 72°$. We observe that e/f is a common multiplier and omit it, to simplify the entries, in Table 7.5. We use the angle $(90–72) = 18°$ in preference to the angle $72°$, as $72°$ corresponds to the angle ϕ in the spherical coordinate system. We also note that all points of intersection that lie on lines parallel to the delta edge AB lie at the same level and therefore have the same z coordinate in common.

Table 7.5

Pt.	x	y	z
0 0	0	0	$f \sin \tau$
1 0	1 cos 18	1 sin 18	$(f-1) \sin \tau$
2 0	2 cos 18	2 sin 18	$(f-2) \sin \tau$
3 0	3 cos 18	3 sin 18	$(f-3) \sin \tau$
4 0	4 cos 18	4 sin 18	$(f-4) \sin \tau$
1 1	0	1	$(f-1) \sin \tau$
2 2	0	2	$(f-2) \sin \tau$
3 3	0	3	$(f-3) \sin \tau$
4 4	0	4	$(f-4) \sin \tau$
2 1	1 cos 18	(1 + 1 sin 18)	$(f-2) \sin \tau$
3 2	1 cos 18	(2 + 1 sin 18)	$(f-3) \sin \tau$
4 3	1 cos 18	(3 + 1 sin 18)	$(f-4) \sin \tau$
3 1	2 cos 18	(1 + 2 sin 18)	$(f-3) \sin \tau$
4 2	2 cos 18	(2 + 2 sin 18)	$(f-4) \sin \tau$
4 1	3 cos 18	(1 + 3 sin 18)	$(f-4) \sin \tau$

Although Table 7.5 has been compiled for a four frequency breakdown, we observe that cos 18 and sin 18 are constants and the tabulation may be extended to higher frequencies, simply by following the established sequence of terms. For instance, the coordinates of the point (6, 0), in a six frequency breakdown are clearly (6 cos 18, 6 sin 18, $(f - 6)$ and the z coordinate checks with the known fact that in a six frequency breakdown the point (6, 0) corresponds to the point B in the parent delta face and this point lies at zero level, i.e. if $(f-6) = z$, then $z = 0$ when $f = 6$. Table 7.5 has, in fact, been printed at length to make the ready extension of the coordinate table, perfectly clear.

If we evaluate all the terms containing sines and cosines, not forgetting to include the effect of τ, we obtain the much simplified numerical Table 7.6, which has been compiled for a three frequency breakdown, just to show that terms may be excluded as readily as added when smaller or larger frequencies are required.

Table 7.6

Pt.		$x = \bar{x}$	$y = \bar{y}$	z	$\bar{z}$
0	0	0	0	1.5772	2.003
1	0	0.9511	0.3090	1.0515	1.4768
2	0	1.9021	0.6180	0.5257	0.9510
3	0	2.8532	0.9271	0	0.4253
1	1	0	1.0000	1.0515	1.4768
2	2	0	2.0000	0.5215	0.9510
3	3	0	3.0000	0	0.4253
2	1	0.9511	1.3090	0.5215	0.47745
3	2	0.9511	2.6180	0	0.4253
3	1	1.9021	1.6180	0	0.4253

(N.B. We must not forget that all entries in Tables 7.5 and 7.6 are to be multiplied by $e/f = e/4$ in the case of Table 7.5, and $e/3$ in the case of Table 7.6.)

The z coordinates listed in both these tables are referred to axes sited in the plane containing the delta apex points A, B. The height of AB above the centroidal plane, which contains the axes $\bar{x}o\bar{y}$, is 0.4253 and hence the $\bar{z}$ coordinates are as listed in Table 7.5.

Once the coordinates of an f frequency breakdown have been obtained for a single face, the coordinates of nodes on all other faces may be readily obtained by applying the appropriate transformations.

In the case of the three frequency icosahedron, we know that five typical faces surround the vertex sited on oz and hence the coordinates of the remaining four faces, which surround this vertex, may be obtained by transforming the coordinates of the first face, by rotations about oz of amounts 72°, 144°, 216°, 288° respectively.

If the coordinates of points on the first face are denoted by $(0, 0)_1, (1, 0)_1, (2, 0)_1 \ldots$ etc., with coordinates as listed, then points in the adjacent face may be obtained by applying a rotational transformation through $\theta_z = 72°$. If the points in the second face are denoted by $(0, 0)_2, (1, 0)_2, (2, 0)_2 \ldots$ etc., then:

$$\begin{bmatrix} \lambda(0,0)_2 \\ \lambda(1,0)_2 \\ \lambda(2,0)_2 \\ \cdot\ \cdot \\ \lambda(4,4)_2 \end{bmatrix} = \begin{bmatrix} \lambda(0,0)_1 \\ \lambda(1,0)_1 \\ \lambda(2,0)_1 \\ \cdot\ \cdot \\ \lambda(4,4)_1 \end{bmatrix} \begin{bmatrix} \cos 72 & \sin 72 & 0 & 0 \\ -\sin 72 & \cos 72 & 0 & 0 \\ 0 & 0 & 1 & 0 \\ 0 & 0 & 0 & 1 \\ & & & \end{bmatrix}$$

where $\lambda(0, 0)$ represents the $(\bar{x}, \bar{y}, \bar{z})$ coordinates of point (0, 0) and $\lambda(1, 0)$ represents the $(\bar{x}, \bar{y}, \bar{z})$ coordinates of point (1, 0) etc.

If we consider a 72° rotation applied to the point $(3, 0)_1$ then it transforms to the position of the point $(3, 3)_1$. The point $(3, 3)_1$ is however in the same position as the point $(3, 0)_2$ and hence the two points $(3, 3)_1$ and $(3, 0)_2$ are clearly one and the same. This is because the line OB_1 in Fig. 7.72, is the common edge between the two faces.

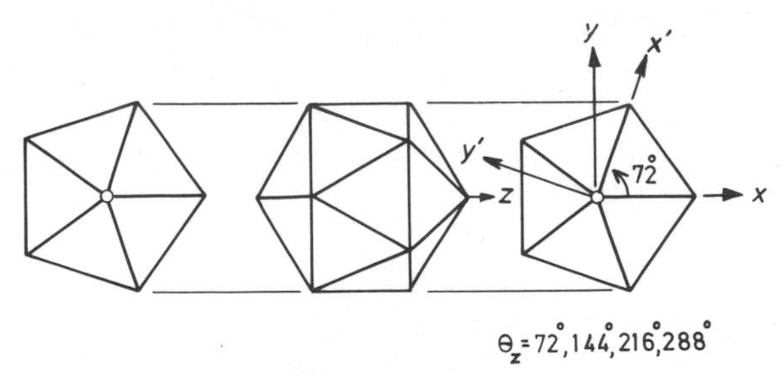

Figure 7.72

The coordinates of the point $(3, 3)_1$ are known from Table 7.6, and if all is correct we should find that the transformed coordinates of $(3, 0)_1$ are equal to the original coordinates of $(3, 3)_1$.

The coordinates of the point $(3, 0)_1$ are from Table 7.6, (2.8532, 0.9271, 0.4253), hence the coordinates of $(3, 0)_2$ are given by

$$[2.8532 \quad 0.9271 \quad 0.4253]\begin{bmatrix} \cos 72 & \sin 72 & 0 & 0 \\ -\sin 72 & \cos 72 & 0 & 0 \\ 0 & 0 & 1 & 0 \\ 0 & 0 & 0 & 1 \end{bmatrix}$$

that is

$$(2.8532 \cos 72 - 0.9271 \sin 72, 2.8532 \sin 72 + 0.9271 \cos 72, 0.4253)$$

hence the coordinates are (0, 3.000, 0.4253) and we note from Table 7.6 that these coordinates are those of the point $(3, 3)_1$ which is as they should be.

All points contained by the five delta faces sited around C, may thus be obtained by this route and all points on the diametrically opposite cap may then be obtained by rotating the full set of points through 180° about the axis oz and then reflecting the full set of points about the plane $\overline{xoy}$ through an angle of 180°.

The method whereby points are rotated is as just given and the required reflection is obtained by applying the reflection matrix.

$$F = \begin{bmatrix} 1 & 0 & 0 & 0 \\ 0 & 1 & 0 & 0 \\ 0 & 0 & -1 & 0 \\ 0 & 0 & 0 & 1 \end{bmatrix}_{F_{xoy}}$$

We note the negative sign in the third row of the third column, as this is the only active element, and the $\bar{z}$ coordinates for a 180° reflection may clearly be written directly.

The methods of rotation, reflection and translation are clearly a considerable asset when performing this type of work.

7.9 DOUBLE LAYER DOMES

It is an inherent geometrical characteristic of domes that as the size increases so more and more frequency divisions are required (both to preserve practical member lengths and as a means of transferring the loads). With increasing frequency the dihedral angle between both member (edges) and panel (faces) is steadily eroded and the inherent structural advantage of lower frequency domes is lost. Due to their low dihedral angles high frequency domes are sensitive to local concentrations of load and as illustrated in Fig. 7.73 are prone to oil-can.

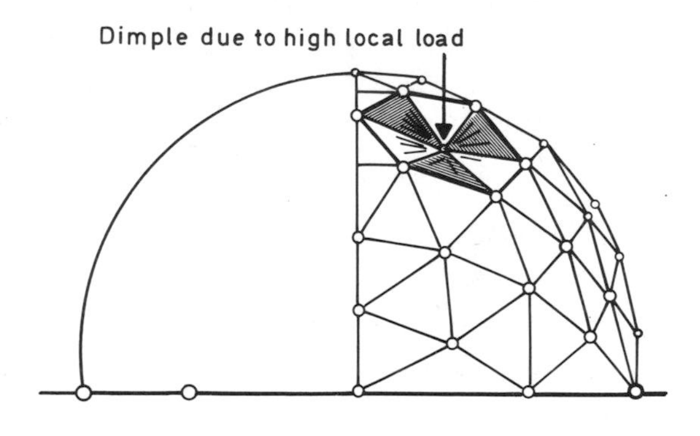

Figure 7.72

One solution to this particular problem is to produce a double layer dome consisting of two concentric circumspheres on which inner and outer sets of vertices are placed. As we have seen the chord factors for a given form of triangulation are invariant – the product of chord factor × radius. The chordal distances for and between inner and outer circumspheres may thus be easily calculated and since in a well-designed high frequency dome all members are of a closely similar length, all face angles are sensibly close to 60° and hence a two layer dome can be made to emulate the octet truss.

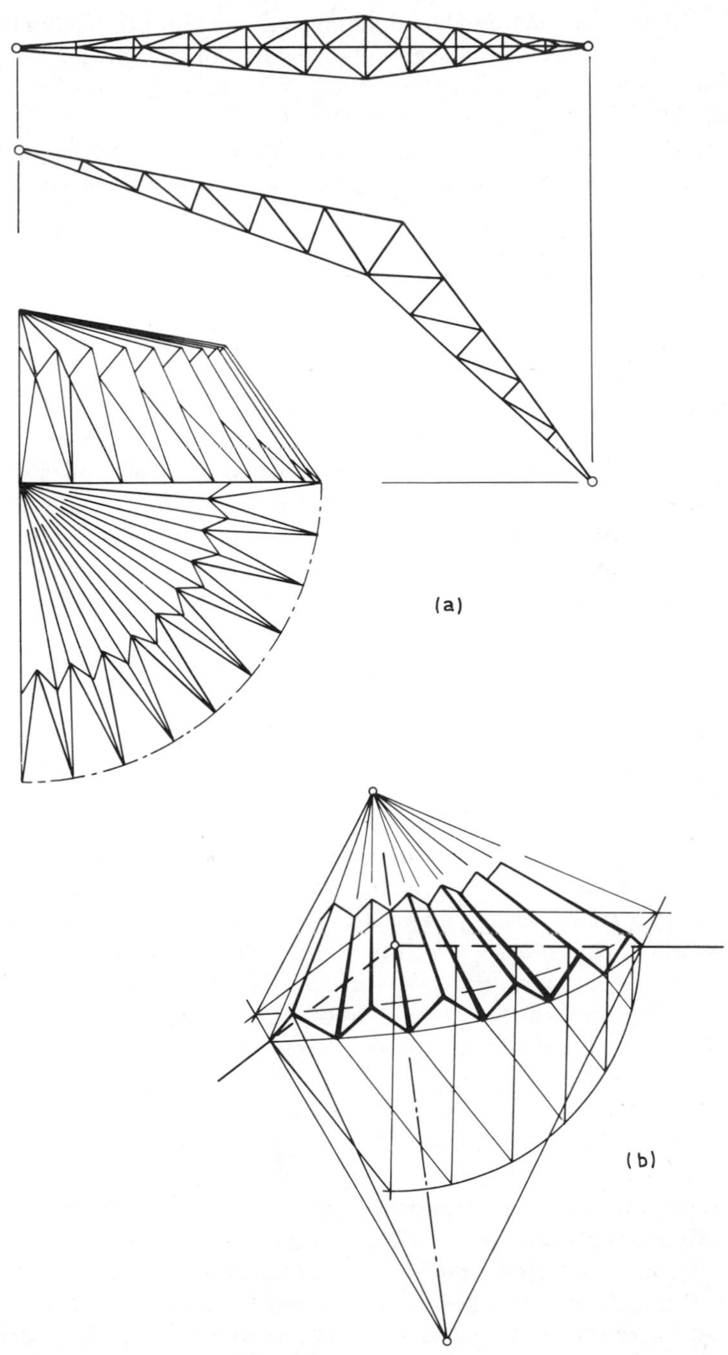

Figure 7.74

Folding flat plate into the third dimension is another highly efficient means of building radial stiffness into an otherwise floppy high frequency dome, and one such method employed by Fuller is based on the use of creased diamond panels.

The fact that two complete shells have been brought into play to cure a relatively local problem means that greatly increased inventory of members or panel has been created and in specific instances it is possible to omit a number of carefully chosen members. This, however, is an advanced structural problem and it must suffice here, to give but one example.

7.10 IMAGINARY STRUCTURES

Throughout this chapter we have been dealing with the placement of real material members which we have endeavoured to arrange in the most expedient way. The success of any structure depends (a) on the appropriateness of its geometry and (b) on the allocation of the right amount and kind of material to carry and withstand the stresses and strains which the structure is designed to carry and resist.

The calculation of stress and strain is of course a subject for separate study but it is pertinent to state that the finite element stress analysis technique is the most powerful method so far devised.

7.10.1 Finite element meshes

The finite element technique is now widely used to calculate (with the aid of a computer) the stresses and strains in complex continuous structures, machine elements and fluids. The method entails a theoretical consideration of the behaviour (under load) of a relatively large number of small, basic and simple elements – called finite elements – into which the parent body is conceptually divided. By considering the behaviour of these simple conceptual elements it is possible to arrive at the behaviour of the real continuum.

Whilst it is obvious that a considerable amount of professional expertise is required to work this basically simple operation effectively, a brief introduction to finite element mesh generation and the all important associate operation of book-keeping is of direct relevance to our subject. The need to define a mesh which leads to a reasonably accurate idealisation of a particular problem is clearly of paramount importance but even so, one's book-keeping as in all computer applications must be immaculate.

In order to define, refer to, call up and recall all nodal coordinates, edge lines and facelets it is necessary to use logical and basically simple numbering systems which are unambiguous which the computer can understand and, with this end in view, we consider a few very simple cases.

7.10.2 Uniform rectangular mesh

With the intent of numbering a rectangular mesh of finite elements in a logical way, acceptable to the computer, we consider the simple case of a 3×2 array as shown in Fig. 7.75.

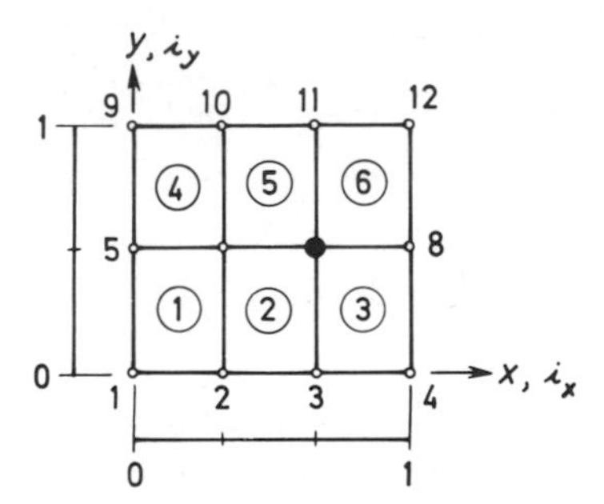

Figure 7.75

We number all nodes from left to right, starting from the bottom left hand corner and working upwards. The nodal numbers are thus as shown in Fig. 7.75. We number each planar element using the same procedure.

We observe that if there are n_x nodes in the x direction and n_y nodes in the y direction, there are $(n_x \cdot n_y)$ nodes in total. n_x nodes in a line yield $(n_x - 1)$ 'squares' in a line and similarly n_y nodes in a line yield $(n_y - 1)$ squares in a line. There are thus $(n_x - 1)(n_y - 1)$ plane square elements in total.

The number of any chosen node, the black dot node for example, is given by $i = i_x + (i_y - 1)\, n_x$.

The black dot node is clearly in line 3 of x and line 2 of y and since n_x in this instance is 4,

$$i \text{ for the black dot} = 3 + (2 - 1)\,4 = 7$$

Now if we normalise the x and y length scales of our boundary element we have a ready means of writing the coordinates of all the nodes, no matter how sizeable the array.

It is clearly the case that

$$X_i = \frac{i_x - 1}{n_x - 1}, \quad Y_i = \frac{i_y - 1}{n_y - 1}$$

Thus the coordinates of the black dot node 7 are

$$X_7 = \frac{3 - 1}{4 - 1} = \frac{2}{3}, \quad Y_7 = \frac{2 - 1}{4 - 1} = \frac{1}{3}$$

(2/3, 1/3) are thus the coordinates of the point 7 in normalised form.

The procedure for numbering the planar elements (facelets) follows a similar route, for if n_R is the number given to a rectangular element then

$$n_R = i_x + (i_y - 1)(n_y - 1)$$

Where i_x and i_y refer to the first corner of the facelet, the i_x and i_y values of the black dot, Fig. 7.75, for instance, are $i_x = 3, i_y = 2$ and these values serve to call up facelet 6.

$$n_R - n_6 = 3 + (2 - 1)(4 - 1) = 6$$

One to root two triangular meshes

If the quadrilateral mesh is divided diagonally as in Fig. 7.76, then we clearly have twice the number of facelets but the same number of nodes, and hence the numbers denoting triangles grow at twice the rate of the original rectangles.

We observed from Fig. 7.76, that, rectangle number n_R divides into two triangles numbered n_T and n_{T+1}.

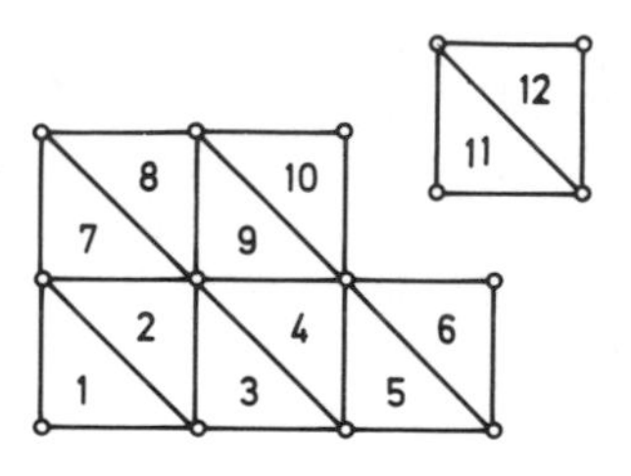

Figure 7.76

For the first and second rectangles we have

$$n_R = 1 \quad n_{T_1} = 1, \quad n_T = 2$$

$$n_R = 2 \quad n_T = 3, \quad n_T = 4$$

and it is easy to see that

$$n_T = 2n_R - 1 \quad n_{T+1} = 2n_R$$

where n_R is the number allotted to the original rectangle. If we again take the rectangular element 6 we see that its two triangular subdivisions are $n_T = (12 - 1) = 11$ and $n_{T+1} = 12$, and as illustrated by Fig. 7.76, this is indeed the case.

7.10.3 Uniform delta mesh

With reference to Fig. 7.77, we see that if the number of nodes per side is n_s, then the average number of nodes per row is $(n_x + 1)/2$ and since there are n_s rows, the total number of nodes (in a triangular stack) is $n_s(n_s + 1)/2$. The average number of triangles per row is $(n_s - 1)$ and since there are $(n_s - 1)$ rows, there are $(n_s - 1)^2$ triangular elements in all.

When numbering this type of mesh we find it convenient to number the partitions (triangular facelets) first and thence deduce the nodal numbers. It is also the practice to count the upright facelets before the inverted facelets are numbered. If the upright triangles are numbered $\Delta_1, \Delta_2, \Delta_3 \ldots \Delta_n$, then the first inverted triangle will be denoted $\Delta_n + \Delta_1$. Inspection of Fig. 7.77, shows that the numbering of nodes defiining a facelet follows the counterclock rule.

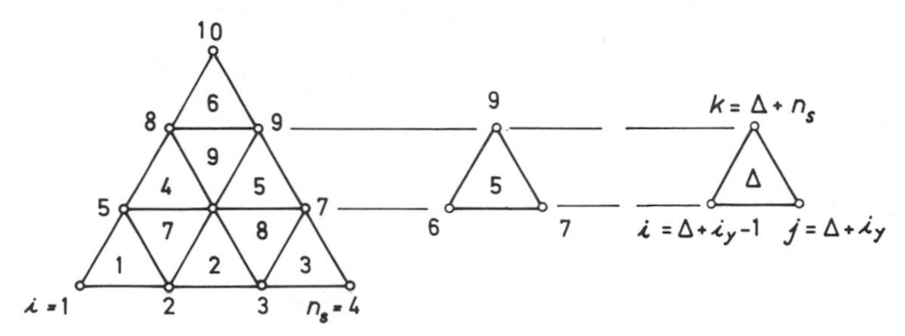

Figure 7.77

First node number $i = (\Delta + i_y - 1)$, second node number $j = (\Delta + i_y)$, third node number $k = (\Delta + n_s)$. For example, if we consider facelet 5, Fig. 7.77, we see that the three node numbers are $i = (5 + 2 - 1) = 6$, $j = (5 + 2) = 7$, $k = (5 + 4) = 9$. Hence the three nodes of facelet 5 are (6, 7, 9) in anticlockwise order.

If the parent figure is of unit side (edge length) then each internal segment is $1/n_s - 1$ units long, and the perpendicular height of each delta facelet is $\sqrt{3}/2$ $(n_s - 1)$ units long. The normalised coordinates of a typical node are

$$X_i = \frac{i_x - 1}{n_s - 1} + \frac{i_y - 1}{2(n_s - 1)}, \quad Y_i = \left(\frac{i_y - 1}{n_s - 1}\right)\frac{\sqrt{3}}{2}$$

where i_x and i_y are the total number of nodes in the particular row and column in which the point $(X_i Y_i)$ is housed.

Node 7, Fig. 7.77, for instance is the third node in the second column whence $i_x = 3$, $i_y = 3$ and since $n_s = 4$

$$X_i = X_7 = \frac{3-1}{4-1} + \frac{2-1}{2(4-1)} = \frac{5}{6}$$

$$Y_i = Y_7 = \frac{2-1}{4-1} \cdot \frac{\sqrt{3}}{2} = \frac{1}{2\sqrt{3}}$$

7.10.4 Triangular elements in a circle

If n_c delta triangles are arranged to fill the inner centre of a circular profile, the remaining area may be (more or less) filled with non-delta triangles. If delta triangles are used to fill the inner centre, then six and six only such triangles are required in plane. If there are n_c deltas about the centre then there are also n_c nodes about the centre. If there are n_r nodes dividing the radius of the parent circle, then we may account the total as follows.

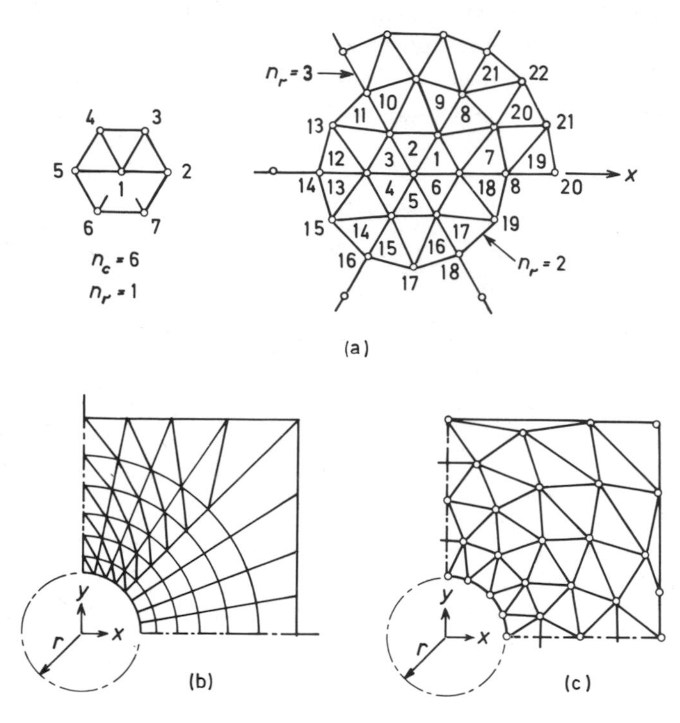

Figure 7.78

The innermost ring contains n_c triangles and n_c nodes. The outermost ring contains $n_c(n_r - 1)$ nodes and so the average number of nodes per ring is $\frac{1}{2}n_c n_r$, and since there are $(n_r - 1)$ rings there are a total of $\frac{1}{2}n_c n_r(n_r - 1) + 1$ nodes in all.

The reader should note the layout, Fig. 7.78, to which these counts refer, and observe how the mesh is built up. There are six true deltas at the centre and these are referred to as the inner ring. The radial edge lines of these deltas are extended outwards. The radius of the second ring is chosen to make the triangles in the second and subsequent rings as near equilateral as possible.

(N.B. Triangles in ring 2 and outwards cannot all be deltas).

Nodes are numbered counter-clockwise ring by ring starting with 1 at the centre. Numbering of rings follows a counter-clock order. Starting with the element in the inner ring immediately above the *ox* axis we number all elements in the inner ring 1-6. We next number all apex inwards elements in the second row 7-18 and similarly starting at 19 we number all apex inwards elements in the third row, etc., before returning to the second row and numbering all apex outwards elements.

The angular coordinate θ of a typical node is given by:

$$\theta = (i_\theta - 1)\ \frac{2\pi}{n_c n_r}$$

For example, node 3 is located at the intersection of radial 2 and lies at a radius 1. Whence $i_\theta = 2, i_r = 1$ and since $n_c = 6$

$$\theta_3 = (2 - 1)\ \frac{2\pi}{6.1} = \frac{\pi}{3} = 60°$$

Similarly node 13 is located at the intersection of radial 6 (we count the radial in relation to the ring to which we refer) and lies at a radius 2. Whence it follows that

$$\theta_{13} = (6 - 1)\ \frac{2\pi}{6.2} = \frac{5\pi}{6} = 150°$$

The Cartesian coordinates are obtained by substituting the appropriate r and θ into the expressions

$$X_i = r\cos\theta,\ \ Y_i = r\sin\theta$$

7.10.5 Iso-parametric elements

We have seen that the essence of the finite element method is to conceptually divide a given structure (or body or fluid) into a number of smaller elements, called finite elements. In areas of pronounced curvature the segmented locality resembles a polyhedron and for the purpose of stress and strain calculations local coordinate axes need to be set in the plane of each separate panel or element, and the best that can be done is to align one axis of a local triad of axes along one edge of the panel, generally as indicated in Fig. 7.79.

If we consider a plane quadrilateral 1234, Fig. 7.79, of arbitrary shape to begin with then it is clear that

$$L' = \left(\frac{x_1 + x_2}{2}, \frac{y_1 + y_2}{2}\right) \qquad M' = \left(\frac{x_3 + x_4}{2}, \frac{y_3 + y_4}{2}\right)$$

$$L'' = \left(\frac{x_2 + x_3}{2}, \frac{y_2 + y_3}{2}\right) \qquad M'' = \left(\frac{x_4 + x_1}{2}, \frac{y_4 + y_1}{2}\right)$$

and it follows that if LML′M′ are the mid-points of their respective sides and the coordinates of O are denoted by O′ and O″

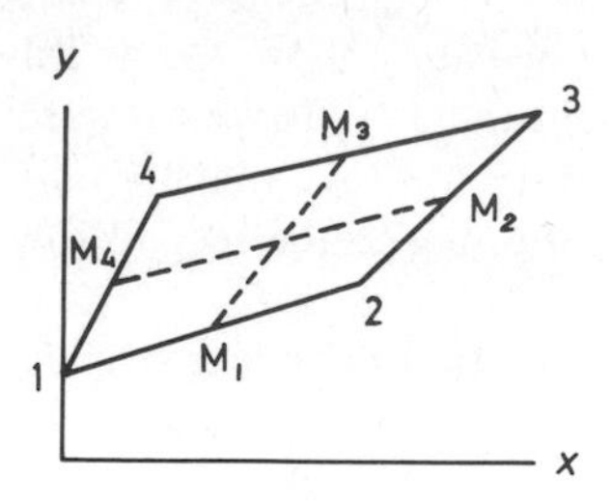

Figure 7.79

$$O' = \left(\frac{\frac{x_1 + x_2}{2} + \frac{x_3 + x_4}{2}}{2}, \frac{\frac{y_1 + y_2}{2} + \frac{y_3 + y_4}{2}}{2} \right)$$

whence

$$\left(\frac{x_1 + x_2 + x_3 + x_4}{4}, \frac{y_1 + y_2 + y_3 + y_4}{4} \right)$$

and similarly

$$O'' = \left(\frac{\frac{x_2 + x_3}{2} + \frac{x_4 + x_1}{2}}{2}, \frac{\frac{y_2 + y_3}{2} + \frac{y_4 + y_1}{2}}{2} \right)$$

$$= \left(\frac{x_1 + x_2 + x_3 + x_4}{4}, \frac{y_1 + y_2 + y_3 + y_4}{4} \right)$$

hence $O' = O'' = O$, the centre.

(N.B. Since LM and L′M′ are the mid-points of their respective sides, LM and L′M′ are the medians of the quadrilateral and therefore bisect each other (at P).)

Having reduced our structure or 3-space body to an amenable calculable multi facetted figure, each plane element is considered in turn and adjacent elements conceptually rejoined, one to the other on the basis of displacement and/or force compatibility. For instance, the displacements at the nodal junctions 1, 3, 5, 7, Fig. 7.80, must match the displacements in adjoining elements, otherwise the structure or body would open up, crack or fall apart. In a plane quadrilateral panel there are clearly eight displacements u_1, u_2, u_3, u_4, v_1, v_2, v_3, v_4, to be considered by the stress man. We, however, are only concerned with how the geometry is defined. In line with the usual convention we define a pair of non dimensional axes ξ, η.

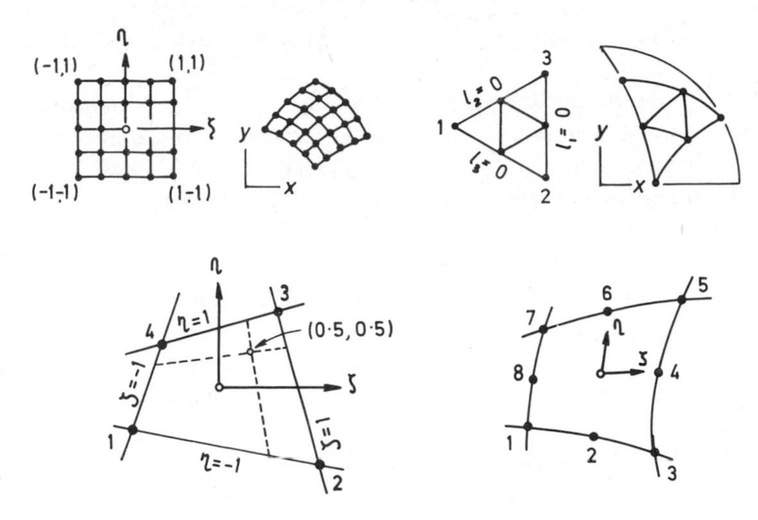

Figure 7.30

If ξ, η are a pair of non-dimensional iso-parametric coordinates and 1, 2, 3, 4, is a plane quadrilateral with nodes at

Node	ξ	η
1	−1	−1
3	+1	−1
5	+1	+1
7	−1	+1

Then it can be shown that the inplane local Cartesian (x', y') and the iso-parametric coordinates (ξ, η) are related as follows:

$$
\begin{aligned}
x &= e_1 + e_2\,\xi + e_3\,\eta + e_4\,\xi\eta \\
y &= f_1 + f_2\,\xi + f_3\,\eta + f_4\,\xi\eta
\end{aligned}
\tag{7.23}
$$

and

$$
\begin{aligned}
e_1 &= \tfrac{1}{4}(x_1 + x_2 + x_3 + x_4) \\
e_2 &= \tfrac{1}{4}(-x_1 + x_2 + x_3 - x_4) \\
e_3 &= \tfrac{1}{4}(x_1 - x_2 + x_3 + x_4) \\
e_4 &= \tfrac{1}{4}(x_1 - x_2 + x_3 - x_4)
\end{aligned}
$$

$$
\begin{aligned}
f_1 &= \tfrac{1}{4}(y_1 + y_2 + y_3 + y_4) \\
f_2 &= \tfrac{1}{4}(-y_1 + y_2 + y_3 - y_4) \\
f_3 &= \tfrac{1}{4}(y_1 - y_2 + y_3 + y_4) \\
f_4 &= \tfrac{1}{4}(y_1 - y_2 + y_3 - y_4)
\end{aligned}
$$

and it follows as a trivial case that when (ξ, η) and (xoy) are sited at 0, $\xi = x$, $\eta = y$ and $e_1 = f_1 = 0$.

The isoparametric quadrilateral thus has edges which correspond to $\xi = 0$, $\xi = 1$, $\eta = 0$, $\eta = 1$, and points within the quadrilateral are readily specified by taking intermediate values of ξ, η, within the range $-1 \leqslant$ or $\leqslant +1$.

The isoparametric concept described above is also applicable to a curvilinear coordinate system in both two and three dimensions, and computing systems now make widespread use of such elements.

There are many well established finite element packages on the market and one such system is PATRAN-G.

PATRAN-G supports all known types of finite elements and features automated grid generation, mesh refinement, element interconnectivity, plus shrunken element display.

The two illustrations, figure 7.81 and 7.82 show a steam turbine generator

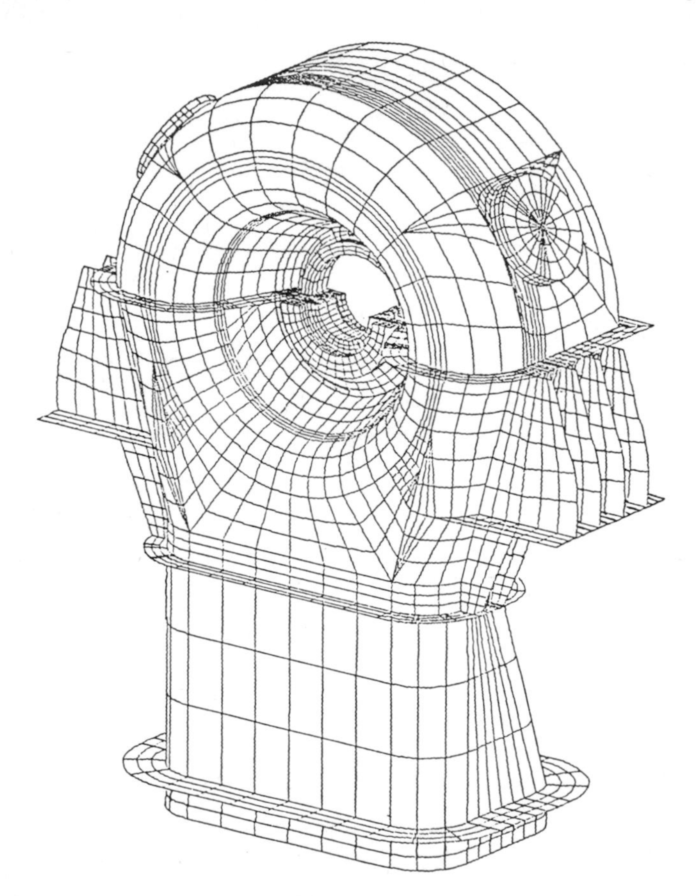

Figure 7.81

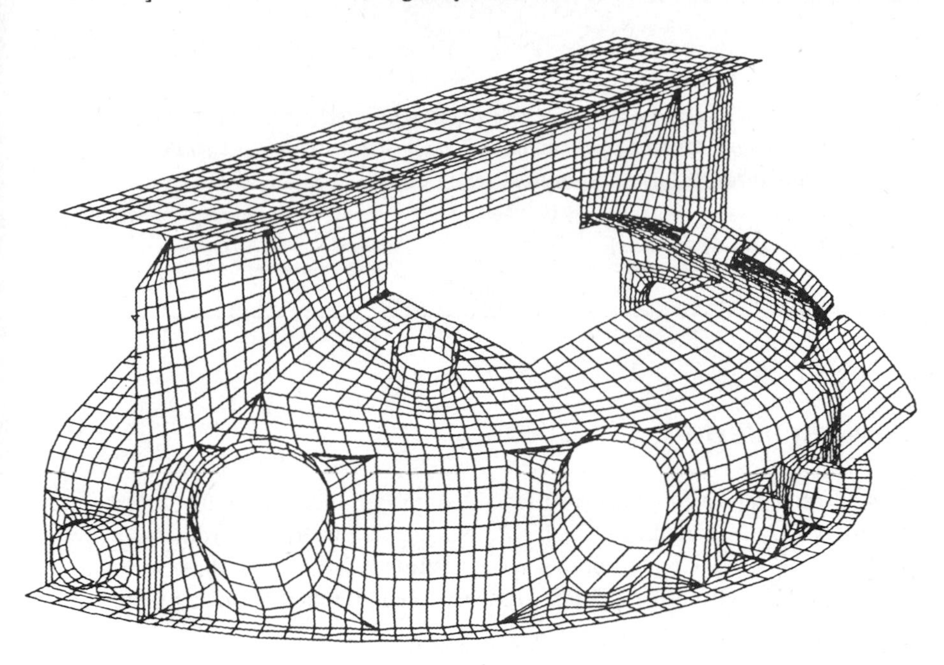

Figure 7.82

housing and the top cap from a nuclear reactor vessel, both are representative examples produced by the PATRAN-G finite element modeller are included here by kind permission of Kins Developments Ltd., Epsom.

Suffice to record that PATRAN-G is a general purpose interactive graphics pre and post processor, designed to construct, display and edit three dimensional finite element models. An important feature of PATRAN-G is that it facilitates the creation of a continuous geometrical model, which is independent of the finite element model and this in turn allows a design data base to be established in a form best suited to each particular type of analysis.

Elements can be colour coded to accord with physical and/or material properties of the structure, whilst the same distinctions may be made on a non-colour device by means of a variety of line types.

8 Classical surfaces

Every curved surface may be regarded as generated by the movement of a line, either constant in form while it changes its position or variable at the same time both in form and in position.

Gaspard Monge

8.1 INTRODUCTION

The position of any point on a surface is completely defined by its Cartesian co-ordinates (x, y, z). A specific point P_1 is identified by the coordinates $x = x_1$, $y = y_1$, $z = z_1$. The equation to a surface is thus a function of three variables x, y and z and we write the equation to the surface in implicit form as follows:

$$F(x, y, z) = 0$$

Thus, the canonical equation to a sphere, by which we mean the equation to the surface of a sphere, about centroidal axes is:

$$x^2 + y^2 + z^2 - a^2 = 0$$

where a is the radius of the sphere.

Similarly the canonical equation to an ellipsoid is:

$$\frac{x^2}{a^2} + \frac{y^2}{b^2} + \frac{z^2}{c^2} - 1 = 0$$

and the canonical equation to a hyperboloid is:

$$\frac{x^2}{a^2} + \frac{y^2}{b^2} - \frac{z^2}{c^2} - 1 = 0$$

and it is clear that although the equation to an ellipsoid and a hyperboloid are surprisingly similar in form, the two surfaces are entirely different.

8.2 GENERATION OF SURFACES

There are two ways in which a smooth curved surface can be generated by the motion of a plane curve along the line of one or two plane guiding curves placed at right angles to it.

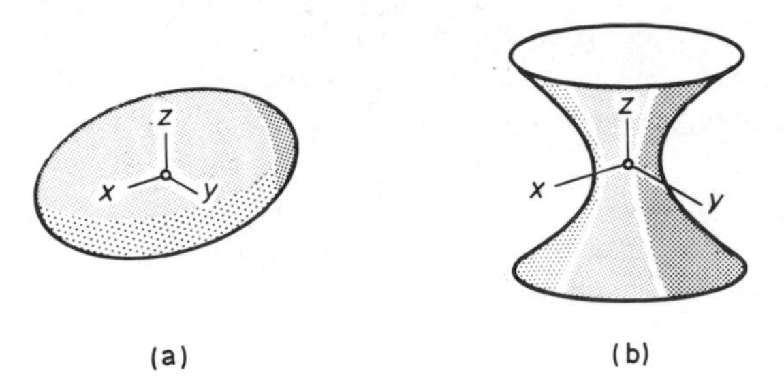

Figure 8.1

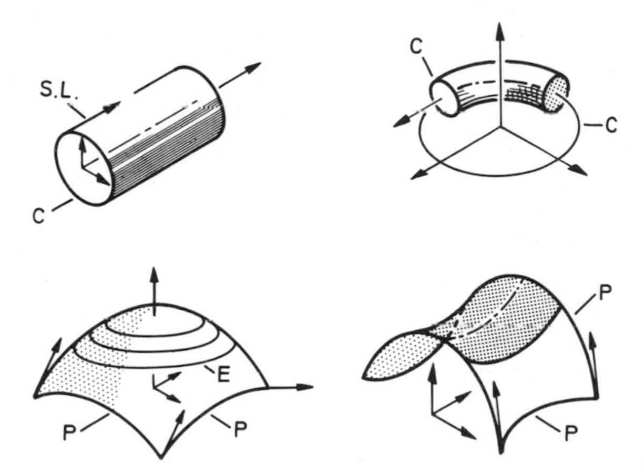

Figure 8.2

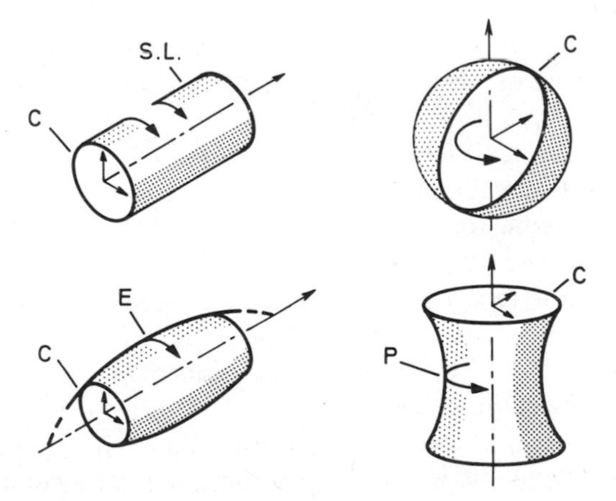

Figure 8.3

The guiding curve along which a generating curve passes is called the directrix and in some instances one such curve suffices. The generating curve which spreads the surface is called the generatrix.

8.2.1 Curve translation

The process of sliding a generating curve – the generatrix – along one or two plane gliding curves is called translation. Four examples of surface generation via translation are shown in Fig. 8.2.

8.2.2 Curve rotation

The process of revolving a plane curve about an inplane axis is called rotation. Four examples of surface generation via rotation are shown in Fig. 8.3, and we note that all cross-sections of a figure of revolution are necessarily circles.

8.3 CLASSIFICATION OF SURFACES

If we consider a point on the surface of a circular cylinder we see that the curvature in a circumferential direction is inversely proportional to its radius. The smaller the radius the larger the curvature: the larger the radius the smaller the curvature, and hence for a radius of infinite length the curvature is zero. A curve of zero curvature is, of course, a straight line. A line parallel to the axis of a circular cylinder is a straight line and hence of zero curvature. A line drawn around the true circumference of a circular cylinder is normal to the axis and its curvature is equal to the inverse of its radius.

A circular cylinder is thus a surface of zero curvature in the longitudinal direction and finite curvature in the circumferential direction.

A sphere, by contrast, has a surface which curves by the same amount in all directions and has but a single centre.

An ellipsoid is an egg-like surface in which the curvature varies from point to point, and an elliptic paraboloid has a similar local characteristic.

The hyperbolic paraboloid is a twisted surface in which the two curvatures originate from opposite sides.

From the brief selection of surfaces listed above, it is easy to identify two mutually perpendicular cross sections (planes) that serve to define the extent of the maximum profile. These cross sections are called the principal directions and are typically as shown in Fig. 8.4.

All curved surfaces of the type summarily discussed are characterised by a maximum and a minimum curvature. The maximum and minimum curvatures are always aligned with the two principal planes and are always mutually perpendicular.

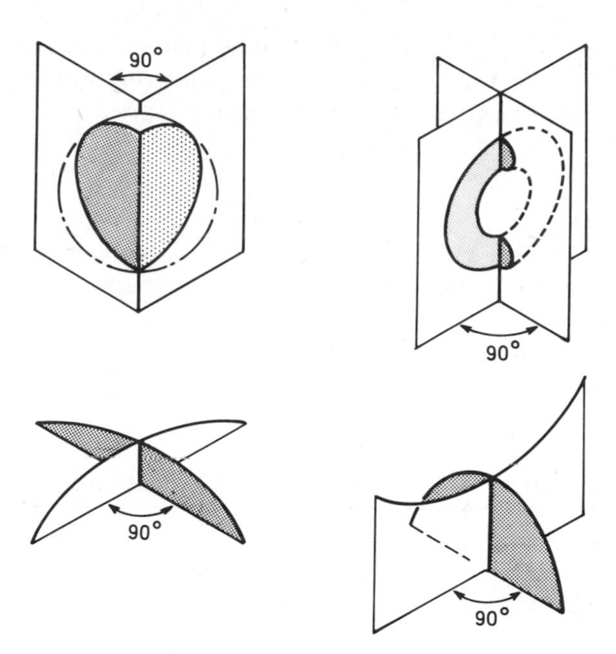

Figure 8.4

8.3.1 Gaussian curvature

By adopting the convention that a normal through a surface has a positive and a negative end, we may allot a positive or negative sign to curvatures struck from opposite sides. The product of the two principal curvatures defined by Gauss (1777–1855) and known as the total or Gaussian curvature, may then be used to distinguish the three possible surface types.

As maysbe observed from Fig. 8.5, the following properties are clearly in evidence.

1. The local surface of a circular cylinder has a ridge-like appearance, with circumferential and longitudinal curvatures respectively $1/r$ and zero. Points on a ridge-like surface are called parabolic points and are always of total curvature zero.
2. The local surface of an ellipsoid is dome-like in appearance and is composed of two positive curvatures $1/r_1$ and $1/r_2$. All points on a dome-like surface are called elliptic points and are characterised by a positive total curvature.
3. The local surface of a hyperboloid is saddle-like in appearance and is composed of two principal curvatures, $1/r_1$, $1/r_2$ of opposite sign. Points on a saddle-like surface are called hyperbolic points and are characterised by a negative total curvature.

The appearance and details relating to these three different surface forms is summarised in Fig. 8.5. C_1 denotes the maximum curvature, C_2 denotes the minimum curvature, C_1 and C_2 are always at right angles.

$$C_1 = \frac{1}{r_1}, \quad C_2 = \frac{1}{r_2}$$

Another measure of local form is given by its mean curvature H, which is simply the average of the two principal curvatures summarily discussed.

$$K = \frac{1}{r_1 r_2}, \quad H = \frac{1}{2}\left(\frac{1}{r_1} + \frac{1}{r_2}\right)$$

Table 8.1

Point surface	Gaussian curvature K	Mean curvature H
Plane	0	0
Parabolic	0	$\frac{1}{2} \cdot \frac{1}{r}$
Elliptic	$+\frac{1}{r_1 r_2}$	$\frac{1}{2}\left(\frac{1}{r_1} + \frac{1}{r_2}\right)$
Hyperbolic	$-\frac{1}{r_1 r_2}$	$\frac{1}{2}\left(\frac{1}{r_1} - \frac{1}{r_2}\right)$

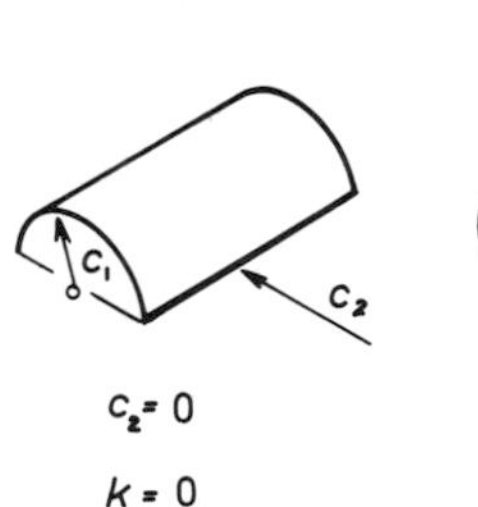

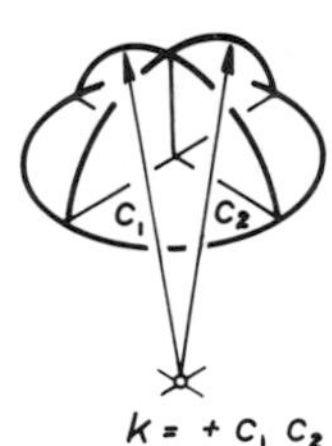

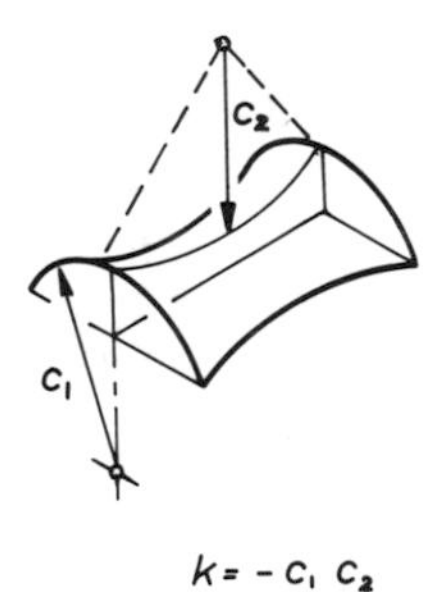

Figure 8.5

8.4 DEVELOPABLE, SYNCLASTIC AND ANTICLASTIC SURFACES

The parabolic, elliptic and hyperbolic point surfaces are also classified as: Developable, Synclastic and Anticlastic surfaces respectively.

8.4.1 Developable surfaces

A developable surface is one which may be produced by bending or rolling an initially flat template into the form required.

It is an intrinsic property of a developable surface that all corresponding geodesic distances on both the curved and the fluttered form are invariant, and this property is true of any surface for which the Gaussian curvature is zero. As explained in Chapter 1 a development of a curved surface is the true shape of its surface, laid out on a two dimensional plane.

8.4.2 Circular cylinder

It is common knowledge that a circular cylinder may be rolled from a single flat sheet, without noticeable stretching, and this is possible because the Gaussian curvature of both the flat and the curved sheet is equal to zero.

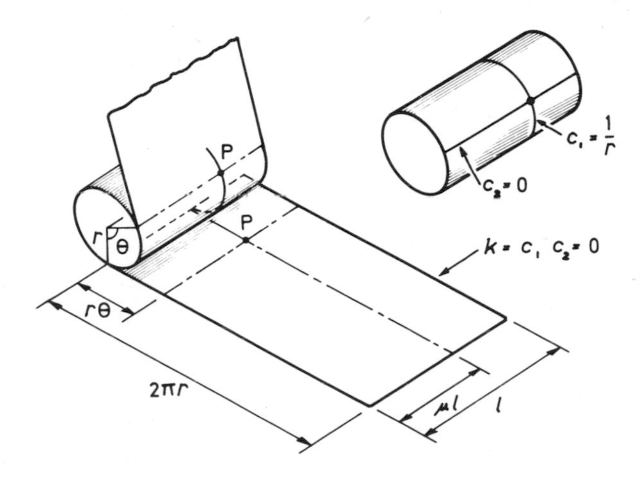

Figure 8.6

8.4.3 Elliptic, parabolic and sine cylinders

Since the surface of all cylinders is comprised of straight line generators, the Gaussian curvature of all cylinders is zero and hence all cylinders are developable forms. Three cylindrical forms commonly used for roofing applications are shown in Fig. 8.7.

The equation to a cylinder is that of its cross section for all values of its length coordinate and the width of its development is equal to the arc length of its surface. The method used to calculate arc length s will therefore be briefly discussed. We define the direction of s, so that as x increases, or if the equation to the curve is in parametric form, namely if $x = f(t), y = g(t)$, we let s increase as t increases. The two procedures are shown in Fig. 8.8.

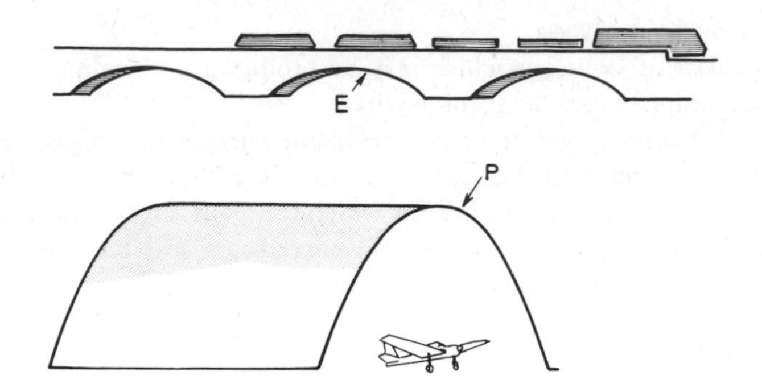

Figure 8.7

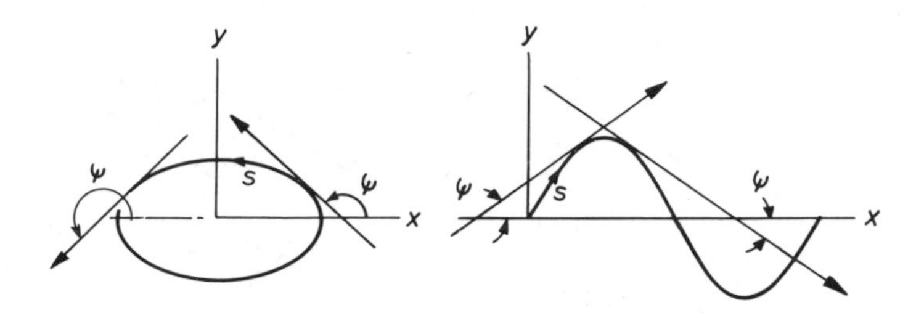

Figure 8.8

From which

$$\sin \psi = \frac{dy}{ds} \quad \cos \psi = \frac{dx}{ds}$$

and

$$\frac{ds}{dx} = \left[1 + \left(\frac{dy}{dx}\right)^2\right]^{1/2}$$

whence

$$s = \int_a^b \left[1 + \left(\frac{dy}{dx}\right)^2\right]^{1/2} dx \tag{8.1}$$

When the parametric form is used and the curve increases from a condition t_1 at A to a condition t_2 at B we have:

$$s = \int_{t_1}^{t_2} \left[1 + \left(\frac{dy}{dx}\right)^2\right]^{1/2} \frac{dx}{dt}\, dt$$

which leads to the result

$$s = \int_{t_1}^{t_2} \left[\left(\frac{dx}{dt}\right)^2 + \left(\frac{dy}{dt}\right)^2\right]^{1/2} dt$$

when dx/dt is positive, x and t increase with s.

Example
Determine the developed length of a catenary arch, the equation to which is $y = c \cosh x/c$.

Solution

$$\frac{dy}{dx} = c \frac{1}{c} \sinh \frac{x}{c} = \sinh \frac{x}{c}$$

$$1 + \left(\frac{dy}{dx}\right)^2 = 1 + \sinh^2 \frac{x}{c} = \cosh^2 \frac{x}{c}$$

$$\left[1 + \left(\frac{dy}{dx}\right)^2\right]^{1/2} = \cosh \frac{x}{c}$$

but the arc length s is given by equation 8.1 and the developed length of the catenary arch is therefore

$$s = \int_{x_1}^{x_2} \cosh \frac{x}{c}\, dx = \left[c \sinh \frac{x}{c}\right]_{x_1}^{x_2}$$

8.4.4 The cone

The equation of a cone may be considered in terms of the parameters that together define a general point on its surface. If the apex of a cone is placed at o, where o is the origin of axes ox, oy, oz, then any point on the surface of a cone with axis coincident with the reference axis oz may be expressed in the form:

$$x^2 + y^2 = r^2 = z^2 \tan^2 \alpha$$

but since $y^2 = x^2 \tan^2 \phi$

$$z^2 \tan^2 \alpha = x^2 \sec^2 \theta$$

whence

$$x = z \tan \alpha \cos \phi, \quad y = z \tan \alpha \sin \phi$$

where α is the half angle of the cone and ϕ the angular displacement of the point referred to a plane parallel to the reference plane xoy as illustrated in Fig. 8.9.

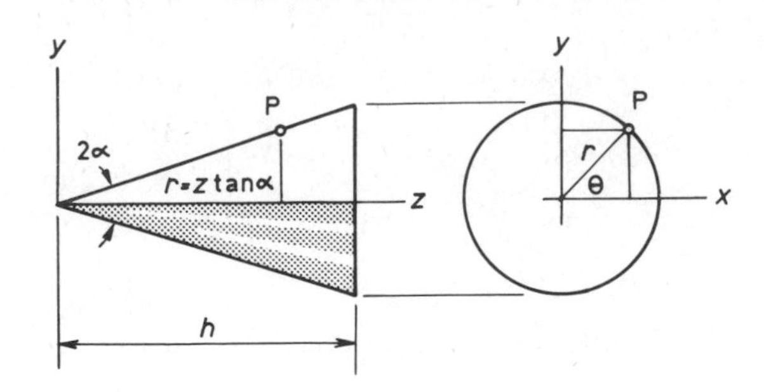

Figure 8.9

8.4.5 Development of a cone

Readers who have studied practical chemistry will know that a conical filter can be produced from a complete circle of filter paper, as shown in Fig. 8.10.

The circular pattern, Fig. 8.10, is not (on account of the overlap OABC) the development of the conical filter.

If the overlapping folds are removed, however, the circular pattern becomes a sector and the sector angle θ determines the apex angle of the cone. The sector angle θ is determined by the condition that no part of the pattern must shrink or extend when rolled to form the cone. The necessary condition is that the circumference of the sector and the cone must be equal.

Whence

$$2\pi r = s\theta$$

and

$$\frac{r}{s} = \frac{\theta}{2\pi} = \sin\alpha$$

where α is the half angle at the apex of the cone.

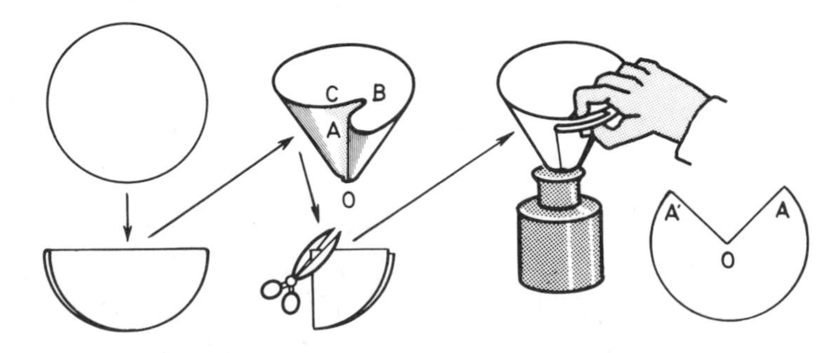

Figure 8.10

8.4.6 The conoid

A conoid is a developable surface produced by ruled straight lines guided by one curved and one straight line directrix. One particular conoid based on a straight line and a circular directrix is shown in Fig. 8.11. Another conoid employing a straight line and a parabolic directrix is shown at (b), and a double sine-wave conoid with a common directing line is shown at (c).

We have seen that a right conoid is a ruled surface comprising one straight line and one curved directrix. The line and the plane of the curve being parallel.

The Cartesian equation to the right conoid, Fig. 8.11(a) may be derived as follows. The conoid is redrawn with appropriate axes as in Fig. 8.12. The curved directrix may be of any suitable form, such as $z = ky^n$ for example.

The curve $z = ky^2$ is a cup curve for which $z = o$ at $y = o$, and symmetric about the oz axis, with both y and z increasing. The curve $z = -ky^2$ is a similar curve but in which z is negative for y increasing. The curve $z = -ky^2 + h$ is a similar curve but has an offset oz axis as shown in Fig. 8.12(c).

Now since it is required that $z = o$, when $y = \pm b$; $k = h/b^2$ whence

$$z = -\frac{h}{b^2}y^2 + h$$

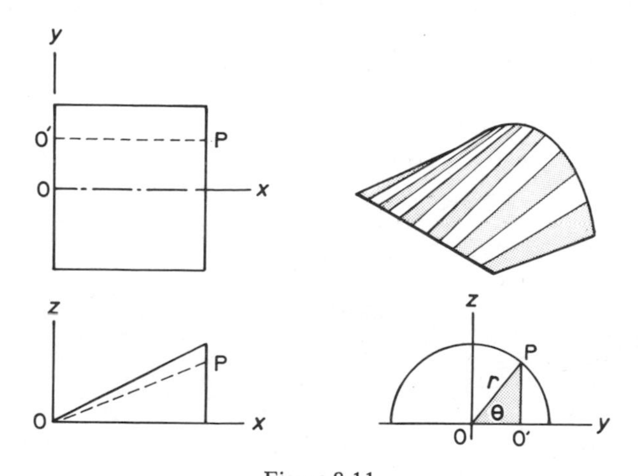

Figure 8.11

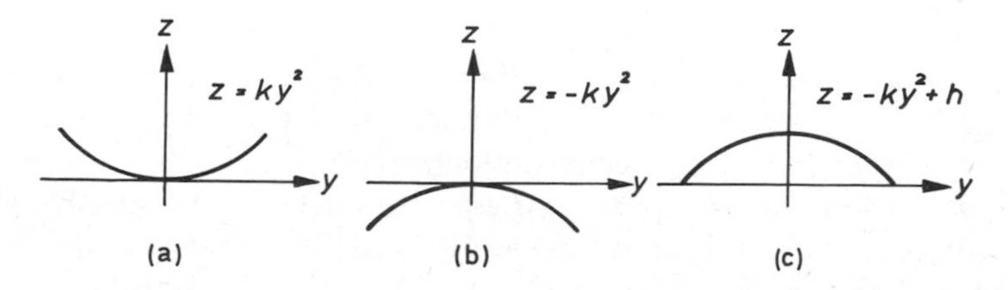

Figure 8.12

The plane section in the plane xoz, Fig. 8.12, shows that the ordinate z increases linearly with increasing x and hence the equation to the generator in the plane xoz is of the form $z = k'x$. We also observe that all generators of a conoid lie parallel to the ox axis and normal to the plane of the curved directrix. Hence the equation to the surface is written:

$$z = \left(-\frac{h}{b^2}\,y^2 + h\right)\frac{x}{a}$$

$$z = -\frac{hx}{ab^2}\,y^2 + \frac{hx}{a}$$

The slopes $dz/dx = (-hy^2/ab^2 + h/a)$ and $dz/dy = -\,2hxy/ab^2$ are obtained by differentiation.

8.4.7 Development of a right conoid

The surface development of a right conoid is easily expressed mathematically when the shape and size of the curved generatrix is given. The information given in Fig. 8.13, is self-explanatory.

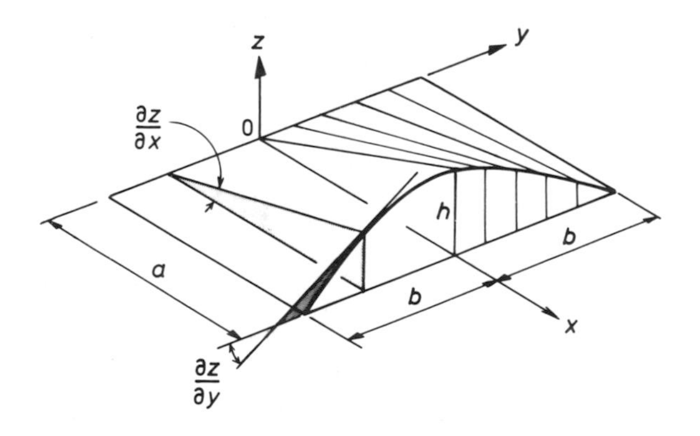

Figure 8.13

Right conoids are widely used in roofing applications and another example of this type of surface occurs in the transition from a square to a circular cross-section – a geometrical feature common to ducting.

In Fig. 8.14(a), the size of square is equal to the escribed circle. The curved generatrix is a quadrant. If $2s$ is the side of the square the radius of the escribed circles is $s\sqrt{2}$ and the rise is $s(\sqrt{2} - 1)$.

The development shown may be calculated as previously explained.

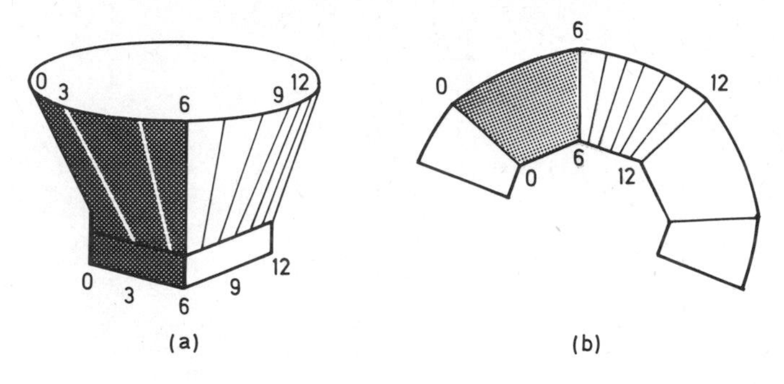

Figure 8.14

Coupled conoids. An interesting roofing surface may be formed by a petal type cluster of conoids, for instance. If adjacent petals intersect in the plane $y = \pm b/a\, x$, we may calculate the curve of intersection as follows.

Since $x = u \cos \phi$ the equation to the line of intersection in the plane uoz is obtained by substituting $u \cos \phi$ for x in the equation of the rectangular based conoid previously derived. The equation:

$$z = -\frac{hx^3}{a^3} + \frac{hx}{a}$$

becomes

$$z = -\frac{h}{a^3}\, u^3 \cos^3 \phi + \frac{h}{a}\, u \cos \phi$$

where $\cos \phi = a/\sqrt{a^2 + b^2}$.

Example
Calculate the line of intersection between two petal conoids for which $a = 4$, $b = 3, h = 2$.

Solution
Since $\cos \phi = 4/5$

$$z = -\frac{2}{64}\, u^3\, \frac{64}{125} + \frac{2}{4} \cdot \frac{4}{5}\, u$$

and

$$z = \frac{1}{125}\,(50u - 2u^3)$$

whence

u	z
0	0
1	$\frac{48}{125}$
2	$\frac{84}{125}$
3	$\frac{96}{125}$
4	$\frac{72}{125}$
5	0

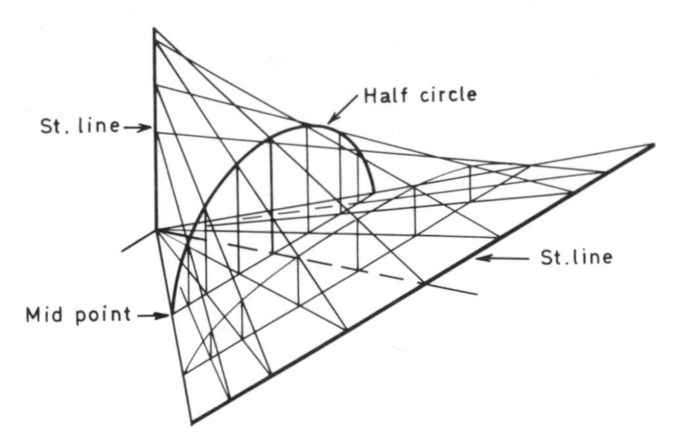

Figure 8.15

8.4.8 A conoid model

A surface composed of two conoids blended end to end, is shown in Fig. 8.16, and a model proposed by Cundy and constructed from card wood or perspex forms; with cotton, cord or wire strings makes an excellent demonstration.

The surface is formed from two outer conoids of $3a{:}a$ elliptical base, set at 90° in planes $z = \pm 2b$. The $a{:}3a$ vertical ellipse at $z = +2b$ transforms to a vertical line, $x = o$, at $z = b$, and opens to a circular cross-section of diameter a, at $z = o$. The circular conoid closes to a horizontal line, $y = o$ at $z = -b$ and then opens to $3a{:}a$ horizontal ellipse at $z = -2b$.

The stringing of the surface is self explanatory.

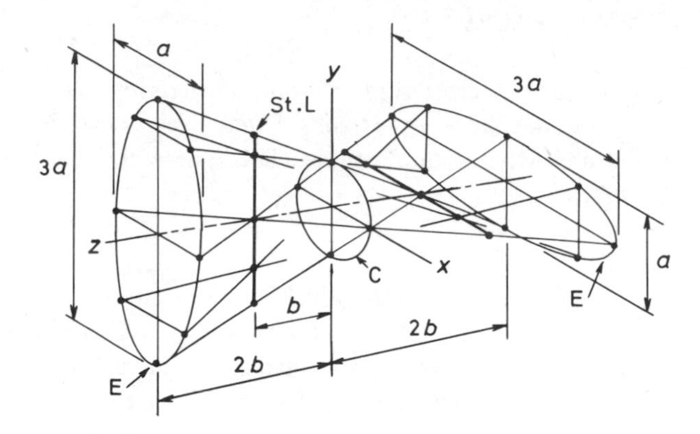

Figure 8.16

8.4.9 Nested sections through a right conoid

A set of superimposed sections through parallel planes is called a nest of sections. A set of nested sections parallel to the xoy plane of a right conoid $z = hxy^2$ may be calculated by setting the coordinate z to a succession of constant values.

If for the purpose of example we elect to put $h = \frac{1}{2}$ we may obtain the corresponding values of x, y and z by denoting y for a given value of z and thence calculating the value of x. A set of such values is shown in the table below and sections illustrated in Fig. 8.17. N.B. The values given apply from the origin to the centre line.

$z = 1$		$z = 2$		$z = 4$	
y	x	y	x	y	x
0	∞	0	∞	0	∞
1	2	1	4	1	8
$\sqrt{2}$	1	$\sqrt{2}$	2	$\sqrt{2}$	4
2	0.5	2	1.0	2	2
4	0.125	4	0.25	4	0.5
∞	0	∞	0	∞	0

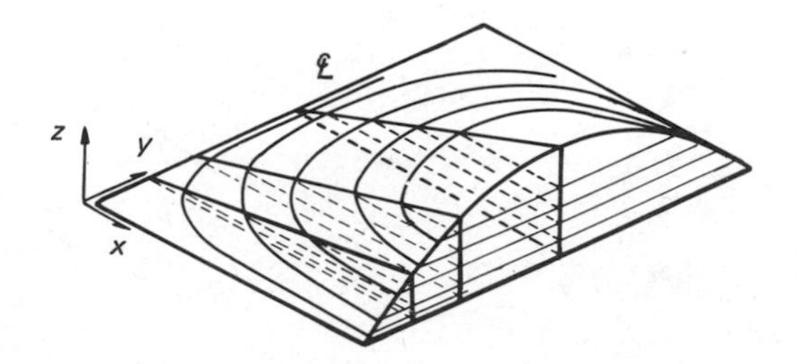

Figure 8.17

8.5 SYNCLASTIC SURFACES

When the centre of both principal curvatures lie on the same side of a surface, the surface is bowl or dome-like in appearance. Dome-like – synclastic surfaces – are doubly curved and are not developable. The sphere and the ellipsoid are two common examples.

8.5.1 The sphere

The sphere has but a single centre from which a single radius is struck. All normal sections, that is all sections that cut the sphere across a full diameter are identical great circles. All other sections are small circles of varying size.

The surface of a sphere is identically symmetric about every diametrial axis and this being so the Gaussian or total curvature is non-zero and the surface is not developable.

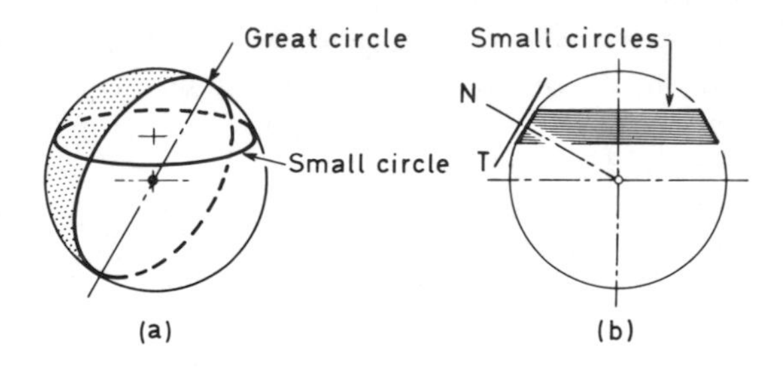

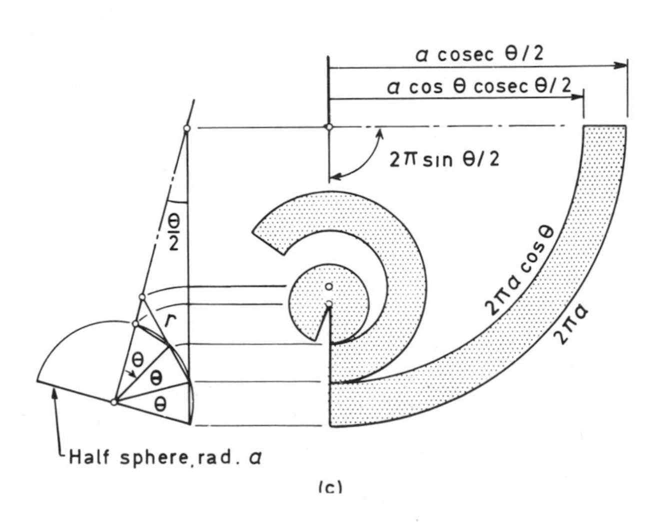

Figure 8.18

8.5.2 Equations of a sphere

The equation of a sphere referred to centroidal axes is:

$$x^2 + y^2 + z^2 = a^2$$

and the equation to the normal (great circle) section in the plane xoy is obtained by setting $z = 0$ and similarly, by setting y and x equal to zero for the other two reference planes.

Normal section xoy $(z = 0)$

$$x^2 + y^2 = a^2$$

Normal section xoz $(y = 0)$

$$x^2 + z^2 = a^2$$

Normal section yoz $(x = 0)$

$$y^2 + z^2 = a^2$$

In the study of surfaces it is sometimes necessary to site the origin of axes on the surface and align two of the axes in a tangential plane.

If the equation of a sphere, with respect to centroidal axes is:

$$x^2 + y^2 + z^2 = a^2$$

The equation with respect to axes situated at $(-u, -v, -w,)$ is:

$$(x + u)^2 + (y + v)^2 + (z + w)^2 = a^2$$

and the expansion

$$x^2 + y^2 + z^2 + 2ux + 2vy + 2wz + u^2 + v^2 + w^2 = a^2$$

is the general equation to a sphere.

For the particular case, Fig. 8.19, in which the origin of axis is placed at (o, o, a). The axis oz' passes through the centroid and hence the mutually perpendicular axes $o'x'$ and $o'y'$ lie in a plane tangential to the sphere through o'.

The equation reflecting this conditions is:

$$x^2 + y^2 + z^2 - 2az = 0$$

Normal sections of a sphere. The normal section in the plane $x'o'z'$ corresponds to the condition $y = o$, whence the equation to the normal section $x'o'z'$ is:

$$x^2 + z^2 - 2az = 0$$

Similarly the normal section $y'oz'$ is:

$$y^2 + z^2 - 2az = 0$$

Since the plane $x'o'y'$ is tangential to the sphere the normal section in the plane $x'o'y'$ is reduced to a single point.

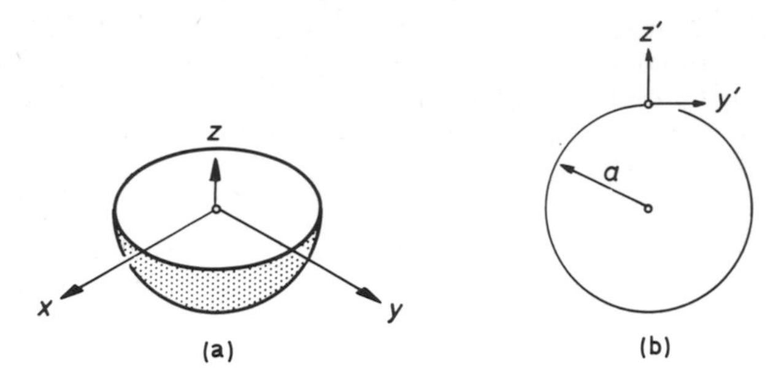

Figure 8.19

8.5.3 Tangent plane to a sphere

The axes $x'o'y'$, Fig. 8.19, define a plane which is tangential to the surface of a sphere through the point where the axis oz' cuts the surface. The tangent plane through any other point on the surface of a sphere may be obtained as follows.

If the equation to the tangent plane with reference to an origin O situated at $(-u, -v, -w)$ is required to pass through an arbitrary point P with coordinates (x_1, y_1, z_1) the direction ratios of the line OP are $(x_1 + u)$, $(y_1 + v)$, $(z_1 + w)$, the direction ratios of a general line in the tangent plane are $(x - x_1)$, $(y - y_1)$, $(z - z_1)$ and since the line OP is normal to the tangent plane, the sum of their products is zero and hence the equation through the point P is:

$$(x_1 + u)(x - x_1) + (y_1 + v)(y - y_1) + (z_1 + w)(z - z_1) = 0$$

(N.B. The reader should note the form of this equation which is the same as that for the analogous case in two dimensions). Whence

$$xx_1 + yy_1 + zz_1 + u(x - x_1) + v(y - y_1) + w(z - z_1) + d = 0$$

where

$$d = -(x_1^2 + y_1^2 + z_1^2)$$

Example

Determine the equation to the tangent plane through the point (3, 4, 5) on the surface of a sphere, with respect to centroidal axes. Also determine the length of the normal from plane to centre of sphere i.e. the radius of the sphere.

Solution

The equation of a tangent plane referred to a set of general axes at $(-u, -v - w)$

$$xx_1 + yy_1 + zz_1 + u(x - x_1) + v(y - y_1) + w(z - z_1) - (x_1^2 + y_1^2 + z_1^2) = 0$$

The equation referred to controidal axes corresponds to the condition

$$u = 0, v = 0, w = 0$$

and hence the equation to a tangent plane with respect to centroidal axes is

$$xx_1 + yy_1 + zz_1 - x_1^2 - y_1^2 - z_1^2 = 0$$

In this particular case $x_1 = 3, y_1 = 4, z_1 = 5$, whence the equation to the tangent plane through the point (3, 4, 5) is

$$3x + 4y + 5z - 50 = 0$$

The radius of the sphere is $\sqrt{3^2 + 4^2 + 5^2} = \sqrt{50} = 5\sqrt{2}$.

N.B. We also note that $xx_1 + yy_1 + zz_1 = r^2$.

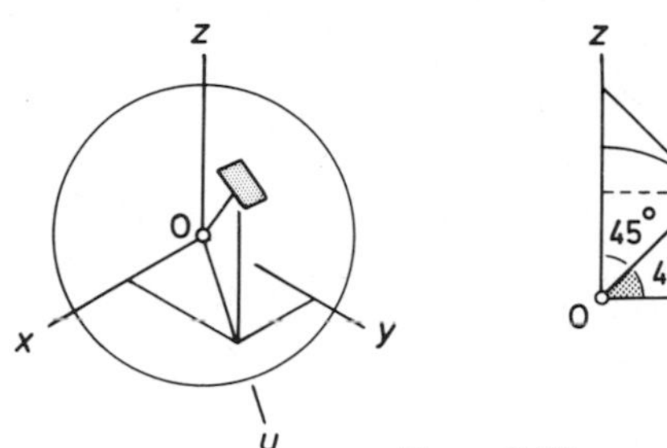

Figure 8.20

8.5.4 The ellipsoid

The equation to an ellipsoid about centroidal axes is of the form:

$$\frac{x^2}{a^2} + \frac{y^2}{b^2} + \frac{z^2}{c^2} = 1$$

and hence its normal sections are:

$$\frac{x^2}{a^2} + \frac{y^2}{b^2} = 1 \quad \text{in plane } xoy$$

$$\frac{x^2}{a^2} + \frac{z^2}{c^2} = 1 \quad \text{in plane } xoz$$

$$\frac{y^2}{b^2} + \frac{z^2}{c^2} = 1 \quad \text{in plane } yoz$$

The ellipsoid is a synclastic surface but unlike the sphere no two normal sections are in general the same.

It is clear, however, that all sections parallel to the coordinate reference planes are also ellipses, as illustrated in fig. 8.21.

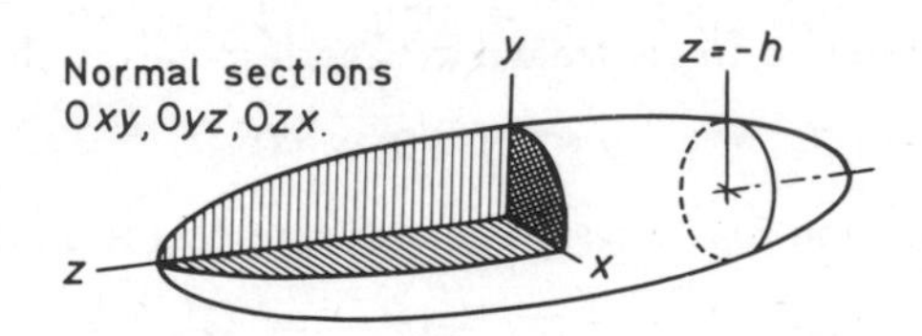

Figure 8.21

8.5.5 Equations of sections parallel to reference planes

The equation to the section on any plane parallel to a coordinate reference plane is obtained by setting the appropriate coordinate to a constant value *f, g, h*.

If we required the equation of the section parallel to the *xoy* plane, through the point $z = h$, we write the surface equation in the form:

$$\frac{x^2}{a^2} + \frac{y^2}{b^2} = 1 - \frac{h^2}{c^2}$$

which yields an ellipse, since the R.H.S. reduces to a constant k^2 say. The major and minor axes of the section are then $(a/k)^2$ and $(b/k)^2$.

The corresponding equation for planes parallel to the *xoz* reference plane is:

$$\frac{x^2}{a^2} + \frac{z^2}{c^2} = 1 - \frac{g^2}{c^2}$$

and the equation for planes parallel to the *yoz* reference plane is:

$$\frac{y^2}{b^2} + \frac{z^2}{c^2} = 1 - \frac{f^2}{c^2}$$

8.5.6 Equation to sections inclined to reference planes

The equation to a section inclined to a reference plane may be obtained by observing that the surface cuts an inclined plane as shown in Fig. 8.22.

The equation to an ellipsoid is:

$$\frac{x^2}{a^2} + \frac{y^2}{b^2} + \frac{z^2}{c^2} = 1$$

and since the inclined plane is defined by:

$$x = u \cos\phi, \ \ y = u \sin\phi$$

we may obtain the equation to the inclined section by direct substitution.

$$\frac{u^2 \cos^2\phi}{a^2} + \frac{u^2 \sin^2\phi}{b^2} = 1 - \frac{z^2}{c^2}$$

or

$$z = c\left[1 - u^2\left(\frac{\cos^2\phi}{a^2} + \frac{\sin^2\phi}{b^2}\right)\right]^{1/2}$$

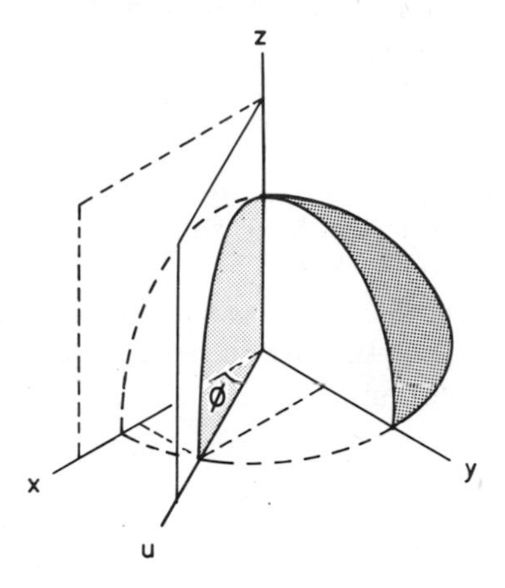

Figure 8.22

Example
The equation of an ellipsoid is as given below. Calculate the equation to the cross section in the plane *uoz*, where the plane *uoz* is given by $\tan\phi = \frac{3}{4}$.

The equation to the ellipsoid is:

$$\frac{x^2}{3} + \frac{y^2}{4} + \frac{z^2}{1} = 1$$

Solution
The equation to the section in the plane *uoz*, where $\tan\phi = \frac{3}{4}$, is from the preceding discussion known to be:

$$u^2\left(\frac{\cos^2\phi}{a^2} + \frac{\sin^2\phi}{b^2}\right) = 1 - \frac{z^2}{c^2}$$

In this particular case $\cos\phi = \frac{3}{5}$, $\sin\phi = \frac{4}{5}$ whence

$$\frac{\cos^2\phi}{a^2} + \frac{\sin^2\phi}{b^2} = \frac{9}{25}\cdot\frac{1}{3} + \frac{16}{25}\cdot\frac{1}{4} = \frac{7}{25}$$

and since $c = 1$

$$z^2 = 1 - \frac{7}{25}u^2$$

8.5.7 Elliptic paraboloid

An elliptic paraboloid is a synclastic surface produced by translating a plane parabola P1 along the profile of another parabola P2, the centres of curvature of both parabolae being on the same side of the surface.

The equation of an elliptical paraboloid is of the form:

$$\frac{x^2}{a^2} + \frac{y^2}{b^2} = 2cz$$

and we see that for any constant value of z, every section parallel to the *xoy* plane is an ellipse, as shown in Fig. 8.23, and all sections parallel to the planes *xoz* and *yoz* are by definition parabolic.

The equation to sections inclined to the reference planes is obtained by the method of article 8.5.6. The plane *uoz* inclined at an angle ϕ to the *ox* axis is given by the parametric equations,

$$x = u \cos \phi, \; y = u \sin \phi$$

and since the section required is the intersection of the plane *uoz* and the surface, we have only to substitute the relevant expressions for x and y into the equation to the surface.

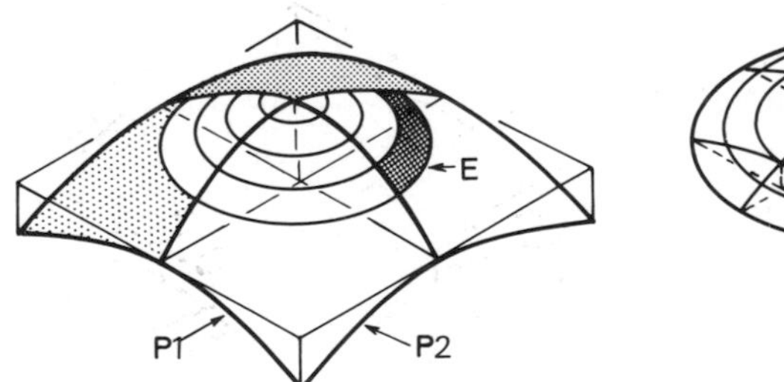

Figure 8.23

The equation to the elliptic paraboloid surface is:

$$\frac{x^2}{a^2} + \frac{y^2}{b^2} = 2cz$$

and hence the equation to an inclined section is:

$$\frac{u^2 \cos^2 \phi}{a^2} + \frac{u^2 \sin^2 \phi}{b^2} = 2cz$$

or expressed in explicit form the equation to the section is:

$$z = \frac{u^2}{2c} \left(\frac{\cos^2 \phi}{a^2} + \frac{\sin^2 \phi}{b^2} \right)$$

8.5.8 Surface slope of an elliptic paraboloid

Given the equation to an elliptic paraboloid surface we may determine the slopes $\partial z/\partial x$ and $\partial z/\partial y$ by direct differentiation. If

$$z = \left(\frac{x^2}{a^2} + \frac{y^2}{b^2}\right)\frac{1}{2c}$$

$$\frac{dz}{dx} = \frac{x}{a^2 c} + \frac{y}{b^2 c} \cdot \frac{dz}{dx}$$

whence

$$\frac{dz}{dx} = \frac{b^2 x}{a^2(b^2 c - y)}$$

and by similar route

$$\frac{dz}{dy} = \frac{a^2 y}{b^2(a^2 c - x)}$$

(N.B. x and y are not independent and hence the value of x depends on the value of y and vice versa.) The surface slopes in the two principal directions are obtained by putting $y = 0$ and $x = 0$ in the above expressions. Whence

$$\frac{\partial z}{\partial x_{y=o}} = \frac{x}{a^2 c}, \quad \frac{\partial x}{\partial y_{x=o}} = \frac{y}{b^2 c}$$

The surface slope $\partial z/\partial u$ may be obtained by differentiation in the usual way. Thus for ϕ equal a constant, the direction ou is fixed, whence

$$z = \frac{u^2}{2c}\left(\frac{\cos^2\phi}{a^2} + \frac{\sin^2\phi}{b^2}\right)$$

and

$$\frac{dz}{du} = \frac{u}{c}\left(\frac{\cos^2\phi}{a^2} + \frac{\sin^2\phi}{b^2}\right)$$

$$\text{when } \phi = o,\ u = x,\ \frac{\partial z}{\partial u_{u=x}} = \frac{u}{a^2 c} = \frac{x}{a^2 c}$$

$$\text{when } \phi = 90^\circ,\ u = y,\ \frac{\partial z}{\partial u_{u=y}} = \frac{u}{b^2 c} = \frac{y}{b^2 c}$$

All cross sections in planes parallel to the xoy reference plane are ellipses, and all cross sections in planes parallel to planes containing the axis oz are parabolae.

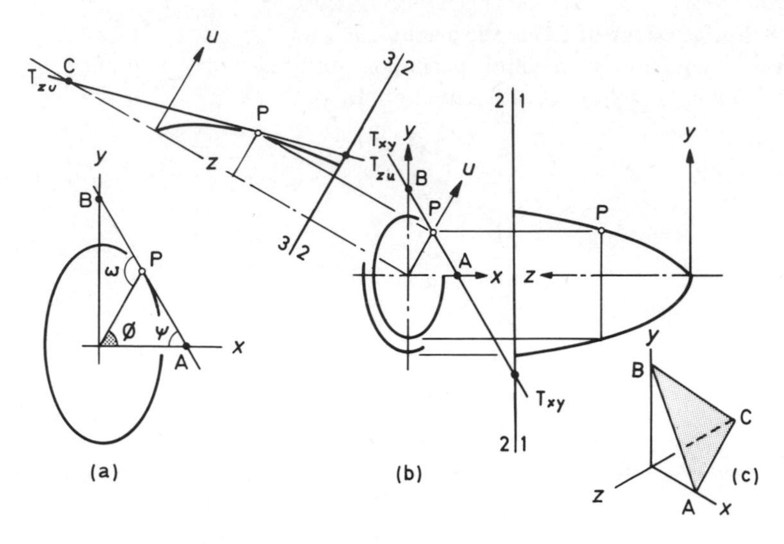

Figure 8.24

The equation to the section at $z = z_1$ is:

$$\frac{x^2}{a^2} + \frac{y^2}{b^2} = 2cz_1 = \text{a constant}$$

$$\frac{2x}{a^2} + \frac{2y}{b^2}\,\frac{dy}{dx} = 0$$

whence

$$\frac{dy}{dx} = -\frac{x}{a^2} \cdot \frac{b^2}{y}$$

but since $x/y = \cot\phi$, for all values of z

$$\frac{dy}{dx} = -\frac{b^2}{a^2}\cot\phi = -\tan\psi \text{ say}$$

We observe that since $\omega = \phi + \psi$

$$\tan\omega = \frac{\tan\phi + \tan\psi}{1 - \tan\phi\tan\psi}$$

The angle ω between the line *ou* and the tangent line TT to the surface is given by:

$$\tan\omega = \frac{\tan\phi + \dfrac{b^2}{a^2}\cot\phi}{1 - \dfrac{b^2}{a^2}}$$

Rider

It is frequently necessary and often desirable to check the validity of equations such as the one above. This may be done in this instance by putting $a = b$ and checking whether the equation gives the correct value of $\tan\omega$ for a circle

$$\text{when } a = b, \quad \tan\omega = \infty, \quad \text{that is } \omega = 90^\circ$$

The radius line *ou* is thus at right angles for all values of ϕ which confirms that by putting $a = b$ the equation is true for a circle.

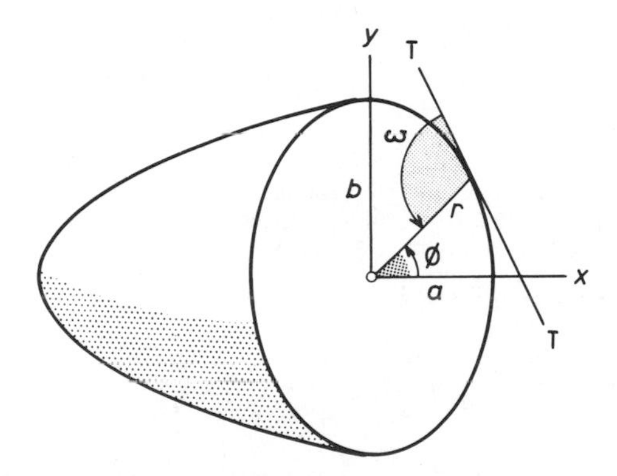

Figure 8.25

Example

A typical elliptical cross-section of an elliptic paraboloid is shown in Fig. 8.24(b). Calculate the angle ω of the tangent line TT, with respect to the line (r, ϕ). Take $\phi = 45^\circ$, $a = 1$, $b = 2$.

Solution

The angle ω is given by the equation:

$$\tan\omega = \frac{\tan\phi + \dfrac{b^2}{a^2}\cot\phi}{1 - \dfrac{b^2}{a^2}}$$

whence

$$\tan \omega = \frac{1+2}{1-2} = -3$$

$$\text{arc tan } 3 = 71.567^\circ$$

From which

$$\omega = (180 - 71.567) = 108.433^\circ$$

Example

Determine the surface slope $\partial z/\partial u$ of an elliptic paraboloid shell at the intersection of two planes $\pi 1$ and $\pi 2$. The plane $\pi 1$ is defined by the axes *uoz* inclined at an angle $\phi = 45^\circ$ to the reference axis *ox*. The plane $\pi 2$ lies parallel to the *xoy* reference plane at level $z = 4$. The equation to the elliptic paraboloid shell is:

$$z = \left(\frac{x^2}{2} + \frac{y^2}{4}\right)\frac{1}{6}$$

whence

$$z = \frac{u^2}{6}\left(\frac{\cos^2\phi}{2} + \frac{\sin^2\phi}{4}\right)$$

and for $\phi = 45^\circ$, $\sin^2\phi = \cos^2\phi = \frac{1}{2}$ and

$$z = \frac{u^2}{6}\left(\frac{1}{4} + \frac{1}{8}\right) = \frac{u^2}{16}$$

whence

$$\frac{dz}{du} = \frac{u}{8}$$

The point at which the slope is required is at a level $z = 4$ and this point lies in the plane *uoz*, defined by $\phi = 45^\circ$. The value of u is therefore given by:

$$u^2 = 4.16$$

whence

$$u = 8$$

and since

$$\frac{dz}{du} = \frac{u}{8}$$

The slope of the surface in the plane uoz at level $z = 1$ is obtained by substituting $u = 8$, whence

$$\frac{dz}{du} = 1$$

and the angle in the uoz plane is 45°

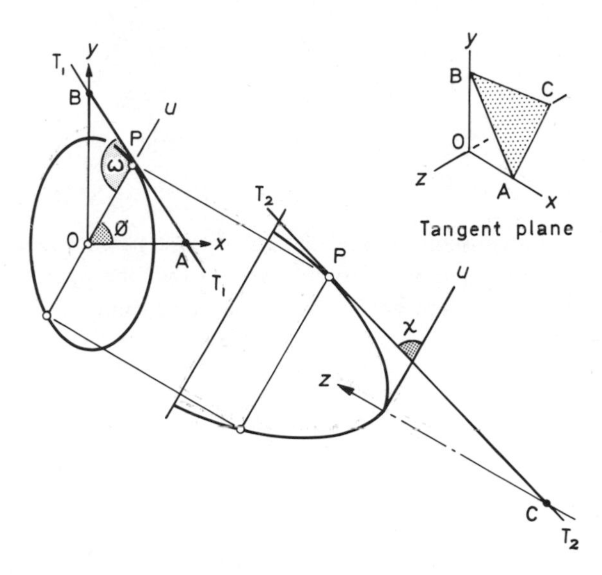

Figure 8.26

8.6 ANTICLASTIC SURFACES

When the centres of the two principal curvatures lie on opposite sides of the surface, the surface is saddle-like in appearance. Anticlastic surfaces are doubly curved and are not developable and it will be shown that anticlastic surfaces are twisted. The hyperboloid is one particular example.

8.6.1 The hyperboloid

The equation to an hyperboloid of one sheet is of the form:

$$\frac{x^2}{a^2} + \frac{y^2}{b^2} - \frac{z^2}{c^2} = 1$$

The equation to an hyperboloid of two sheets is of the form:

$$-\frac{x^2}{a^2} - \frac{y^2}{b^2} + \frac{z^2}{c^2} = 1$$

The two surfaces are shown in Fig. 8.27(a) and (b).

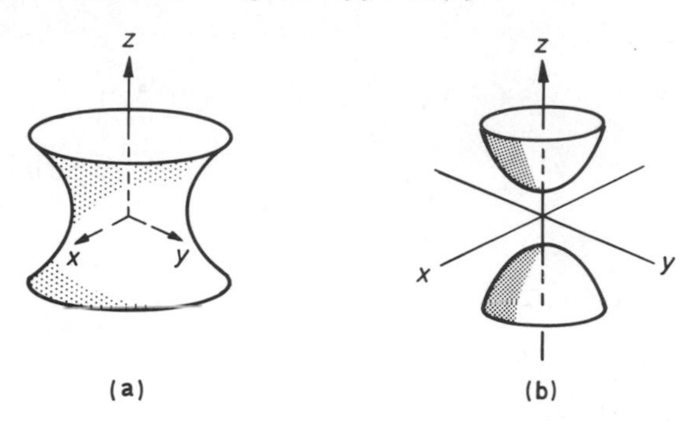

Figure 8.27

If we let z equal a constant we see that all sections parallel to the xoy plane are ellipses, and by putting x equal to zero we see that the form of all sections parallel to the reference plane yoz are parabolic and similarly for all sections parallel to the reference plane xoz.

For the plane yoz

$$z = \frac{c}{b}\,(y^2 - b^2)^{1/2}$$

For the plane xoz

$$z = \frac{c}{a}\,(x^2 - a^2)^{1/2}$$

8.6.2 The hyperbolic paraboloid

The hyperbolic paraboloid of translation is the surface produced by translating a plane hogging parabola 1 along two sagging plane parabola 2 as shown in Fig. 8.26.

The equation to the hyperbolic paraboloid is of the form:

$$\frac{x^2}{a^2} - \frac{y^2}{b^2} = 2cz - h$$

where h is the minimum value of z corresponding to the station $x = y = o$. All sections parallel to the reference planes xoz and yoz are clearly parabolic in form and by letting z equal a constant, we see that:

$$y = \left(\frac{b^2}{a^2}\,x^2 - k\right)^{1/2}$$

and this equation confirms that all sections parallel to the xoy reference plane are also parabolic.

Ruled anticlastic surfaces. The nature of twist in, and the ruling of, an anticlastic surface may be demonstrated by means of a model. If two discs are set at a fixed distance apart and coupled around their periphery by longitudinal elastic strings, forming a squirrel cage, the envelope profile, Fig. 8.27(b), is produced when each disc is counter rotated and the more the two discs are rotated the more acute the waisting.

8.6.3 Hyperboloid of revolution

The equation to a plane hyperbola in the *yoz* plane is written:

$$\frac{y^2}{b^2} - \frac{z^2}{c^2} = 1$$

and the corresponding surface of revolution is obtained from the general equation:

$$\frac{x^2}{a^2} + \frac{y^2}{b^2} - \frac{z^2}{c^2} = 1$$

We note that all cross sections in the *xoy* plane are circles and put $a = b$ where a is the radius of the root directing circle.

The equation to the surface of revolution is

$$x^2 + y^2 - \frac{a^2}{c^2} z^2 - a^2 = o$$

but since all cross sections are circular $x^2 + y^2 = r^2$ and we may express $x^2 + y^2$ in cylindrical coordinates. Whence

$$x^2 + y^2 = (x \cos \alpha + y \sin \alpha)^2 + (y \cos \alpha - x \sin \alpha)^2$$

(N.B. as may be readily verified by evaluating the R.H.S.).

On making the above substitution we find that:

$$(x \cos \alpha + y \sin \alpha)^2 + (y \cos \alpha - x \sin \alpha)^2 - \frac{a^2 z^2}{c^2} - a^2 = 0$$

and since none of these terms are in general zero

$$(x \cos \alpha + y \sin \alpha) = a \quad \text{and} \quad (y \cos \alpha - x \sin \alpha) = \frac{az}{c}$$

The first equation is that of a system of straight lines expressed in perpendicular form.

The second equation is also a system of straight lines, the envelope to which is a hyperbola.

These lines lie on the hyperboloid surface and are the ruled straight line generators of the surface. A corresponding system of straight line generators is formed by changing the sign of z.

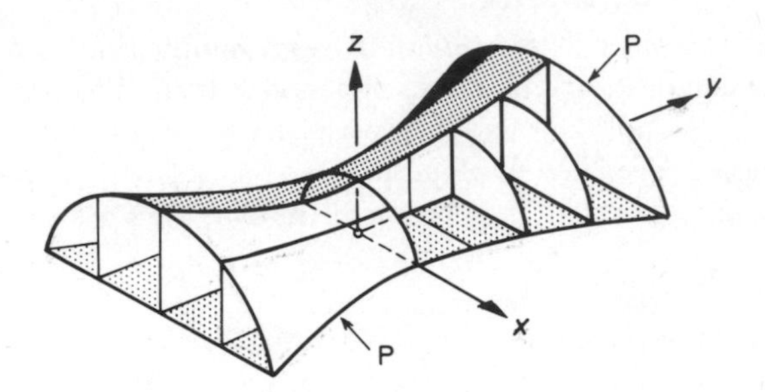

Figure 8.28

In order to plot these lines in the zoy plane we need to eliminate x from the equation.

$$y \cos \alpha - x \sin \alpha = \frac{az}{c}$$

and this we do by substituting the expression for x obtained from the lines in plane xoy.

It is easily verified that:

$$y = \frac{az}{c} \cos \alpha + a \sin \alpha$$

is the equation required. A simple ruling would be obtained by putting α to the following values.

$$\alpha = 0 \quad y = \frac{az}{c}$$

$$\alpha = 30^\circ \quad y = \frac{az}{c} \frac{\sqrt{3}}{2} + \frac{a}{2}$$

$$\alpha = 60^\circ \quad y = \frac{az}{c} \frac{1}{2} + \frac{a\sqrt{3}}{2}$$

$$\alpha = 90^\circ \quad y = a$$

The ruling is shown in Fig. 8.28.

The two systems of lines ruled on the surface of a hyperbola of revolution may be alternatively expressed in terms of two variable parameters λ, μ.

Given the equation:

$$\frac{x^2}{a^2} + \frac{y^2}{b^2} - \frac{z^2}{c^2} = 1$$

$$\frac{x^2}{a^2} - \frac{z^2}{c^2} = 1 - \frac{y^2}{b^2}$$

whence

$$\left(\frac{x}{a} + \frac{z}{c}\right)\left(\frac{x}{a} - \frac{z}{c}\right) = \left(1 + \frac{y}{b}\right)\left(1 - \frac{y}{b}\right)$$

The left hand side may be expressed in the form of two equations

$$\frac{x}{a} + \frac{z}{c} = \lambda\left(1 + \frac{y}{b}\right), \quad \frac{x}{a} - \frac{z}{c} = \frac{1}{\lambda}\left(1 - \frac{y}{b}\right)$$

and similarly

$$\frac{x}{a} - \frac{z}{c} = \mu\left(1 + \frac{y}{b}\right), \quad \frac{x}{a} + \frac{z}{c} = \frac{1}{\mu}\left(1 - \frac{y}{b}\right)$$

and these equations represent two systems of straight (ruled) lines which lie in the surface. The complete surface can be laced as finely as desired by allotting a small incremental change to the parameters λ and μ.

8.6.4 Paraboloid of revolution

The equation to a parabola of revolution is written:

$$x^2 + y^2 = \frac{2a}{c} z + a$$

where a is the root or throat radius and c is the semi-distance between the two extreme circles.

Making the following substitution for $x^2 + y^2$ we may express z in terms of the angle α

$$(x \cos \alpha + y \sin \alpha)^2 + (y \cos \alpha - x \sin \alpha)^2 = \frac{2a}{c} z + a$$

and since none of these terms is zero,

$$(x \cos \alpha + y \sin \alpha) = a \quad \text{and} \quad (y \cos \alpha - x \sin \alpha) = \frac{2a}{c} z$$

The first equation is that of a system of straight lines expressed in perpendicular form and is thus the equation to the envelope of a circle.

The second equation also represents a system of straight lines, the envelope to which is a parabola.

By eliminating x from the second equation we obtain the equation to the parabola of revolution.

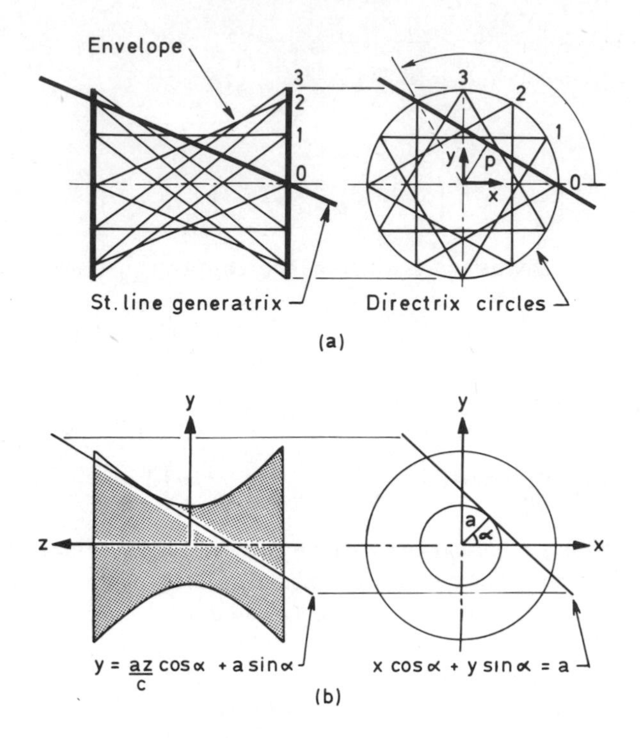
Envelope
3
2
1
0
3
2
1
0
y
p
x
St. line generatrix
Directrix circles
(a)
y
y
z
a
x
y = az/c cos α + a sin α
x cos α + y sin α = a
(b)

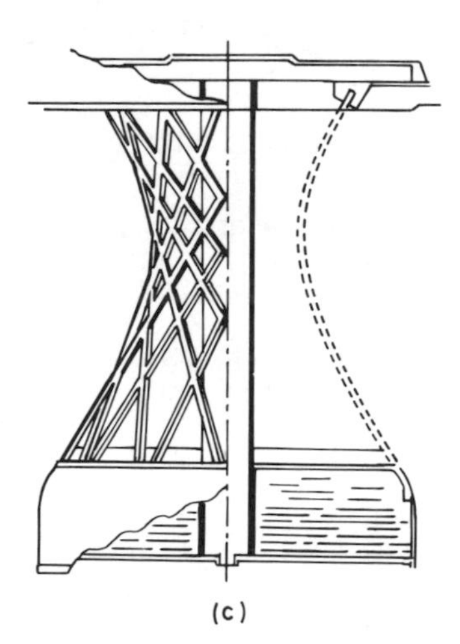
(c)

Figure 8.29

From the first equation:

$$x = \frac{a - y \sin \alpha}{\cos \alpha}$$

and on substitution into the second equation

$$y \cos \alpha - \left(\frac{a - y \sin \alpha}{\cos \alpha} \right) \sin \alpha = \frac{2az}{c}$$

whence

$$y = \frac{2az}{c} \cos \alpha + a \sin \alpha$$

This is the equation used to produce the surface of revolution, Fig. 8.29, on which a few ruled lines have been drawn. A cooling tower is shown at c.

8.6.5 The right helicoid

A helix is a curve drawn on the surface of a right circular cylinder, in such a way that each successive point advances through a constant axial displacement for each successive degree of rotation. The curve of a standard screw thread is a helix.

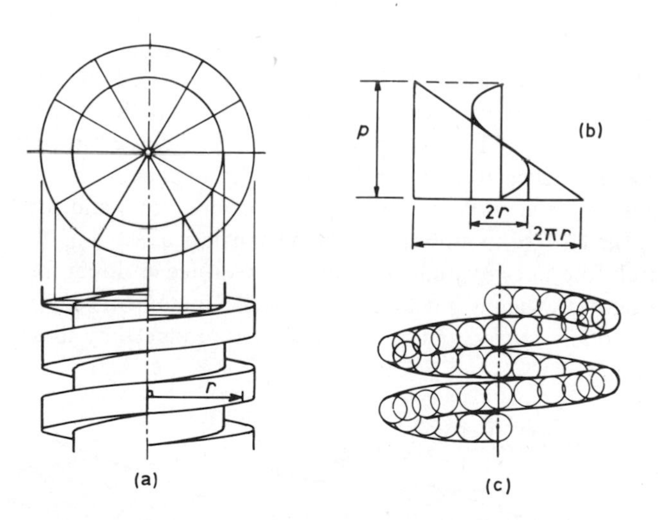

Figure 8.30

If the radius of the cylinder is r and the axial advance per complete revolution is p, then the helix angle is arc tan $p/2\pi r$. One pitch of unwrapped circumference showing the true helix angle is shown at (b).

The helical path may be alternatively considered as the curve produced by the tip of a radius vector, which is at all times normal to the virtual axis of the cylinder and advances axially a fixed distance per unit angle of rotation, as shown in Fig. 8.30.

The ruled helicoid is the basis of square section screw threads, spiral staircases, wireless aerials and numerous forms of screw feeder as used in process plants.

8.7 TWISTED SURFACE ELEMENTS

The geometrical nature of a twisted surface is clearly shown in Fig. 8.31, and its qualities further demonstrated by means of a tangible model.

If (x_0, y_0), $(-x_0, -y_0)$, are stationary pivot centres about which the lines PQ, QR, RS, SP are free to rotate without restraint at their ends, then a small counter rotation of opposite sides transforms the flat plane PQRS into the twisted form P′Q′R′S′.

The point P′ is above the point P by the same amount as the point S′ is below the point S, and a similar asymmetry applies to the points RR′ and QQ′. Since the points $(x_0, -x_0)$ and $(y_0, -y_0)$ do not translate the mid-point O lies on both the flat and twisted surface. But since the points P and R are above the point O and the points Q and S are below the point O, the join between POR and QOS are oppositely curved as shown in Fig. 8.31.

The four quadrants are oppositely curved in pairs and hence the surface is anticlastic.

As can be seen from Fig. 8.31(b) each quadrant is based on four lines, two of which intersect in a right angle at O and are directed through the co-plannar points $x_0 = a$, $y_0 = b$. The other two lines aP′ and bP′ are directed through P′, where PP′ is parallel to the reference axis oz and the angle aPb is a right angle.

The lines ox_0 and y_0P′ may be regarded as directrices and the line aP′ as a generator which is free to move in parallel increments of x.

Conversely the lines oy_0 and x_0P′ may be regarded as directrices and the line bP′ as a generator which is free to move in parallel increments of y.

A single doubly curved and twisted surfaçe is produced by these two alternative procedures.

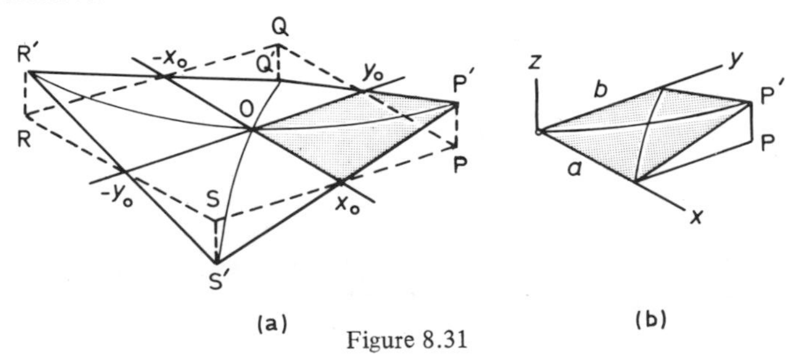

Figure 8.31

A surface of the type under discussion is called a ruled surface and a model to illustrate its counter curvatures is readily constructed from card and cotton. A square base O, aP, b, with two inclined planes aPP′ and bPP′ can be folded from card and cottons spaced at regular intervals of x and y.

The equation to the quadrant surface considered above is:

$$z = hxy$$

with axes sited as in Fig. 8.31.

The wedge angles $(P'x_0P) = \alpha$ and $(P'y_0P) = \beta$ are obtained directly by partial differentiation.

$$\frac{\partial z}{\partial x} = hy = \tan\alpha \quad \text{for } y = \text{ a constant}$$

$$\frac{\partial z}{\partial y} = hx = \tan\beta \quad \text{for } x = \text{a constant}$$

and the complete surface of Fig. 8.31(a) may be readily obtained by taking both positive and negative values of the variable x and y.

The equation to the section of the surface through the plane OPP′, Fig. 8.31(b) is obtained by setting

$$y = \frac{b}{a}x$$

in the equation of the surface.

The equation to the section OPP′ is therefore:

$$z = \frac{h \cdot b}{a} x^2$$

or in terms of an oblique coordinate u, through the points OP.

$$\tan\theta = \frac{b}{a}$$

$$\cos\theta = \frac{a}{\sqrt{a^2 + b^2}}$$

$$x = u\cos\theta$$

$$x^2 = u^2\cos^2\theta$$

whence

$$z = \frac{ha^3 b\, u^2}{a^2 + b^2}$$

z varies as u^2 and the section is thus a parabola.

If the twisted surface is square $a = b$ and the equation to the section through ROP becomes:

$$z = \frac{ha^2}{2} u^2$$

where h is the full rise of the surface as measured parallel to oz.

The four quadrant elements discussed above may clearly be reorientated to form the alternative surface shown in Fig. 8.32, and numerous other orientations of ruled and twisted surfaces are widely used in building.

8.7.1 Duality of curvature and twist

Consider the continuous surface, Fig. 8.32, for which the normal section in the plane xoz is different from the normal section in the plane yoz.

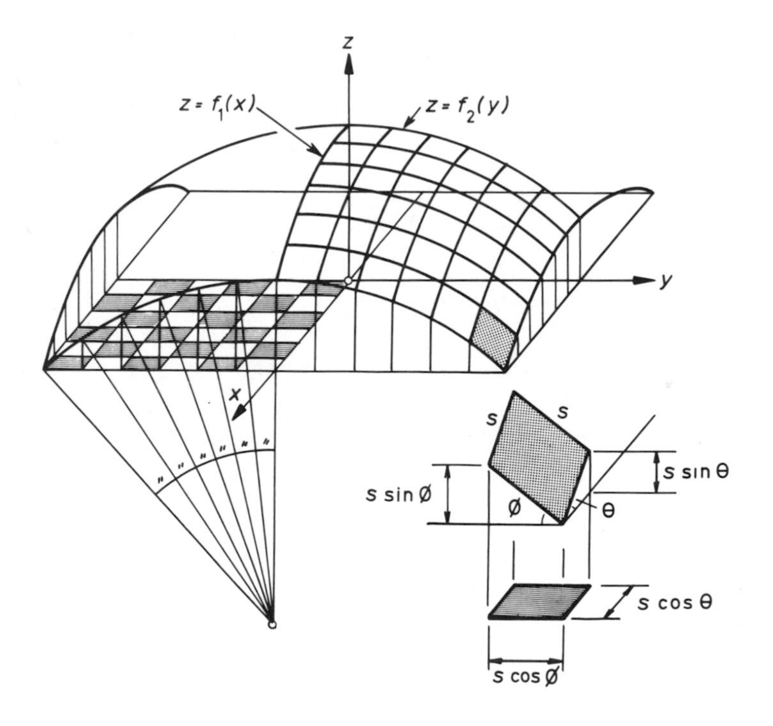

Figure 8.32

Let $z = f_1(x)$ and $z = f_2(y)$ define the normal sections in the planes xoz and yoz respectively.

The rate of change of slope of the surface in the x direction is called the curvature in that direction and is denoted by Cx, where

$$Cx = \frac{d^2 f}{dx^2}$$

The curvature in the y direction is denoted by Cy, where

$$Cy = \frac{d^2 f}{dy^2}$$

The complementary nature of twist in a surface may be further explained by considering the rectangular element of surface OPQR shown in Fig. 8.33.

We define the twist t_{xy} in a surface as the rate of change of slope in the x direction with respect to an elemental length Δy measured in the y direction or as the rate of change of slope in the y direction with respect to an elemental length Δx measured in the x direction.

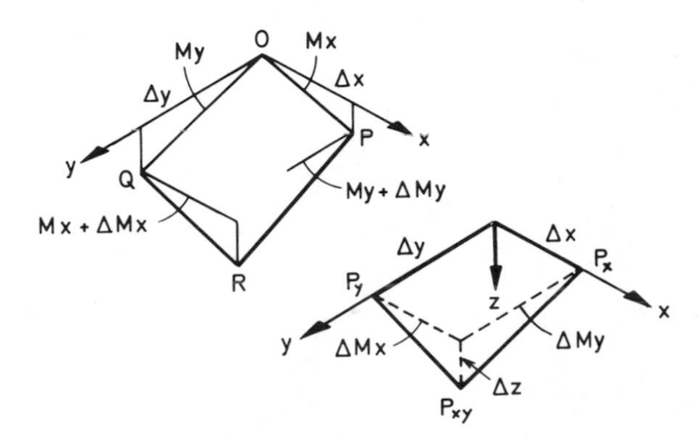

Figure 8.33

Since any three points lie flat in a plane the element of surface OPQR, Fig. 8.33, may be raised and rotated in space until the points P and Q lie on the ox and oy axes respectively, or conversely the reference axes xoy may be aligned with the element edges OP and OQ respectively.

$$t_{xy} = \frac{\Delta mx}{\Delta y} = \frac{\Delta z}{\Delta x} \cdot \frac{1}{\Delta y}$$

$$t_{yx} = \frac{\Delta my}{\Delta x} = \frac{\Delta z}{\Delta y} \; \frac{1}{\Delta x}$$

and we see that the two twists are identical. That is

$$t_{xy} = t_{yx}.$$

8.8 SYNCLASTIC AND ANTICLASTIC SURFACES COMPARED

With the exception of a spherical surface the twist varies from point to point but is always zero for all points that lie along the principal directions of curvature. As may be deduced from the ellipsoid, Fig. 8.34, the maximum and/or minimum twist occurs on a line which bisects the two principal directions of curvature.

The essential difference between a synclastic and an anticlastic surface can be readily appreciated by comparing the surface characteristics of the two surfaces formed by ruled straight lines produced by (1) folding all four parabolic generatrices up, (2) folding two opposite parabolic generatrices up and the other two parabolic generatrices down. The geometric nature of the two types of surface is clearly seen in Fig. 8.34.

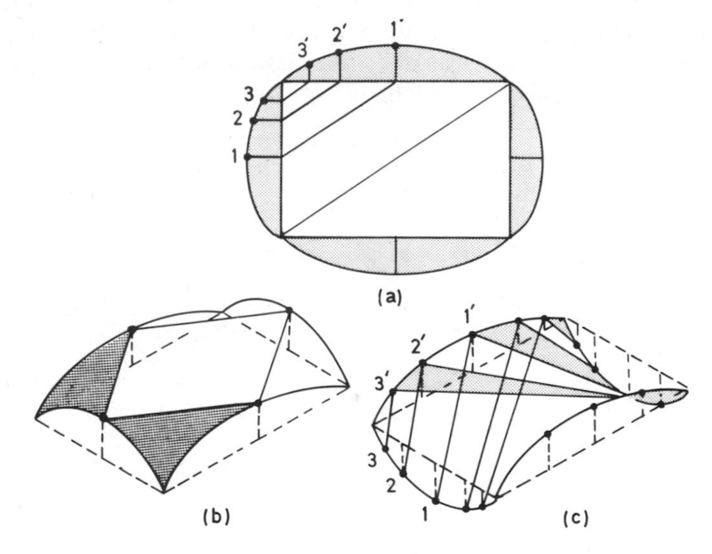

Figure 8.34

The relationship between curvature and twist for developable, synclastic and anti-clastic surfaces is given in Table 8.2.

Table 8.2

Surface type	Curvature/twist	Relationships
Plane	$c_1c_2 - t_{12} = 0$	$c_1 = c_2 = t_{12} = 0$
Developable	$c_1c_2 - t^2_{12} = 0$	$c_1 = 0$ or $c_2 = 0$
Synclastic	$c_1c_2 - t_{12} > 0$	$c_1 > 0, c_2 > 0, t_{12} = 0$
Anticlastic	$c_1c_2 - t^2_{12} < 0$	$c_1 > 0, c_2 < 0$

A visual inspection of any synclastic surface is sufficient to show that a small rectangular element of surface which is symmetrically placed about a principal plane is free from twist but the absence of twist in an element centered on a non-principal plane is not so readily apparent but this will become clear with an example.

Example

The four node element P, Q, R, S, on the surface of an ellipsoid is formed by the intersection of two parallel sections E1 and E2 with two diverging planes $\pi 1$ and $\pi 2$, as shown in Fig. 8.35, with other details as given. Show that the points P, Q, R, S, lie in a plane and determine the equation to the plane in which the points P, Q, R, S, lie.

Problem data, with reference to Fig. 8.35.

The equation to the cllipsoidal surface is:

$$\frac{x^2}{6} + \frac{y^2}{3} + \frac{z^2}{9} = 1$$

The equation of the elliptical section in the xy plane at level $z = 0$ is:

$$\frac{x^2}{6} + \frac{y^2}{3} = 1 \qquad \text{– ellipse 1}$$

The equation of the elliptical section in the xoy plane at level $z = 2$ is:

$$\frac{x^2}{6} + \frac{y^2}{3} = \frac{5}{9} \qquad \text{– ellipse 2}$$

The equation of the plane through the points OAC is:

$$y^2 = 3 \times 2 \qquad \text{– line 1}$$

The equation of the plane through the points OBD is:

$$3y^2 = x^2 \qquad \text{– line 2}$$

The cross sections through the points COA and DOB are clearly different and the coordinates of the points A, B, C, D are as listed in the table.

	x	y	z
P	0.6901	1.1952	2
Q	1.4142	0.8164	2
R	0.9258	1.6036	0
S	1.8974	1.0954	0

Since the coordinates of four points on the surface are known we may consider any two sets of three points to form a plane.

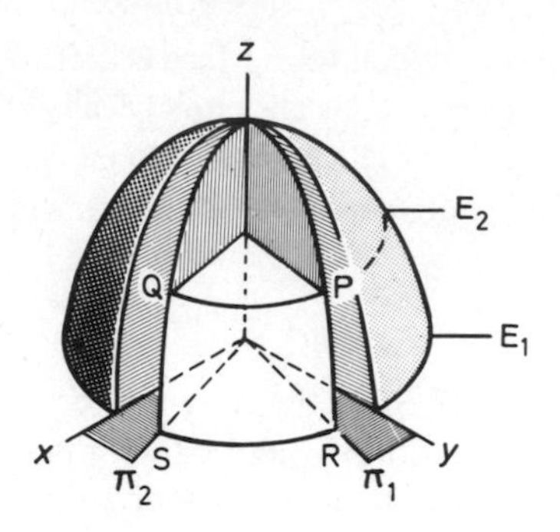

Figure 8.35

If we divide the element PQRS into two sub-planes formed by the points PQR, QRS, we may demonstrate the absence of twist by showing that the two sub-planes form part of the plane PQRS. This we may do by proving that the direction cosines and the normal for the two cases are equal.

The equation to a plane in normal perpendicular form is:

$$x \cos \alpha + y \cos \beta + z \cos \gamma = p$$

where α, β, γ, are the direction cosines of the normal to the plane and p is the normal distance of the plane from the origin.

We define each sub-plane in terms of its known coordinate dimensions:
For plane PQR we have:

$$0.6901 \cos \alpha + 1.1952 \cos \beta + 2 \cos \gamma = p$$
$$1.4142 \cos \alpha + 0.8164 \cos \beta + 2 \cos \gamma = p$$
$$0.9258 \cos \alpha + 1.6036 \cos \beta + 0 \qquad = p$$

We subtract the first two equations and obtain:

$$-0.7241 \cos \alpha + 0.3788 \cos \beta = 0$$

whence

$$\cos \beta = 1.9117 \cos \alpha$$

Substitution into the third equation yields:

$$\cos \alpha = 0.2504\, p$$
$$\cos \beta = 0.4790\, p$$

and substitution into the first equation yields:

$$\cos \gamma = 0.1273\, p$$

The direction cosines of the normal to the plane PQR are:

$$0.2504\,p, \quad 0.4790\,p, \quad 0.1273\,p$$

We next write the equations to the sub-plane formed by the points QRS

$$1.4142 \cos \alpha_1 + 0.8164 \cos \beta_1 + 2 \cos \gamma_1 = p_1$$
$$0.9252 \cos \alpha_1 + 1.6036 \cos \beta_1 + 0 = p_1$$
$$1.8974 \cos \alpha_1 + 1.0954 \cos \beta_1 + 0 = p_1$$

We multiply the second equation by 1.0954 and the third equation by 1.6036 and subtract. Whence

$$\cos \alpha_1 = 0.2505\, p_1$$
$$\cos \beta_1 = 0.4790\, p_1$$

and by substitution into the first equation

$$\cos \gamma_1 = 0.1273\, p_1$$

The direction cosines of the normal to the plane QRS are:

$$0.2505\, p_1, \quad 0.4790\, p_1, \quad 0.1273\, p_1$$

and since p and p_1 cancel in both normal equations we conclude that:

$$\cos \alpha = \cos \alpha_1, \quad \cos \beta = \cos \beta_1, \quad \cos \gamma = \cos \gamma_1, \quad p = p_1$$

The element formed by the chords PQ, QS, SR, RP is thus a flat plane and contains no twist. The absolute values of α, β, γ, may be obtained as follows. The perpendicular (normal) distance from the origin to the plane PQRS is readily calculated as follows:

We know that:

$$\cos \alpha = \frac{x}{p} = 0.2505\, p$$

whence $x = 0.2505\, p^2$

$$\cos \beta = \frac{y}{p} = 0.4790\, p$$

whence $y = 0.4790\, p^2$

$$\cos \gamma = \frac{z}{p} = 0.1273\, p$$

whence $z = 0.1273\, p^2$.

We also know that:

$$p^2 = x^2 + y^2 + z^2$$
$$= (0.2505^2 + 0.4790^2 + 0.1273^2)\,p^4$$

from which

$$p = 1.8009$$

whence

$$\cos\alpha = 0.4509, \quad \cos\beta = 0.8626, \quad \cos\gamma = 0.2293$$

and the equation to the plane formed by the chords PQ, QS, SR, RP is:

$$0.4509\,x + 0.8626\,y + 0.2293\,z = 1.8009$$

or

$$x + 1.9131\,y + 5.0853\,z - 3.9940 = 0$$

as preferred.

(N.B. The equation above is the general form in which $Ax + By + Cz + D = 0$, where)

$$A = 1, B = 1.9131, C = 5.0853, D = 3.9940$$

8.9 SURFACE TANGENTS BY PARTIAL DIFFERENTIATION

If $z = f(x, y)$, $dz = (\partial z/\partial x)dx + (\partial z/\partial y)dy$, so that for y equal a constant $dy = 0$ and for x equal a constant $dx = 0$. Hence

$$dz_{y=c} = \frac{\partial z}{\partial x}\,dx, \quad dz_{x=c} = \frac{\partial z}{\partial y}\,dy$$

and

$$\frac{dz}{dx} = \frac{\partial z}{\partial x}, \quad \frac{dz}{dy} = \frac{\partial z}{\partial y}$$

under these conditions. (N.B. The terms $\partial z/\partial x$ and $\partial z/\partial y$ are known as partial derivatives).

Now if x and y are both functions of two independent variables u and v then

$$\frac{\partial z}{\partial u} = \frac{\partial z}{\partial x}\frac{\partial x}{\partial u} + \frac{\partial z}{\partial y}\frac{\partial y}{\partial u}$$

and

$$\frac{\partial z}{\partial v} = \frac{\partial z}{\partial x}\frac{\partial x}{\partial v} + \frac{\partial z}{\partial y}\frac{\partial y}{\partial v}$$

and it follows that if V is a function of the three variables x, y, z, for all values of the independent variable t

$$\frac{dV}{dt} = \frac{\partial V}{\partial x}\frac{dx}{dt} + \frac{\partial V}{\partial y}\frac{dy}{dt} + \frac{\partial V}{\partial z}\frac{dz}{dt}$$

Example

Suppose we wish to determine the partial derivatives $\partial x/\partial u$, $\partial y/\partial u$, $\partial x/\partial v$, $\partial y/\partial v$, for the case when $u = x + y$ and $v = xy$.

Solution

Since $du = dx + dy$ and $dv = y\ dx + x\ dy$ we multiply the first equation by x and subtract from the second equation, thereby illuminating dy. Whence

$$dv - x\ du = (y - x)\ dx$$

Now we obtain $dx/du = \partial x/\partial u$ by setting dv to zero, hence

$$\frac{\partial x}{\partial u} = \frac{-x}{(y - x)} = \frac{x}{x - y}$$

and

$$\frac{\partial y}{\partial u} = \frac{y}{y - x}$$

The values of $\partial x/\partial v$ and $\partial y/\partial v$ are obtained by multiplying the first equation by y and subtracting from the second equation, before setting $du = 0$.

N.B. The method of partial derivatives may also be used to express variables x and y in terms of u and v when $f(x, y) = 0$, $dy/dx = df = 0$) and it follows that:

$$fx\ dx + fy\ dy = 0$$

hence

$$\frac{dy}{dx} = -\frac{fx}{fy}$$

and similarly for the three variable case, when $f(x, y, z) = 0$ the values of $\partial z/\partial x$, $\partial z/\partial y$, are given by setting $y = a$ constant, hence $dy = 0$, and $x = a$ constant, hence $dx = 0$ in the expression

$$fx\ dx + fy\ dy + fz\ dz = 0$$

When $dy = 0$

$$\frac{\partial z}{\partial x} = -\frac{fx}{fz}$$

When $dx = 0$

$$\frac{\partial z}{\partial y} = -\frac{fy}{fz}$$

For instance, the equation to the elliptic paraboloid, article 8.5.8 is

$$z = \left(\frac{x^2}{a^2} + \frac{y^2}{b^2}\right)\frac{1}{2c}$$

and since

$$\frac{x^2}{a^2} + \frac{y^2}{b^2} - 2cz = 0$$

$$fx = \frac{2x}{a^2}, \quad fy = \frac{2y}{b^2}, \quad fz = -2c$$

and since

$$\frac{\partial z}{\partial x} = -\frac{fx}{fz} \quad \text{and} \quad \frac{\partial z}{\partial y} = -\frac{fy}{fz}$$

$$\frac{\partial z}{\partial x} = \frac{x}{a^2c} \quad \text{and} \quad \frac{\partial z}{\partial y} = \frac{y}{b^2c}$$

as previously determined in article 8.5.8.

8.10 PAPPAS VOLUME AND SURFACE AREA THEOREMS

If a plane area A is revolved about an axis in its own plane (the axis being outside or coincident with an edge of the area) the volume swept out by the area or cross section is equal to cross sectional area times the length of the path traced by its centroid. That is, volume = $2\pi\bar{y}$A, where $\bar{y}$ is the radius of the centroid of area and A the area of cross-section.

Similarly, if a plane curve of arc length l is revolved around an axis in its own plane, as shown in Fig. 8.38, then the surface area swept out by the curve is the arc length l times the length of the path traced by its centroid. Hence surface area = $2\pi l\bar{y}$.

The need to calculate areas and volumes occurs in many applications. In most technical applications the contours are such that the problem is relatively straightforward. In the case of aesthetically fashioned bottles and canisters, however, in which volumetric content needs to be calculated in advance of manufacture, rigorous calculation can be quite complex. The reader is referred to Faux and Pratt.

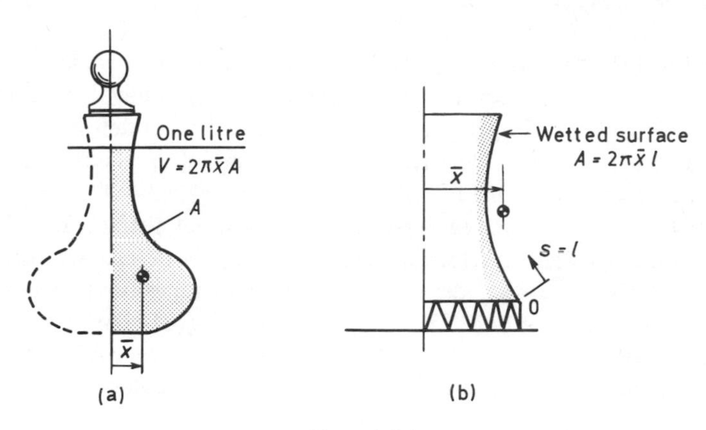

Figure 8.36

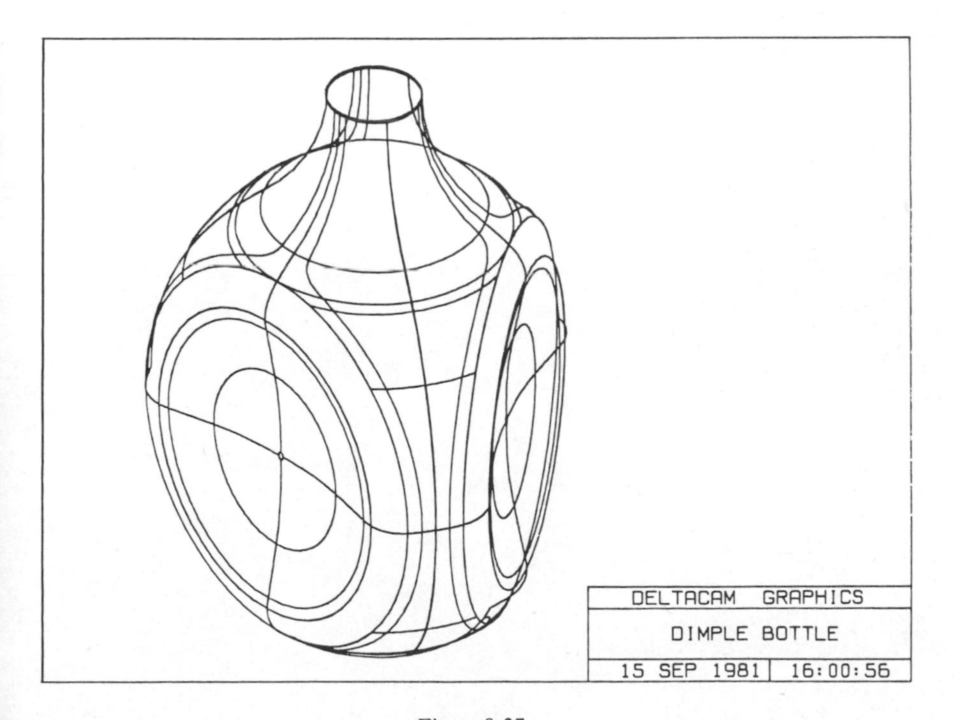

Figure 8.37

An interesting bottle design produced by the DUCT computer aided design program and reproduced here by courtesy of Delta Computer Aided Engineering Ltd. Birmingham are shown with caption in figure 8.37. This is not however a classical surface.

8.11 Surface topology

According to Firby and Gardiner topology is an abstraction in which continuity and closeness are prime considerations. Because we consider the geometrical properties of figures as though they were drawn on sheets of deformable rubber, the subject is sometimes crudely referred to as rubber sheet geometry. The subject was introduced by Listing in 1847.

In article 8.4 we recognised three different types of surface and classified them according to their developable, synclastic, anticlastic characteristics. The surface topologist, however, identifies four different types of surface, namely, Torus T, Sphere S, Klein bottle K, Projective plane P. (N.B. the twisted Möbius band which has but a single side is a K surface. Readers who wish to pursue the study of Surface Topology are referred to Surface Topology by Firby and Gardiner.

9 Lofted lines

Everything turned on a miracle of a theorem that Pascal called 'L'hexagramme mystique', commonly acknowledged to be the greatest theorem in mediaeval geometry.

H. W. Turnbull

9.1 INTRODUCTION

Until quite recently, the lines of ships, cars and aeroplanes were all laid down using a manual process called lofting. As traditionally employed lofting is a largely graphical procedure in which the cross-sections and thence the lines were developed full scale; not on the traditional double elephant drawing board but on the floor of a shed or more congenially in the loft above the works, set aside for that purpose. The word lofting is clearly derived from use of this amenity.

The shape of the transverse bulkheads, frames, rings and ribs is first detemined to meet functional and/or aesthetic requirements, then the longitudinal lines (spanwise lines in the case of a wing) are drawn as 'fair' lines through a number of transverse cardinal stations.

As seen in Chapter 8, a classical surface requries but one directrix and one generatrix to define it theoretically but at least two directrices are commonly required in practice. Indeed the method, as traditionally employed, entails plotting a series of nested cross sections: which in the case of ships, cars and aeroplanes are typically set at approximately equal intervals in the transverse vertical plane and these sections do in fact correspond to the actual bulkheads, frames, rings and ribs which go to make up the real structure of the ship or aeroplane. A small ship or aeroplane may contain from four to ten such formers, a large ship or aeroplane may contain one hundred or more.

The need for exceptionally high graphical accuracy is obvious and one can appreciate that in the case of many long shallow curves, typical of ships and aeroplanes, the task of drawing a 'perfectly' smooth curve through a distance of hundreds of feet is nigh on impossible and hence the qualitative term – fair – is used to describe the smoothness of all such curves. (N.B. Prior to the advent of computer-aided methods the draughtsman's hand was guided by a wood or metal spline bent to pass through a given set of plotted points.

9.2 THEOREMS OF PASCAL AND BRIANCHON

Since the early 1940s there has been an ever increasing use of analytic methods, principal among which are conic lofting and proportional development.

9.2.1 Pascal's theorem

If six vertices of a hexagon (of any shape) lie on a conic, the three pairs of opposite sides intersect in three points which lie on a straight line. Thus in Fig. 9.1, the curve 1, 2, 3, 4, 5, 6, is a circle. A circle is a conic and hence the theorem should hold.

(12, 45) (13, 56) (26, 34) are opposite sides, and as may be seen from the figure P, Q, R, lie on a straight line.

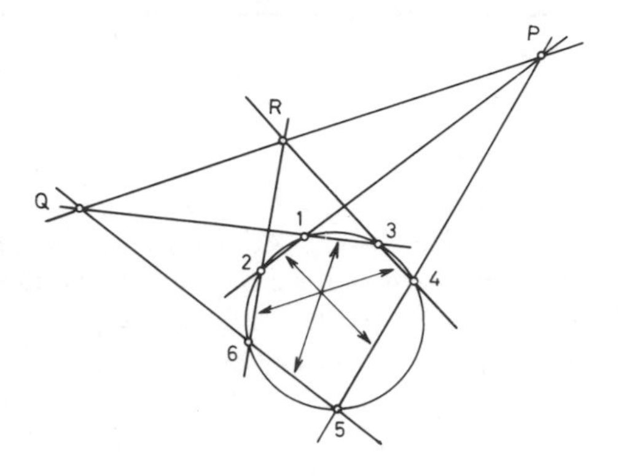

Figure 9.1

9.2.2. Proof of Pascal's theorem

A full analytic proof of Pascal's theorem is lengthy, and use of the cross-ratio theorem will be made here.

Let ABCDEF be any hexagon inscribed in any conic. The geometrical construction, Fig. 9.2, may be reproduced for any conic, but a circle is used for convenience. We need to prove that the points of intersection of opposite sides of the hexagon ABCDEF, lie on the straight line, LMN. We assume the cross ratio theorem to be true in respect of the construction given, in Fig. 9.2, and proceed with the proof.

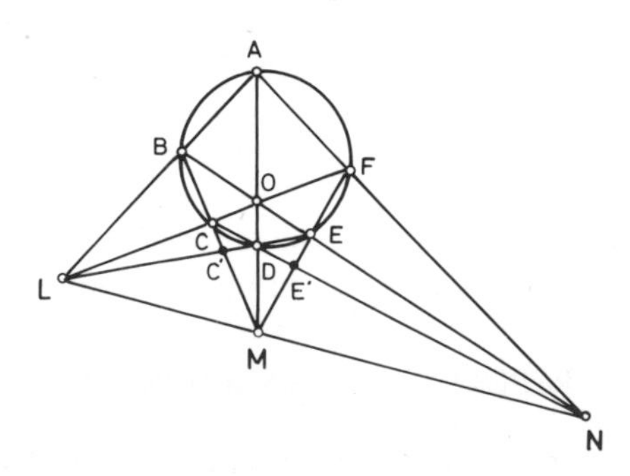

Figure 9.2

The cross (or anharmonic) ratio of the line divisions L to E, with respect to M is written

$$M(LCDE) = M(LC'DE)$$

This statement is true, for C and C′ are on a common ray.

We may likewise write the cross ratio of the line divisions L to E, with reference to the point B, whence

$$B(LC'DE) = B(ACDE)$$

Now, it is known, from the theory of cross ratios, that the value of a cross ratio is unaltered by a change from one vertex point to another. Thus if LC′DE are in cross ratio with respect to M they are also in cross ratio with respect to B.

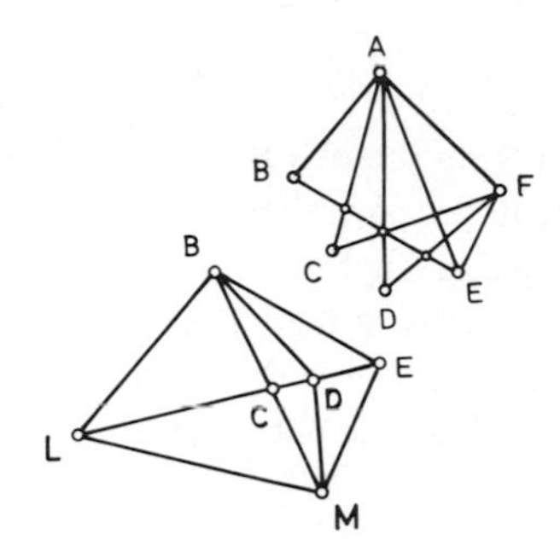

Figure 9.3

It follows from Fig, 9.3, that:

$$M(LCDE) = M(LC'DE) = B(LC'DE) = B(ACDE)$$

and similarly

$$M(NCDE) = M(NCDE') = F(NCDE') = F(ACDE)$$

But

$$B(ACDE) = F(ACDE)$$

and therefore

$$M(LCDE) = M(NCDE)$$

We note that points C, D, E, are common and hence LM, MN are identical and if this is true, the points L, M, N are collinear.

It may also be shown that a circle may be projected (conically) into any other conic, namely a parabola, hyperbola or ellipse, and conversely any conic may be so projected into a circle. Any property in cross ratio which is true of a circle is thus true of any conic section. Hence we may say that the theorem of Pascal is proved.

9.2.3 Application of Pascal's theorem

Pascal's theorem provides a ready means of drawing a conic through five given points and hence constitutes an accurate method of plotting the full extent of any conic, once five suitable points are given.

Suppose that only the points ABCDE, Fig. 9.2, are given, then we know that the point F lies on the intersection of the lines ME (produced) and the line AN. Hence we may obtain the position of F. Having obtained the position of F we may produce another Pascal diagram through the points BCDEF and hence obtain another point G, internal or external to F, and this procedure may be continued until sufficient points to plot the entire curve have been obtained. The power of the method lies in the repetitive use of the straight edge, no other tool apart from a pencil being required.

9.2.4 Brianchon's theorem

A direct analogue of Pascal's theorem is Brianchon's theorem, which states:

If the sides of a hexagon are tangential to a conic, the lines joining pairs of opposite vertices pass through a single point.

The truth of this statement is immediately apparent from the circumscribed ellipse, Fig. 9.4(a), but great care is required in setting the tangent lines when applied to an open conic, such as the parabola, Fig. 9.4(b).

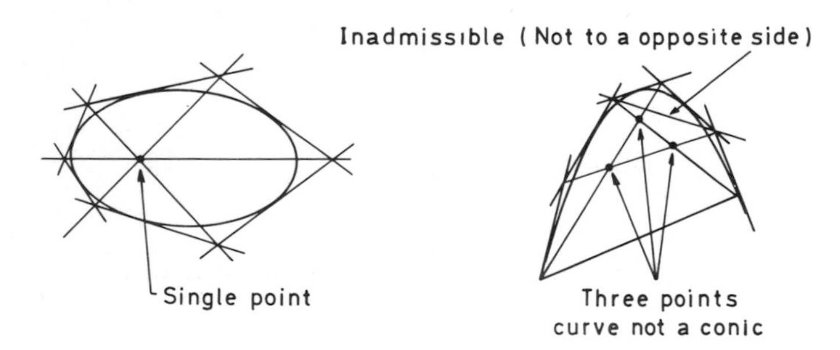

Figure 9.4

9.2.5 Proof of Brianchon's theorem

The points of contact of any circumscribing hexagon are the vertex points of an inscribed hexagon. Each vertex of the circumscribing hexagon is thus the pole of the corresponding side of the inscribed hexagon and the diagonal of the circumscribing hexagon is the polar of the point of intersection of a pair of opposite sides of the inscribed hexagon. We know from Pascal's theorem that the three points of intersection of opposite sides of the inscribed hexagon lie on a straight line and hence the three diagonals of the circumscribing hexagon meet at a point.

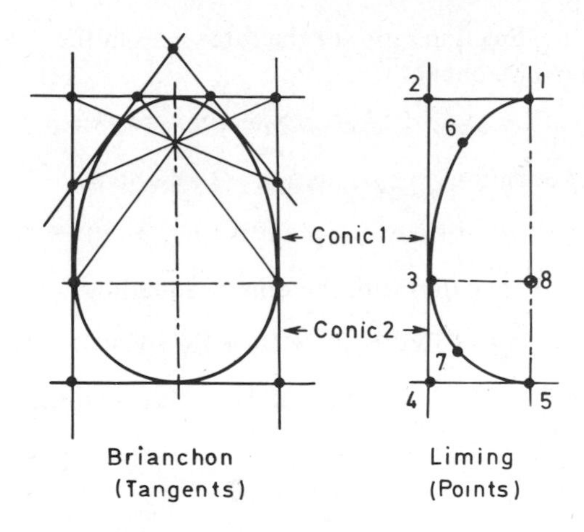

Figure 9.5

9.1.6 Application of Brianchon's theorem

If we are given five tangents to a conic we may obtain their five points of intersection. Having thus obtained five points of intersection we may obtain the sixth point and thence the sixth tangent by the application of Brianchon's theorem, as shown in Fig. 9.5.

9.1.7 Pole, polar definition of a conic

If (x_1y_1), (x_3y_3) are two points on a conic, such that a line (α) through (x_1y_1), (x_2y_2) and another line (β) through (x_3y_3), (x_2y_2) are tangential to the conic at (x_1y_1) and (x_3y_3), then (x_2y_2) is the pole and the line $(x_1y_1)\,(x_3y_3)$ denoted by (γ) is its polar.

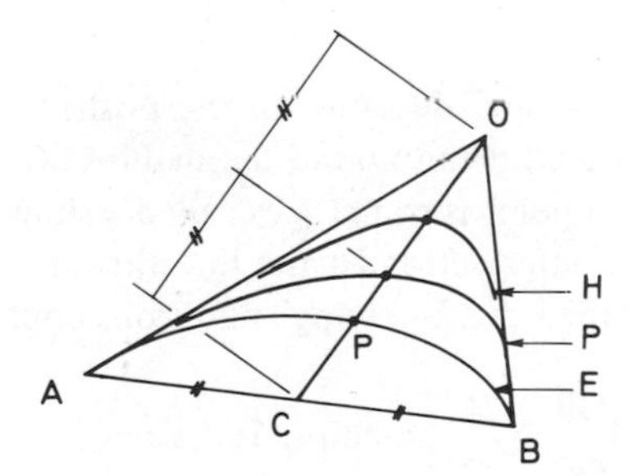

Figure 9.6

If the three lines (α), (β), (γ) are expressed in the form

$$y - \lambda x - a = 0 \qquad (\alpha)$$

$$y - \mu x - b = 0 \qquad (\beta)$$

$$y - \nu x - c = 0 \qquad (\gamma)$$

where λ, μ, ν are the direction ratios of the three lines in the xoy plane. Then by squaring equation (α) we obtain

$$y^2 - 2\lambda xy - \lambda^2 x^2 - 2ay + 2a\lambda x + a^2 = 0$$

and by multiplying equation (β) by equation (γ) we obtain

$$y^2 - (\mu + \nu)xy + \mu\nu x^2 - (\mu + \nu)y + (\mu b + c\nu)x + bc = 0$$

and by comparing $(\alpha)^2$ and $(\beta\gamma)$ with the general equation

$$\mathrm{A}y^2 + \mathrm{B}xy + \mathrm{C}x^2 + \mathrm{D}y + \mathrm{E}x + \mathrm{F} = 0$$

we see that both $(\alpha)^2$ and $(\beta\gamma)$ are conics, and it is clear that since $(\alpha)^2 = 0 = (\beta\gamma)$, the equation

$$\alpha^2 \pm k\beta\gamma = 0$$

is also a conic. Whence

$$k = \frac{\alpha^2}{\beta\gamma}$$

We note that

$$\frac{x - x_1}{x_2 - x_1} = \frac{y - y_1}{y_2 - y_1}$$

$$y = x\,\frac{(y_2 - y_1)}{(x_2 - x_1)} + (x_2 y_1 - x_1 y_2)$$

and by cycling sub-scripts equations to all lines $\alpha\beta\gamma\delta$ may be written,

9.1.8 Conic control point

The form of a pole/polar conic depends on the position of a control point P, which may lie anywhere within the bound defined by the two tangent lines and the polar. NB. The control point is sometimes called a shoulder point

If, as in Fig. 9.6, the intersection of the line through OP, divides the polar into two equal segments then the following conditions apply

$$\text{If } \frac{\mathrm{OP}}{\mathrm{OC}} > \frac{1}{2} \text{ an ellipse is defined}$$

$$\text{If } \frac{\mathrm{OP}}{\mathrm{OC}} = \frac{1}{2} \text{ a parabola is defined}$$

$$\text{If } \frac{\mathrm{OP}}{\mathrm{OC}} < \frac{1}{2} \text{ an hyperbola is defined}$$

9.2.9 Intersection conics

Two overlaid conics intersect at four points, real or imaginary, depending on their type and orientation.

If we denote two conics by $Ay^2 + Bxy + Cx^2 + Dy + Ex + F = 0$ and $A'y^2 + B'xy + C'x^2 + D'y + E'x + F' = 0$ by S and S′ we see that $S + S' = 0$ and $S + kS' = 0$ for all constant values of k. The first two conics intersect in four points and the equation $S + kS' = 0$ defines a whole family of conics which pass through the four same points. k is thus an arbitrary constant which may be used to direct the conic through a fifth point, as shown in Fig. 9.7.

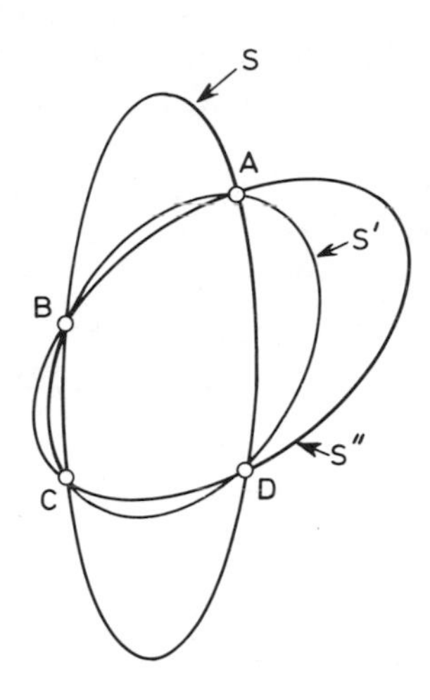

Figure 9.7

Example

Two ellipse S and S′ intersect in four points which form a square. Use the equation $S + kS' = 0$ to determine the equation of the circle that shares these same four points.

$$\frac{x^2}{a^2} + \frac{y^2}{2a^2} = 1 \qquad (S)$$

$$\frac{x^2}{2a^2} + \frac{y^2}{a^2} = 1 \qquad (S')$$

Solution

$$(2x^2 + y^2 - 2a^2) + k(x^2 + 2y^2 - 2a^2) = 0$$

so that

$$x^2(2 + k) + y^2(1 + 2k) = 4a^2$$

and for a circle we must have $2 + k = 1 + 2k$ hence $k = 1$, and the equation to the circle is

$$x^2 + y^2 = \frac{4}{3} a^2$$

9.2.10 Four point definition of a conic

In so far that the two points of tangency, Fig. 9.6, are technically double points, the pole/polar expression, article 9.1.7, is a four point definition. However, as shown in article 9.1.9, a conic may be directed through four separate points, defined by the intersections of four straight lines $\alpha, \beta, \gamma, \delta$, here taken in cyclic (anticlockwise) order. See Fig. 9.8.

The expression

$$\alpha\delta \pm k\beta\gamma = 0$$

defines a conic through these four points, the exact form of which depends on the value allotted to the arbitrary constant k. See Fig. 9.9 in which 1, 2 and 5, 6 are double points.

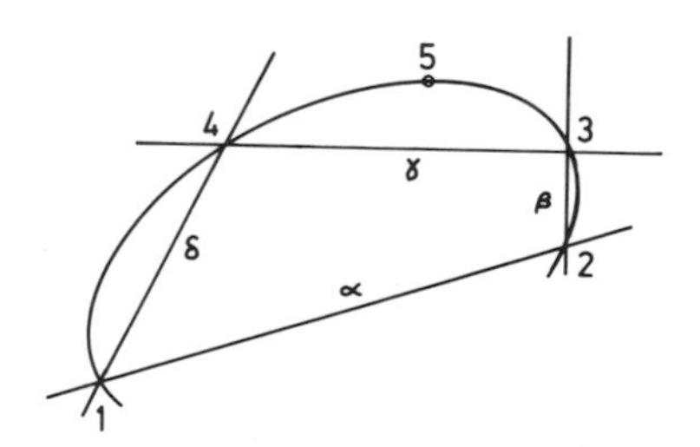

Figure 9.8

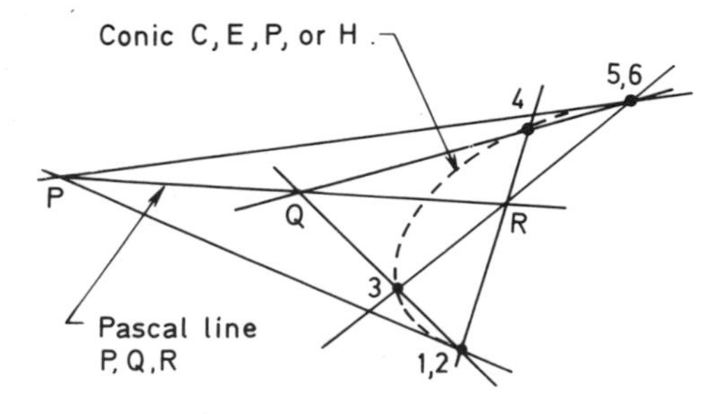

Figure 9.9

9.2.11 Five point definition of a conic

The general equation to a conic contains six constants but these six constants may be reduced to give constants by dividing through by any one of them. The general equation to a conic is

$$\mathrm{A}y^2 + \mathrm{B}xy + \mathrm{C}x^2 + \mathrm{D}y + \mathrm{E}x + \mathrm{F} = 0$$

and it is clear that the equation

$$\frac{\mathrm{A}}{\mathrm{F}}y^2 + \frac{\mathrm{B}}{\mathrm{F}}xy + \frac{\mathrm{C}}{\mathrm{F}}x^2 + \frac{\mathrm{D}}{\mathrm{F}}y + \frac{\mathrm{E}}{\mathrm{F}}x + 1 = 0$$

has only five variable coefficients and hence there are only five degrees of freedom. The equation $\alpha\delta \pm k\beta\gamma = 0$ may thus be used to fit a conic through a maximum of five arbitrary chosen points.

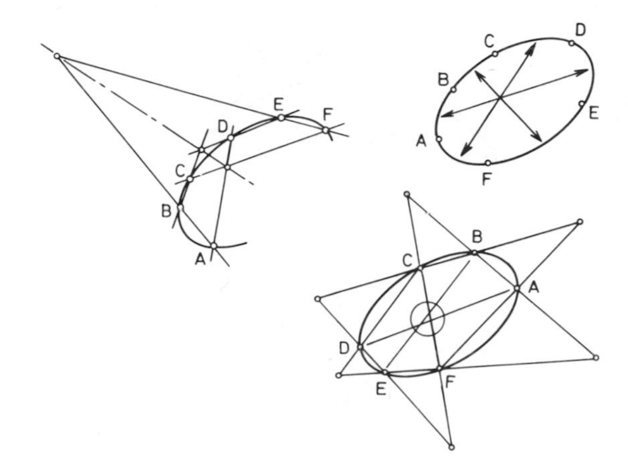

Figure 9.10

Example

The profile shape of a wing tip, or vertical stabilizer is defined by two double points (12), (56) and two single points 3 and 4. These four points are not, as may be verified from Fig. 9.11, independent. If points (12), 4, (56) are chosen then the position of point 3 is set by Pascal's theorem: which states that if the curve is a conic the points of intersection P, Q, R, lie on a straight line. If the points (P, Q, R), (12, 3, R) do not lie on straight lines which intersect at R, the curve through the points 12, 3, 4, 56, will not be a conic.

Example

A limiting case of some practical importance occurs when two cross-sections – a square and a circle – are to blend as shown in Fig. 9.12. For this particular case we have the two tangent lines at right angles and the conic curve is a circle.

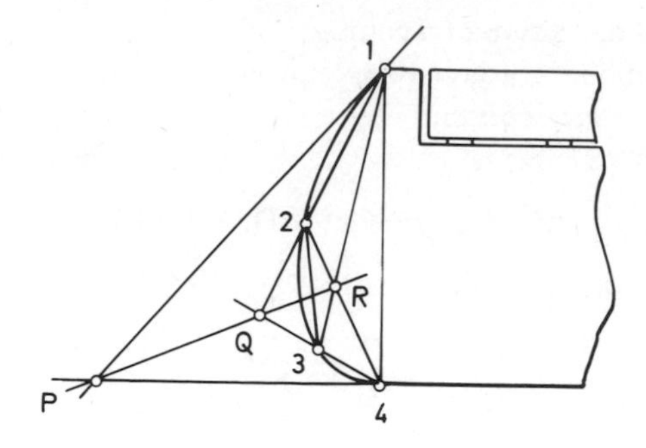

Figure 9.11

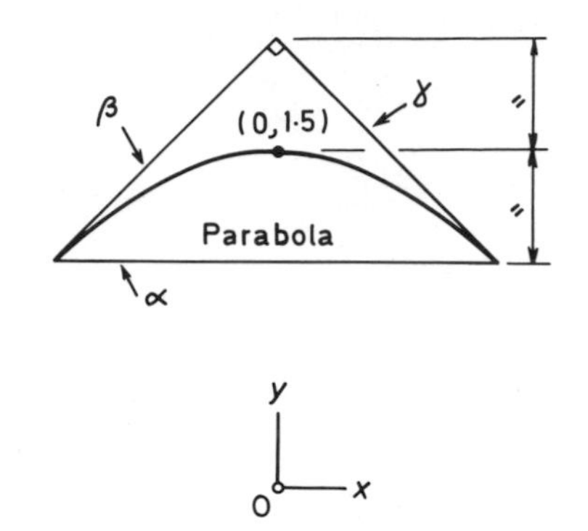

Figure 9.12

First, however, consider the geometry of Fig. 9.12.

The point P(0, 1.5) is mid-way between the chord and the apex and hence the blending conic through P is a parabola, the equation to which may be calculated as follows:

We apply the following theorem:

$$\alpha^2 - k\beta\gamma = 0$$

where α, β, γ are the homogeneous equations to the lines BC, AB and AC respectively.

For the particular case considered we have:

$$(\alpha) \qquad y - 1 = 0$$

$$(\beta) \qquad y - x - 2 = 0$$

$$(\gamma) \qquad y + x - 2 = 0$$

Hence by the equation above:

$$(y-1)^2 - k(y-x-2)(y+x-2) = 0$$

from which

$$k = \frac{y^2 - 2y + 1}{y^2 - x^2 - 4y + 4}$$

If we require the equation of the blend conic through the point P (0, 1.5) we merely substitute the coordinates (0, 1.5) into the equation above. In this instance we obtained $k = 1$ and hence the equation of the blending conic through the point P(0, 1.5) is:

$$x^2 + 2y - 3 = 0$$

This equation is a parabola as expected.

If, on the other hand, we require the blend curve corresponding to $k = 4$ we substitute $k = 4$ into the original equation:

$$(y^2 - 2y + 1) - 4(y^2 - x^2 - 4y + 4) = 0$$

from which

$$3y^2 - 4x^2 - 14y + 15 = 0$$

On substituting selected values of x or y we obtain a quadratic equation in the variable y or x. The values, Table 9.1, apply to the case of $k = 4$.

Table 9.1

x	0	0.265	0.356	0.433	0.500	0.866	1.000
y	1.667	1.600	1.550	1.500	1.451	1.132	1.000

The blend curve $k = 4$ is shown in Fig. 9.13.

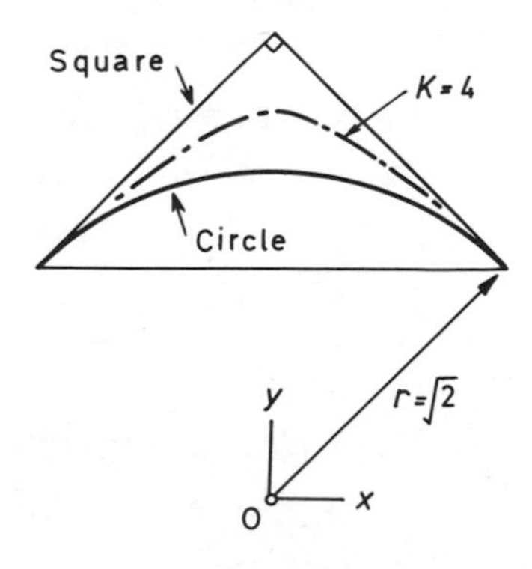

Figure 9.13

Example

Suppose that when laying down the lines of a boat we have known cross sectional coordinates for gunnel and keel. For the purpose of example, let these two points be located at the xy positions $(8, -4)$, $(0 + 4)$, denote these two points by G and K respectively. From a consideration of hydrodynamic and buoyancy requirements we supply two further points H and J, as shown in Fig. 9.14, on which a control point P is also shown.

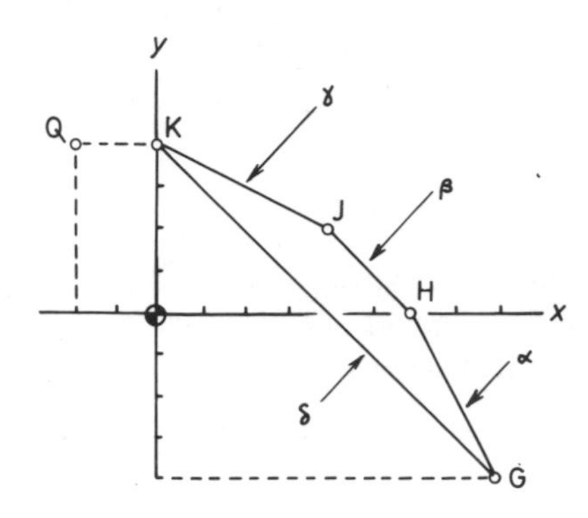

Figure 9.14

Given the coordinates of the points G, H, J, K, Q, the problem is to determine the straight line equations for the lines (GH), (HJ), (JK), (GK), and it is emphasised that the points and the lines should be plotted on squared paper as a check on the straight line equations obtained. Label the straight lines α, β, γ, δ, as indicated in Fig. 9.14.

We adopt the equation:

$$k\beta\delta - \alpha\gamma = 0$$

If

$$y + 2x - 12 = 0 \qquad (\alpha)$$
$$y + x - 6 = 0 \qquad (\beta)$$
$$2y + x - 8 = 0 \qquad (\gamma)$$
$$y + x - 4 = 0 \qquad (\delta)$$

then

$$k(y + x - 6)(y + x - 4) - (y + 2x - 12)(2y + x - 8) = 0$$

whence

$$y^2(2 - k) + xy(5 - 2k) + x^2(2 - k) - y(32 + 10k) - x(28 + 10k) + 96 - 24k = 0$$

The value of k may be determined by substituting the coordinates of the fifth point, but first it is as well to check the truth of the equation, before progressing.

A check on the truth of this equation may be obtained by substituting the coordinates of any one of the points G, H, J, K, already used.

Thus if we take the point H(6, 0) we obtain:

$$(72 - 36K) - 168 - 60K + 96 - 24K = 0$$

hence

$$0 = 0$$

and the truth of the equation is proved.

In order to establish the value of k we must substitute the coordinates of any known or assumed point which has not previously been used.

Take the point Q(–2, 4).

When the coordinates (–2, 4) are substituted we obtain:

$$32 - 16k - 40 + 16k + 8 - 4k - 128 - 40k - 56 + 20k + 96 - 24k = 0$$

which reduces to:

$$-48k + 24 = 0$$

whence

$$k = +\tfrac{1}{2}$$

Substituting this value of k into the equation we obtain the equation to the conic required. Hence

$$3y^2 + 8xy + 3x^2 - 54y - 46x + 168 = 0$$

and this equation should also be checked, this time by the reader, using the procedure above.

We now have the equation to a unique conic – the only conic – which passes through the five given points.

The method is based on the use of point locations and does not specify the direction of terminal tangents.

Thus, by placing point K, Fig. 9.15, on the y axis and giving the line JK a finite negative slope, we impose a negative, but smaller slope, on the conic at K and the same is true of the (larger) slope through G. If we had required the slope through K (or G) to be parallel to the ox (or oy) axis we could have achieved this for the point K by choosing two new points Q′, K′, symmetrically disposed about the centre line, as shown in Fig. 9.15.

The terminal tangents of the curve as specified may, of course, be obtained by differentiation.

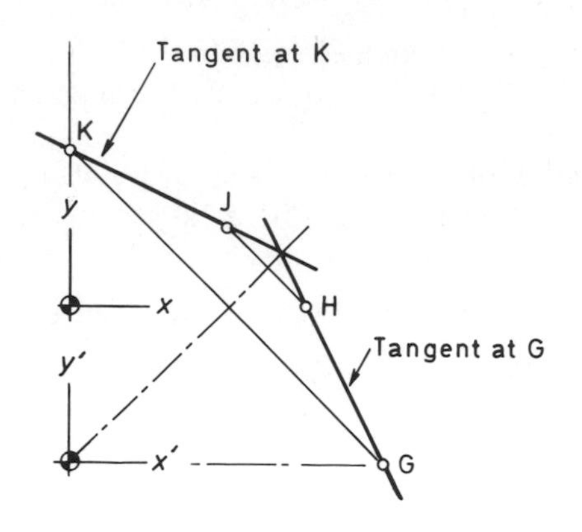

Figure 9.15

Given the equation

$$3y^2 + 8xy + 3x^2 - 54y - 46x + 168 = 0$$

we may obtain the slope at any point as follows: we note that the term $8xy$ is a product and therefore we differentiate this term in accord with the product rule.

$$\frac{dy}{dx} \text{ of } xy \text{ is of the form: } \frac{dy}{dx} = u\frac{dv}{dx} + v\frac{du}{dx}$$

where $u = x$ and $v = y$, hence

$$\frac{dy}{dx} \text{ of } xy = x\frac{dy}{dx} + y\frac{dx}{dx} = x\frac{dy}{dx} + y$$

and

$$\frac{dy}{dx} \text{ of } 8xy = 8x\frac{dy}{dx} + 8y$$

Differentiation of the other terms is straightforward. Thus we may write:

$$6y\frac{dy}{dx} + 8x\frac{dy}{dx} + 8y + 6x - 54\frac{dy}{dx} - 46 = 0$$

hence

$$\frac{dy}{dx} = \frac{-8y - 6x + 46}{6y + 8x - 54}$$

$$\frac{dy}{dx} \text{ at G} = \frac{32 - 48 + 46}{-24 + 64 - 54} = \frac{-30}{14} = \frac{-15}{7}$$

$$\frac{dy}{dx} \text{ at K} = \frac{-32 - 0 + 46}{24 + 0 - 54} = \frac{-14}{30} = \frac{-7}{15}$$

The tangent at K may be written as:

$$y = \frac{-7}{15} x + 4$$

and the tangent at G

$$y = \frac{-15}{7} x + C$$

where the constant C remains to be determined.

For this particular conic, the slope at G is the reciprocal of the slope at K. If we take the point 0, −4, as origin O′, the two tangent lines intersect at a point on the line $y' = x'$. This is so because the points G and K lie at equal distances along the x' and y' axes.

By solving the two equations

$$y' = \frac{-7}{15} x' + 8$$

$$y' = x'$$

we obtain the intersection of the two tangent lines.

We find that $x' = 5.4545 = y'$ and since $y = (y' - 4)$, $y = 1.4545$. The intersection of the two tangent lines lies at the point $x = 5.4545, y = 1.4545$. The y intercept of the tangent through G is equal to C. Where

$$y = \frac{-15}{7} x + C$$

hence

$$1.4545 = -5.4545.15/7 + C$$

$$C = 13.142$$

and hence the two tangents are:

$$y = \frac{-15}{7} x + 13.142$$

and

$$y = \frac{-7}{15} x + 4$$

with respect to the original axes, or

$$y' = \frac{-15}{7} x' + 17.142$$

$$y' = \frac{-7}{15} x' + 8$$

with respect to the axes $x'oy'$.

Example
In this example we accept the data derived above and seek to determine the equation to the same conic in terms of two tangents and a control point.

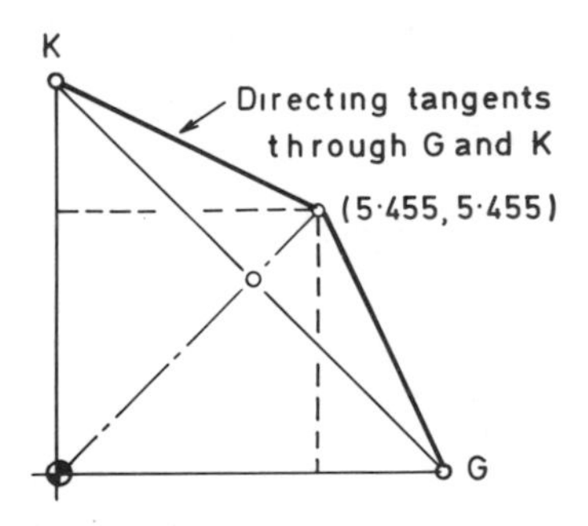

Figure 9.16

The conic with respect to the axes xoy, is:

$$3y^2 + 8xy + 3x^2 - 54y - 46x + 168 = 0$$

and this may be checked as follows:
when $x = 6, y = 0$,

$$+108 - 276 + 168 = 0$$
$$0 = 0 \quad \text{correct}$$

when $x = 0, y = 4$,

$$+48 - 216 + 168 = 0$$
$$0 = 0 \quad \text{correct}$$

In order to express the conic in terms of the axes $x'oy'$ we merely substitute the coordinates of the point 0, –4, into the equation above.

Thus we substitute $(y - 4)$ for y.

$$3(y - 4)^2 + 8x(y - 4) + 3x^2 - 54(y - 4) - 46x + 168 = 0$$

and this yields the equation with axes $x'\,oy'$,

$$3y^2 + 8xy + 3x^2 - 78y - 78x + 432 = 0$$

we drop the prime for convenience.

This equation may be checked in the usual way. Thus when $x = 8$, $y = 0$

$$192 - 624 + 432 = 0$$

$$0 = 0 \quad \text{correct}$$

If we now equate the equation of the conic $x'oy'$, to the straight line $y' = x'$ we obtain the point on the conic which is on the bisector of the line GK.

Thus if we write x' for y' and again drop the prime for convenience we obtain:

$$3x^2 + 8x^2 + 3x^2 - 78x - 78x + 432 = 0$$

$$14x^2 - 156x + 432 = 0$$

whence

$$x = \frac{144}{28} = 5.1429 = y$$

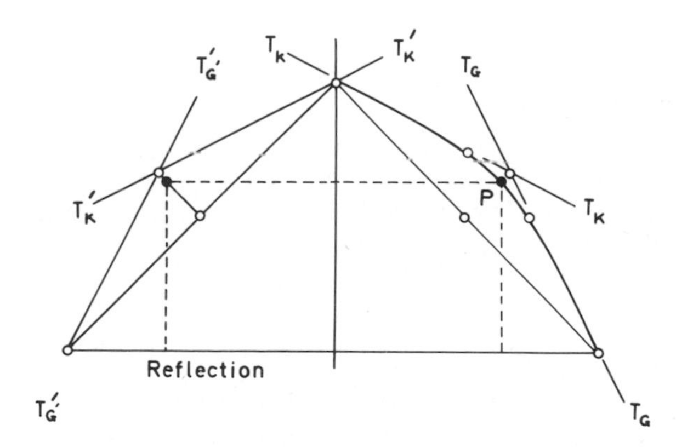

Figure 9.17

With the above information to hand we may express the conic in terms of its two terminal tangents and a control point. We have seen that once the two terminal tangents are given, the ratio PD/CP = ρ determines the shape of the conic,

$$PD = \sqrt{2.1.1429^2} = 1.6163$$
$$CP = \sqrt{2.0.3116^2} = 0.1942$$
$$\rho = \frac{PD}{CP} = 8.3233$$

We note in conclusion that:

$$\text{arc tan } \frac{7}{15} = 25.017^\circ \qquad 25.017^\circ$$

$$\text{arc tan } \frac{15}{7} = \underline{64.983^\circ} \qquad \underline{64.983}$$

$$89.999^\circ = 90^\circ \text{ correct}$$

Using the control point method above the angles 25° and 65° would serve for most practical purposes.

9.3 THE QUADRANT PARABOLA METHOD

Another useful curve fitting technique entails fitting a quadrant parabola between two tangent lines, as shown in Fig. 9.18.

The general equation to a parabola is

$$v = au^2 + bu + c$$

where a, b, c, are constants.

For a parabola passing through the origin, $c = 0$, and if we write: $v = y, u = \sqrt{x}$ the equation becomes:

$$y = ax + b\sqrt{x}$$

If we define the bounding lines as in Fig. 9.18, we may describe a quadrant parabola as follows:

$$y = (s - 2t)\frac{x}{r} + 2t\frac{\sqrt{x}}{\sqrt{r}}$$

Two such quadrant parabolae may be used, face to face, to form a discontinuous curve, such as the ship bulk-head cross-section, Fig. 9.19.

The method as described above produces a known slope at the keel and a vertical tangent at the gunnels.

If in the above equation $s = t$

$$y = -s\frac{x}{r} + 2s\frac{\sqrt{x}}{r}$$

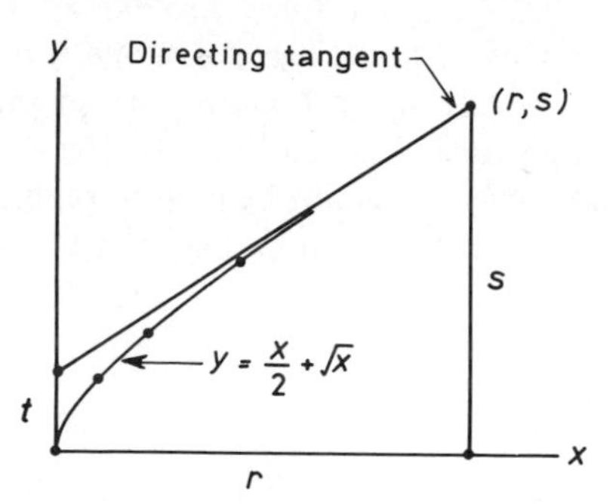

Figure 9.18

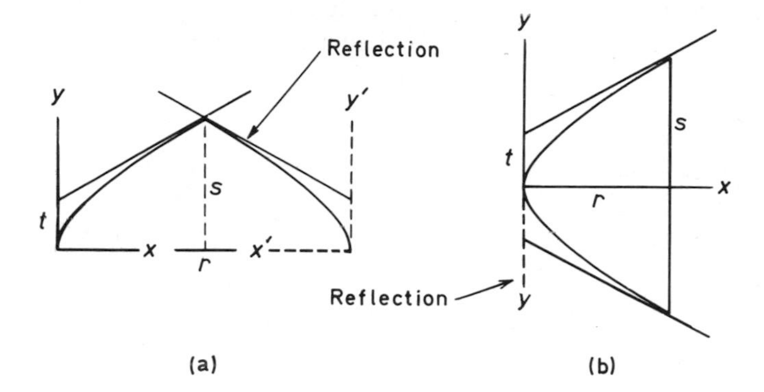

Figure 9.19

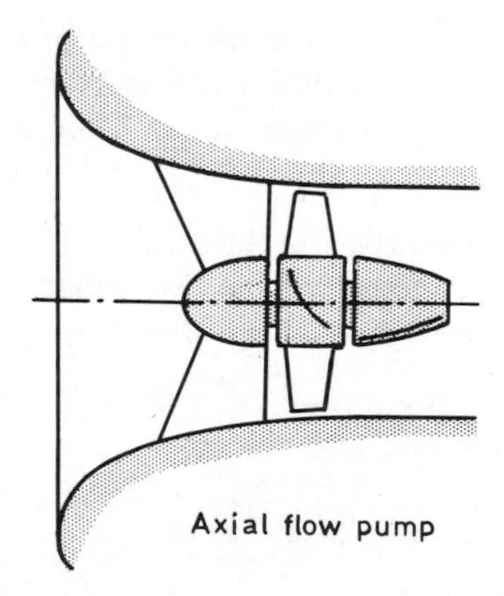

Figure 9.20

and when this is the case the parabola has zero slope through K and the origin of axes is better placed on the vertical centre line so that the more common symmetrical equation to a parabola may be used.

The quadrant parabola may, however, be used to generate a hollow orifice or solid cone of revolution an axial flow pump, Fig. 9.20.

9.4 MATCHING CONICS

Aerodynamic, hydrodynamic and aesthetic considerations are often best served when there is continuity of curvature across the join in two adjacent parts of a line.

Whilst it is obvious that there is no continuity of curvature across the join of the two lines, Fig. 9.21 (a), it is not so readily apparent that the blend between the two circular arcs, Fig. 9.21 (b), is not mathematically smooth.

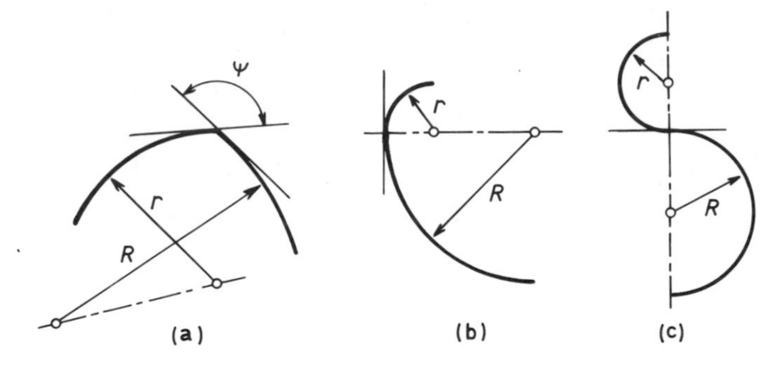

Figure 9.21

The curvature of a circle is defined as the inverse of its radius and hence no two circles blend smoothly together, unless their radii are equal.

The requirement that a second conic should blend smoothly with the first may be stated as follows:

1. Conic 2 must be tangent to conic 1 at the origin.
2. Conic 2 must be tangent to conic 1 at the given point.
3. Conic 1 and conic 2 must be of equal curvature at the join.

Suppose we have a conic of the form: $Ax^2 + Bxy + Cy^2 + Dx = 0$ we assume a suitable circle of curvature with equation:

$$x^2 + y^2 - \frac{D}{C}x = 0$$

as shown in Fig. 9.22.

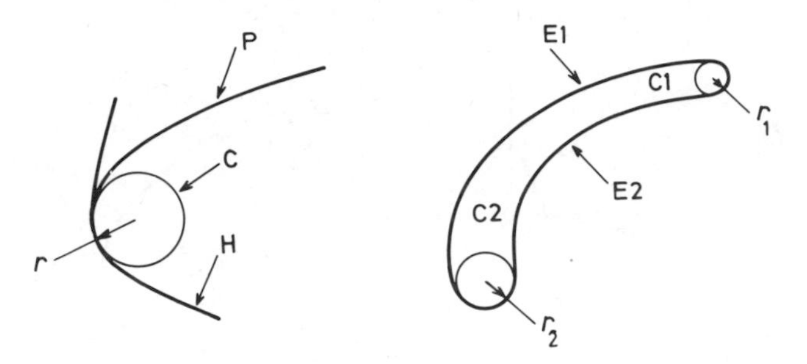

Figure 9.22

For the purpose of example suppose D = –2, C = 9. Let P(3, 2) be the point on the conic for which the tangent to the conic through the point P(3, 2) is either known or zero. We assume that the tangent is parallel to the ox axis in this example.

We write the equation to the conic thus:

$$Ax^2 + Bxy + 9y^2 - 2x = 0$$

we know that $dy/dx = 0$ at $x = 3, y = 2$,

$$\frac{dy}{dx} = 2Ax + By - 2 = 0$$

We also know that the point $x = 3$, $y = 2$, lies on the curve hence; we obtain a second equation by direct substitution:

From the parent equation we have:

$$3A + 2B = -10$$

From the slope equation we have:

$$6A + 2B = +2$$

From these two equations we obtain:

$$A = 4 \quad B = -11$$

and the required conic has the equation:

$$4x^2 - 11xy + 9y^2 - 2x = 0$$

Any two conics with a common circle of curvature will blend smoothly at the point (or points) to which the common circle of curvature applies.

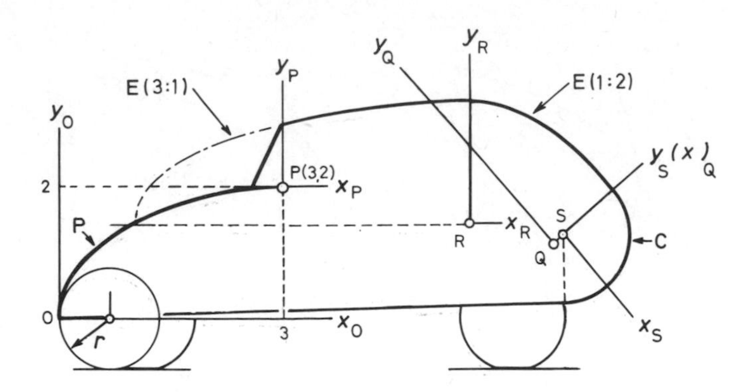

Figure 9.23

9.5 NESTED SECTIONS

The equation $S + kS' = 0$, which expresses the summation of two conics, has been established.

If we let $k = \mu$ and note that $(1 - \mu) + \mu = 1$, we may write:

$$(1 - \mu)S + \mu S = S$$

and hence

$$(1 - \mu)S + \mu S + \mu S' = 0$$

and

$$(1 - \mu)S + \mu(S + S') = 0$$

where $(S + S')$ is another conic S'', say.

The form of the equation:

$$(1 - \mu)S + \mu S'' = 0$$

should be noted. When $\mu = 0$, $S = 0$, and when $\mu = 1$, $S'' = 0$ and this property makes the equation of direct application to nested sections.

Example

A problem frequently encountered in aeroplane fuselage design is that of blending an elliptical cross section into a circle.

Suppose two cross sections of an aircraft fuselage are defined by the equations:

$$\frac{x^2}{1} + \frac{y^2}{4} - 1 = 0 \text{ (an ellipse)}$$

and

$$x^2 + y^2 - \frac{1}{4} = 0 \text{ (a circle)}$$

We know that:

$$(1 - \lambda)\left(x^2 + \frac{y^2}{4} - 1\right) + \lambda\left(x^2 + y^2 - \frac{1}{4}\right) = 0$$

(N.B. Some authors use μ, others use λ, it makes no difference).

Hence

$$(1 - \lambda)(4x^2 + y^2 - 4) + 4\lambda(x^2 + y^2) - \lambda = 0$$

$$4x^2 + y^2(1 + 3\lambda) + 3\lambda - 4 = 0$$

when $\lambda = 0$

$$4x^2 + y^2 - 4 = 0$$

when $x = 0$, $y = 2$. When $x = 1$, $y = 0$ and the major and minor axes for other values of λ are shown in Table 9.2.

Table 9.2

λ	0		1/3		2/3		1	
x	0	1	0	$\sqrt{3/4}$	0	$\sqrt{1/2}$	0	1/2
y	2	0	$\sqrt{3/2}$	0	$\sqrt{2/3}$	0	1/2	0

when $\lambda = 0$ we have an elliptical cross section for which the major and minor axes are $x = \pm 1$, $y = \pm 2$. When $\lambda = 1$ we have a circular cross section for which the radii, and diameters in x and y are equal to $\pm 1/2$.

When $\lambda = 1$ we have a circular cross-section for which the radii, and diameters in x and y are equal to $\pm 1/2$.

When λ has a value between 0 and 1 the curve is one of a nested family of ellipses, which tend to the ellipse and circle originally defined.

The nest of sections is shown in Fig. 9.24.

In an aircraft fuselage each curve in the nest would exist either as an open frame or as a bulk head. The spacing of these sections obeys the parameter λ, which may assume any positive fractional value between 0 and 1.

The spacing of the two extreme cross-sections in practice depends on the requirements set by the designer, but whatever distance is set, intermediate sections are spaced according to the assigned value of λ.

Example

The two equations used to define the bounds of the nested sections derived above are both very compact, indeed one of the boundary equations is a circle.

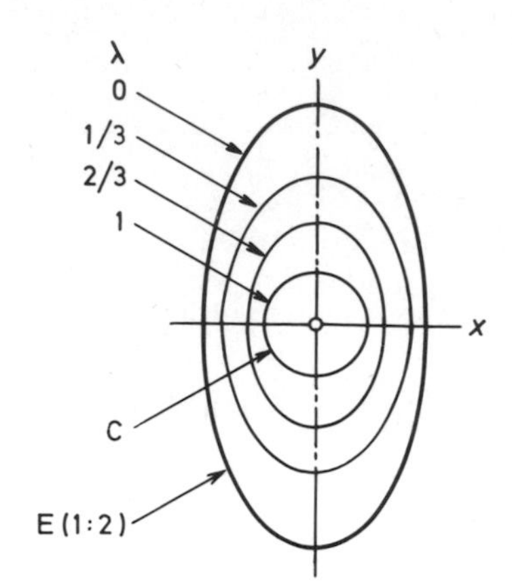

Figure 9.24

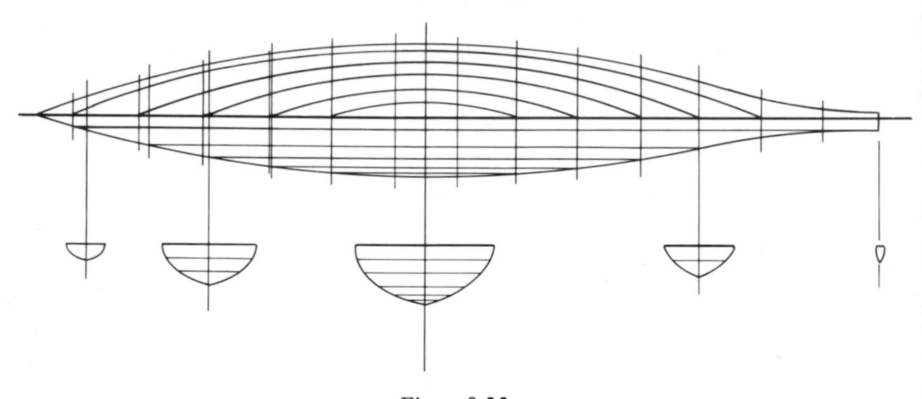

Figure 9.25

Suppose we take a second example in which the two boundary curves are much more oblate, for which purpose the horizontal sections of a rowing eight serve as a ready example. Typical length and girth dimensions of a rowing eight are shown in Fig. 9.25.

Let the plan of the gunwale be represented by the conic:

$$\frac{x^2}{100^2} + \frac{y^2}{1^2} - 1 = 0$$

Let the keel line be represented by the conic:

$$\frac{x^2}{50^2} + \frac{y^2}{0.1^2} - 1 = 0$$

Both these curves are, of course, ellipses.

The nest of horizontal sections lying between these two extremes may be represented by the equation:

$$(1-\lambda)\left(\frac{x^2}{100^2}+\frac{y^2}{1^2}-1\right)+\lambda\left(\frac{x^2}{50^2}+\frac{y^2}{0.1^2}-1\right)=0$$

To obtain these intermediate sections we substitute suitable values of λ.

Thus if we require the horizontal section corresponding to $\lambda = 1/5$ we substitute this value of λ into the above equation. Hence we obtain the equation:

$$8x^2 + y^2(4.100^2 + 100^4) - 5.100^2 = 0$$

when $y = 0, \quad x = 79.057$

when $x = 0, \quad y = 0.2192$

when $x = 50, \quad y = 0.1732$

In practice we would compute many more coordinates than this but these three pairs of coordinates suffice to map the section here.

We may obtain any intermediate section by re-writing the equation in terms of a new value of λ.

Thus when $\lambda = 1/20$ we obtain the equation:

$$23x^2 + y^2(19.100^2 + 100^4) - 20.100^2 = 0$$

and hence the equation to the horizontal section corresponding to $\lambda = 1/20$, for which the coordinates are:

when $y = 0, \quad x = 93.25$

when $x = 0, \quad y = 0.4468$

when $x = 50, \quad y = 0.3738$

Hence this horizontal section may also be plotted.

We may likewise obtain the coordinates of the section corresponding to $\lambda = 1/1000$, for which the equation is:

$$999x^2 + 999.100^2y^2 - 999.100^2 + 4x^2 + 100^4y^2 - 100^2 = 0$$

and hence

$$1.1003x^2 + 19999y^2 - 10000 = 0$$

which equation yields the coordinates:

x	0	25	50	75	90	95	99.985
y	0.7071	0.6847	0.6124	0.4676	0.3080	0.2208	0

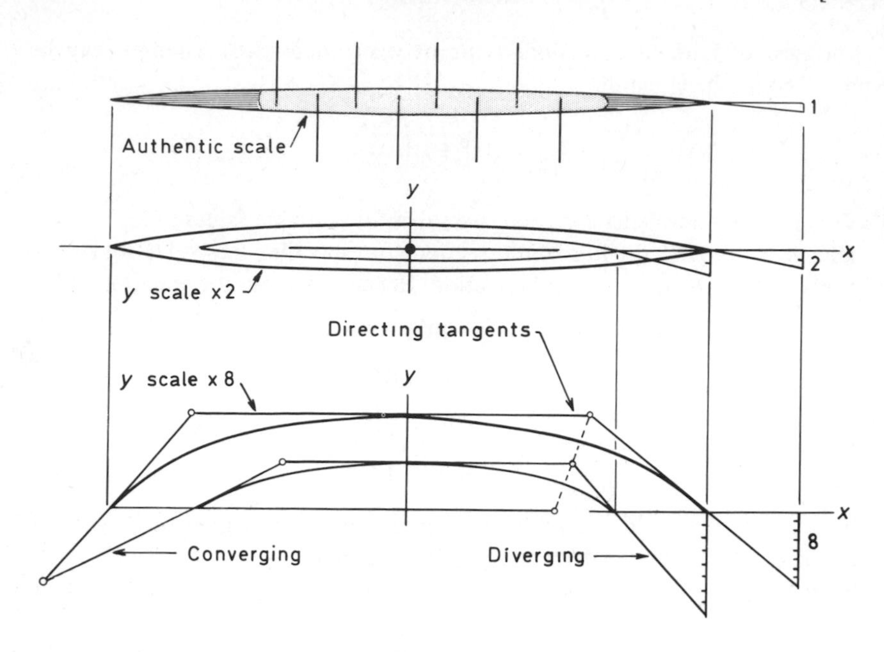

Figure 9.26

9.6 THE STRAIGHT LINE AS A DEGENERATE CONIC

The format used above produces a very slender elliptical keel line, as defined by the equation:

$$\frac{x^2}{50^2} + \frac{y^2}{0.1^2} - 1 = 0$$

Let us, however, investigate what happens when the very slender ellipse degenerates to a line.

If we use the equation:

$$\frac{x^2}{e^2} + \frac{y^2}{f^2} - 1 = 0$$

to represent the keel ellipse. We see that for $y = f = 0$.

$$x^2 - e^2 = 0$$

$$(x + e)(x - e) = 0$$

$x = \pm e$ is the equation to a straight line, which extends along the x axis between the points $\pm e$.

The final equation appropriate to a nest of sections lying between an elliptical gunnel line and a straight keel line is therefore:

$$(1-\lambda)\left(\frac{x^2}{a^2}+\frac{y^2}{b^2}-1\right)+\lambda(x+e)(x-e)=0$$

or simply

$$\frac{x^2}{a^2}+\frac{y^2}{b^2}-1+k(x^2-e^2)=0$$

where $k=\lambda/1-\lambda$.

9.7 THE MEANING OF THE PARAMETER λ

As demonstrated above the parameter λ is a weighting factor which facilitates the progressive transformation of one conic curve into another. When $\lambda=0$ we have the contour of the conic 1, when $\lambda=1$ we have the contour of the conic 2, and for intermediate values of λ we have an infinity of blend conics between these two extremes. The parameter $\lambda=\frac{1}{2}$ does not, however, produce a mean section. For instance, if $y=2x$ and $y=x/4$ are two straight lines $(1-\lambda)(y-2x)+(\lambda)(y-x/4)$ represents a family of straight lines, but the line corresponding to $\lambda=\frac{1}{2}$ does not bisect the angle between them. Moreover if we consider two circular concentric circles $x^2+y^2-r_1^2=0$ and $x^2+y^2-r_2^2=0$ situated at a distance of one unit apart then $(1-\lambda)x^2+y^2-r_1^2+(\lambda)x^2+y^2-r_2^2=0$ and when $\lambda=\frac{1}{2}$, we have an intermediate conic:

$$x^2+y^2-\frac{(r_1^2+r_2^2)}{2}=0$$

This intermediate conic is, of course, a circle of radius r_3, where

$$r_3=\sqrt{\frac{r_1^2+r_2^2}{2}}$$

and this indicates that the conic corresponding to $\lambda=\frac{1}{2}$ is not the mean diameter conic.

Example

Let the lines l_1, l_2, l_3, l_4, be represented by their Cartesian equations:

$$x-1=0 \qquad (l_1)$$

$$y-1=0 \qquad (l_2)$$

$$x+1=0 \qquad (l_3)$$

$$y+1=0 \qquad (l_4)$$

Since we may treat a straight line as a degenerate conic we may solve this problem by applying the equation:

$$k\alpha\gamma + \beta\delta = 0$$

We must, however, supply a suitable control point, for if not, an infinite number of conics passing through the given points A, B, C, D, may be drawn.

We will consider two cases.

If we specify the point $(0, \sqrt{2})$ as being on the required conic, we use co-ordinates of this point to determine the required value of k.

We know that:

$$k(x-1)(x+1) + (y-1)(y+1) = 0$$

that is:

$$k(x^2-1) + (y^2-1) = 0$$

If the point $(0, \sqrt{2})$ lies on the curve:

$$k(0-1) + (2-1) = 0$$

hence

$$k = 1$$

The equation to the conic which passes through all five points is therefore:

$$x^2 + y^2 - 2 = 0$$

and this conic is clearly a circle.

If, however, we specify the point (0, 2) as being on the required curve we obtain a different equation:

We find that when the point (0, 2) lies on the curve $k = 3$ and the equation to the required conic is:

$$3x^2 + y^2 - 4 = 0$$

$$\text{when } x = 0 \qquad y = \pm 2$$

$$\text{when } y = 0 \qquad x = \pm\sqrt{3}/2$$

This latter curve is clearly an ellipse.

9.7 REDUCTION OF COMPLEX CURVES

A problem commonly encountered in ducting is the translation of a circle into a square, and whilst this particular application may not require a high order of mathematical accuracy, the problem is one of fundamental interest.

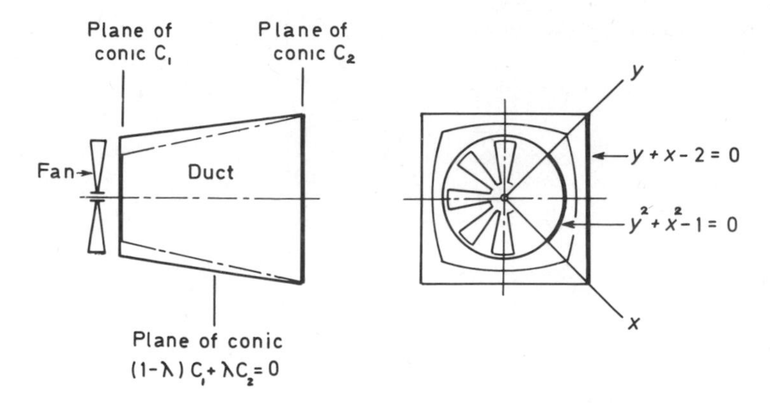

Figure 9.27

Consider the nest of sections lying between the circle $x^2 + y^2 - 1 = 0$ and the square quadrant defined by the axes xoy, and the line $y + x - 2 = 0$.

The equation to the nest of sections is:

$$(1 - \lambda)(x^2 + y^2 - 1) + \lambda(y + x - 2) = 0$$

and if for the purpose of example we take $\lambda = \frac{1}{2}$

$$x^2 + y^2 + x + y - 3 = 0$$

and similarly when $\lambda = \frac{3}{4}$

$$x^2 + y^2 + 3x + 3y - 7 = 0$$

Solving these two equations we obtain the coordinates of two different sections. When $\lambda = \frac{1}{2}$

x	0	0.4	0.8229	1.0	1.156	1.3028
y	1.3028	1.156	0.8229	0.618	0.4	0

when $\lambda = \frac{3}{4}$

x	0	0.4	0.8979	1.0	1.3089	1.5414
y	1.5414	1.3089	0.8979	0.7913	0.4	0

and these two intermediate sections are as plotted in Fig. 9.28.

In practical duct work a transverse curvature discontinuity is not usually important and a triangulated development, would probably be used. In the context of the present problem, however, we observe that the nested sections do not intersect the x and y axes at right angles and hence there is a curvature discontinuity as each nested section passes into an adjacent quadrant. The slope of each conic may be obtained by differentiating the equation of the conic for each value of λ.

If we consider the case when $\lambda = \frac{1}{2}$, we have:

$$x^2 + y^2 + x + y - 3 = 0$$

and

$$2x + 2y \frac{dy}{dx} + 1 + \frac{dy}{dx} = 0$$

hence

$$\frac{dy}{dx} \text{ when } \lambda = \tfrac{1}{2} = -\frac{(2x + 1)}{(2y + 1)}$$

when $x = 0, y = 1, dy/dx = \frac{1}{3}$

when $x = 1, y = 0, dy/dx = \frac{3}{1}$

Hence there is curvature discontinuity.

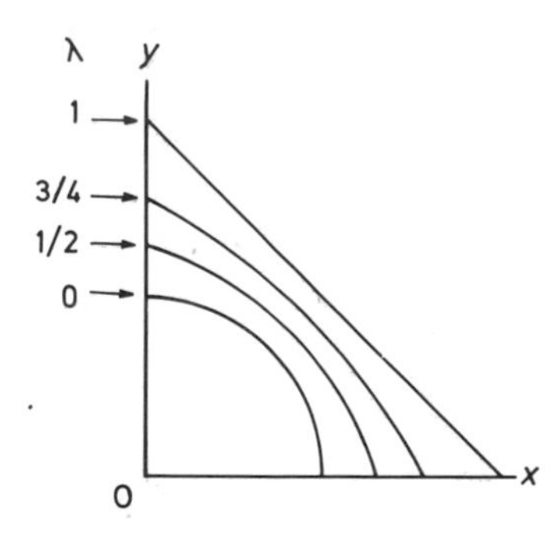

Figure 9.28

Whilst it is easy to verify (using a compass and scale drawing) that the nest of sections in the problem above are indeed circular arcs, it is instructive to prove this observation and moreover to obtain an equation which enables the infinity of intermediate sections to be drawn and indeed manufactured as truly circular arcs.

In order to prove that all intermediate sections are circular arcs we proceed as follows:

$$(1 - \lambda)(x^2 + y^2 - 1) + \lambda(y + x - 2) = 0$$

and for simplicity we write $1 - \lambda/\lambda = k$, hence the equation:

$$k(x^2 + y^2 - 1) + x + y - 2 = 0$$

represents the nest of sections as defined at the beginning of this example.

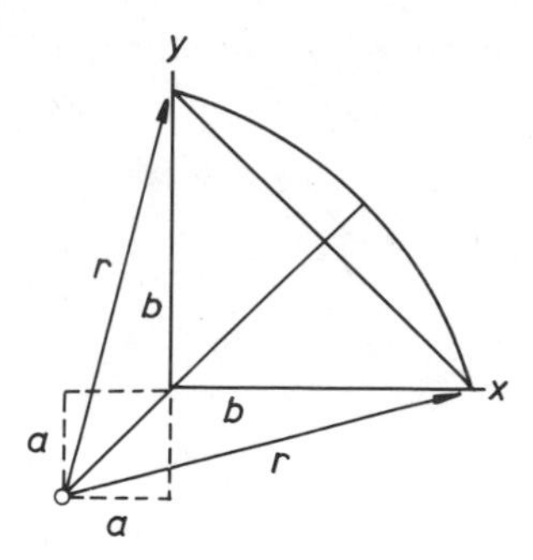

Figure 9.29

We observe that the intercepts in x and y are equal and hence the centre of any circle which passes through the general points x and y must lie on a line at 45° which passes through the origin O, as shown in Fig. 9.29.

We begin by translating the origin of axes to a point $(-a, -a)$. A translation which is reached by writing $(x + a)$ for x and $(y + a)$ for y in the original equation. Making this substitution we obtain the equation:

$$\mathrm{k}\,[(x + a)^2 + (y + a)^2 - 1] + (x + a) + (y + a) - 2 = 0$$

$$\mathrm{k}\,[x^2 + 2ax + a^2 + y^2 + 2ay + a^2 - 1] + 2a + (x + y - 2) = 0$$

we note that $(x + y - 2) = -\mathrm{k}(x^2 + y^2 - 1)$ and

hence

$$\mathrm{k}\,[2ax + 2ay + a^2] + 2a = 0$$

$$2a\mathrm{k}(x + y) + \mathrm{k}a^2 + 2a = 0$$

we note that $(x + y) = -\mathrm{k}(x^2 + y^2 + 1) + 2$, hence

$$-2a\mathrm{k}^2(x^2 + y^2 - 1) + 4a\mathrm{k} + 2a = 0$$

and

$$x^2 + y^2 - \frac{2\mathrm{k} + 1 + \mathrm{k}^2}{\mathrm{k}^2} = 0$$

and this is the general equation to a circle, for which:

$$\mathrm{R}^2 = \left(\frac{\mathrm{k} + 1}{\mathrm{k}}\right)^2$$

and hence

$$\mathrm{R} = \frac{(\mathrm{k} + 1)}{\mathrm{k}}$$

In order to calculate the offset a, we solve the triangle, Fig. 9.32, for which b equals the intercept of the circle with the x and y axes.

$$a^2 + (a+b)^2 = R^2$$

and hence

$$a = \frac{-b \pm \sqrt{2R^2 - b^2}}{2}$$

$$\text{when } \lambda = \tfrac{1}{2} \quad \text{and} \quad b = 1.3028$$

$$k = \frac{1-\lambda}{\lambda} = 1$$

$$R = \frac{k+1}{k}$$

hence R = 2, and

$$a = \frac{-1.3028 \pm 2.5105}{2} = 0.60386$$

We check this result by direct substitution into the equation:

$$a^2 + (a+b)^2 = R^2$$

whence

$$0.60386^2 + 1.9067^2 = 4.0$$

which yields the correct result, R = 2. Each quadrant of every intermediate section may thus be described as a pure circular arc.

We also note that since $k = (1 - \lambda)/\lambda$, the radius R may be expressed in terms of λ. We have:

$$R = \frac{1}{1-\lambda}$$

which tells us that when $\lambda = 0$, R = 1, when $\lambda = 1$, $R = \infty$. R = 1 is of course a circle and $R = \infty$ a straight line. When $\lambda = \frac{1}{2}$, R = 2 which is the particular conic we have been considering.

(N.B. The equation to the outer dome of St. Paul's is unknown and no drawings of it remain. However, Barnes Wallis believed that Wren struck two circular arcs from two separate points on the base line and this is indeed a close approximation.)

10 Computational geometry

Computational geometry is the computer representation, analysis and synthesis of shape information.

A. R. Forrest

10.1 INTRODUCTION

The classical curves and surfaces described in Chapter 8 were all characterised by a single equation and herein lies the elegance and attraction of their formulation. A spherical shell makes an excellent play-ball, as (from a stress point of view) it is also the ideal shape of a pressure vessel. An ellipsoidal envelope makes a grand airship and a hyperparaboloid makes a stylistic roof. There are, however, many practical requirements that cannot be met by classical forms.

The technical requirements of ships, automobiles, aircraft, bottles, cups, plates, knives, forks, spoons and turbine blades, etc., etc., are such that other forms must necessarily be invoked. The use of conic lofting in the shape design of ships, autos and aircraft has already been discussed in Chapter 9 and the use of piecewise curve building will shortly be considered.

Although Barnes Wallis made considerable use of the mathematical equations of the ellipsoid, when designing the R.100 airship in 1929, the widespread use of numerical geometry did not materialise until the early 1940s. Up to that date, the shape definition of ships, autos and aircraft depended largely on the shape designers' manual lofting skills and the myriad of calculations required to define the surface form of an aircraft were, until the 1960s, performed on mechanical desk top calculators, of which the Monroe was perhaps most highly regarded. Since the advent of the much faster electronic computer it has beome increasingly possible to define and modify shapes and forms much more quickly and more accurately and it is with these latter-day methods of shape and form definition that we are now concerned.

The methods currently used and subject to on-going development are collectively classified as computational geometry, and the book by Faux and Pratt is perhaps the most complete text on the subject. The use of vector methods feature predominantly in the literature and practice of the subject.

We therefore open this chapter by considering the more elementary formulations and uses of vectors and go on to consider the more elementary problems associated with the piecewise definition of three dimensional curves and surface patches.

10.2 VECTOR CALCULUS

Vector methods feature prominently in the literature and practice of computational geometry. There are, as we have seen, two ways in which we may travel to a point situated in three-space. The indirect line path along the Cartesian coordinate directions has been used extensively and the notion of a direct line path has been introduced principally in terms of the spherical coordinate system. We now need to consider the properties ascribable to direct line paths in greater detail.

10.2.1 The concept of a vector

The shortest distance between two points P and Q may be represented by a straight line segment of length equal to or proportional to the length of the direct line path PQ. The sequence PQ may be used to denote the direction $P \rightarrow Q$ and the sequence QP may be used to denote the direction $Q \rightarrow P$. If the sense of PQ is positive then the sense of QP is negative. The negation of PQ = –PQ and –PQ = +QP when a length –PQ is added to a length +PQ the result is clearly zero and when –PQ and +PQ are regarded as displacements the geometrical interpretation of (–PQ +PQ) is zero movement.

Quantities that depend on both magnitude and direction are called vectors.

10.2.2 Addition and subtraction of vectors

If a number of positive vectors are joined end to end then a continuous chain of positive vectors is produced.

$$\overrightarrow{MN} + \overrightarrow{NS} = \overrightarrow{MS}, \quad \overrightarrow{PQ} + \overrightarrow{QR} + \overrightarrow{RS} = \overrightarrow{PS}$$

and $\overrightarrow{MS}$ is said to be the resultant of the two vectors $\overrightarrow{MN} + \overrightarrow{NS}$, where $\overline{MN}$ and $\overline{NS}$ are joined in accord with the parallelogram rule.

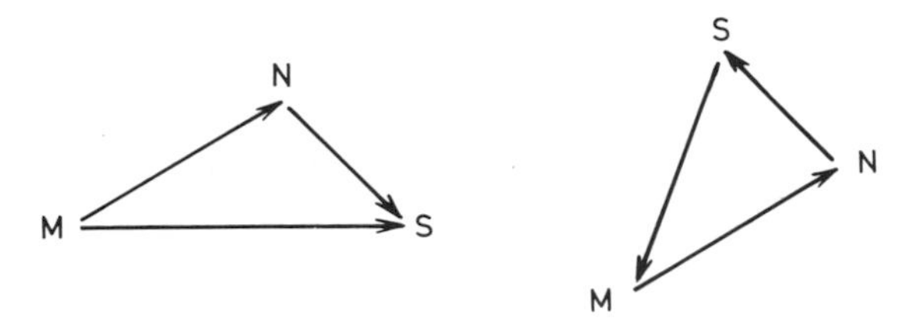

Figure 10.1

10.2.3 Modulus of a vector

If three points P, Q, R, are in a straight line and PQ = QR = a, then PQ + QR = $2a$. The numerical multiplier 2 is a scalar quantity which indicates that the length, that is the magnitude of PR is two units of a. In vector algebra numerical multipliers such as this are known as the modulus of the vector. In this case the modulus of the vector $2a$ is 2.

If the three points P, Q, R, are not in a straight line then the direct line path PR is less than $2a$, and in this event the numerical multiplier, i.e. the modulus, would be obtained by applying the cosine rule.

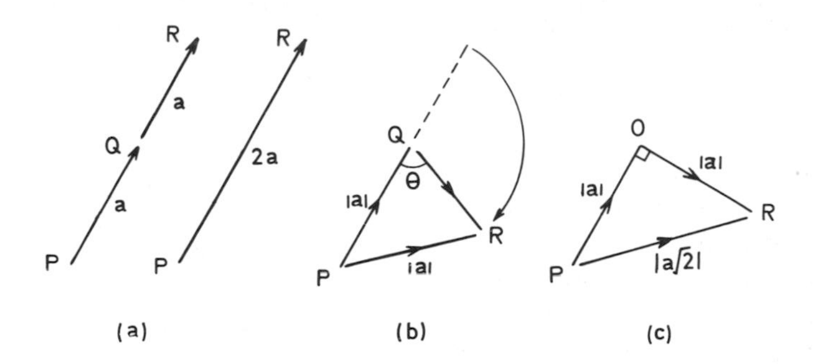

Figure 10.2

If, as in Fig. 10.2, the included angle is 60° then the magnitude of PR is a and if the included angle is 90° then the magnitude of PQ is $\sqrt{2}a$ and since we have not specified the direction of PQ, a and $\sqrt{2}a$ are scalar quantities, which merely tell us the length (i.e. the magnitude) but not the direction of the vector PQ.

In order to distinguish between pure scalar lengths, such as the hypotenuse $\sqrt{2}a$, Fig. 10.2, and a fully defined vector quantity, we adopt the convention of printing vectors in bold type or for convenience in handwriting place a bar or arrow above the character used.

Characters such as the a in Fig. 10.2, which merely indicates the magnitude of the distance from P to Q, are enclosed between vertical bars. Thus the magnitude of the vector $\overrightarrow{PQ} = a$ is written as $|a|$.

It is therefore understood that: **a** (bold type), $\bar{a}$ (ordinary a with a bar), $\vec{a}$ (ordinary a with an arrow) are vectors, whilst $|a|$ (ordinary a between vertical bars) is a scalar. The modulus of a vector is its undirected length and undirected length is always a scalar.

Some authors always use lower case letters to indicate vectors but this practice is by no means universal, we shall use upper and lower case letters according to mood and in accord with common usage. When dealing with small specific problems and proofs it is sometimes helpful to use capital letters for points and the corresponding lower case letters for their position vectors, see article 10.3.

If $\overline{PQ} + \overline{QR}$ are two positive vectors at right angles which lie in the xoy plane, then $\overline{PQ} + \overline{QR} = \overline{PR}$. Then the Cartesian components of the resultant vector $\overline{PR}$ are

$$x_{PR} = (x_2 - x_1) \quad \text{and} \quad y_{PR} = (y_2 - y_1)$$

whence the length (that is the magnitude) of the vector $\overline{PR}$ is equal to $|PR|$ where

$$|PR| = ((x_2 - x_1)^2 + (y_2 - y_1)^2)^{1/2}$$

If $\overline{PR}$ is a three dimensional vector then its length

$$|PR| = [(x_2 - x_1)^2 + (y_2 - y_1)^2 + (z_2 - z_1)^2]^{1/2}$$

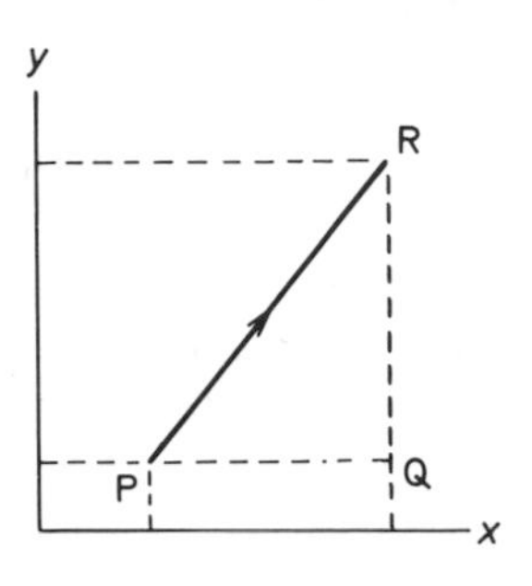

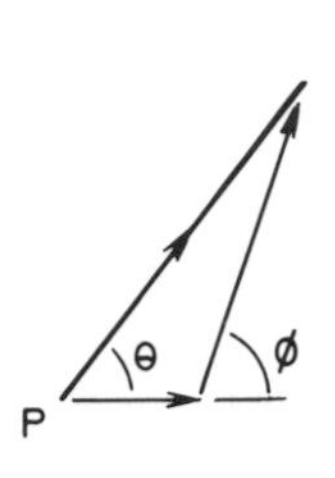

Figure 10.3

10.2.4 Unit vectors

In order to specify the direction of a vector we introduce the concept of a unit vector the direction of which is known. Any vector such as $\overline{PQ}$, Fig. 10.3, which lies parallel to the *ox* axis may be written in the form $x\mathbf{i}$ and any vector such as $\overline{QR}$ which is parallel to the *oy* axis may be written in the form $y\mathbf{j}$, where **i** and **j** are defined as unit vectors (that is vectors of unit length) which lie parallel to the *ox* and *oy* axes respectively.

Since both **i** and **j** are of unit magnitude they do not affect the magnitude of x and y but serve merely to indicate direction.

If the vector $\overline{PQ} = \mathbf{a}$

$$\mathbf{a} = x\mathbf{i} + y\mathbf{j}$$

and if the vector a has 3-space freedom then its magnitude and direction is completely defined by the vector equation

$$\mathbf{a} = x\mathbf{i} + y\mathbf{j} + z\mathbf{k}$$

in which **i**, **j**, **k**, are unit vectors directed along the *ox*, *oy*, *oz*, axes respectively.

This vector equation concisely states that the vector **a** is comprised of three mutually perpendicular components, each of which is parallel to the reference axis to which it is referred.

(N.B. A unit vector **u** is often written $\hat{\mathbf{u}}$ to signify that it is a vector of unit length.)

10.2.5 Scalar product

A vector **b** inclined at an angle θ to another vector **a** may be resolved into two components, one parallel to and the other perpendicular to the line of vector **a**.

The component of vector **b** which is parallel to the vector **a** is equal to the magnitude of **b** times cos θ, that is $|\mathbf{b}| \cos \theta$.

The product obtained by multiplying the magnitude of **a**, that is, $|\mathbf{a}|$ by $|\mathbf{b}|$ cos θ is defined as the scalar product. A scalar product sometimes called the dot product is distinguished from a vector product, article 10.2.6, by the use of a bold dot to indicate multiplication.

$$\mathbf{a} \cdot \mathbf{b} = |\mathbf{ab}| \cos \theta = \mathbf{b} \cdot \mathbf{a}$$

We observe that $|\mathbf{ab}| \cos \theta$ is a pure number which explains why this particular operation is called a scalar product.

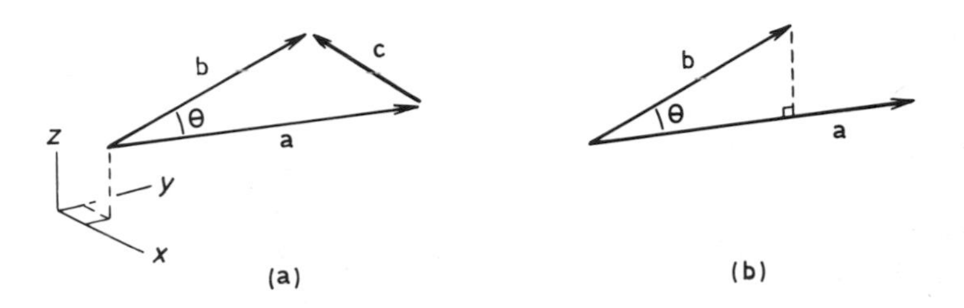

Figure 10.4

(N.B. The scalar product is sometimes written in the simplified form)

$$\mathbf{a}\,\mathbf{b} = ab \cos \theta$$

Bearing in mind that we are dealing with magnitudes, we see that

$$|\mathbf{a} - \mathbf{b}|^2 = |\mathbf{a}|^2 + |\mathbf{b}|^2 - 2|\mathbf{a}|\,|\mathbf{b}| \cos \theta$$

so that if the Cartesian components of **a** are $(\mathbf{a}_1, \mathbf{a}_2, \mathbf{a}_3)$ and those of **b** are $(\mathbf{b}_1, \mathbf{b}_2, \mathbf{b}_3)$ then

$$(a_1 - b_1)^2 + (a_2 - b_2)^2 + (a_3 - b_3)^2 = (a_1^2 + a_2^2 + a_3^2)$$
$$+ (b_1^2 + b_2^2 + b_3^2) - 2 \cdot |\mathbf{a}|\,|\mathbf{b}| \cos \theta$$

expanding and cancelling terms we obtain

$$2a_1 b_1 + 2a_2 b_2 + 2a_3 b_3 = 2\,|\mathbf{a}|\,|\mathbf{b}| \cos \theta$$

and hence

$$\cos\theta = \frac{a_1b_1 + a_2b_2 + a_3b_3}{|\mathbf{a}|\,|\mathbf{b}|}$$

and since $|b| \cos\theta$ = the component of b along a, the component of b along a is

$$\frac{a_1b_1 + a_2b_2 + a_3b_3}{|\mathbf{a}|}$$

and since $a_1/|\mathbf{a}| = l_a$, $a_2/|\mathbf{a}| = m_a$, $b_1/|\mathbf{b}| = l_b$, $b_2/|\mathbf{b}| = m_b$, where $l_a{:}m_a$, $l_b{:}m_b$ are the direction cosines of the vectors a and b whence we obtain

$$\cos\theta = l_al_b + m_am_b + n_an_b$$

as previously.

(N.B. The scalar dot product is also referred to as the inner product).

When the x, y, z, components of a vector A are Ax, Ay, Az and the x, y, z, components of a vector B are Bx, By, Bz. The scalar product

$$\begin{aligned}\mathbf{A} \cdot \mathbf{B} &= (\mathrm{A}_x\mathbf{i} + \mathrm{A}_y\mathbf{j} + \mathrm{A}_z\mathbf{k}) \cdot (\mathrm{B}_x\mathbf{i} + \mathrm{B}_y\mathbf{j} + \mathrm{B}_z\mathbf{k})\\ &= \mathrm{A}x\,\mathrm{B}x\,(\mathbf{i} \cdot \mathbf{i}) + \mathrm{A}x\,\mathrm{B}y\,(\mathbf{i} \cdot \mathbf{j}) + \mathrm{A}x\,\mathrm{B}z\,(\mathbf{i} \cdot \mathbf{k})\\ &+ \mathrm{A}y\,\mathrm{B}x\,(\mathbf{j} \cdot \mathbf{i}) + \mathrm{A}y\,\mathrm{B}y\,(\mathbf{j} \cdot \mathbf{j}) + \mathrm{A}y\,\mathrm{B}z\,(\mathbf{j} \cdot \mathbf{k})\\ &+ \mathrm{A}z\,\mathrm{B}x\,(\mathbf{k} \cdot \mathbf{i}) + \mathrm{A}z\,\mathrm{B}y\,(\mathbf{k} \cdot \mathbf{j}) + \mathrm{A}z\,\mathrm{B}z\,(\mathbf{k} \cdot \mathbf{k})\end{aligned}$$

(N.B. The above expansion is obtained by ordinary algebraic multiplication of the two bracketted terms).

However, the angle between two identical vectors is zero and since $\cos\theta = 1$ it follows that $(\mathbf{i} \cdot \mathbf{i}) = (\mathbf{j} \cdot \mathbf{j}) = (\mathbf{k} \cdot \mathbf{k}) = 1$. Similarly the angle between two vectors at right angles is 90° and since $\cos 90 = 0$, it follows that $(\mathbf{i} \cdot \mathbf{y}) = (\mathbf{i} \cdot \mathbf{k}) = (\mathbf{y} \cdot \mathbf{i}) = (\mathbf{j} \cdot \mathbf{k}) = (\mathbf{k} \cdot \mathbf{i}) = (\mathbf{k} \cdot \mathbf{j}) = 0$. Hence the scalar product **A . B** is simply

$$\mathbf{A} \cdot \mathbf{B} = \mathrm{A}_x\mathrm{B}_x + \mathrm{A}_y\mathrm{B}_y + \mathrm{A}_z\mathrm{B}_z$$

That is the scalar product of two vectors is the sum of their component products.

The relationships between unit vectors at right angles play an important part in the solution of many problems and are tabulated for ease of reference.

$$\begin{array}{lll}\mathbf{i} \cdot \mathbf{i} = 1 & \mathbf{i} \cdot \mathbf{j} = 0 & \mathbf{i} \cdot \mathbf{k} = 0\\ \mathbf{j} \times \mathbf{i} = 0 & \mathbf{j} \cdot \mathbf{j} = 1 & \mathbf{j} \cdot \mathbf{k} = 0\\ \mathbf{k} \times \mathbf{i} = 0 & \mathbf{k} \cdot \mathbf{j} = 0 & \mathbf{k} \cdot \mathbf{k} = 1\end{array}$$

(N.B. a_1, a_2, a_3 and Ax, Ay, Az are both commonly used to represent the components of the vector **A**.)

Example
If $\mathbf{u} = 3\mathbf{i} + 4\mathbf{j}$ is a vector prove that its magnitude $|\mathbf{u}| = (\mathbf{u} \cdot \mathbf{u})^{1/2}$ and hence determine the magnitude of $\mathbf{u}$.

Solution
The general form of a vector is $\mathbf{u} = x\mathbf{i} + y\mathbf{j}$, which means that x and y are the Cartesian coordinates of $\mathbf{u}$. Now since

$$\mathbf{u} \cdot \mathbf{u} = xx + yy = x^2 + y^2 = |\mathbf{u}|^2$$

$$|\mathbf{u}| = (\mathbf{u} \cdot \mathbf{u})^{1/2} = (x^2 + y^2)^{1/2}$$

Thus the length, or magnitude, of the vector $\mathbf{u} = 3\mathbf{i} + 4\mathbf{j}$ is equal to $(3^2 + 4^2)^{1/2} = 5$.

Example
Use the scalar (dot) product method to determine the angle between two vectors

$$\mathbf{u}_1 = \sqrt{3}\,\mathbf{i} + \mathbf{j} \quad \text{and} \quad \mathbf{u}_2 = -\mathbf{i} + \sqrt{3}\,\mathbf{j}$$

Solution
We know that since this is a two dimensional problem $\cos\theta = (l_1 l_2 + m_1 m_2)$, where $l_1 : m_1$, $l_2 : m_2$ are the direction cosines of the vectors $\mathbf{u}_1$ and $\mathbf{u}_2$ respectively. Now since $x_1 = |\mathbf{u}_1| l_1$, $y_1 = |\mathbf{u}_1| m_1$, $x_2 = |\mathbf{u}_2| l_2$, $y_2 = |\mathbf{u}_2| m_2$ and $\mathbf{u}_1 \cdot \mathbf{u}_2 = x_1 x_2 + y_1 y_2$ (as shown in the previous example).

$$\begin{aligned} \mathbf{u}_1 \cdot \mathbf{u}_2 &= |\mathbf{u}_1||\mathbf{u}_2|(l_1 l_2 + m_1 m_2) \\ &= |\mathbf{u}_1||\mathbf{u}_2| \cos\theta \end{aligned}$$

That is

$$\cos\theta = \frac{\mathbf{u}_1 \cdot \mathbf{u}_2}{|\mathbf{u}_1||\mathbf{u}_2|}$$

For this particular problem we have

$$\mathbf{u}_1 \cdot \mathbf{u}_2 = -\sqrt{3}\,\mathbf{i}^2 - \mathbf{ij} + 3\mathbf{ji} + \sqrt{3}\,\mathbf{j}^2$$

but $\mathbf{i}^2 = \mathbf{j}^2 = 1$, $\mathbf{ij} = \mathbf{ji} = 0$ and hence $\mathbf{u}_1 \cdot \mathbf{u}_2 = 0$, and since $\cos 90° = 0$, the vectors $\mathbf{u}_1$ and $\mathbf{u}_2$ are at right angles.

(N.B. The solution of the corresponding three-dimensional problem follows a similar route).

10.2.6 The vector product

It happens quite often in geometrical work that we need to define a vector which is normal to the plane with which we are dealing and the vector product fulfils this requirement.

The vector product of two vectors **a** and **b** is distinguished from the scalar dot product by using a cross and for this reason the vector product is often referred to as the cross product of the two vectors. The magnitude of $\mathbf{a} \times \mathbf{b}$ is $|\mathbf{a}|\,|\mathbf{b}|\sin\theta$, where θ is the smallest angle between them. In order to distinguish between the two possible cases, illustrated in Fig. 10.5(a) and (b), we apply the right-handed screw rule. Thus if **a** is thought to rotate towards **b** in an anticlockwise direction, as in Fig. 10.5(a) then the positive direction of the vector **c** is out of the page towards the eye. If as in Fig. 10.5(b), **a** rotates towards **b** in a clockwise direction, then the positive direction of the vector **c** is into the page. That is it follows the path of a right handed screw and we note that $\mathbf{a} \times \mathbf{b} = -\mathbf{b} \times \mathbf{a}$.

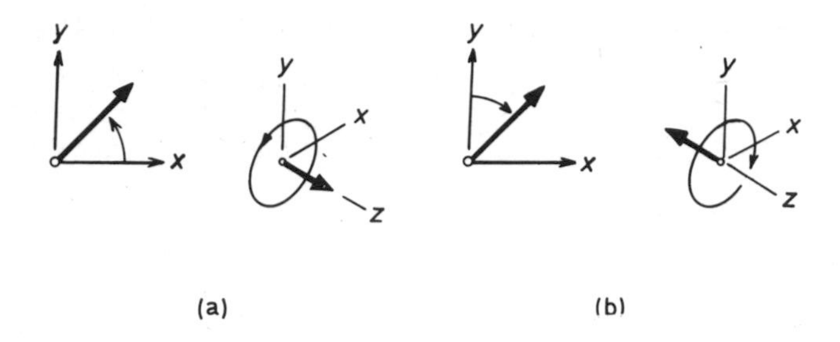

Figure 10.5

Now since **c** is a vector of magnitude $|\mathbf{a}|\,|\mathbf{b}|\sin\theta$ we write the vector or cross product of the vectors **a** and **b** as follows:

$$\mathbf{a} \times \mathbf{b} = |\mathbf{a}|\,|\mathbf{b}|\sin\theta\ \mathbf{c}$$

or simply as in some books

$$\mathbf{a} \times \mathbf{b} = \mathbf{a}\,\mathbf{b}\sin\theta\ \mathbf{c}$$

When the x, y, z, components of a vector **A** are Ax, Ay, Az and the x, y, z, components of a vector **B** are Bx, By, Bz, the vector product

$$\begin{aligned}\mathbf{A} \times \mathbf{B} &= (A_x\mathbf{i} + A_y\mathbf{j} + A_z\mathbf{k}) \times (B_x\mathbf{i} + B_y\mathbf{j} + B_z\mathbf{k})\\ &= A_xB_x\,(\mathbf{i} \times \mathbf{i}) + A_xB_y\,(\mathbf{i} \times \mathbf{j}) + A_xA_z\,(\mathbf{i} \times \mathbf{k})\\ &= +\,A_yB_x\,(\mathbf{j} \times \mathbf{i}) + A_yB_y\,(\mathbf{j} \times \mathbf{j}) + B_yB_z\,(\mathbf{j} \times \mathbf{k})\\ &\quad +\,A_zB_x\,(\mathbf{k} \times \mathbf{i}) + A_zB_y\,(\mathbf{k} \times \mathbf{j}) + A_zB_z\,(\mathbf{k} \times \mathbf{k})\end{aligned}$$

(N.B. The above expression is obtained by ordinary algebraic multiplication of the two bracketed terms).

However, the angle between two identical vectors is zero and since $\sin 0° = 0$ it follows that $(\mathbf{i} \times \mathbf{i}) = (\mathbf{j} \times \mathbf{j}) = (\mathbf{k} \times \mathbf{k}) = 0$. Similarly the angle between two vectors at right angles is 90° and since sin 90° = 1 it follows that (**i** × **j**) = (**i** × **k**) = (**j** × **i**) = (**j** × **k**) = (**k** × **i**) = (**k** × **j**) = 1. Moreover (**i** × **j**) = **k**, (**i** × **k**) = **j**, (**j** × **i**) = –**k**, (**j** × **k**) = **i**, (**k** × **i**) = **k**, (**k** × **j**) = –**i**. The vector product **A** × **B** is thus given by

$$\mathbf{A} \times \mathbf{B} = (A_y B_z - A_z B_y)\,\mathbf{i} + (A_z B_x - A_x B_z)\,\mathbf{j} + (A_x B_y - A_y B_x)\,\mathbf{k}$$

or if written in determinant form

$$\mathbf{A} \times \mathbf{B} = \begin{vmatrix} \mathbf{i} & \mathbf{j} & \mathbf{k} \\ A_x & A_y & A_z \\ B_x & B_y & B_z \end{vmatrix}$$

The relationships between unit vectors at right angles play an important part in the solution of many problems and are tabulated for easy reference:

$$\begin{array}{lll} \mathbf{i}\times\mathbf{i} = \ \ 0 & \mathbf{i}\times\mathbf{j} = \ \ \mathbf{k} & \mathbf{i}\times\mathbf{k} = -\mathbf{j} \\ \mathbf{j}\times\mathbf{i} = -\mathbf{k} & \mathbf{j}\times\mathbf{j} = \ \ 0 & \mathbf{j}\times\mathbf{k} = \ \ \mathbf{i} \\ \mathbf{k}\times\mathbf{i} = \ \ \mathbf{j} & \mathbf{k}\times\mathbf{j} = -\mathbf{i} & \mathbf{k}\times\mathbf{k} = \ \ 0 \end{array}$$

(N.B. The reader who is new to vector methods should take care not to confuse the results of scalar and vector multiplication). For example (**i** · **i**) = 1, (**i** × **i**) = 0, (**i** · **j**) = 0, (**i** × **j**) = 1, (**j** · **i**) = (**i** · **j**) but (**j** × **i**) = –**k** whereas (**i** × **j**) = +**k**. The difference is of course traceable to the use of cosine in the scalar product and the use of sine in the vector product.

10.2.7 The triple scalar product

A vector multiplication **a** × **b** followed by a scalar multiplication such that

$$\mathbf{a} \times \mathbf{b} \cdot \mathbf{c} = (a_y b_z - a_z b_y)\,c_x + (a_z b_x - a_x b_z)\,c_y + (a_x b_y - a_y b_x)\,c_z$$

is called a triple scalar product.

Example

Show that the volume of the parallelopiped, Fig. 10.6, is given by the triple scalar product **a** × **b** . **c**.

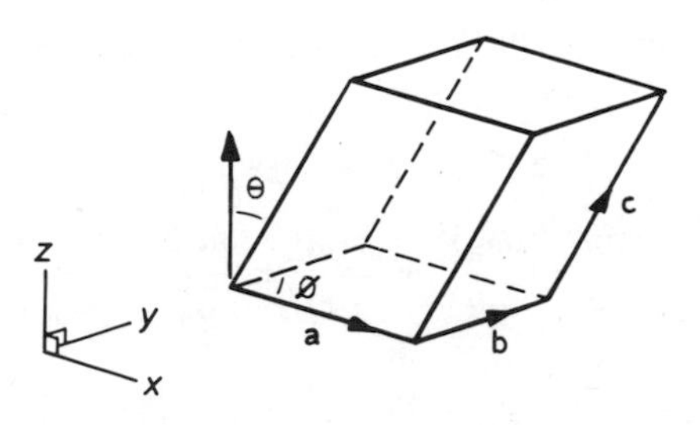

Figure 10.6

Solution

The area of a parallelopiped is given by area of base × perpendicular height. In terms of vectors the area of base is the vector product **a** × **b**, the magnitude of which is $ab \sin\theta$, in a direction perpendicular to the base. If **p** is a unit vector perpendicular to the base then $\mathbf{a} \times \mathbf{b} = ab \sin\theta\, \mathbf{p}$. Now the projection of **c** on **p** is $c \cos\phi$ and the scalar product of **p** and $\mathbf{c} \cos\phi$ is $pc \cos\phi$, whence

$$\mathbf{a} \times \mathbf{b} \cdot \mathbf{c} = ab \sin\theta\, \mathbf{p} \cdot \mathbf{c} = ab \sin\theta\, c \cos\phi$$

We see that $ab \sin\theta\, c \cos\phi$ is a magnitude, and since $ab \sin\theta$ is the area of the base and $c \cos\phi$ the perpendicular height it follows that $\mathbf{a} \times \mathbf{b} \cdot \mathbf{c}$ = the volume of the parallelopiped.

The volume of an actual parallelopiped is obtained by writing:

$$\mathbf{a} \times \mathbf{b} \cdot \mathbf{c} = (a_y b_z - a_z b_y)\, c_x + (a_z b_x - a_x b_z)\, c_y + (a_x b_y - a_y b_x)\, c_z$$

(N.B. If it so happens that $\mathbf{a} \times \mathbf{b} \cdot \mathbf{c} = 0$, then the volume is zero and **a**, **b**, **c**, must be coplanar. We also observe that if any two of the vectors **a**, **b**, **c**, are equal then $\mathbf{a} \times \mathbf{b} \cdot \mathbf{c} = 0$).

10.2.8 The triple vector product

A vector multiplication $\mathbf{a} \times \mathbf{b} \times \mathbf{c}$ is called a triple vector product, and entails the double application of the cross product equation. As always with vector multiplication the order of multiplication is important and in the present case

$$(\mathbf{a} \times \mathbf{b}) \times \mathbf{c} \neq \mathbf{a} \times (\mathbf{b} \times \mathbf{c})$$

Suffice to record that

$$\mathbf{a} \times (\mathbf{b} \times \mathbf{c}) = (\mathbf{a} \cdot \mathbf{c})\mathbf{b} - (\mathbf{a} \cdot \mathbf{b})\mathbf{c}$$

but we shall not pursue the triple vector product further.

10.2.9 Differentiation of vectors

Since vectors are not always constant in terms of magnitude and direction we need to confirm that if

$$\mathbf{a} = a_x\mathbf{i} + a_y\mathbf{j} + a_z\mathbf{k}$$

$$\frac{d\mathbf{a}}{dt} = \frac{da_x\mathbf{i}}{dt} + \frac{da_y\mathbf{j}}{dt} + \frac{da_z\mathbf{k}}{dt}$$

and if

$$\mathbf{b} = b_x\mathbf{i} + b_y\mathbf{j} + b_z\mathbf{k}$$

the differential of the scalar product is

$$\frac{d}{dt}(\mathbf{a}\,.\,\mathbf{b}) = \frac{d}{dt}(a_x b_x + a_y b_y + a_z b_z)$$

$$= \left(\mathbf{a} \cdot \frac{d\hat{b}}{dt}\right) + \left(\frac{d\hat{a}}{dt} \cdot \mathbf{b}\right)$$

The differential of the vector product is

$$\frac{d}{dt}(\mathbf{a} \times \mathbf{b}) = \left(\mathbf{a} \times \frac{d\hat{b}}{dt}\right) + \left(\frac{d\hat{a}}{dt} \times \mathbf{b}\right)$$

10.2.10 Vector usage

We have seen that a vector equation is very concise and offers a considerable economy in terms of writing. The vector **A**, for instance, represents three indirect line path Cartesian components. The single symbol **A** may thus be used during the process of analysis in place of $(A_x\mathbf{i} + A_y\mathbf{j} + A_z\mathbf{k})$ by writing $\mathbf{A} = (x, y, z)$.

We may also appreciate that scalar and vector products possess properties that make them powerful analytic tools. We do however need to revert to algebraic methods when making specific numerical evaluations.

10.3 VECTOR EQUATIONS TO POINT, LINE AND PLANE

The position of a point in 3-space may be concisely specified, relative to a known origin, by means of a single position vector. The vectors $\overrightarrow{OP}$ and $\overrightarrow{OQ}$ position the points P and Q in relation to the origin O. Vectors of the type $\overrightarrow{OP}$ and $\overrightarrow{OQ}$ are called absolute position vectors whilst vectors of the type $\overrightarrow{PQ}$ are called relative position vectors (since the position of P is known indirectly).

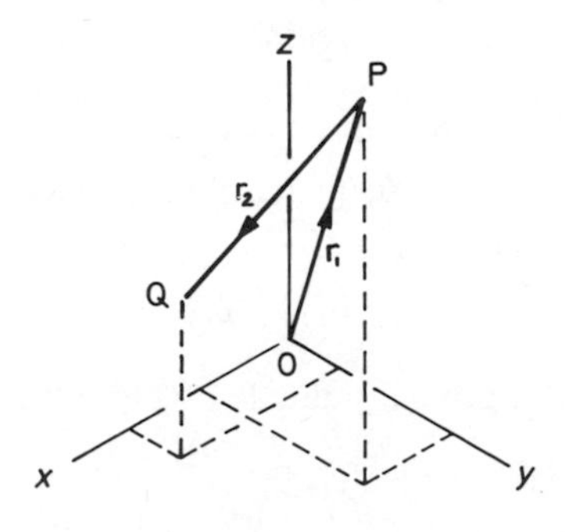

Figure 10.7

(N.B. The terms base vector and direction vector are also used).

The relative position vector $\overrightarrow{PQ}$ may be expressed in terms of the two absolute position vectors $\overrightarrow{OP}$ and $\overrightarrow{OQ}$ by straightforward vector addition.

Bearing in mind that the vector $\overrightarrow{PO}$ equals the negation of the vector $\overrightarrow{OP}$ we accomplish this addition as follows:

$$\overrightarrow{PQ} = -\overrightarrow{OP} + \overrightarrow{OQ}$$

or in terms of the vectors $\mathbf{r}_1$ and $\mathbf{r}_2$

$$\overrightarrow{PQ} = (\mathbf{r}_2 - \mathbf{r}_1)$$

where $\mathbf{r}_1$ and $\mathbf{r}_2$ are the two absolute position vectors, which position the points P and Q.

The position vectors $\mathbf{r}_1$ and $\mathbf{r}_2$ may clearly be expressed in terms of their Cartesian components as follows:

$$\overrightarrow{PQ} = (x_2\mathbf{i} + y_2\mathbf{j} + z_2\mathbf{k}) - (x_1\mathbf{i} + y_1\mathbf{j} + z_1\mathbf{k})$$

$$\overrightarrow{PQ} = (x_2 - x_1)\mathbf{i} + (y_2 - y_1)\mathbf{j} + (z_2 - z_1)\mathbf{k}$$

where

$$(x_2 - x_1), (y_2 - y_1), (z_2 - z_1)$$

are the direction ratios of the directed line vector $\overrightarrow{PQ}$ and since

$$|\mathbf{PQ}| = [(x_2 - x_1)^2 + (y_2 - y_1)^2 + (z_2 - z_1)^2]^{1/2}$$

$$\frac{(x_2 - x_1)}{|\mathbf{PQ}|}, \frac{(y_2 - y_1)}{|\mathbf{PQ}|}, \frac{(z_2 - z_1)}{|\mathbf{PQ}|}$$

are the corresponding direction cosines.

10.3.1 Vector equation to a straight line

If P is a point on a straight line with a position vector $\mathbf{r}_1$ and Q is a second point on the line with a position vector $\mathbf{r}_2$ and V is a vector of known magnitude which defines the direction of the line, as shown in Fig. 10.8, then the position of all other points (that is a general point) on the line may be specified as follows.

We see from Fig. 10.8, that

$$\overrightarrow{OQ} = \overrightarrow{OP} + \overrightarrow{PQ}$$

and hence the position of a general point R is given by a vector equation of the form

$$\mathbf{r} = \mathbf{r}_1 + \mu\mathbf{V}$$

where $\mu\mathbf{V}$ is some specified multiple of the given vector $\mathbf{V}$.

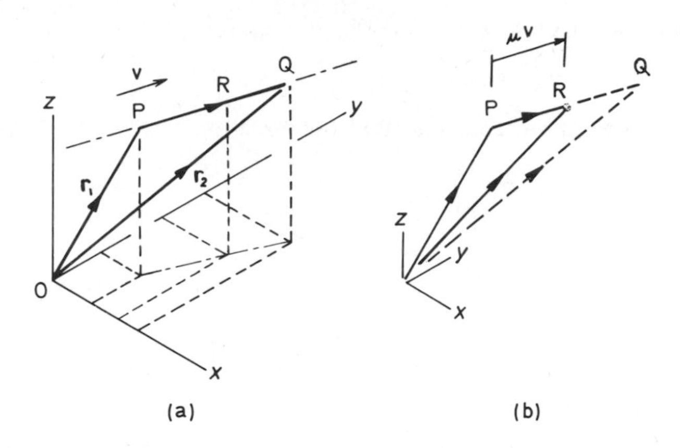

Figure 10.8

If $\mathbf{V}$ is a unit vector, $\hat{\mathbf{u}}$ say, then $(\mathbf{r} - \mathbf{r}_1) = \lambda\mathbf{u}$ and the equation to the general point on a straight line is written

$$\mathbf{r} = \mathbf{r}_1 + \lambda\mathbf{u}$$

The distance PR is a scalar and is therefore written as follows

$$|\mathbf{r} - \mathbf{r}_1| = |\lambda\mathbf{u}| = \lambda$$

If $\hat{\mathbf{u}}$ is of unit length then

$$\begin{aligned}\mathbf{u} &= \cos\alpha\,\mathbf{i} + \cos\beta\,\mathbf{j} + \cos\gamma\,\mathbf{k}\\ &= l\mathbf{i} + m\mathbf{j} + n\mathbf{k}\end{aligned}$$

where $\cos\alpha = l$, $\cos\beta = m$, $\cos\gamma = n$ are the direction cosines which fix the direction of $\mathbf{u}$.

(N.B. The direction of a non unit vector V may similarly be specified in terms of its direction ratios).

Example

If the direction of a straight line is given by $2\mathbf{i} + 3\mathbf{j} + 4\mathbf{k}$ (the components of its direction vector) and the position of a known point on the line is known to be $\mathbf{i} + \mathbf{j} + 2\mathbf{k}$, write the vector equation to the line.

The vector equation to a straight line is of the form:

$$\mathbf{r} = \mathbf{r}_1 + \mu\mathbf{V}$$

whence

$$\begin{aligned}\mathbf{r} &= (\mathbf{i} + \mathbf{j} + 2\mathbf{k}) + \mu(2\mathbf{i} + 3\mathbf{j} + 4\mathbf{k})\\ &= (1 + 2\mu)\mathbf{i} + (1 + 3\mu)\mathbf{j} + (2 + 4\mu)\mathbf{k}\end{aligned}$$

The Cartesian coordinates of any point Q are therefore:

$$x\ (1 + 2\mu),\ y = (1 + 3\mu),\ z = (2 + 4\mu)$$

and on eliminating μ we conclude that the equation connecting x, y and z is:

$$\frac{x-1}{2} = \frac{y-1}{3} = \frac{z-2}{4} = \mu$$

whence

μ	x	y	z
0	1	1	2
1	3	4	6

Other points on the line may be readily obtained by specifying different values of μ. The slope of the line is confirmed by the direction ratios obtained from both the equation and the table. The direction ratios are seen to be $(3 - 1)$, $(4 - 1)$, $(6 - 2) = 2{:}3{:}4$ and these ratios clearly agree with the direction vector originally specified.

10.3.2 Vector equation to a line through two known points

The direction of a straight line is uniquely defined by the position vectors of any two points on a line.

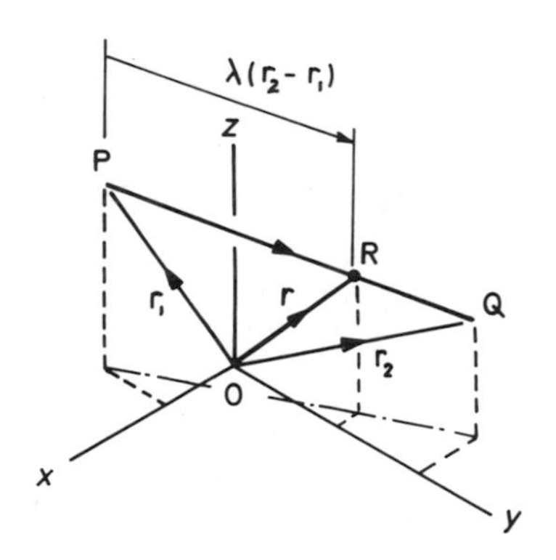

Figure 10.9

If P and Q are two points on a line the position vectors of which are $\mathbf{r}_1$ and $\mathbf{r}_2$, then a third point R on the line is located as follows:

$$\overrightarrow{PQ} = (\mathbf{r}_2 - \mathbf{r}_1)$$

$\overrightarrow{PR}$ is some proportion λ of $\overrightarrow{PQ}$, whence

$$\overrightarrow{PR} = \lambda(\mathbf{r}_2 - \mathbf{r}_1)$$

$$\overrightarrow{OR} = \overrightarrow{OP} + \overrightarrow{PR}$$

therefore

$$\mathbf{r} = \mathbf{r}_1 + \lambda(\mathbf{r}_2 - \mathbf{r}_1)$$

alternatively

$$\mathbf{r} = \mathbf{r}_1(1 - \lambda) + \lambda\mathbf{r}_2$$

The position of any point R on the line is controlled by the numerical parameter λ or μ. The expression

$$\mathbf{r}(\mu) = \mathbf{r}_1 + \mu(\mathbf{r}_2 - \mathbf{r}_1)$$

is frequently used to signify that the vector $\mathbf{r}$ is the outcome of applying the constant μ. The expression

$$\mathbf{r}(\mu) = (1 - \mu)x_1 + \mu x_2, (1 - \mu)y_1 + \mu y_2, (1 - \mu)z_1 + \mu z_2$$

indicates that the vector $\mathbf{r}(u)$ is the sum of its Cartesian coordinates.

Example

If the coordinates of two points P and Q are (6, 7, 8), (−4, 3, −2) write the vector equation to the straight line through P and Q.

Solution

The vector equation to a straight line is of the form:

$$\mathbf{r} = \mathbf{r}_1 + \lambda(\mathbf{r}_2 - \mathbf{r}_1)$$

The position vector of P is: $6\mathbf{i} + 7\mathbf{j} + 8\mathbf{k}$.

The position vector of Q is: $-4\mathbf{i} + 3\mathbf{j} - 2\mathbf{k}$.

Hence the position vector of R is given by the equation:

$$\begin{aligned} r &= 6\mathbf{i} + 7\mathbf{j} + 8\mathbf{k} + \lambda(-4\mathbf{i} + 3\mathbf{j} - 2\mathbf{k} - 6\mathbf{i} - 7\mathbf{j} - 8\mathbf{k}) \\ &= 6\mathbf{i} + 7\mathbf{j} + 8\mathbf{k} - 10\lambda\mathbf{i} - 4\lambda\mathbf{j} - 10\lambda\mathbf{k} \\ &= (6 - 10\lambda)\mathbf{i} + (7 - 4\lambda)\mathbf{j} + (8 - 10\lambda)\mathbf{k} \end{aligned}$$

and this is the vector equation to the line through P and Q.

The coordinates of any point R on the line are

$$x = (6 - 10\lambda),\ y = (7 - 4\lambda),\ z = (8 - 10\lambda)$$

and the equation connecting x, y and z is:

$$\frac{(6-x)}{10} = \frac{(7-y)}{4} = \frac{(8-z)}{10} = \lambda$$

As demonstrated previously the denominators in the above equation are the direction ratios to the line. Hence the direction ratios are (10: 4:10) = (5:2:5).

When $\lambda = 0$, $x = 6$, $y = 7$, $z = 8$, that is, $\lambda = 0$ corresponds to the point P.

When $\lambda = 1$, $x = -4$, $y = 3$, $z = -2$, that is, $\lambda = 1$ corresponds to the point Q.

Any point R between P and Q may thus be defined by settling λ to any value $0 < \lambda < 1$.

If $\lambda = \frac{1}{2}$, $x_R = 1, y_R = 5, z_R = 3$ and the point R is at a distance of

$$\sqrt{(6-1)^2 + (7-5)^2 + (8-3)^2} = \sqrt{29} \text{ from P}$$

10.3.3 Division of a line in given ratio

Any point on the line PQ may be specified in terms of a position vector. If the point R divides the line PQ in the ratio $\lambda:\mu$

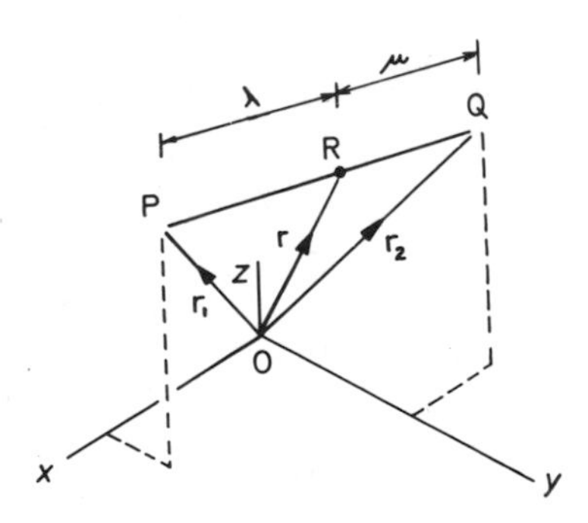

Figure 10.10

$$\overrightarrow{OR} = \overrightarrow{OP} + \frac{PR}{PQ}\overrightarrow{PQ}$$

$$= \mathbf{r}_1 + \frac{\lambda}{\lambda+\mu}(\mathbf{r}_2 - \mathbf{r}_1)$$

$$= \frac{\lambda\mathbf{r}_2 + \mu\mathbf{r}_1}{\lambda+\mu}$$

If $\lambda = \mu$, R is the mid-point, M say, whence the mid-point is specified as follows:

$$\overrightarrow{OM} = \frac{\mathbf{r}_2 + \mathbf{r}_1}{2} = \frac{\mathbf{r}_1 + \mathbf{r}_2}{2}$$

The position vectors $\mathbf{r}_1$ and $\mathbf{r}_2$ may be expressed in terms of Cartesian coordinates as in the preceding section:

$$\overrightarrow{OM} = \tfrac{1}{2}[(x_1 + x_2)\mathbf{i} + (y_1 + y_2)\mathbf{j} + (z_1 + z_2)\mathbf{k}]$$

where

$$\frac{(x_1 + x_2)}{2}, \frac{(y_1 + y_2)}{2}, \frac{(z_1 + z_2)}{2}$$

are the Cartesian coordinates of the mid-point R = M.

The direction cosines of the lines $\overrightarrow{OR}$ and line $\overrightarrow{OM}$ are equal.

$$\frac{(x_2 - x_1)}{|OR|}, \frac{(y_2 - y_1)}{|OR|}, \frac{(z_2 - z_1)}{|OR|}$$

as previously.

10.3.4 Vector equation to the perpendicular distance between two skew lines

If the position and direction of two skew lines P_1R_1, P_2R_2 is defined by position vectors $\mathbf{r}_1$ and $\mathbf{r}_2$ and unit direction vectors $\hat{\mathbf{u}}_1$ and $\hat{\mathbf{u}}_2$ and the orientation is such that Q_1Q_2 in direction $\hat{u}$ is the unknown perpendicular between them. Then with the aid of Fig. 10.11, we may write the vector equation to the line Q_1Q_2 by equating the path OP_1Q_1 to the path $OP_2Q_2Q_1$.

$$\overrightarrow{OQ_1} = \mathbf{r}_1 + P_1Q_1\hat{\mathbf{u}}_1 = \mathbf{r}_2 P_2Q_2\hat{\mathbf{u}}_2 - Q_1Q_2\hat{\mathbf{u}}$$

Since the direction of Q_1Q_2 is unknown the direction of the unit vector $\hat{\mathbf{u}}$ is unknown. We may, however, eliminate $\hat{\mathbf{u}}$ by applying the scalar product rule. We note that $\hat{\mathbf{u}} \cdot \hat{\mathbf{u}} = 1$ and hence we may write

$$\mathbf{r}_1\hat{\mathbf{u}} + P_1Q_1\hat{\mathbf{u}}_1\hat{\mathbf{u}} = \mathbf{r}_2 P_2Q_2\hat{\mathbf{u}}_2\hat{\mathbf{u}} - Q_1Q_2$$

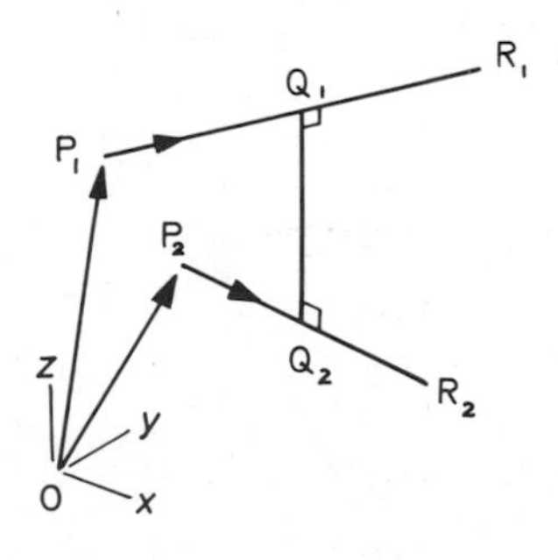

Figure 10.11

However, since the line Q_1Q_2 is perpendicular to both P_1R_1 and P_2R_2, $\hat{\mathbf{u}}_1\hat{\mathbf{u}} = \hat{\mathbf{u}}_2\hat{\mathbf{u}} = 0$, and this means that

$$\overrightarrow{Q_1Q_2} = (\mathbf{r}_2 - \mathbf{r}_1)\,\hat{\mathbf{u}}$$

But $\hat{\mathbf{u}}$ is perpendicular to both $\hat{\mathbf{u}}_1$ and $\hat{\mathbf{u}}_2$ and hence proportional to the vector (cross) product $\hat{\mathbf{u}}_1 \times \hat{\mathbf{u}}_2$. Q_1Q_2 is a length and hence

$$\overrightarrow{Q_1Q_2} = \frac{(\mathbf{r}_2 - \mathbf{r}_1)\cdot(\mathbf{u}_1 \times \mathbf{u}_2)}{|\mathbf{u}_1 \times \mathbf{u}_2|}$$

10.3.5 Vector equation to a plane

If ON = **p** is the perpendicular distance from the origin to a plane and $\hat{\mathbf{u}}$ is a unit vector in the direction $\overrightarrow{ON}$ and P is a point on the plane with position vector **r** making an angle θ with ON, as shown in Fig. 10.12.

Then ON = **p** is the component of $\overrightarrow{OP}$ in the direction $\overrightarrow{ON}$. ON = OP cos θ which means that $\mathbf{p} = \mathbf{r}\hat{\mathbf{u}}$.

$\mathbf{p} = \mathbf{r}\hat{\mathbf{u}}$ is the vector equation to the plane.

Alternatively we can make use of the scalar (dot) product rule, for since the angle ONP is a right angle

$$\overrightarrow{ON}\cdot\overrightarrow{NP} = 0$$

and since

$$\overrightarrow{ON} = \mathbf{p}\hat{\mathbf{u}} \text{ and } \overrightarrow{NP} = \mathbf{r} - \mathbf{p}\hat{\mathbf{u}}$$

$$\mathbf{p}\hat{\mathbf{u}}\cdot(\mathbf{r} - \mathbf{p}\hat{\mathbf{u}}) = 0$$

whence

$$\mathbf{r}\mathbf{p}\hat{\mathbf{u}} - \mathbf{p}^2\hat{\mathbf{u}}^2 = 0$$

but since

$$\hat{\mathbf{u}}\cdot\hat{\mathbf{u}} = 1$$

$$\mathbf{r}\cdot\hat{\mathbf{u}} = \mathbf{p}$$

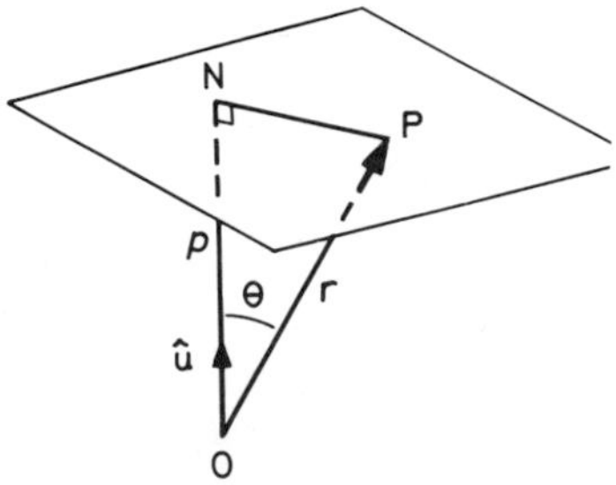

Figure 10.12

N.B. If a plane is defined by $\hat{\mathbf{u}} = (\mathbf{i} - 2\mathbf{j} + 3\mathbf{k})$ and $\mathbf{p} = 4$, the Cartesian equation is $x - 2y + 3z = 4$.

If P(x, y, z) is the point with position vector $\mathbf{r}$ then

$$\mathbf{r} = x\mathbf{i} + y\mathbf{j} + z\mathbf{k}$$

and if the direction cosines of the perpendicular are l, m, n, then

$$\hat{\mathbf{u}} = l\mathbf{i} + m\mathbf{j} + n\mathbf{k}$$

and hence

$$\mathbf{r}\hat{\mathbf{u}} = (x\mathbf{i} + y\mathbf{k} + z\mathbf{k})\,(l\mathbf{i} + m\mathbf{k} + n\mathbf{k}) = \mathbf{p}$$

whence

$$\mathbf{p} = l\mathrm{x} + m\mathrm{y} + n\mathrm{z}$$

which is, as previously explained the Cartesian equation to the perpendicular.

10.3.6 Vector equation to a plane expressed in parametric form

Let the lay of a plane be defined by two non-parallel vectors **d** and **e**, which pass through a given point A, as shown in Fig. 10.13.

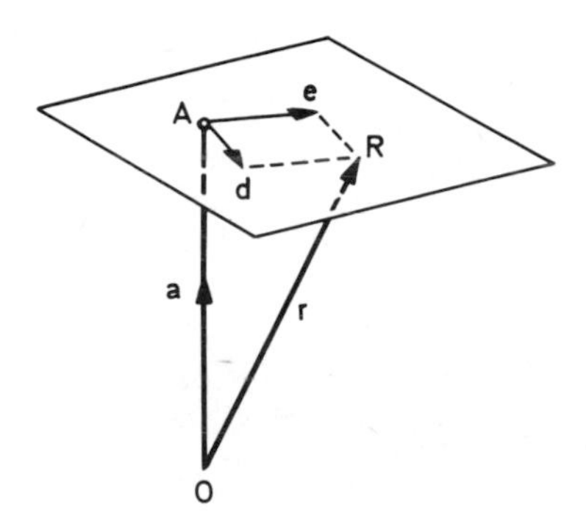

Figure 10.13

If R is any point on the plane

$$\overrightarrow{\mathrm{AR}} = \lambda\mathbf{d} + \mu\mathbf{e}$$

where λ and μ are independent parameters which fix the position of the general point R.

By simple vector addition

$$\mathbf{r} = \mathbf{a} + \lambda\mathbf{d} + \mu\mathbf{e}$$

For instance if A, B and C are three points on a plane then

$$\mathbf{r} = \mathbf{a} + \lambda\overrightarrow{\mathrm{AB}} + \mu\overrightarrow{\mathrm{AC}}$$

$$= \mathbf{a} + \lambda(\mathbf{b} - \mathbf{a}) + \mu(\mathbf{c} - \mathbf{a})$$

whence

$$\mathbf{r} = (1 - \lambda - \mu)\mathbf{a} + \lambda\mathbf{b} + \mu\mathbf{c}$$

or

$$\mathbf{r} = k\mathbf{a} + \lambda\mathbf{b} + \mu\mathbf{c}$$

where

$$k = (1 - \lambda - \mu)$$

whilst the parameters λ and μ may be independently chosen, the parameter k is clearly dependent on the values of λ and μ.

We also observe that since $\mathbf{r} = \mathbf{a} + \lambda\mathbf{d} + \mu\mathbf{e}$ where $\mathbf{a} = \mathbf{r}_0$ is regarded as a base vector from which points in the plane are measured. Then

$$\mathbf{r} = \mathbf{r}_0 + \lambda\mathbf{d} + \mu\mathbf{e}$$

$$x = x_0 + \lambda d_x + \mu e_x$$

$$y = y_0 + \lambda d_y + \mu e_y$$

$$z = z_0 + \lambda d_z + \mu e_z$$

and hence on eliminating λ and μ

$$(\mathbf{r} - \mathbf{r}_0) \cdot (\mathbf{d} \times \mathbf{e}) = 0$$

or

$$(\mathbf{r} - \mathbf{r}_0) \cdot (\mathbf{e}_1 \times \mathbf{e}_2) = 0$$

whichever notation is preferred.

Example

If three vertices A, B, C, define a triangular plate element in a structure and A(2, 2, 7), B(3, 6, 6), C(0, 2, 8), write the equation to the plane containing the plate element (a) in Cartesian form, (b) in vector form.

The position vectors of A, B and C are obtained directly from the given coordinates

$$\overrightarrow{\mathrm{OA}} = 2\mathbf{i} + 2\mathbf{j} + 7\mathbf{k} = \mathbf{a}$$

$$\overrightarrow{\mathrm{OB}} = 3\mathbf{i} + 6\mathbf{j} + 6\mathbf{k} = \mathbf{b}$$

$$\overrightarrow{\mathrm{OC}} = 2\mathbf{j} + 8\mathbf{k} \qquad = \mathbf{c}$$

The direction vectors $\overrightarrow{AB}$ and $\overrightarrow{AC}$ are each obtained by vector addition, that is, we subtract the position vector of A from the position vector of B and similarly subtract the position vector of A from the position vector of C. Thus we obtain:

$$\overrightarrow{AB} = \mathbf{i} + 4\mathbf{j} - \mathbf{k} = \mathbf{d}$$

$$\overrightarrow{AC} = -2\mathbf{i} + \mathbf{k} \quad = \mathbf{e}$$

Now the vector equation to a plane is

$$\mathbf{r} = \mathbf{a} + \lambda\mathbf{d} + \mu\mathbf{e}$$

and on making the appropriate substitutions we obtain

$$\mathbf{r} = 2\mathbf{i} + 2\mathbf{j} + 7\mathbf{k} + \lambda(\mathbf{i} + 4\mathbf{j} - \mathbf{k}) + \mu(-2\mathbf{i} + \mathbf{k})$$

$$= (2 + \lambda - 2\mu)\,\mathbf{i} + (2 + 4\lambda)\,\mathbf{j} + (7 - \lambda + \mu)\,\mathbf{k}$$

whence

$$x = (2 + \lambda - 2\mu), y = (2 + 4\lambda), z = (7 - \lambda + \mu)$$

If we multiply x by 4 and subtract y we eliminate λ. If we multiply z by 4 and subtract y we again eliminate λ. If we then multiply the first equation by 2 we eliminate μ and hence obtain the equation to the plane in Cartesian form.

$$8x - y + 4z = 42$$

which is equivalent to

$$\mathbf{r} \cdot (8\mathbf{i} - \mathbf{j} + 4\mathbf{k}) = 42$$

in vector form.

10.3.7 The angle between a line and a plane

The vector equation to a plane is of the form $\mathbf{r}\hat{\mathbf{n}} = \mathbf{d}$ and that of a line $\mathbf{r} = \mathbf{a} + \lambda\mathbf{b}$.

The projection of **b** onto NM is $|\mathbf{b}|\cos\phi$ and hence application of the scalar product rule yields

$$\cos\phi = \frac{\hat{b} \cdot \hat{n}}{|\mathbf{b}|} = \sin\theta$$

where θ is the acute angle between the line and the plane.

Example

If the equation to a line is $\mathbf{r} = (\mathbf{i} + \mathbf{j} + \mathbf{k}) + \lambda(4\mathbf{i} + 3\mathbf{k})$ and the equation to a plane is $\mathbf{r}\hat{\mathbf{n}} = \mathbf{r}(3\mathbf{i} - 2\mathbf{j} + 6\mathbf{k})$. Determine the true angle between line and plane.

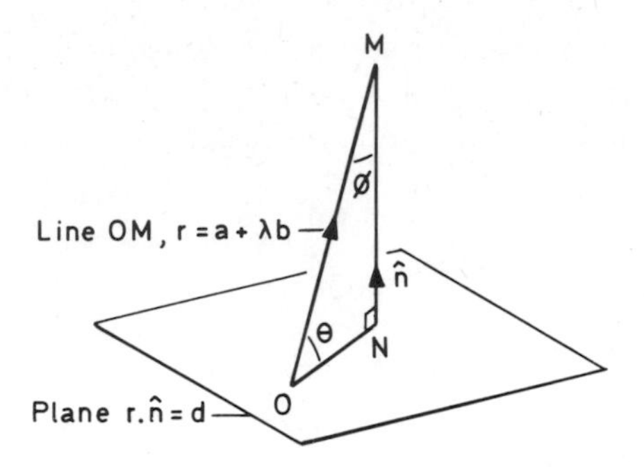

Figure 10.14

Solution

$$\sin\theta = \frac{\hat{b}\hat{n}}{|\mathbf{b}|} = \frac{\hat{b}}{|\mathbf{b}|} \cdot \frac{\hat{n}}{|\mathbf{n}|}$$

$$|\mathbf{b}| = (3^2 + 4^2)^{1/2} = 5 \quad |\mathbf{n}| = (3^2 + 2^2 + 6^2)^{1/2} = 7$$

hence

$$\sin\theta = \frac{(4\mathbf{i} + 3\mathbf{k})}{5} \cdot \frac{(3\mathbf{i} - 2\mathbf{j} + 6\mathbf{k})}{7} = \frac{6}{7}$$

10.3.8 The distance of a point from a plane

If Q is a point the location of which is given by the position vector **q** and π_1 a plane given by the equation $\mathbf{r} \cdot \hat{\mathbf{n}} = d$. Then the equation of a second plane π_2 which passes through the point Q and is parallel to the plane π_1 may be written as follows:

$$\mathbf{r} \cdot \hat{\mathbf{n}} = \mathrm{Q} \cdot \hat{\mathbf{n}}$$

When Q and O are on opposite sides of the plane π_1 (as shown in the figure) the perpendicular distance $\overrightarrow{\mathrm{PN}} = \mathbf{p}$ between the two planes is $\mathbf{a} \cdot \hat{\mathbf{n}} - \mathbf{d}$ and this is the distance from the point Q to the plane π_1. Clearly if the point Q is on the same side as O, equation 10.1, gives a negative result:

$$\mathbf{p} = \mathbf{q} \cdot \hat{\mathbf{n}} - \mathbf{d} \qquad (10.1)$$

Example

If the position vector of a point A is $\mathbf{i} + 2\mathbf{j} + \mathbf{k}$ and the equation to a plane is $\mathbf{r}(3\mathbf{i} - 2\mathbf{j} + 6\mathbf{k}) = 2$. Determine the perpendicular distance between point and plane.

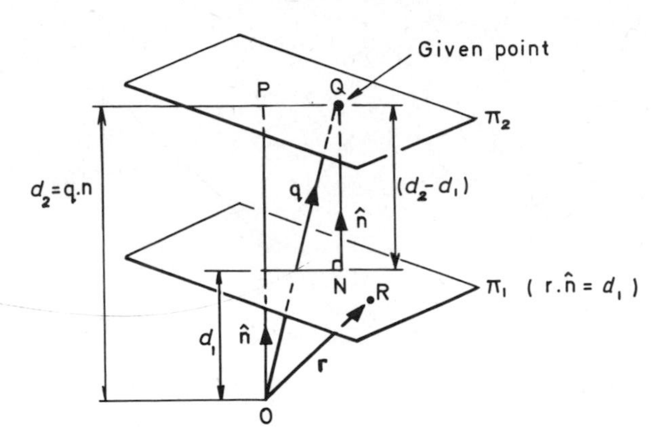

Figure 10.15

Solution

Let the perpendicular distance between point and plane equal p then $\mathbf{p} = \mathbf{a}\hat{\mathbf{n}} - \mathbf{d}$.

The unit normal vector

$$\hat{\mathbf{n}} = \frac{\hat{\mathbf{n}}}{|\mathbf{n}|} = \frac{3\mathbf{i} - 2\mathbf{j} + 6\mathbf{k}}{7} .$$

Hence

$$\mathbf{p} = (\mathbf{i} + 2\mathbf{j} + \mathbf{k})\ \frac{(3\mathbf{i} - 2\mathbf{j} + 6\mathbf{k})}{7} - \frac{2}{7}$$

and since $\mathbf{i} \cdot \mathbf{i} = \mathbf{j} \cdot \mathbf{j} = \mathbf{k} \cdot \mathbf{k}$ and mixed dot products of $\mathbf{i}, \mathbf{j}, \mathbf{k} = 0$

$$p = \frac{3 - 4 + 6}{7} - \frac{2}{7} = \frac{3}{7}$$

10.3.9 The angle between two planes

If two planes are defined by the vector equations

$$\mathbf{r} \cdot \hat{\mathbf{n}}_1 = \mathbf{d}_1 \quad \text{and} \quad \mathbf{r} \cdot \hat{\mathbf{n}}_2 = \mathbf{d}_2$$

the angle between the two planes may be determined by calculating the angle between their two normals.

The cosine of the angle between two lines is obtained directly from the definition of a scalar product.

If $\hat{\mathbf{n}}_1$ and $\hat{\mathbf{n}}_2$ are unit vectors along the normals and θ the angle between them

$$\hat{\mathbf{n}}_1 \cdot \hat{\mathbf{n}}_2 = \cos\theta$$

(N.B. The two planes are perpendicular when $\cos\theta = 0$, that is, when the scalar product $\hat{\mathbf{n}}_1 \cdot \hat{\mathbf{n}}_2 = 0$ and similarly if $\hat{\mathbf{n}}_1 = \hat{\mathbf{n}}_2$ the scalar product $\hat{\mathbf{n}}_1 \cdot \hat{\mathbf{n}}_2 = 1$ and hence the two planes are parallel.

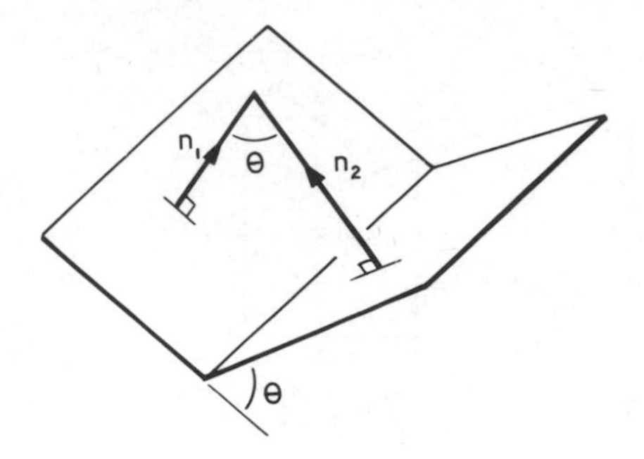

Figure 10.16

10.3.10 The line of intersection between two planes

If two planes π_1 and π_2 are defined by the vector equations

$$\mathbf{r} \cdot \hat{\mathbf{n}}_1 = d_1 \quad \text{and} \quad \mathbf{r} \cdot \hat{\mathbf{n}}_2 = d_2$$

then their line of intersection is the one and only line common to both planes π_1 and π_2.

The line of intersection is moreover perpendicular to both $\hat{\mathbf{n}}_1$ and $\hat{\mathbf{n}}_2$ where $\hat{\mathbf{n}}_1$ and $\hat{\mathbf{n}}_2$ are the unit vectors defining the direction of each normal respectively.

If the line of intersection is normal to both $\hat{\mathbf{n}}_1$ and $\hat{\mathbf{n}}_2$ then we know from the (cross) product rule that the line of intersection is parallel to $\mathbf{n}_1 \times \mathbf{n}_2$.

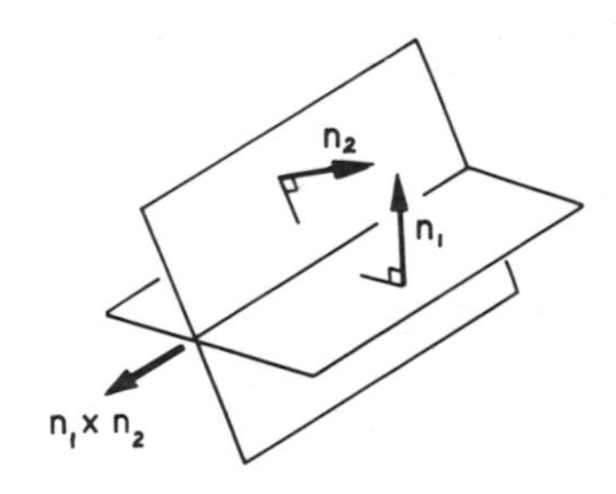

Figure 10.17

10.3.11 Equation to a plane through intersection of two other planes

If the vector equations of two intersecting planes are $\mathbf{r}\hat{\mathbf{n}}_1 = \mathbf{d}_1$ and $\mathbf{r}\hat{\mathbf{n}}_2 = \mathbf{d}_2$ then their line of intersection is the one and only line common to both π_1 and π_2. If a is the position vector of a point on the common line then

$$\mathbf{a} \cdot \hat{\mathbf{n}}_1 = \mathbf{d}_1 \quad \text{and} \quad \mathbf{a} \cdot \hat{\mathbf{n}}_2 = \mathbf{d}_2$$

whence

$$\mathbf{a} \cdot \hat{\mathbf{n}}_1 - ka \cdot \hat{\mathbf{n}}_2 = \mathbf{d}_1 - k\mathbf{d}_2$$

$$\mathbf{a}(\hat{\mathbf{n}}_1 - k\hat{\mathbf{n}}_2) = \mathbf{d}_1 - k\mathbf{d}_2$$

If we compare this equation with that of article 10.3.10 we see that they are of identical form, and hence if a is replaced by r, and r satisfies the equations of both π_1 and π_2, then r is the general point on a third plane π_3 which passes through the line of intersection of planes π_1 and π_2, as shown in Fig. 10.18. The equation to the third plane is

$$\mathbf{r}(\hat{\mathbf{n}}_1 - k\hat{\mathbf{n}}_2) = \mathbf{d}_1 - k\mathbf{d}_2$$

where k is a constant which sets the inclination of π_3 in relation to planes π_1 and π_2.

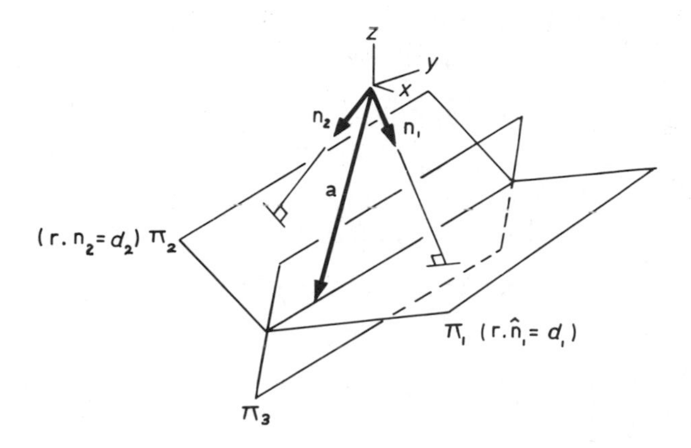

Figure 10.18

10.4 VECTOR EQUATIONS TO PLANE CURVES

The methods of 10.3 are directly applicable to the description of plane curves and we consider the circle and ellipse as representative examples.

10.4.1 The vector equation to a circle

The Cartesian equation to a circle with centre at the origin is $x^2 + y^2 = a^2$ and if $\mathbf{r}$ is the position vector of a point P on the circumference then $\mathbf{a} = |\mathbf{r}|$. The projection of OP = $\mathbf{r}$ onto ox yields the component $a \cos \theta\, i$ and the projection of OP = $\mathbf{r}$ onto oy yields the component $\mathbf{a} \sin \theta\, \mathbf{j}$.

Whence the vector equation to a circle is

$$\mathbf{r} = a \cos \theta\, \mathbf{i} + a \sin \theta\, \mathbf{j}$$

where (r, θ) is as defined in Fig. 10.19(a).

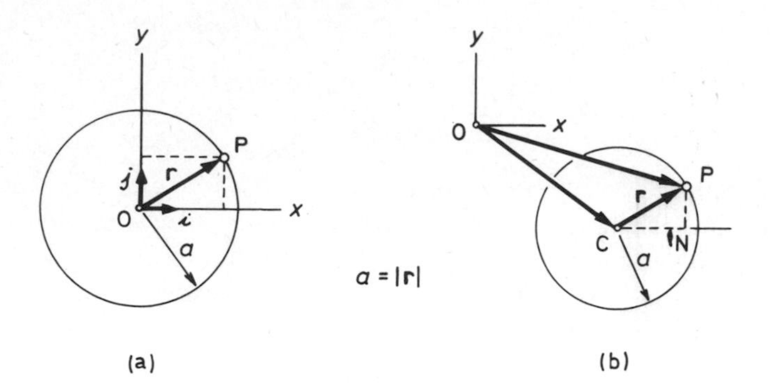

Figure 10.19

If the circle centre is not sited at the origin then the vector equation is derived as follows. With reference to Fig. 10.19(b)

$$\mathbf{r} = \overrightarrow{OC} + \overrightarrow{CP} = \overrightarrow{OC} + \overrightarrow{CN} + \overrightarrow{NP}$$

$\overrightarrow{OC}$ is simply the position vector of the circle centre C and $\overrightarrow{CN} + \overrightarrow{NP}$ are the rectangular components of the circle as previously given.

If the circle centre is at (x_c, y_c) then the positive vector is $x_c\mathbf{i} + y_c\mathbf{j}$ and

$$\mathbf{r} = x_c\mathbf{i} + y_c\mathbf{j} + a \cos\theta\ \mathbf{i} + b \sin\theta\ \mathbf{j}$$

Example

If the centre of a circle of radius a is sited at $x_c = 3a$, $y_c = 4a$, write the vector equation to a general point P.

$$\mathbf{r} = 3a\mathbf{i} + 4a\mathbf{j} + a \cos\theta\ \mathbf{i} + a \sin\theta\ \mathbf{j}$$

$$\mathbf{r} = a\ [(3 + \cos\theta)\,\mathbf{i} + (4 + \sin\theta)\,\mathbf{j}]$$

10.4.2 The vector equation to an ellipse

The Cartesian equation to an ellipse is $x^2/a^2 + y^2/b^2 = 1$ and the parametric equation is $x = a \cos\theta$, $y = b \sin\theta$. Hence the vector equation to an on-centre ellipse is

$$\mathbf{r} = a \cos\theta\ \mathbf{i} + b \sin\theta\ \mathbf{j}$$

An ellipse with major axis rotated through a positive or negative angle ψ may similarly be defined, whence

$$\mathbf{r} = a \cos(\theta \mp \psi)\ \mathbf{i} + b \sin(\theta \mp \psi)\ \mathbf{j}$$

and if the centre of the ellipse is displaced from the zero position, then we include its position vector as a vectorially additive term.

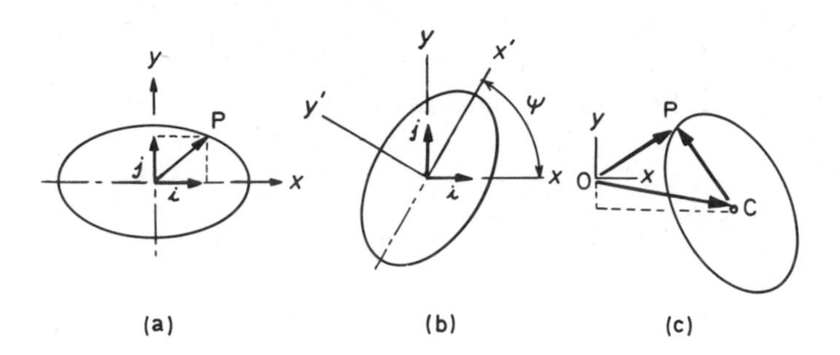

Figure 10.20

If the centre of the ellipse is at (x_c, y_c) then

$$\mathbf{r} = x_c\mathbf{i} + y_c\mathbf{j} + a\cos(\theta \mp \psi) + b\sin(\theta \mp \psi)$$

10.5 SPACE CURVES AND SURFACES

Consideration of the easily recognised properties of a sphere provides an easy way into the study of differential geometry and piecewise shape design, but before doing so, we will consider the vectorial properties of the circular cylinder and helix.

10.5.1 Vector equation to a point on the surface of a circular cylinder

We have seen that the vector equation to a circle with centre coincident with the origin is of the form

$$\mathbf{r} = \mathbf{r}(\theta) = a\cos\theta\,\mathbf{i} + a\sin\theta\,\mathbf{j}$$

and we can translate any general point P into the third dimension and thus make a cylinder, simply by including the coordinate z

$$\mathbf{r} = \mathbf{r}(\theta, z) = (a\cos\theta, a\sin\theta, z)$$

is thus a circular cylinder with axis aligned with oz and similarly

$$\mathbf{r} = \mathbf{r}(\theta, y) = (a\cos\theta, y, a\sin\theta)$$

and

$$\mathbf{r} = \mathbf{r}(\theta, x) = (x, a\cos\theta, a\sin\theta)$$

are the equations of circular cylinders with axes aligned with oy and ox respectively.

Equations to cylinders not centred on zero may be written by including the centre line off-sets, and the coordinates of cylinders inclined to the reference

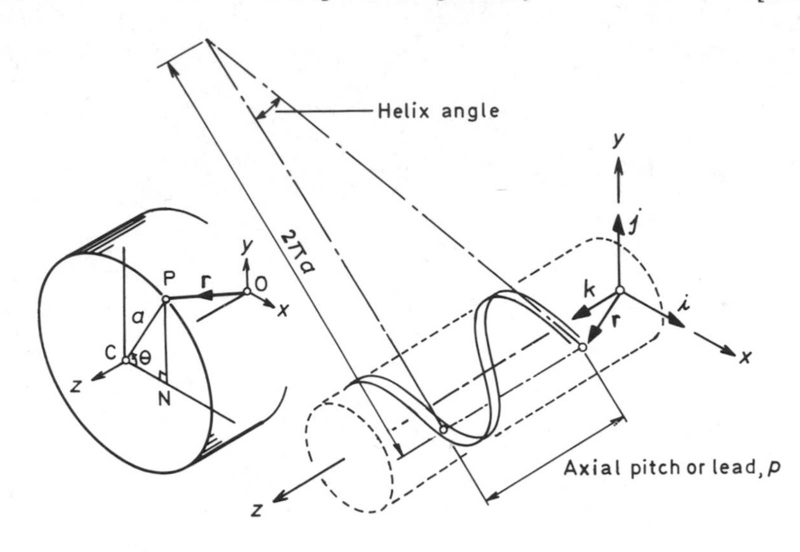

Figure 10.21

axes may be obtained by defining the lay on the axial centre line, using the appropriate transformation, see Chapter 5.

10.5.2 Vector equation to a circular helix

The circular helix is perhaps the best known space curve and no introduction to its geometrical appearance is needed. Suffice to say we consider the constant pitch single start thread in this case.

If the line AP, Fig. 10.21, makes a constant angle $(90 - \phi) = \alpha = \arctan P/\pi D$ with a normal section and θ is the angular rotation relative to the xoy plane.

Then

$$CP = a,\ CN = a\cos\theta,\ NP = a\sin\theta$$

Since the helical curve advances a distance p in the axial (z direction) a rotation θ produces an axial advance of $p\theta/2\pi = OC$.

Now the vector $\overrightarrow{CN}$ is parallel to ox, the vector $\overrightarrow{NP}$ is parallel to oy and the vector $\overrightarrow{OC}$ is along oz, and hence

$$\overrightarrow{CN} = a\cos\theta\ \mathbf{i},\ \overrightarrow{NP} = a\sin\theta\ \mathbf{j},\ OC = \frac{p\theta}{2\pi}\ \mathbf{k}$$

and since

$$\overrightarrow{OP} = \overrightarrow{OC} + \overrightarrow{CN} + \overrightarrow{NP}$$

$$\mathbf{r} = a\cos\theta\ \mathbf{i} + a\sin\theta\ \mathbf{j} + \frac{p\theta\mathbf{k}}{2\pi}$$

and this is the vector equation to the helix which advances in the positive direction of oz and is synonymous with a right handed screw. The equation to the helix which recedes in the negative direction of oz is evidently

$$\mathbf{r} = a \cos \theta \, \mathbf{i} + a \sin \theta \, \mathbf{j} + \frac{-p\theta \mathbf{k}}{2\pi}$$

and this is synonymous with a left handed screw.

Care must therefore be taken to observe that positive rotations (angles) are measured from ox to oy, from oy to oz, and oz to ox, and hence the all, sine, tan, cos convention must be strictly observed.

(N.B. A right handed screw turned anticlockwise advances towards the eye and conversely. A left handed screw turned anticlockwise advances away from the eye and conversely.)

10.5.3 The vector sphere

The line of intersection produced by a plane which 'cuts' a sphere is always a circle. When the plane of intersection passes through the north/south pole points the section is a great circle and is said to be a meridian of longitude. When the plane of intersection lies at right angles to the north/south axis it is a small circle except at the equator and is said to be a parallel of latitude.

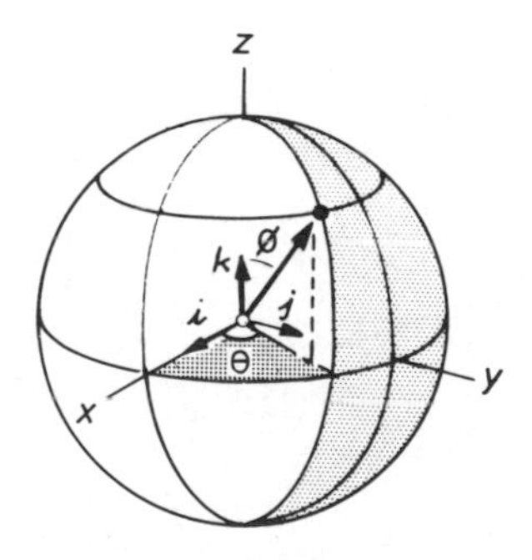

Figure 10.22

If central axes ox, oy, oz are introduced as in Fig. 10.22, then θ and ϕ are the angles in a spherical coordinate system in which

$$x = \cos \phi \sin \theta, \; y = \sin \phi \sin \theta, \; z = \cos \theta$$

Planes containing the meridians of longitude are defined by the equation $y = mx$, where $x = \tan \phi = \sin \phi / \cos \phi$ whence the equation which defines all longitudinal meridians is

$$y \cos \phi - x \sin \phi = 0$$

All parallels of latitude are defined by the equation

$$z = r \cos \theta = c$$

where c is the intercept of the plane on the oz axis.

The line of intersection between any meridian is a circle and since we are here considering a unit sphere, the equation to the line of meridian intersection is obtained by solving $y \cos \phi - x \sin \phi$ and $x^2 + y^2 + z^2 = 1$, simultaneously. Similarly, the line of intersection between any parallels of latitude and the sphere is a circle, the equation to which is obtained by solving the equations $z \cos \theta = c$ and $x^2 + y^2 + z^2 = 1$, simultaneously.

An element of spherical surface is clearly defined by any two meridians of longitude and any two parallels of latitude. If we consider a full quadrant in ϕ and take $\theta = (60°, 90°)$ as in Fig. 10.23, then the four boundary curves are $x^2 + y^2 = 1, y^2 + z^2 = 1, x^2 + z^2 = 1, x^2 + y^2 + c^2 = 1$, where c equals $r \cos \theta = 0.5$. The equation to the small bounding circle through P and Q is thus $x^2 + y^2 = \frac{3}{4}$.

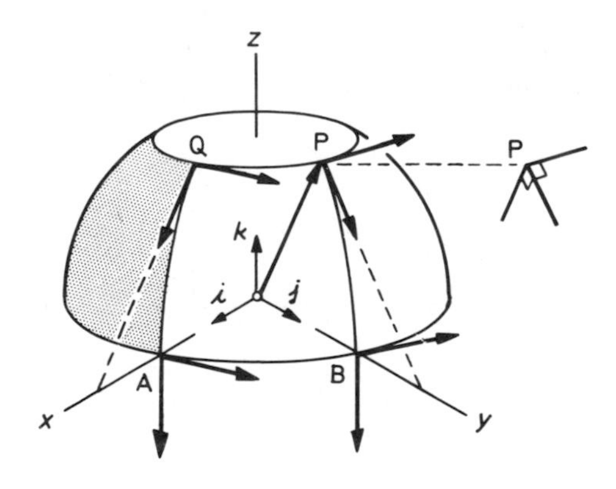

Figure 10.23

Now if, as in the present instance, the analytic form of the surface is known then it is clearly possible to define the surface in terms of its four 'corner' coordinates A, B, P, Q, and eight corner tangents. The coordinates of A, B, P, Q, are of course calculated in the usual way and the tangents found by differentiation.

The canonical equation to a sphere $x^2 + y^2 + z^2 = 1$, may be expressed in parametric form using the spherical coordinate system.

$$\mathbf{T}_\mathrm{N} = \mathbf{r} = \sin \theta \cos \phi \, \mathbf{i} + \sin \theta \sin \phi \, \mathbf{j} + \cos \theta \, \mathbf{k}$$

where *i, j, k*, are unit vectors used to define the *ox, oy, oz* directions.

$$\mathbf{T}_\phi = \frac{\partial \hat{r}}{\partial \phi} = -\sin\theta \sin\phi\,\mathbf{i} + \sin\theta\cos\phi\,\mathbf{j}$$

$$\mathbf{T}_\theta = \frac{\partial \hat{r}}{\partial \theta} = \cos\theta\cos\phi\,\mathbf{i} + \cos\theta\sin\phi\,\mathbf{j} - \sin\theta\,\mathbf{k}$$

The magnitude of the vector $\partial r/\partial\phi = |\mathbf{T}_\phi|$ where $|\mathbf{T}_\phi|$ is the square root of the sum of the squares.

$$|\mathbf{T}_\phi| = [\sin^2\theta\sin^2\phi + \sin^2\theta\cos^2\phi]^{1/2} = \sin\theta$$

and since the unit tangent $\hat{\mathbf{T}}_\phi = \mathbf{T}_\phi/|\mathbf{T}_\phi|$

$$\hat{\mathbf{T}}_\phi = -\sin\phi\,\mathbf{i} + \cos\phi\,\mathbf{j}$$

The magnitude of the vector $\partial\hat{r}/\partial\theta = |\mathbf{T}_\theta|$, where $|\mathbf{T}_\theta|$ is the square root of the sum of the squares

$$|\mathbf{T}_\theta| = [\cos^2\theta\cos^2\phi + \cos^2\theta\sin^2\phi + \sin^2\theta]^{1/2} = 1$$

whence

$$\hat{\mathbf{T}}_\theta = \cos\theta\cos\phi\,\mathbf{i} + \cos\theta\sin\phi\,\mathbf{j} - \sin\theta\,\mathbf{k}$$

(N.B. We may confirm that $\hat{\mathbf{T}}_\phi$ and $\hat{\mathbf{T}}_\theta$ are unit tangents by taking the Pythagoran sum of their respective components).

The position of A, Fig. 10.23, is $(0, \pi/2)$ and the unit tangents at A are therefore given by $\partial\hat{r}/\partial\phi = \mathbf{i}$, $\partial\hat{r}/\partial\theta = -\mathbf{k}$, and similarly the position of P is $(\pi/2, \pi/3)$ and the unit tangents at P are $\partial/\partial\theta = (\sqrt{3}/2)\mathbf{i}$, $\partial/\partial\phi = \frac{1}{2}\mathbf{j} - (\sqrt{3}/2)\mathbf{k}$ and other tangents are as tabulated.

	$\partial\hat{\mathbf{r}}/\partial\phi$	$\partial\hat{\mathbf{r}}/\partial\theta$	$\lvert\partial\hat{\mathbf{r}}/\partial\phi\rvert$	$\lvert\partial\hat{\mathbf{r}}/\partial\theta\rvert$
A	$\mathbf{i}$	$-\mathbf{k}$	1	1
B	$-\mathbf{i}$	$-\mathbf{k}$	1	1
P	$-\frac{\sqrt{3}}{2}\mathbf{i}$	$\frac{1}{2}\mathbf{j} - \frac{\sqrt{3}}{2}\mathbf{k}$	$-\frac{\sqrt{3}}{2}$	1
Q	$\frac{\sqrt{3}}{2}\mathbf{j}$	$\frac{1}{2}\mathbf{j} - \frac{\sqrt{3}}{2}\mathbf{k}$	$\frac{\sqrt{3}}{2}$	1

In certain cases we may need to specify surface normals and these are expressed as follows.

Now since $\mathbf{r} = x\mathbf{i} + y\mathbf{k} + z\mathbf{k}$

$$\frac{\partial \hat{r}}{\partial \theta} = \cos\theta \cos\phi \, \mathbf{i} + \cos\theta \sin\phi \, \mathbf{j} - \sin\theta \, \mathbf{k}$$

$$\frac{\partial \hat{r}}{\partial \phi} = -\sin\theta \sin\phi \, \mathbf{i} + \sin\theta \cos\phi \, \mathbf{j}$$

and since both $\partial \mathbf{r}/\partial\theta$ and $\partial \mathbf{r}/\partial\phi$ are mutually perpendicular tangents the normal to the surface is given by the vector (cross) product of the two tangents $\partial \mathbf{r}/\partial\theta$, $\partial \mathbf{r}/\partial\phi$.

Now bearing in mind the importance of order (sequence) in a vector product we multiply $\partial \mathbf{r}/\partial\phi$ by $\partial \mathbf{r}/\partial\theta$ in accord with the expression

$$\mathbf{T_N} = \frac{\partial \hat{r}}{\partial \phi} \times \frac{\partial \hat{r}}{\partial \theta} = \mathbf{N} \text{ say}$$

whence

$$\begin{aligned} \mathbf{N} = &-(\sin\theta \cos\theta \cos\phi)\, \mathbf{i} \times \mathbf{i} - (\sin\theta \cos\theta \sin^2\phi)\, \mathbf{i} \times \mathbf{j} \\ &+ (\sin^2\theta \sin\phi)\, \mathbf{i} \times \mathbf{k} + (\sin\theta \cos\theta \cos^2\phi)\, \mathbf{j} \times \mathbf{i} \\ &+ (\sin\theta \cos\theta \sin\phi \cos\phi)\, \mathbf{j} \times \mathbf{j} - (\sin^2\theta \cos\phi)\, \mathbf{j} \times \mathbf{k} \end{aligned}$$

and since $\mathbf{i} \times \mathbf{i} = \mathbf{j} \times \mathbf{j} = 0$ and $\mathbf{i} \times \mathbf{j} = \mathbf{k}$, $\mathbf{i} \times \mathbf{k} = -\mathbf{j}$, $\mathbf{j} \times \mathbf{i} = -\mathbf{k}$, $\mathbf{j} \times \mathbf{k} = \mathbf{i}$, the above expression reduces to

$$\mathbf{N} = -(\sin^2\theta \cos\phi)\, \mathbf{i} - (\sin^2\theta \sin\phi)\, \mathbf{j} - (\sin\theta \cos\theta)\, \mathbf{k}$$

Now the magnitude of the normal vector $|\mathbf{N}|$ is the square root of the sum of squares of its components. Hence

$$\begin{aligned} |\mathbf{N}| &= [\sin^2\theta \cos^2\phi + \sin^4\theta \sin^2\phi + \sin^2\theta \cos^2\theta]^{1/2} \\ &= [\sin^2\theta + \sin^2\theta \cos^2\theta]^{1/2} = \sin\theta \end{aligned}$$

and since the unit normal is equal to $\hat{\mathbf{N}} = \mathbf{N}/|\mathbf{N}|$

$$\hat{\mathbf{N}} = -(\sin\theta \cos\phi)\, \mathbf{i} - (\sin\theta \sin\phi)\, \mathbf{j} - \cos\theta \, \mathbf{k}$$

(N.B. The components $(\sin\theta \cos\phi)$, $(\sin\theta \sin\phi)$, $(\cos\theta)$ are identically the same as those in the spherical coordinate system, they are the components of a point (ϕ, θ) on a unit sphere to which $\hat{\mathbf{N}}$ is the inward unit normal).

A twist vector may also be defined by taking the differential $\partial\hat{r}/\partial\phi$ with respect to θ and the differential of $\partial\hat{r}/\partial\theta$ with respect to ϕ

$$\frac{\partial \hat{r}}{\partial \phi} = -(\sin\theta \sin\phi)\, \mathbf{i} + \sin\theta \cos\phi)\, \mathbf{j}$$

$$\frac{\partial \hat{r}}{\partial \theta} = (\cos\theta \cos\phi)\, \mathbf{i} + (\cos\theta \sin\phi)\, \mathbf{j} - (\sin\theta)\, \mathbf{k}$$

Now if we differentiate the first equation with respect to θ we obtain

$$\frac{\partial}{\partial\theta}\frac{\partial\hat{r}}{\partial\phi} = -(\cos\theta\sin\phi)\,\mathbf{i} + (\cos\theta\cos\phi)\,\mathbf{j}$$

and if we differentiate the second equation with respect to ϕ we obtain

$$\frac{\partial}{\partial\phi}\frac{\partial\hat{r}}{\partial\theta} = -(\cos\theta\sin\phi)\,\mathbf{i} + (\cos\theta\cos\phi)\,\mathbf{j}$$

and we see that $\partial/\partial\theta\ \partial\hat{r}/\partial\phi = \partial/\partial\phi\ \partial\hat{r}/\partial\theta$ and it matters not which of the two we take.

The magnitude of the twist vector is given by the square root of the sum of the squares of its components, hence

$$|\mathbf{V}| = [\cos^2\theta\sin^2\phi + \cos^2\theta\cos^2\phi]^{1/2} = \cos\theta$$

The unit twist vector is $\hat{V}/|\mathbf{V}|$ and

$$\frac{\hat{V}}{|\mathbf{V}|} = -\frac{(\cos\theta\sin\phi)\,\mathbf{i} + (\cos\theta\cos\phi)\,\mathbf{j}}{\cos\theta}$$

$$= -\sin\phi\,\mathbf{i} + \cos\phi\,\mathbf{j} = \hat{\mathbf{V}}$$

The twist vectors for the four points A, B, P, Q, on our unit sphere, Fig. 10.23, are obtained by substituting the relevant coordinates.

The coordinates of A are $(0, \pi/2)$ and $\cos\theta = 0$ and since

$$\frac{\partial}{\partial\theta}\cdot\frac{\partial\hat{r}}{\partial\phi} = -(\cos\theta\sin\phi)\,\mathbf{i} + (\cos\theta\cos\phi)\,\mathbf{j} = 0$$

the twist of A is

$$\left(\frac{\partial}{\partial\theta}\frac{\partial\hat{r}}{\partial\phi}\right)_A = \left(\frac{\partial}{\partial\phi}\frac{\partial\hat{r}}{\partial\theta}\right)_A = 0$$

Similarly the coordinates of P are $(\pi/2, \pi/3)$, hence

$$\left(\frac{\partial}{\partial\theta}\frac{\partial\hat{r}}{\partial\phi}\right)_P = -\frac{1}{2}\,\mathbf{i}$$

and the coordinates of Q are $(0, \pi/3)$, hence

$$\left(\frac{\partial}{\partial\theta}\frac{\partial\hat{r}}{\partial\phi}\right)_Q = +\frac{1}{2}\,\mathbf{j}$$

The four twist vectors at ABPQ are

	$\dfrac{\partial}{\partial\theta}\dfrac{\partial\hat{r}}{\partial\phi}$	$\left\lvert\dfrac{\partial}{\partial\theta}\dfrac{\partial\hat{r}}{\partial\phi}\right\rvert$
A	0	0
B	0	0
P	$-\dfrac{1}{2}\mathbf{i}$	$-\dfrac{1}{2}$
Q	$+\dfrac{1}{2}\mathbf{j}$	$\dfrac{1}{2}$

(N.B. The unit vectors $\hat{\mathbf{T}}_\theta$, $\mathbf{T}_\phi$, $\mathbf{N}$ more commonly denoted by $\hat{\mathbf{B}}$, $\hat{\mathbf{T}}$, $\hat{\mathbf{N}}$ in that order form as right handed (orthonormal) triad of axes.)

10.6 DIFFERENTIAL GEOMETRY

If a space curve is defined by the parametric equations $\mathbf{x} = x(u)$, $\mathbf{y} = y(u)$, $\mathbf{z} = z(u)$ then a small element of length dl has Cartesian components $dx, dy, dz,$ whence

$$dl^2 = dx^2 + dy^2 + dz^2 = \left\{\left(\frac{dx}{du}\right)^2 + \left(\frac{dy}{du}\right)^2 + \left(\frac{dz}{du}\right)^2\right\}du$$

and the arc length s from u_0 to u is given by the intregal

$$s = \int_0^l ds = \int_{u_0}^u \left\{\left(\frac{dx}{du}\right)^2 + \left(\frac{dy}{du}\right)^2 + \left(\frac{dz}{du}\right)^2\right\}^{1/2} du$$

If this equation can be solved for u in terms of s so that $\dot{u} = \mathrm{F}(l)$ then the parametric equations

$$\mathbf{x} = x(u),\ y = y(u),\ z = z(u)$$

may be replaced by the parametric equations

$$\mathbf{x} = x(s),\ \mathbf{y} = y(s),\ \mathbf{z} = z(s)$$

The necessary and sufficient condition which ensures that u is indeed the arc length s is that

$$\left(\frac{dx}{du}\right)^2 + \left(\frac{dy}{du}\right)^2 + \left(\frac{dz}{du}\right)^2 = 1$$

and if this is true

$$\frac{dx}{du} = \frac{dx}{ds}, \frac{dy}{du} = \frac{dy}{ds}, \frac{dz}{du} = \frac{dz}{ds}$$

and these are the direction ratios of the space curve at the point considered.

10.6.1 The unit tangent vector

If the position vector of a point P is **r**, where P is some distance s from A along the curve as shown in Fig. 10.24.

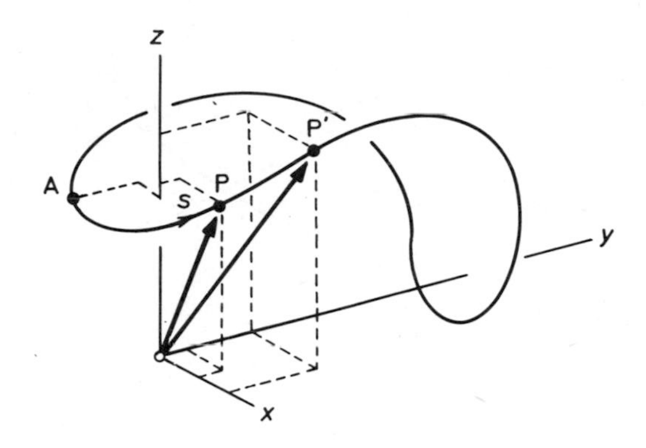

Figure 10.24

Then the position vector of a point P′ close to P is $\mathbf{r} + \delta\mathbf{r}$ and hence as P′ approaches P, $\delta\mathbf{r}/\delta s$ becomes $d\mathbf{r}/ds$ in the limit.

The vector equation to any point on the curve is of the form

$$\mathbf{r} = x\mathbf{i} + y\mathbf{j} + z\mathbf{k}$$

and

$$\frac{d\hat{r}}{ds} = \frac{dx}{ds}\mathbf{i} + \frac{dy}{ds}\mathbf{j} + \frac{dz}{ds}\mathbf{k} = \mathbf{T} \text{ say}$$

and since

$$\left(\frac{dx}{ds}\right)^2 + \left(\frac{dy}{ds}\right)^2 + \left(\frac{dz}{ds}\right)^2 = 1$$

T is a unit vector, which is tangential to the curve.

$\mathbf{T} = d\mathbf{r}/ds = \dot{\mathbf{r}}$ is called the unit tangent.

The equation to the line through the point P is thus

$$\mathbf{r} = \mathbf{r}_1 + k\mathbf{T} = \mathbf{r}_1 + k\dot{\mathbf{r}}$$

where k is a variable scalar (positive or negative) and **r** is the position vector of a general point on the line.

Since $\dot{\mathbf{r}}$ is a unit vector the scalar product $\dot{\mathbf{r}} \cdot \dot{\mathbf{r}} = 1$ and its differential $\dot{\mathbf{r}} \cdot \ddot{\mathbf{r}} = 0$. The vector $\ddot{\mathbf{r}}$ is thus perpendicular to the vector $\dot{\mathbf{r}}$.

The plane through P which is perpendicular to the tangent to the curve at P is called the normal plane. The equation of the normal plane is $(\mathbf{r} - \mathbf{r}_1) \cdot \mathbf{T} = 0$.

As shown in Fig. 10.25, there are any number of lines in the normal plane which are perpendicular to the unit tangent **T**. One such line lies in the direction of the unit vector **N**.

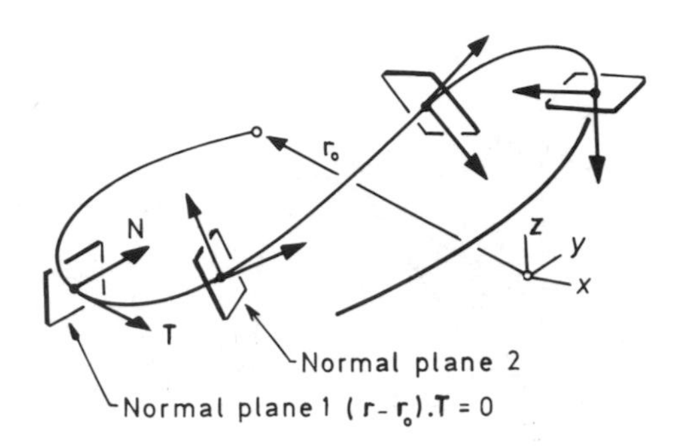

Figure 10.25

10.6.2 The unit normal vector

In a plane two dimensional curve the direction of **N** is along the radius of the osculating circle, see article 8.6.4, whence the plane containing **T** and **N** is termed the osculating plane. For a plane curve the osculating plane is clearly perpendicular to the normal plane previously defined.

In a three dimensional curve the vector **N** likewise points towards the instantaneous centre of curvature and we write

$$\frac{d\hat{T}}{ds} = \mathrm{K}\mathbf{N} = \dot{\mathbf{T}} = \ddot{\mathbf{r}}$$

where k is a positive scalar of magnitude

$$\mathrm{K} = \left| \frac{d\hat{T}}{ds} \right|$$

Since arc $= \theta^\circ \mathbf{r}$ the arc rate of turning $d\theta/ds$ of the tangent at P is called the curvature and we see from Fig. 10.26, that in the limit

$$\mathrm{K} = \left| \frac{d\hat{T}}{ds} \right| = \frac{d\theta}{ds}$$

and this indicates that K is the reciprocal of the radius of curvature ρ. The vector $\overrightarrow{PC} = \rho N$ and since $K = 1/\rho$

$$\dot{T} = \ddot{r} = \frac{1}{\rho} N = KN$$

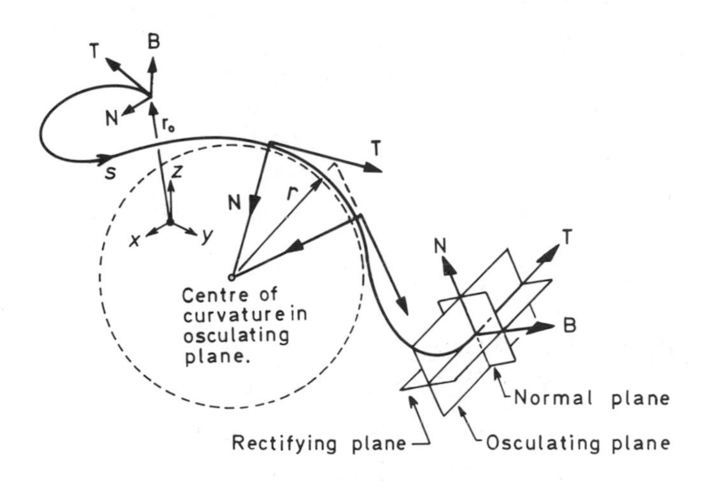

Figure 10.26

10.6.3 The unit binormal vector

The other important line through P is the so-called binormal **B**. Now as we have proved for a sphere $\hat{\mathbf{B}} = \hat{\mathbf{T}} \times \hat{\mathbf{N}}$ and this equality indicates that $\hat{\mathbf{B}}$, $\hat{\mathbf{T}}$ and $\hat{\mathbf{N}}$ form a right-handed triad as previously explained.

10.6.4 The osculating, normal and rectifying planes

An illustration of the importance of tangent normal and binormal planes is provided by consideration of a simple problem in automated manufacture, for which a system such as APT (Automatically programmed tools) is used.

Now, whilst it is possible to machine a $2\frac{1}{2}$D (cylindrical object) such as the parabolic wing, Fig. 10.27, using a vertical bull-nosed cutter and making longitudinal passes, the machining of truly 3D shapes is, in many ways, better served by allowing the cutter to align with the surface normal. Suffice to point out that not all machines have this facility and the user then has no option but to resolve the problem of cutter offset by other means. (N.B. The centre line tip of the cutter and the cutting point are not, in the latter case, one and the same.)

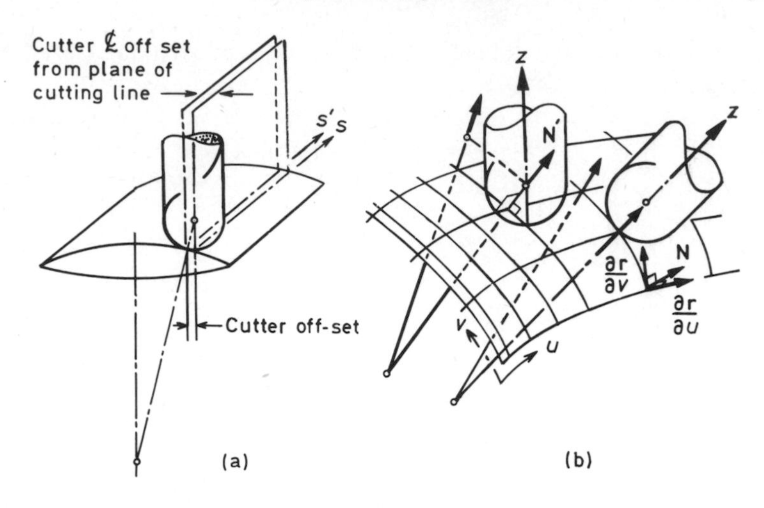

Figure 10.27

10.6.5 The Frenet Serret formulae

Since the three unit vectors **B**, **T** and **N** form an orthonormal right-handed triad

$$\mathbf{B} = \mathbf{T} \times \mathbf{N}, \mathbf{T} = \mathbf{N} \times \mathbf{B}, \mathbf{N} = \mathbf{B} \times \mathbf{T}$$

and since these are all vector products the order of multiplication is important.

In a plane curve the binormal **B** does not change direction with increasing or decreasing arc length s and for a plane curve $d\mathbf{B}/ds = 0$. The binormal of a curve which occupies three dimensions has, however, a non-zero rate of change with respect to s, and since $\mathbf{B} \cdot \mathbf{T} = 0$, the differential of $\mathbf{B} \cdot \mathbf{T} = 0$

$$\frac{d\hat{B}}{ds} \cdot \mathbf{T} + \mathbf{B} \cdot \frac{d\hat{T}}{ds} = 0$$

Now we know that $\dot{\mathbf{T}} = d\hat{T}/ds = \mathrm{K}\mathbf{N}$, therefore $\mathbf{B} \cdot d\hat{T}/ds = \mathrm{K}\mathbf{B} : \mathbf{N}$ and since $\mathbf{B} \cdot \mathbf{N} = 0$, $d\hat{B}/ds \cdot \mathbf{T} = 0$ and since **B** is a unit vector $\mathbf{B} \cdot d\hat{B}/ds = 0$. It follows therefore that $d\hat{B}/ds$ is perpendicular to both **T** and **B** and since $d\hat{B}/ds$ is in the direction of **N** we write

$$\frac{d\hat{B}}{ds} = -\tau\mathbf{N}$$

τ is known as the torsion of the curve.

Recalling that $\mathbf{N} = \mathbf{B} \times \mathbf{T}$ we may now determine the rate of change of **N** with respect to s.

$$\frac{d\hat{N}}{ds} = \frac{d}{ds}(\mathbf{B} \times \mathbf{T}) = \frac{d\hat{B}}{ds} \times \mathbf{T} + \mathbf{B} \times \frac{d\hat{T}}{ds}$$

and it follows that

$$\frac{d\hat{N}}{ds} = \tau \mathbf{B} - k\mathbf{T}$$

The four equations on the left-hand side of the table are known as the Frenet-Serret formulae. The four equations on the right-hand side are those given by Faux and Pratt for dealing with a curve $\mathbf{r}(u)$.

$\frac{d\hat{r}}{ds} = \mathbf{T}$	$\mathbf{T} = \dot{\mathbf{r}}/\dot{\mathbf{s}}$
$\frac{d\hat{T}}{ds} = k\mathbf{N}$	$k\mathbf{B} = (\dot{\mathbf{r}} \times \ddot{\mathbf{r}})/\dot{\mathbf{s}}^3$
$\frac{d\hat{B}}{ds} = -\tau\mathbf{N}$	$\tau = \dot{\mathbf{r}} \cdot (\ddot{\mathbf{r}} \times \dddot{\mathbf{r}})/\dot{\mathbf{s}}^6 k^2$
$\frac{d\hat{N}}{ds} = \tau\mathbf{B} - k\mathbf{T}$	$\dot{\mathbf{s}} = \lvert \dot{\mathbf{r}} \rvert$
	$\mathbf{N} = \mathbf{B} \times \mathbf{T}$

10.7 CURVILINEAR COORDINATES

Since we often need to calculate the geometries of bodies which have curved boundary edges and surfaces we need to consider a system of coordinates in which an orthonormal triad of axes follow the curve.

If (u, v, w) are a set of orthogonal curvilinear coordinates which are related to the Cartesian coordinates then

$$\mathbf{x} = x(u, v, w) \quad \mathbf{y} = y(u, v, w) \quad \mathbf{z} = z(u, v, w)$$

or in the vector form

$$\mathbf{r} = \mathbf{r}(u, v, w) \text{ where } \mathbf{r} = x\mathbf{i} + y\mathbf{j} + z\mathbf{k}$$

If v and w are constant, the u coordinate describes a single plane curve and this also holds true for the v and w coordinate curves.

If $\mathbf{e}_u$, $\mathbf{e}_v$, $\mathbf{e}_w$ are unit tangents, which relate to the curvilinear coordinate directions u, v, w, then

$$\mathbf{e}_u = \frac{1}{h_1}\frac{\partial \hat{r}}{\partial u} \quad \mathbf{e}_v = \frac{1}{h_2}\frac{\partial \hat{r}}{\partial u} \quad \mathbf{e}_2 = \frac{1}{h_3}\frac{\partial \hat{r}}{\partial u}$$

where

$$h_1 = \frac{\partial \hat{r}}{\partial u} \qquad h_2 = \frac{\partial \hat{r}}{\partial v} \qquad h_3 = \frac{\partial \hat{r}}{\partial w}$$

by definition.

10.7.1 Cylindrical coordinates

If the Cartesian axis *oz* is aligned with the axial direction of a circular cylindrical object, then the unit vector $\mathbf{e}_w$ is directed along the axial direction, the unit vector $\mathbf{e}_u$ is usually written $\mathbf{e}_\mathrm{R}$ (capital R to distinguish from the position vector) and is directed outwards along the radial direction. The unit vector $\mathbf{e}_v$ is usually written $\mathbf{e}_\phi$ because $\mathbf{e}_\phi$ is normal to $\mathbf{e}_\mathrm{R}$ the radius vector and hence lies in a circumferential direction. The cylindrical polar coordinate orientation is as shown in Fig. 10.28.

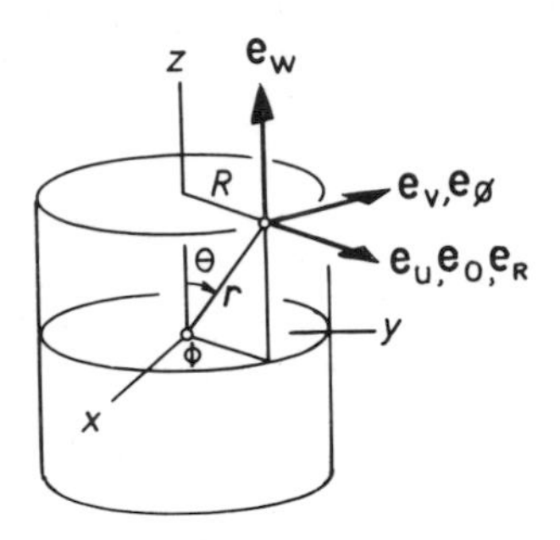

Figure 10.28

In the cylindrical polar coordinate system

$$\mathbf{r} = (\mathrm{R}\cos\phi, \mathrm{R}\sin\phi, z)$$

$$\frac{\partial \hat{r}}{\partial \mathrm{R}} = (\cos\phi, \sin\phi), \qquad \left|\frac{\partial \hat{r}}{\partial \mathrm{R}}\right| = 1 = h_1$$

$$\frac{\partial \hat{r}}{\partial \phi} = \mathrm{R}(-\sin\phi, \cos\phi), \qquad \left|\frac{\partial \hat{r}}{\partial \phi}\right| = \mathrm{R} = h_2$$

$$\frac{\partial \hat{r}}{\partial z} = 1 \qquad \left|\frac{\partial \hat{r}}{\partial z}\right| = 1 = h_3$$

and since

$$\mathbf{e}_\mathrm{R} = \frac{1}{h_1}\frac{\partial \hat{r}}{\partial \mathrm{R}}, \mathbf{e}_\phi = \frac{1}{h_2}\frac{\partial \hat{r}}{\partial \phi}, \mathbf{e}_\theta = \frac{1}{h_2}\frac{\partial \hat{r}}{\partial \theta}$$

and hence the unit tangents in the cylindrical system are

$$\mathbf{e}_R = (\cos\phi, \sin\phi, 0)$$
$$\mathbf{e}_\phi = (-\sin\phi, \cos\phi, 0)$$
$$\mathbf{e}_z = (0, 0, 1)$$

10.7.2 Spherical coordinates

When the unit tangents $\mathbf{e}_1$, $\mathbf{e}_2$, $\mathbf{e}_3$, are referred to a spherical system of coordinates they are usually aligned with the r, ϕ, θ directions and are written as $\mathbf{e}_r$, $\mathbf{e}_\phi$, $\mathbf{e}_\theta$ respectively.

(N.B. A lower case r is used to indicate that $\mathbf{e}_r$ is the unit vector associated with the position vector $\mathbf{r}$. $\mathbf{e}_\phi$ is normal to the OP_ϕ line and $\mathbf{e}_\theta$ lies in the POP_ϕ and is normal to $\mathbf{r}$).

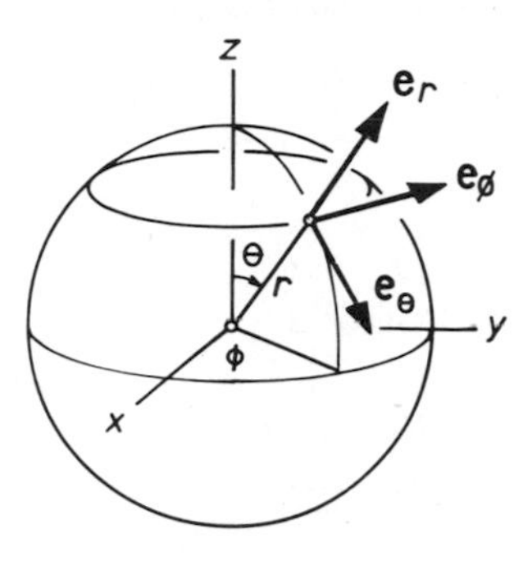

Figure 10.29

We see that

$$\mathbf{r} = r(\sin\theta\cos\phi, \sin\theta\sin\phi, \cos\theta)$$

$$\frac{\partial\hat{r}}{\partial r} = r(\sin\theta\cos\phi, \sin\theta\sin\phi, \cos\theta)$$

$$\frac{\partial\hat{r}}{\partial\theta} = r(\cos\theta\cos\phi, \cos\theta\sin\phi, -\sin\theta)$$

$$\frac{\partial\hat{r}}{\partial\phi} = r(-\sin\theta\sin\phi, \sin\theta\cos\phi, 0)$$

and since

$$h_1 = \left|\frac{\partial\hat{r}}{\partial r}\right| \quad h_2 = \left|\frac{\partial\hat{r}}{\partial\theta}\right| \quad h_3 = \left|\frac{\partial\hat{r}}{\partial\phi}\right|$$

$$h_1^2 = r^2(\sin^2\theta\cos^2\phi + \sin^2\theta\sin^2\phi + \cos^2\theta) = 1$$
$$h_2^2 = r^2(\cos^2\theta\cos^2\phi + \cos^2\theta\sin^2\phi + \sin^2\theta) = r^2$$
$$h_3^2 = r^2(\sin^2\theta\sin^2\phi + \sin^2\theta\cos^2\phi) = r^2\sin^2\theta$$

whence

$$h_1 = 1, \quad h_2 = r, \quad h_3 = r\sin\theta$$

since $0 \leqslant \theta \leqslant \pi$ and $\sin\theta \geqslant 0$.

It follows therefore that the unit tangents in the spherical coordinate system are

$$\mathbf{e}_r = (\sin\theta\cos\phi, \sin\theta\sin\phi, \cos\theta)$$
$$\mathbf{e}_\theta = (\cos\theta\cos\phi, \cos\theta\sin\phi - \sin\theta)$$
$$\mathbf{e}_\phi = (-\sin\phi, \cos\phi, 0)$$

10.8 CURVE FITTING TECHNIQUES

We have seen in Chapter 2 that the graph $y = a_0 + a_1 x$ is a straight line and it is clear that we may choose the constants a_0 and a_1 so that the line passes through any two given points. The graph $y = a_0 + a_1 x + a_2 x^2$ is a second degree polynomial curve and by choosing the three constants a_0, a_1, a_2, we may direct the curve through any three given points. (N.B. If the three points happen to be in a straight line the second degree equation automatically reverts to the straight line form). A second degree polynomial will not however yield a perfect circle, for clearly there is no term in y^2. The four coefficients of a cubic $y = a_0 + a_1 x + a_2 x^2 + a_3 x^3$ may be chosen so that the curve passes through four distinct points, or alternatively may be such that the curve passes through two given points, at predefined end point inclinations.

Polynomial equations of higher degree n may also be used when $n + 1$ prescribed points are involved, but computational effort increases dramatically as the number of predetermined points on the curve is increased. It is principally for this reason that the cubic spine finds widespread application, and as we shall see the method to be developed is amenable to a piecewise approach.

(N.B. A piecewise curve is a curve of the form $r = r(t)$ for which $r(t)$ is continuous over a number of successive intervals $t_0 \leqslant t \leqslant t_1$, $t_1 \leqslant t \leqslant t_2$, such that the curve is smooth over the full range $t_0 \leqslant t \leqslant t_n$.

Let us consider the parametric cubic

$$\mathrm{P}(\mathbf{u}) = a_0 + a_1\mathbf{u} + a_2\mathbf{u}^2 + a_3\mathbf{u}^3$$

which has a differential

$$\dot{P}(u) = a_1 + 2a_2u + 3a_3u^2$$

spanning the range $u = 0$ to $u = 1$.

The end points and slopes corresponding to $u = 0$ and $u = 1$ are

$$\begin{aligned}
P(0) &= a_0 \\
P(1) &= a_0 + a_1 + a_2 + a_3 \\
\dot{P}(0) &= a_1 \\
\dot{P}(1) &= a_1 + 2a_2 + 3a_3
\end{aligned}$$

or in column matrix form

$$\begin{bmatrix} P(0) \\ P(1) \\ \dot{P}(0) \\ \dot{P}(1) \end{bmatrix} = \begin{bmatrix} 1 & 0 & 0 & 0 \\ 1 & 1 & 1 & 1 \\ 0 & 1 & 0 & 0 \\ 0 & 1 & 2 & 3 \end{bmatrix} \begin{bmatrix} a_0 \\ a_1 \\ a_2 \\ a_3 \end{bmatrix}$$

or simply

$$\mathbf{P} = \mathbf{MA}$$

we solve for **A** by writing

$$\mathbf{A} = \mathbf{M}^{-1}\mathbf{P}$$

Now the matrix **M** was inverted in article 4.2.6, and if

$$\mathbf{M} = \begin{bmatrix} 1 & 0 & 0 & 0 \\ 1 & 1 & 1 & 1 \\ 0 & 1 & 0 & 0 \\ 0 & 1 & 2 & 3 \end{bmatrix}$$

we may write $\mathbf{M}^{-1}$ directly

$$\mathbf{M}^{-1} = \begin{bmatrix} 1 & 0 & 0 & 0 \\ 0 & 0 & 1 & 0 \\ -3 & 3 & -2 & -1 \\ 2 & -2 & 1 & 1 \end{bmatrix}$$

We may now express the curve $P(u)$ in terms of $P(0)$, $P(1)$, $\dot{P}(0)$, $\dot{P}(1)$ by writing

$$P(u) = f_0(u)P(0) + f_1(u)P(1) + f_2(u)\dot{P}(0) + f_3(u)\dot{P}(1)$$

where

$$f_0 = 1 - 3u^2 + 2u^3$$
$$f_1 = 3u^2 - 2u^3$$
$$f_2 = u - 2u^2 + u^3$$
$$f_3 = -u^2 + u^3$$

The four function f_0, f_1, f_2, f_3 collectively control the shape of a single polynomial curve, between the boundaries $u = 0$, $u = 1$, and for this reason are widely referred to as blending or shape functions. (N.B. f_0, f_1, f_2, f_3, are in fact the intrinsic equations to any surface curve they may represent).

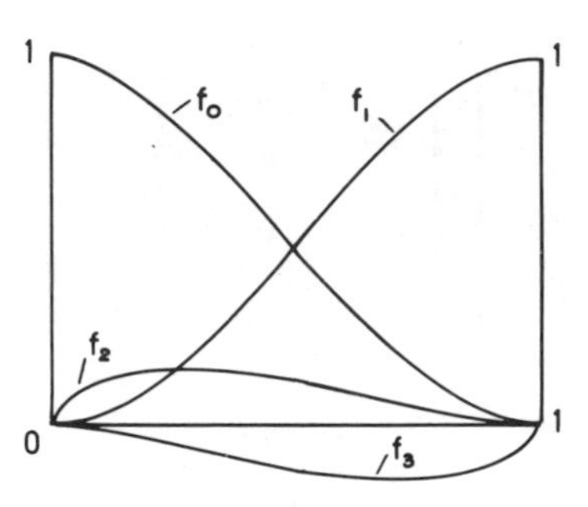

Figure 10.30

Readers of structural mechanics will be aware that the deflected shape of a long slender spline follows a cubic form and the functions f_0 and f_1 are akin to a real draughting spline to which similar moments and a shear force are applied at each end. The functions f_2 and f_3 are akin to a real spline clamped at one end and bent by a moment at the free end. As is visually apparent from Fig. 10.30, the blending function f_0 predominates in the region of $u = 0$, the function f_1 predominates in the region of $u = 1$, whilst the functions f_2 and f_3 predominate at $u = +\frac{1}{3}$ and $u = +\frac{2}{3}$. The straight addition of f_0 and f_1 equals unity and the straight addition of f_2 and f_3 equals zero throughout the entire range of u. Other combinations of the functions f_0, f_1, f_2, f_3, compound to give curves at other stations of u.

We commonly need to define the end point positions and the slopes at intervals along a curve. The intervals $(i, i + 1)$, $(i + 1, i + 2)$, see Fig. 10.31, are indicative of this procedure.

10.8.1 Piecewise definition of curves

If it is our aim to direct a smooth curve through a number of given points P(i), P(i + 1), P(i + 2) . . . etc., we clearly seek three levels of continuity. (i) continuity of position, i.e. there must be no steps in the curve, (ii) continuity of slope, i.e. there must be no abrupt changes in gradient; (iii) continuity of curvature, i.e. there must be no visual impression of flats.

For reasons of computational convenience we commonly divide complete curves into segments, typically as shown in Fig. 10.31.

Each segment is characterised by two end points P_i, P_{i+1} and two end slopes. The first real interval $(i, i+1)$ corresponds to $u = 0, u = 1$ on the intrinsic scale.

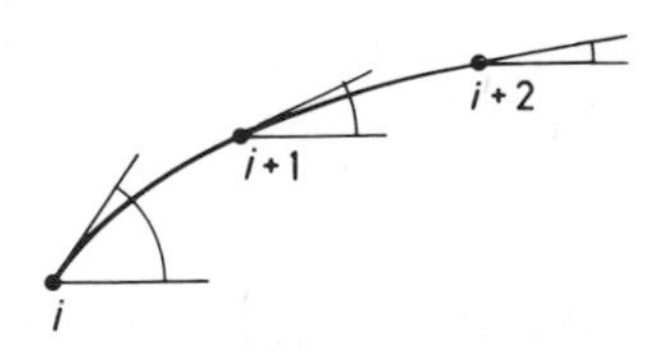

Figure 10.31

On making the substitutions $u = 0, u = 1$ the functions f_0, f_1, f_2, f_3 and their first derivatives become

n		$u = 0$	$u = 1$
0	$f_0 = 1 - 3u^2 + 2u^3$	1	0
1	$f_1 = 3u^2 - 2u^3$	0	1
2	$f_2 = u - 2u^2 + u^3$	0	0
3	$f_3 = -u^2 + u^3$	0	0
0	$\dot{f}_0 = -6u + 6u^2$	0	0
1	$\dot{f}_1 = 6u - 6u^2$	0	0
2	$\dot{f}_2 = 1 - 4u + 3u^2$	1	0
3	$\dot{f}_3 = -2u + 3u^2$	0	1

whence the end point values for the i, $i + 1$ interval are

n	$u = 0$ $fn(i)$	$u = 1$ $fn(i+1)$	$u = 0$ $\dot{fn}(i)$	$u = 1$ $\dot{fn}(i+1)$
0	1	0	0	0
1	0	1	0	0
2	0	0	1	0
3	0	0	0	1

At this stage it is instructive to consider an example, from which the data given has been obtained, from a prior knowledge of the curve.

Example

The three points $z = (0, 2, 14)$ and the three slopes $\dot{z} = (1, 5, 21)$ are known points and slopes on the cubic curve $z = u - u^2 + 2u^3$. The problem is to verify the truth of the given data, by way of the intrinsic functions.

Solution

We know that for interval 1

$$z(u)_1 = f_0(u) + f_1(u) + f_2(u) + f_3(u) \left| \begin{array}{l} z_i \\ z_{i+1} \\ \dot{z}_i \\ \dot{z}_{i+1} \end{array} \right.$$

In this particular example $z_i = 0, z_{i+1} = 2, \dot{z}_i = 1, \dot{z}_{i+1} = 5$, and by evaluating in terms of the intrinsic functions f_0, f_1, f_2, f_3, we should obviously recoup the known equation to the curve.

Thus for the first interval we have

$$z(u)_1 = 0f_0 + 2f_1 + 1f_2 + 5f_3$$

whence

$$\begin{aligned} z(u)_1 &= 2(3u^2 - 2u^3) + 1(u - 2u^2 + u^3) + 5(-u^2 + u^3) \\ &= u - u^2 + 2u^3 \end{aligned}$$

which is the result expected.

If we now move on to consider the second interval $(i + 1, i + 2)$, we have

$$z_{i+1} = 2, z_{i+2} = 14, \dot{z}_{i+1} = 5, \dot{z}_{i+2} = 21$$

from which

$$z(u)_2 = 2f_0 + 14f_1 + 5f_2 + 21f_3$$

whence

$$\begin{aligned} z(u)_2 &= 2(1 - 3u^2 + 2u^3) + 14(3u^2 - 2u^3) + 5(u - 2u^2 + u^3) + 21(-u^2 + u^3) \\ &= 2 + 5u + 5u^2 + 2u^3 \end{aligned}$$

We see that the equation for interval 2 is not the same as the equation for interval 1 and this is because we have written the first expression in relation to the point i and the second expression in relation to the point $i + 1$. The two expressions should, however, yield identical positions, slopes and curvatures at their common junction, i.e. at the point $i + 1$. Now the equations for intervals 1 and 2 are

$$\begin{aligned} z(u)_1 &= u - u^2 + 2u^3 & \qquad z(u)_2 &= 2 + 5u + 5u^2 + 2u^3 \\ \dot{z}(u)_1 &= 1 - 2u + 6u^2 & \qquad \dot{z}(u)_2 &= 5 + 10u + 6u^2 \\ \ddot{z}(u)_1 &= -2 + 12u & \qquad \ddot{z}(u)_2 &= 10 + 12u \end{aligned}$$

and since we require complete continuity across the junction $i + 1$

$$z(u)_1 \text{ at } u = 1 = z(u)_2 \text{ at } u = 0$$
$$\dot{z}(u)_1 \text{ at } u = 1 = \dot{z}(u)_2 \text{ at } u = 0$$
$$\ddot{z}(u)_1 \text{ at } u = 1 = \ddot{z}(u)_2 \text{ at } u = 0$$

In this trivial example we already know that these three conditions are met but in a real design problem we usually need to establish continuity of position, slope and curvature at various common intermediate junctions.

If we wish to write the equation to the second interval in relation to the start of the first interval, we have only to translate the origin of coordinates for the second interval from $i + 1$ to i. That is we translate the origin of $i + 1$ through a distance of $u = -1$ on the intrinsic scale. We therefore substitute $(u - 1)$ for u in the equation to the second interval and observe that $z(u)_2$ becomes

$$\begin{aligned} z(u)_2 &= 2 + 5(u - 1) + 5(u - 1)^2 + 2(u - 1)^3 \\ &= u - u^2 + 2u^3 \end{aligned}$$

The equation $z = u - u^2 + 2u^3$ is clearly the true equation which directs the curve with complete continuity through all three given points. If, however, we introduce two further points beyond point $i + 2$ which taken together are definitely not on the curve, $z = u - u^2 + 2u^3$, then it is clear that we must revert to the use of a piecewise definition.

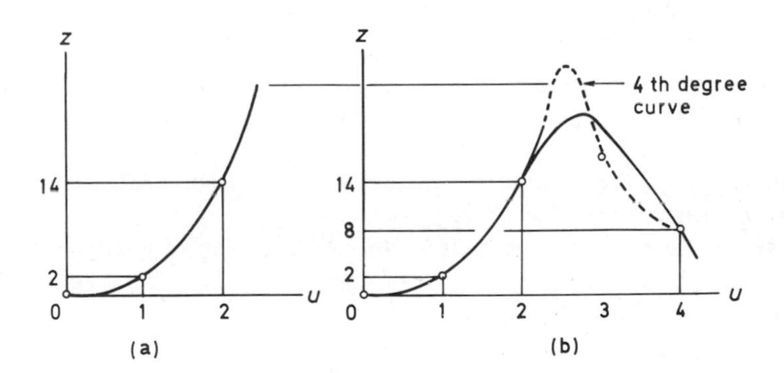

Figure 10.32

If the position and slope of $i + 3$ is specified then we may treat the third interval $i + 2$, $i + 3$ and the fourth interval $i + 3$, $i + 4$, as separate piecewise curves. If, on the other hand, the position and slope at $i + 4$ is specified then we may prefer to include the point $i + 3$, by specifying continuity of position and slope at $i + 3$ in which case the slope $i + 3$ would depend on the requirements at points $i + 2$ and $i + 4$.

A single cubic equation can span a maximum of four arbitrarily chosen points but the curve designer then has no control over slopes.

Example
Consider the problem of directing a cubic polynomial through the four points (0, 0), (1, 2), (2, 14), (4, 8), Fig. 10.32.

Solution

$$z = a_0 + a_1 u + a_2 u^2 + a_3 u^3$$

where the parameters a are to be determined. Since four known points are given we can write four independent equations and solve for a_0, a_1, a_2, a_3 simultaneously.

Now since $z = 0$ when $u = 0$, $a_0 = 0$. With regard to the second, third and fourth points we have

$$\begin{aligned} 2 &= a_1 + a_2 + a_3 \\ 14 &= 2a_1 + 4a_2 + 8a_3 \\ 8 &= 4a_1 + 16a_2 + 64a_3 \end{aligned}$$

and it follows that $a_1 = -16/2$, $a_2 = 25/2$, $a_3 = -5/2$, whence the required equation is

$$2z = -16u + 25u^2 - 5u^3$$

Had we specified end slopes, in addition to the four data points, a fifth degree polynomial would have been required.

It is clear, therefore, that if we are to define the shape of a complex curve (and more so in the case of a surface) a relatively large number of data points are necessarily required, and it is then that a piecewise approach, based on the use of the intrinsic functions becomes essential.

If we wish to direct a smooth curve through a number of equispaced points $(1, z_1)$, $(2, z_2)$, $(3, z_3) \ldots (m, z_m)$, with slopes $\dot{z}_1, \dot{z}_2, \dot{z}_3 \ldots \dot{z}_m$ then the piecewise equation for the ith interval is

$$z(u) = z_i f_0(u) + z_{i+1} f_1(u) + \dot{z}_i f_2(u) + \dot{z}_{i+1} f_3(u)$$

or

$$z(u) = [f_0(u) f_1(u) f_2(u) f_3(u)] \begin{bmatrix} z_i \\ z_{i+1} \\ \dot{z}_i \\ \dot{z}_{i+1} \end{bmatrix}$$

Now since the coefficients of the ith segment are also the coefficients of the $i+1$th segment and the other two coefficients of the ith segment are also the coefficients of the $i-1$th segment, there are $2m$ coefficients ($z_1, z_2, \ldots z_m$. $\dot{z}_1, \dot{z}_2, \ldots \dot{z}_m$) in an m point curve to be determined, and this compares favourably with the $(4m - 4)$ coefficients required for a perfect fit of the cubic $z(u) = a_0 + a_1 u + a_2 u^2 + a_3 u^3$.

In the case of 3 space curves we need to consider components in x, y, z, and we summarise the resulting equations using the vector notation of article 10.2.10.

$$\mathbf{a} = a_x\mathbf{i} + a_y\mathbf{j} + a_z\mathbf{k}$$

The single vector **a** thus represents three Cartesian terms and we observe that an expression such as equation 10.2 contains twelve independent coefficients.

10.8.2 Ferguson cubic curves

We may relate the derivation above to the work of Faux and Pratt by following their example.

If we write

$$\mathbf{r} = \mathbf{r}(u) = \mathbf{a}_0 + u\mathbf{a}_1 + u^2\mathbf{a}_2 + u^3\mathbf{a}_3 \tag{10.2}$$

then

$$\dot{\mathbf{r}} = \dot{\mathbf{r}}(u) = \mathbf{a}_1 + 2u\mathbf{a}_2 + 3u^2\mathbf{a}_3 \tag{10.3}$$

where

$$\dot{\mathbf{r}}(u) = \frac{d\hat{r}}{du}$$

and when we put $u = 0$ and $u = 1$

$$\begin{aligned}
\mathbf{r}(0) &= \mathbf{a}_0 \\
\dot{\mathbf{r}}(0) &= \mathbf{a}_1 \\
\mathbf{r}(1) &= \mathbf{a}_0 + \mathbf{a}_1 + \mathbf{a}_2 + \mathbf{a}_3 \\
\dot{\mathbf{r}}(1) &= \mathbf{a}_1 + 2\mathbf{a}_2 + 3\mathbf{a}_3
\end{aligned}$$

Now on substituting $\mathbf{r}(0) = \mathbf{a}_0$ and $\dot{\mathbf{r}}(0) = \mathbf{a}_1$ into equations 10.2 and 10.3 the equations for $\mathbf{a}_2$ and $\mathbf{a}_3$ are obtained in the following form:

$$\begin{aligned}
\mathbf{a}_0 &= \mathbf{r}(0) \\
\mathbf{a}_1 &= \dot{\mathbf{r}}(0) \\
\mathbf{a}_2 &= [3\,\mathbf{r}(1) - \mathbf{r}(0)] - 2\dot{\mathbf{r}}(0) - \dot{\mathbf{r}}(1) \\
\mathbf{a}_3 &= [2\,\mathbf{r}(0) - \mathbf{r}(1)] + \dot{\mathbf{r}}(0) + \dot{\mathbf{r}}(1)
\end{aligned}$$

and on substituting these values back into equation 10.2, we obtain the following result.

$$\begin{aligned}\mathbf{r}(u) &= \mathbf{r}(0) + u\dot{\mathbf{r}}(0) + 3u^2[\mathbf{r}(1) - \mathbf{r}(0)] - 2u^2\dot{\mathbf{r}}(0) - u^2\dot{\mathbf{r}}(1) \\ &\quad + 2u^3\,[\mathbf{r}(0) - \mathbf{r}(1)] + u^3\dot{\mathbf{r}}(0) + u^3\dot{\mathbf{r}}(1) \\ &= \mathbf{r}(0) + u\dot{\mathbf{r}}(0) + 3u^2\mathbf{r}(1) - 3u^2\mathbf{r}(0) - 2u^2\dot{\mathbf{r}}(0) - u^2\dot{\mathbf{r}}(1) + 2u^3\mathbf{r}(0) \\ &\quad - 2u^3\mathbf{r}(1) + u^3\dot{\mathbf{r}}(0) + u^3\dot{\mathbf{r}}(1) \\ &= \mathbf{r}(0)\,(1 - 3u^2 + 2u^3) + \dot{\mathbf{r}}(0)\,(u - 2u^2 + u^3) + \mathbf{r}(1)\,(3u^2 - 2u^3) \\ &\quad + \dot{\mathbf{r}}(1)\,(-u^2 + u^3)\end{aligned}$$

and hence the source and derivation of the intrinsic functions is clear.

The natural outcome of this derivation is the matrix formulation

$$\mathbf{r}(u) = [1 \quad u \quad u^2 \quad u^3]\begin{bmatrix} 1 & 0 & 0 & 0 \\ 0 & 1 & 0 & 0 \\ -3 & -2 & 3 & -1 \\ 2 & 1 & -2 & 1 \end{bmatrix}\begin{bmatrix} \mathbf{r}(0) \\ \dot{\mathbf{r}}(0) \\ \mathbf{r}(1) \\ \dot{\mathbf{r}}(1) \end{bmatrix}$$

and the reader's attention is drawn to the order in which the $\mathbf{r}$ and $\dot{\mathbf{r}}$ terms appear. The $\mathbf{r}, \dot{\mathbf{r}}$ order used elsewhere in this book and by Faux and Pratt is

$$\mathbf{r}(u) = [1 \quad u \quad u^2 \quad u^3]\begin{bmatrix} 1 & 0 & 0 & 0 \\ 0 & 0 & 1 & 0 \\ -3 & 3 & -2 & -1 \\ 2 & -2 & 1 & 1 \end{bmatrix}\begin{matrix} \mathbf{r}(0) \\ \mathbf{r}(1) \\ \dot{\mathbf{r}}(0) \\ \dot{\mathbf{r}}(1) \end{matrix}$$

The two presentations give identical results but terms and elements must not be inadvertently mixed. The two presentations are

$\mathbf{r}(0),\ \dot{\mathbf{r}}(0),\ \mathbf{r}(1),\ \dot{\mathbf{r}}(1)$ Sabin

$\mathbf{r}(0),\ \mathbf{r}(1),\ \dot{\mathbf{r}}(0),\ \dot{\mathbf{r}}(1)$ Faux and Pratt

10.8.3 Bezier cubic curves

When using the parametric cubic, article 10.8.1, in a piecewise mode we sought to pass the curve through a series of predefined points and this we did by specifying the two end positions and two end slopes of each segment.

An alternative approach, originated by Bezier 1970, is to define the end slopes in terms of two additional points (not on the curve)

If, as in Fig. 10.33, P_0, P_3, are two given points on a curve, the shape of which is to be determined and P_1, P_2, are two points not on the curve which serve to direct the two end tangents. Then it is clear from Fig. 10.33, that te curve through the end points P_1 and P_3 is tangential to the lines P_0P_1, P_3P_2 and lies wholly within the bound of the quadrilateral $P_0P_1P_2P_3$.

The quadrilateral $P_0P_1P_2P_3$, Fig. 10.33, is known as the convex polygon and as can be seen from the figure, the shape of the curve through P_0P_3 can be

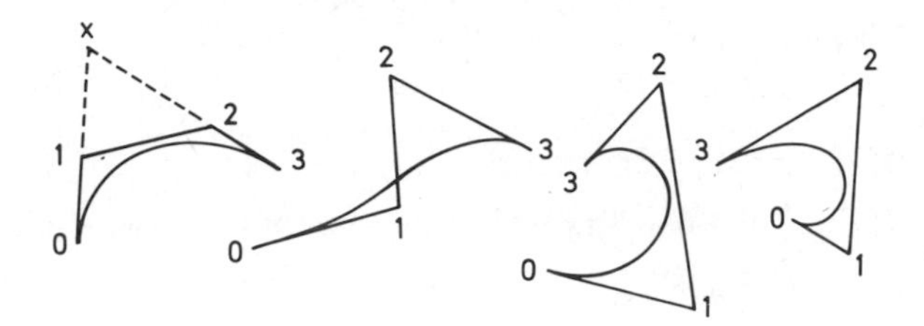

Figure 10.33

radically altered simply by redirecting one or both end tangents. By moving the control points P_1 and P_3 further apart both angles $P_0P_3P_2$, $P_3P_0P_1$, are increased and the curve becomes fuller. If only one of these angles is increased the curve is drawn towards the line of the increased tangent and if the angles $P_0P_3P_2$, $P_3P_0P_1$ are altered differentially, the curve swings to one side or the other, and if the intersection of the tangents P_0P_1, P_3P_2 is internal to P_1P_2 a loop in the curve may be produced. It is however unlikely that a looped curve would be of practical value and in a two dimensional curve loops are avoided by making P_0P_1, P_2P_3 less than the chord length P_0P_3.

The three dimensional Bezier parametrisation of a u curve is of the form

$$\mathbf{r} = \mathbf{r}(u) = (1-u)^3\mathbf{r}_0 + 3u(1-u)^2\mathbf{r}_1 + 3u^2(1-u)\mathbf{r}_2 + u^3\mathbf{r}_3 \tag{10.4}$$

the expansion of which gives

$$\mathbf{r}(u) = (1 - 3u + 3u^2 - u^3)\mathbf{r}_0 + (3u - 6u^2 + 3u^3)\mathbf{r}_1 + (3u^2 - 3u^3)\mathbf{r}_2 + (u^3)\mathbf{r}_3 \tag{10.5}$$

and hence

$$\dot{\mathbf{r}}(u) = (-3 + 6u - 3u^2)\mathbf{r}_0 + (3 - 12u + 9u^2)\mathbf{r}_1 + (6u - 9u^2)\mathbf{r}_2 + (3u^2)\mathbf{r}_3 \tag{10.6}$$

when equation 10.5 is expressed in matrix form

$$\mathbf{r}(u) = [1 \quad u \quad u^2 \quad u^3] \begin{bmatrix} 1 & 0 & 0 & 0 \\ -3 & 3 & 0 & 0 \\ 3 & -6 & 3 & 0 \\ -1 & 3 & -3 & 1 \end{bmatrix} \begin{bmatrix} \mathbf{r}_0 \\ \mathbf{r}_1 \\ \mathbf{r}_2 \\ \mathbf{r}_3 \end{bmatrix}$$

If we set up the intrinsic scale by substituting $u = 0$, $u = 1$ in equations 10.5 and 10.6 we obtain

$$\mathbf{r}(0) = \mathbf{r}_0$$
$$\mathbf{r}(1) = \mathbf{r}_3$$
$$\dot{\mathbf{r}}(0) = (-3\mathbf{r}_0 + 3\mathbf{r}_1) = 3(\mathbf{r}_1 - \mathbf{r}_0)$$
$$\dot{\mathbf{r}}(1) = (-3 + 6 - 3)\mathbf{r}_0 + (3 - 12 + 9)\mathbf{r}_1 + (6 - 9)\mathbf{r}_2 + 3\mathbf{r}_3$$
$$= (-3\mathbf{r}_2 + 3\mathbf{r}_3) = 3(\mathbf{r}_3 - \mathbf{r}_2)$$

that is

u	0	1
$\mathbf{r}(u)$	$\mathbf{r}_0$	$\mathbf{r}_3$
$\dot{\mathbf{r}}(u)$	$3(\mathbf{r}_1 - \mathbf{r}_0)$	$3(\mathbf{r}_3 - \mathbf{r}_2)$

(10.7)

and if we substitute the Bezier values, $\mathbf{r}(0)$, $\mathbf{r}(1)$, $\dot{\mathbf{r}}(0)$, $\dot{\mathbf{r}}(1)$, into the Ferguson equations 10.2, 10.3, we conclude that the values $\mathbf{a}_0, \mathbf{a}_1, \mathbf{a}_2, \mathbf{a}_3$, in the Ferguson cubic are related to the Bezier values as follows

$$\begin{aligned} \mathbf{a}_0 &= \mathbf{r}_0 \\ \mathbf{a}_1 &= 3(\mathbf{r}_1 - \mathbf{r}_0) \\ \mathbf{a}_2 &= 3(\mathbf{r}_2 - 2\mathbf{r}_1 + \mathbf{r}_0) \\ \mathbf{a}_3 &= \mathbf{r}_3 - 3\mathbf{r}_2 + 3\mathbf{r}_1 - \mathbf{r}_0 \end{aligned} \qquad (10.8)$$

Readers who find difficulty in relating the three vector groups ($\mathbf{r}(0)$, $\mathbf{r}(1)$, $\dot{\mathbf{r}}(0)$, $\dot{\mathbf{r}}(1)$), ($\mathbf{r}_0$, $\mathbf{r}_1$, $\mathbf{r}_2$, $\mathbf{r}_3$), ($\mathbf{a}_0$, $\mathbf{a}_1$, $\mathbf{a}_2$, $\mathbf{a}_3$) may check the expressions given by substituting a valid set of numerical values. One simple numerical example is given below.

Example

The equation $y = 2 + 3u + 4u^2 + 5u^3$ is clearly a plane cubic in which $a_0 = 2$, $a_1 = 3, a_2 = 4, a_3 = 5$.

Demonstration

If

$$y = 2 + 3u + 4u^2 + 5u^3$$
$$\dot{y} = 3 + 8u + 15u^2$$

and

$$\mathbf{r}(0) = 2, \quad \mathbf{r}(1) = 14$$
$$\dot{\mathbf{r}}(0) = 3, \quad \dot{\mathbf{r}}(1) = 26$$

Now according to equations 10.7

$$\mathbf{r}_0 = \mathbf{r}(0) = 2, \qquad \mathbf{r}_3 = \mathbf{r}(1) = 14$$
$$3(\mathbf{r}_1 - \mathbf{r}_0) = \dot{\mathbf{r}}(0) = 3, \quad 3(\mathbf{r}_3 - \mathbf{r}_2) = \dot{\mathbf{r}}(1) = 26$$

whence

$$\mathbf{r}_1 = 3, \quad \mathbf{r}_2 = \frac{16}{3}$$

and hence by equations 10.8

$$\begin{aligned}
\mathbf{a}_0 &= \mathbf{r}_0 = 2\\
\mathbf{a}_1 &= 3(\mathbf{r}_1 - \mathbf{r}_0) = 3(3-2) = 3\\
\mathbf{a}_2 &= 3(\mathbf{r}_2 - 2\mathbf{r}_1 + \mathbf{r}_0) = (16 - 18 + 6) = 4\\
\mathbf{a}_3 &= (\mathbf{r}_3 - 3\mathbf{r}_2 + 3\mathbf{r}_1 - \mathbf{r}_0) = (14 - 16 + 9 - 2) = 5
\end{aligned}$$

and this is clearly correct.

The principal virtue of the Bezier method is that it allows control points to be selected which are not tied to the curve and this facility allows the designer much greater flexibility. The second important feature is that Bezier segments may be easily divided into smaller segments, to cater for high local curvatures where necessary, whereas, in a curve fitting technique one is committed to the use of a curve of fixed degree (usually a cubic) throughout the entire range considered.

10.8.4 Bezier unisurf, characteristic polygon

In the Bezier approach the slope of the curve at the end points is identical to the slope of the defining lines P_0P_1, P_2P_3, Fig. 10.34, and this enables both end positions and end slopes to be set quite easily. Since the first and third straight line segments in a four-sided polygon are the tangents to a curve spanning the first and fourth corner points, a Bezier curve lies entirely within the defining polygon and this is also true of Bezier curves and surfaces in three dimensions.

The Bezier formulation for a surface patch is

$$P(u, v) = [(1-u)^3, 3(1-u)^2 u, 3(1-u)u^2, u^3]\ [B] \begin{bmatrix} (1-v)^3 \\ 3(1-v)^2 v \\ 3(1-v)v^2 \\ v^3 \end{bmatrix} \tag{10.9}$$

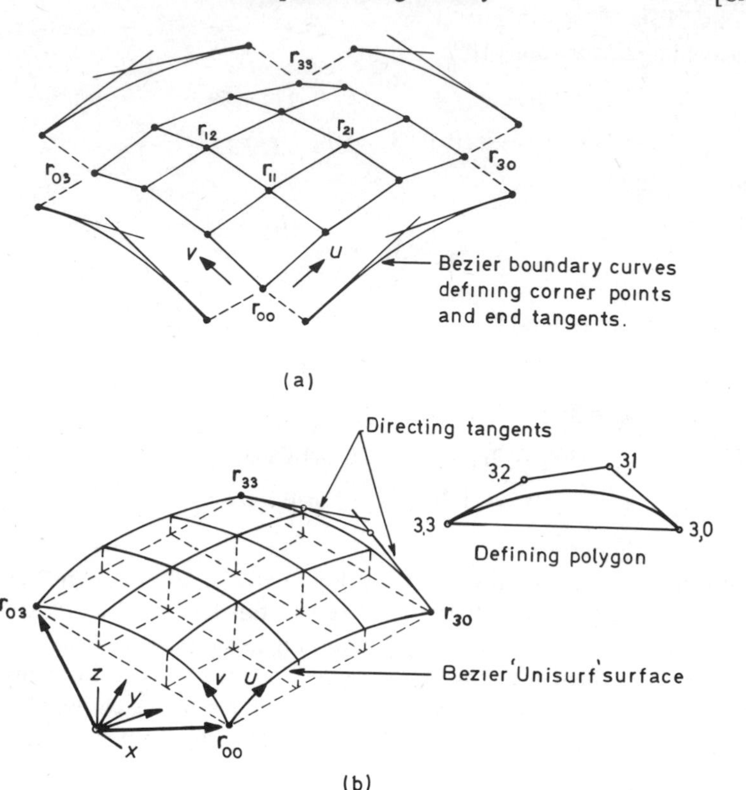

Figure 10.34

where the position vectors of the defining polygon, Fig. 10.34, are given by

$$B = \begin{bmatrix} B(0,0) & B(0,1) & B(0,2) & B(0,3) \\ B(1,0) & B(1,1) & B(1,2) & B(1,3) \\ B(2,0) & B(2,1) & B(2,2) & B(2,3) \\ B(3,0) & B(3,1) & B(3,2) & B(3,3) \end{bmatrix}$$

and it is seen from the figure that only the corner points B(0, 0), B(3, 0) B(0, 3) B(3, 3) lie on the surface of a nine patch configuration.

10.9 PROPORTIONAL DEVELOPMENT

If AB, CD are two plane cross sections set normal to the *oz* axis, as shown in Fig. 10.35, we may generate any number of intermediate cross sections, such as EF, using the process of linear interpolation known – since the early days of manual draughting – as proportional development.

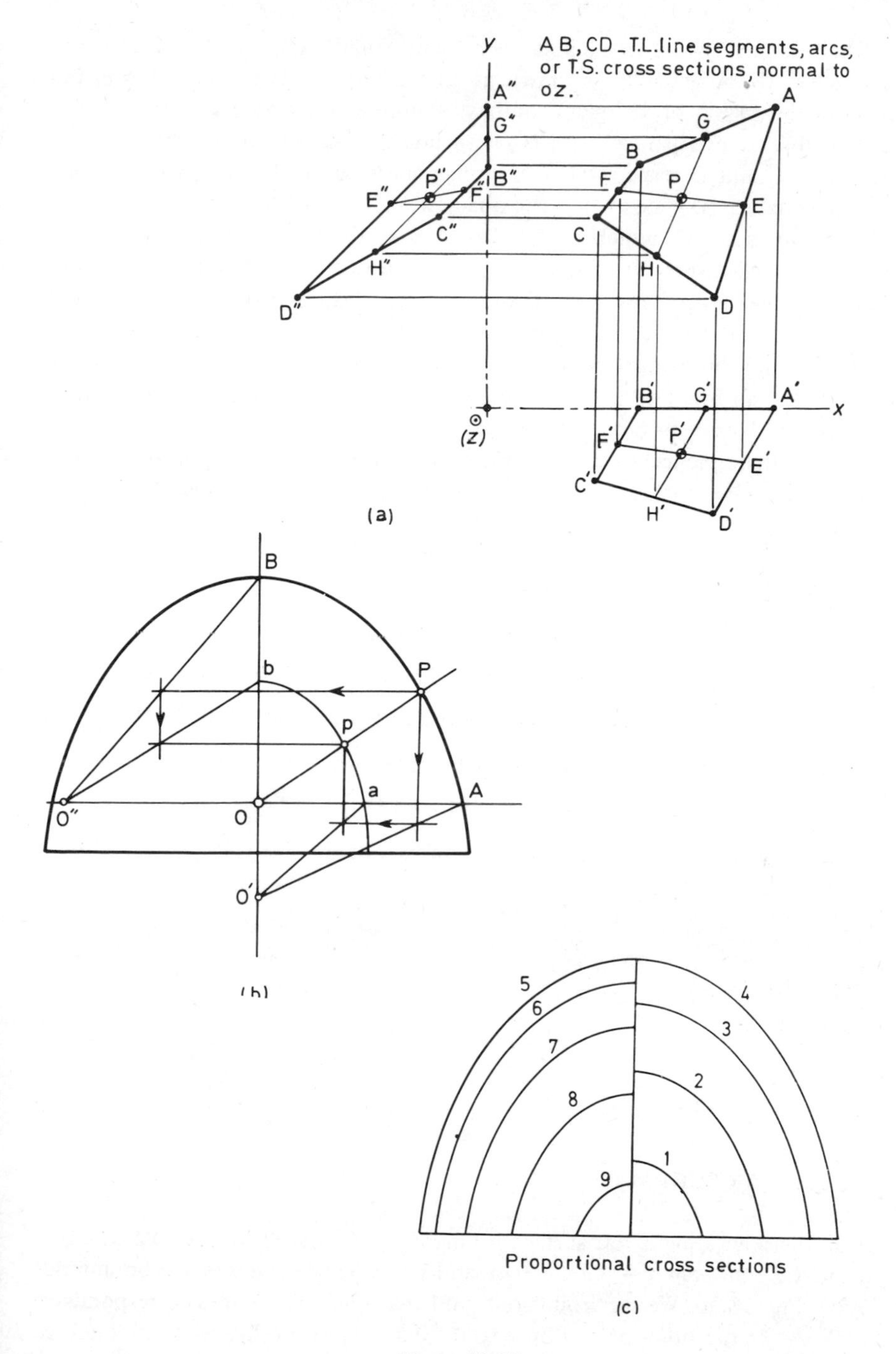

AB, CD _ T.L. line segments, arcs, or T.S. cross sections, normal to oz.
(a)
(b)
Proportional cross sections
(c)

Figure 10.35

The line A′B′ is drawn parallel to ox with length equal to the parallel projection of AB. A line B′C′ is drawn at an arbitrary angle to the line A′B′ (an angle in the region of 45° gives the best definition for graphical work). A line A′D′ is then drawn parallel to B′C′, with points C′ and D′ on the parallel projectors of C and D respectively. A similar construction starting with the line A″D″ parallel to the y axis yields the quadrilateral A″B″C″D″.

Now any point G′ which divides line A′B′ in the ratio of $\alpha : 1 - \alpha$ and any point E″ which divides the line A″D″ in the ratio of $\beta : 1 - \beta$ may be chosen to locate the position of any desired intermediate cross section. Lines G′H′ and E′F′ produce the intersection P′ and the lines E″F″ and G″H″ produce the intersection P″ and the parallel projections of these five points locate the end points EF and P of an intermediate cross section EPF. The Point P is clearly a general point on the 3-space surface.

Now with reference to Fig. 10.35, it is seen that the position vector of a general point G on the curve AB is given by $r_G(\alpha) = r_A + (r_B - r_A)\alpha$

$$r_G(\alpha) = r_A(1 - \alpha) + r_B\alpha$$
$$r_H(\alpha) = r_D(1 - \alpha) + r_C\alpha$$
$$r_E(\beta) = r_A(1 - \beta) + r_B\beta$$
$$r_F(\beta) = r_B(1 - \beta) + r_C\beta$$

and the point P lies at the intersection of the curves through EF and GH.

$$r_P(\mathrm{EF}) = r_E(1 - \alpha) + r_F\alpha$$
$$r_P(\mathrm{GH}) = r_G(1 - \beta) + r_H\beta$$

whence the x, y coordinates of P are x_p and y_p where

$$x_p = x_E(1 - \alpha) + x_F\alpha$$
$$y_p = y_G(1 - \beta) + y_H\beta$$

10.10 LINEAR SURFACES

Let us consider a plane flat surface bounded by four straight lines. Let the two parametric equations $x = 3u + v + uv$ and $y = u + 2v + 1$ define the boundaries OPRQ, Fig. 10.36. We see fromsthe figure that one pair of lines corresponds to $u = 0, u = 1$, the other pair of lines to $v = 0, v = 1$, with other relevant details as given in the table.

u	v	x	y	Pt
0	0	0	1	P
0	1	1	3	Q
1	0	3	2	R
1	1	5	4	S

The Cartesian equations to the lines corresponding to $u = 0, u = 1, v = 0, v = 1$ are obtained by substituting these values into the parametric form, whence the Cartesian equations are as follows

$$y = 2x + 1 \qquad \frac{dy}{dx} = 2 \ (u = 0)$$

$$y = x - 1 \qquad \frac{dy}{dx} = 1 \ (u = 1)$$

$$y = \frac{x}{3} + 1 \qquad \frac{dy}{dx} = \frac{1}{3} \ (v = 0)$$

$$y = \frac{x}{4} + \frac{11}{4}, \qquad \frac{dy}{dx} = \frac{1}{4} \ (v = 1)$$

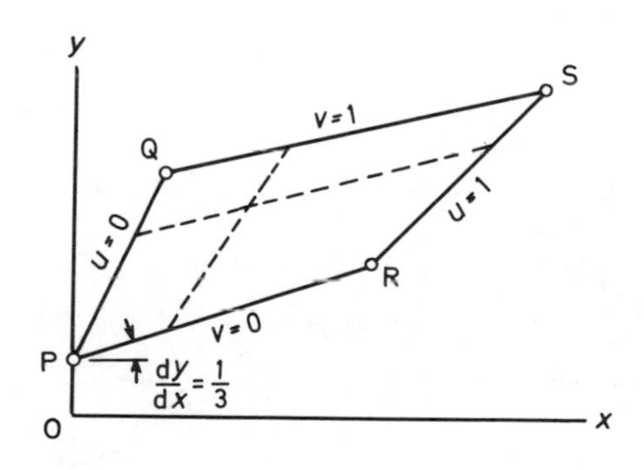

Figure 10.36

The parametric equations $x = 3u + v + uv, y = u + 2v + 1$, are of the form

$$x = f(u, v), \quad y = g(u, v)$$

and the derivatives of these functions are given by

$$dx = \frac{\partial x}{\partial u} du + \frac{\partial x}{\partial v} dv, \quad dy = \frac{\partial y}{\partial u} du + \frac{\partial y}{\partial v} dv$$

Now since

$$\frac{\partial x}{\partial u} = 3 + v,\ \frac{\partial x}{\partial v} = 1 + u,\ \frac{\partial y}{\partial u} = 1,\ \frac{\partial y}{\partial v} = 2$$

it follows that

$$\frac{dy}{dx}_{(v=c)} = \frac{\partial y/\partial u}{\partial x/\partial u} = \frac{1}{3+v} \quad \text{and} \quad \frac{dy}{dx}_{(u=c)} = \frac{\partial y/\partial v}{\partial x/\partial v} = \frac{2}{1+u}$$

and it is seen that when $v = 0$, $dy/dx = 1/3$ as previously determined. Other values follow suit and it is clear that any number of rulings may be made on the surface of the patch and any number of nodal junctions formed by intersecting rulings. A pair of rulings corresponding to $u = 2/3$, $v = 1/3$ are shown on the figure.

10.10.1 Coons patches

A linear surface patch defined by Coons (during the sixties) defines the locus of a point in terms of two variable parameters u, v. All points $P(u, v)$ on a Coons patch are therefore linear functions of the type

$$P(u, v) = [P_x(u, v) + P_y(u, v) + P_z(u, v)] \tag{10.10}$$

where P_x, P_y, P_z are arbitrary functions.

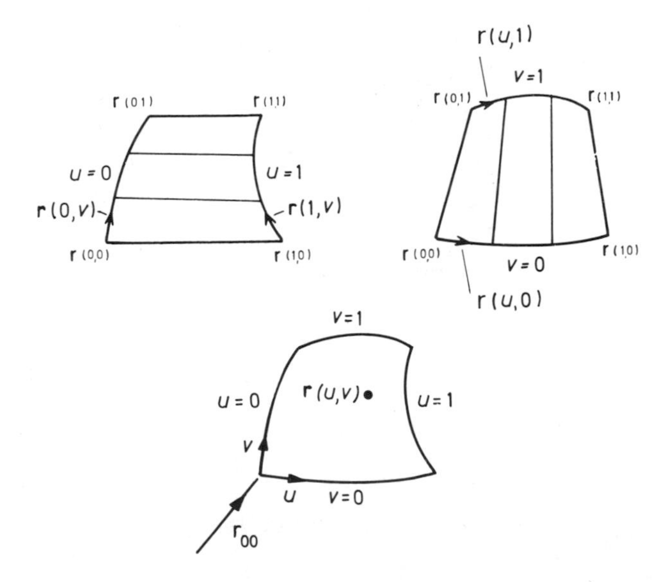

Figure 10.37

If we initially consider two boundary curves of the form

$$\mathbf{r}_1(u, v) = (1 - u)\mathbf{r}(o, v) + u\mathbf{r}(1, v) \tag{10.11a}$$

$$\mathbf{r}_2(u, v) = (1 - v)\mathbf{r}(u, 0) + v\mathbf{r}(u, 1) \tag{10.11b}$$

that is, we limit u to the range $0 < u < 1$ and interpolate linearly between the two boundaries $(0, v)$ and $(1, v)$ and do likewise for the other two boundary curves. We have the means of defining a surface generally as shown in Fig. 10.37.

Now in the case of a surface based on four boundary curves such as that illustrated in Fig. 10.38, we see that if $\mathbf{r}_{00}$ is the position vector to one end of the $v = 0$ line and $\mathbf{r}_{10}$ the position vector to the other end. Then the position vector $\mathbf{r}_u$ of a general point which divides the segment in the ratio $u:1 - u$ is obtained directly from the Fig. 10.38.

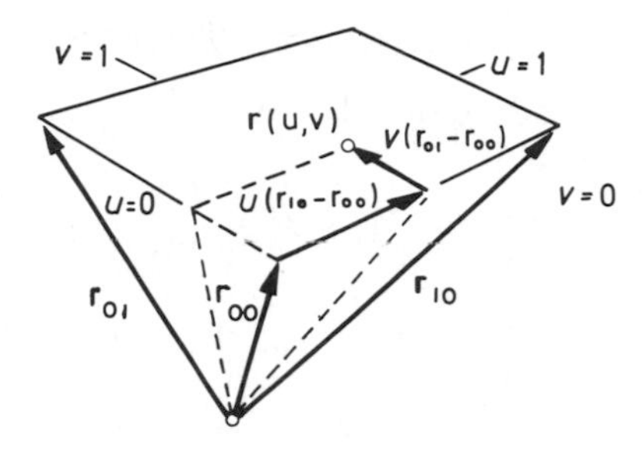

Figure 10.38

The length of the $v = 0$ line is $(\mathbf{r}_{10} - \mathbf{r}_{00})$ and a proportion u of this segment is $u(\mathbf{r}_{10} - \mathbf{r}_{00})$. The position vector of the point from which u is taken is $\mathbf{r}_{00}$ and the general point on the $v = 0$ line is therefore

$$\mathbf{r}_u = \mathbf{r}_{00} + u(\mathbf{r}_{10} - \mathbf{r}_{00})$$

and similarly is the general point on the $u = 0$ line is

$$\mathbf{r}_v = \mathbf{r}_{00} + v(\mathbf{r}_{01} - \mathbf{r}_{00})$$

Now it is clearly necessary to combine this information in such a way that the surface represented by the four boundary curves $u = 0, u = 1, v = 0, v = 1$ is truthfully produced. Our first step is to add equation 10.11a and 10.11b, whence we obtain

$$\begin{aligned} \mathbf{r}_u + \mathbf{r}_v &= \mathbf{r}_{00} + u(\mathbf{r}_{10} - \mathbf{r}_{00}) + \mathbf{r}_{00} + v(\mathbf{r}_{01} - \mathbf{r}_{00}) \\ &= \mathbf{r}_{00}(2 - u - v) + \mathbf{r}_{10}(u) + \mathbf{r}_{01}(v) \end{aligned} \tag{10.11c}$$

However, we see from Fig. 10.38, that the general point $\mathbf{r}(u, v)$ is located by the sum of the vectors from the position (0, 0) to the position (u, v). It follows therefore that the position vector $\mathbf{r}(u, v)$ is given by

$$\mathbf{r}(u, v) = \mathbf{r}_{00} + u(\mathbf{r}_{10} - \mathbf{r}_{00}) + v(\mathbf{r}_{01} - \mathbf{r}_{00})$$
$$= \mathbf{r}_{00}(1 - u - v) + \mathbf{r}_{10}u + \mathbf{r}_{01}v \qquad (10.11d)$$

and on comparing equations 10.11c and 10.11d we see that the straight sum of $\mathbf{r}_u$ and $\mathbf{r}_v$ yields one $\mathbf{r}_{00}$ too many, moreover the position of the general point (u, v) depends on the position of all four corners, and we must therefore subtract the excess contribution.

Now if we write

$$P(u, v) = P(u, 0)(1 - v) + P(u, 1)v + P(0, v)(1 - u) + P(1, v)u$$

it is easy to see that

$$P(00)(1 - u)(1 - v) + P(0, 1)(1 - u)v + P(1, 0)u(1 - v) + P(1, 1)uv$$

is the quantity to be subtracted, whence we may legitimately write the equation to the surface in the form

$$P(u, v) = [1 - u, u, 1] \begin{bmatrix} -P(0,0) & -P(0,1) & P(0,v) \\ -P(1,0) & -P(1,1) & P(1,v) \\ P(u,0) & P(u,1) & 0 \end{bmatrix} \begin{bmatrix} 1 - v \\ v \\ 1 \end{bmatrix} \qquad (10.12)$$

An example of a linear coons surface is given in the next article in which the concept of a twisted or warped patch is developed.

10.10.2 Twisted (Coons) surface

We know from Chapter 8, that if two parallel and oppositely inclined lines OP and QR serve as directrices, a third line LM which slides on OP and OR rules a hyperbolic paraboloid surface. Similarly a line L′M′ which slides on OQ, PR crosses the line LM at the point (uv on the surface. The directrices OP and QR are identified as the lines $(u, 0)$ and $(u, 1)$, and the extreme bound of the generators OQ and PR are identified by $(0, v)$ and $(1, v)$ respectively.

Now since $P(u, v) = [x(u, v), y(u, v), z(u, v)]$ a straight line may be ruled on the surface by setting u = constant and a node on the surface defined by the intersection of two such lines, the first with u = constant, the second with v = constant. Thus $P(u_i, v_i)$ is a node on the surface.

If the generator LM lies at a distance u along the directrices $v = 0, v = 1$, and the generator L′M′ lies at a distance w along the alternative directrices $u = 0$, $u = 1$, then the intersection of LM and L′M′ is the point (u, v) and this is a point on the hyperbolic paraboloid surface.

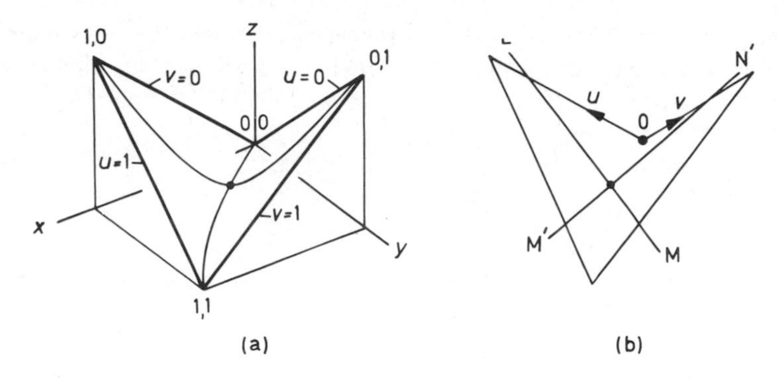

Figure 10.39

The position vector of L is $\mathbf{r}_{00} + u(\mathbf{r}_{10} - \mathbf{r}_{00})$ the position vector of M is $\mathbf{r}_{01} + u(\mathbf{r}_{11} - \mathbf{r}_{01})$. The point (u, v) lies at a distance v along LM. The vector equation to the line LM is

$$[\mathbf{r}_{01} + u(\mathbf{r}_{11} - \mathbf{r}_{01}) - \mathbf{r}_{00} - u(\mathbf{r}_{10} - \mathbf{r}_{00})]$$

and the position vector of the point (u, v) is therefore

$$\mathbf{r}_{00} + u(\mathbf{r}_{10} - \mathbf{r}_{00}) + v[\mathbf{r}_{01} + u(\mathbf{r}_{11} - \mathbf{r}_{01}) - \mathbf{r}_{00} - u(\mathbf{r}_{10} - \mathbf{r}_{00})]$$

N.B. The point (u, v) may also be approached via the alternative vector path, v along $u = 0$ followed by u along L$'$M$'$. The equation to this alternative path is

$$\mathbf{r}_{00} + v(\mathbf{r}_{01} - \mathbf{r}_{00}) + u[\mathbf{r}_{10} + v(\mathbf{r}_{11} - \mathbf{r}_{10}) - \mathbf{r}_{00} - v(\mathbf{r}_{01} - \mathbf{r}_{00})]$$

Expanding and rearranging the equation to the point (u, v) a distance v along LM yields

$$\mathbf{r}_{00}(1 - u - v + uv) + \mathbf{r}_{01}(v - uv) + \mathbf{r}_{10}(u - uv) + \mathbf{r}_{11}(uv) \quad (10.12)$$

If $\mathbf{r}_{00} = (0, 0, 0)$, $\mathbf{r}_{01} = (0, 1, 1)$, $\mathbf{r}_{10} = (1, 0, 1)$, $\mathbf{r}_{11} = (1, 1, 0)$, as in Fig. 10.39. Then we know by symmetry that when $u = v = \frac{1}{2}$ the point (u, v) lies on the surface, and since this point is the middleof the surface $x = y = z = \frac{1}{2}$.

On substituting the relevant values into equation 10.12, we obtain

$$(0,0,0)(0) + (0,1,1)(\tfrac{1}{2} - \tfrac{1}{4}) + (1,0,1)(\tfrac{1}{2} - \tfrac{1}{4}) + (1,1,0)(\tfrac{1}{2} \cdot \tfrac{1}{2})$$

and it follows that the coordinates of (uv) relative to the point $(0, 0, 0)$ are

$$x = 0 + 0 + \tfrac{1}{4} + \tfrac{1}{4} = \tfrac{1}{2}$$
$$y = 0 + \tfrac{1}{4} + 0 + \tfrac{1}{4} = \tfrac{1}{2}$$
$$z = 0 + \tfrac{1}{4} + \tfrac{1}{4} + 0 = \tfrac{1}{2}$$

and these results confirm our expectations.

As may be seen from Fig. 10.40, twisted surface elements are clearly present in the bow of a ship, and a number of such elements may be assembled to describe its total form.

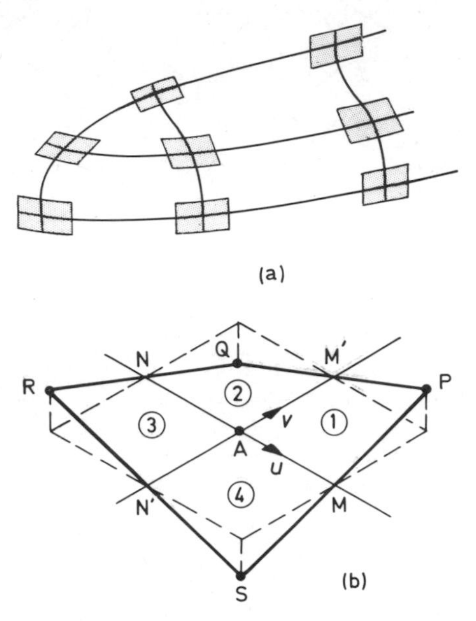

Figure 10.40

A square twisted element PQRS may clearly be divided into four smaller elements (patches) with common corner sited at $\mathbf{A}(x, y, z)$ and vectors $\mathbf{A}_u$, $\mathbf{A}_v$ in the surface. Since $\mathbf{A}$ is a common corner the twist or cross derivative $\mathbf{A}_{uv}$ at A is the same for each adjacent patch and if it can be arranged for the other corner points PQRS to lie in the surface, then one small element of surface can be defined.

If we position MN and M′N′ at suitable distances along $\mathbf{A}_u$, $\mathbf{A}_v$ then the position of the mid-point A is given by

$$\mathrm{A} = \frac{\mathrm{MN}}{2} = \frac{\mathrm{M'N'}}{2}$$

The twist of the surface at A is governed by the ‘elevation’ of the points P, Q, R, S. Now if the sub-patches are square and the twist is the same in each, the ‘elevations’ of the points P, Q, R, S are equal and it follows that

$$\mathrm{M} = \frac{\mathrm{P}+\mathrm{Q}}{2} = \mathrm{N} = \frac{\mathrm{R}+\mathrm{S}}{2} \quad \mathrm{M'} = \frac{\mathrm{S}+\mathrm{P}}{2} = \mathrm{N'} = \frac{\mathrm{Q}+\mathrm{R}}{2}$$

where the signs of P, Q, R, S, from the datum level are to be observed.

The level of the centre point A lies mid-way between the levels M and N. Whence the level of A is given by

$$\frac{\frac{P+Q}{2}+\frac{R+S}{2}}{2}=\frac{(P+Q+R+S)}{4}$$

The slope at A in the direction **A***u* is the difference in levels M′, N′ divided by twice the normalised half-span.

$$\mathbf{A}u=\frac{(P-Q)/2-(R-S)/2}{2}=\frac{P-Q-R+S}{4}$$

The slope at A in the direction **A***w* is by similar route.

$$\mathbf{A}w=\frac{(P-S)/2-(R-Q)/2}{2}=\frac{P+Q-R+S}{4}$$

The twist at A, denoted by **A***uw* is given by

$$\mathbf{A}uw=\frac{(P-Q)/2+(R-S)/2}{2}=\frac{P-Q+R+S}{4}$$

In accord with the recommendations of Yuille we make **A***u*, **A***w* some proportion *ku*, *kw* of the half-span. That is **A***u* = *k*(M – A), **A***w* = *k*(M′ – A). (N.B. *ku* = *kw* = *k* because the panel is square. The position, two slopes and twist at **A** are then given by

$$\mathbf{A}=\tfrac{1}{4}(P+Q+R+S) \qquad \mathbf{A}u=\tfrac{1}{4}k(P-Q-R+S)$$

$$\mathbf{A}_w=\tfrac{1}{4}k(P+Q-R-S) \qquad \mathbf{A}_{uw}=\tfrac{1}{4}k^2(P-Q+R-S)$$

It is found that if $k = 3$, adjacent patches will blend smoothly without the occurrence of overshoot.

The points MN, M′N′ are thus located to match the slopes and the points PQRS chosen in such a way that they control the twist of the surface at the point A. The matrix for patches adjacent to the centre point A is

$$[\mathbf{A}\quad \mathbf{A}_u\quad \mathbf{A}_w\quad \mathbf{A}_{uw}]=\begin{bmatrix}1 & 3 & 3 & 9\\ 1 & -3 & 3 & -9\\ 1 & -3 & -3 & 9\\ 1 & 3 & -3 & -9\end{bmatrix}[\mathbf{P}\quad \mathbf{Q}\quad \mathbf{R}\quad \mathbf{S}]$$

The equivalent column presentation is

$$\begin{bmatrix} \mathbf{A} \\ \mathbf{A}_u \\ \mathbf{A}_w \\ \mathbf{A}_{uw} \end{bmatrix} \begin{bmatrix} 1 & 1 & 1 & 1 \\ 3 & -3 & -3 & 3 \\ 3 & 3 & -3 & -3 \\ 9 & -9 & 9 & -9 \end{bmatrix} \begin{bmatrix} \mathbf{P} \\ \mathbf{Q} \\ \mathbf{R} \\ \mathbf{S} \end{bmatrix} \tag{10.13}$$

10.11 NUMERICAL MASTER GEOMETRY

The shape design and manufacturing system developed by Sabin and others is widely used in the aircraft industry and we shall approach the topic by way of Sabin's original paper.

If P is a generic point on a surface, defined by the vector function P(u, v), in which the parameters u and v provided the necessary two degrees of freedom over the entire surface, any point (i, j) may be represented by the equation

$$\mathrm{P} = \sum_{i=o}^{3} \sum_{i=j}^{3} \mathbf{A}_{ij}\, u^i\, v^j \tag{10.14}$$

The spline curves, Fig. 10.41(a), may be bridged by any number of transverse spline curves, as shown in Fig. 10.41(b), such that the points of crossing, nodes or knots, map the contour of a surface.

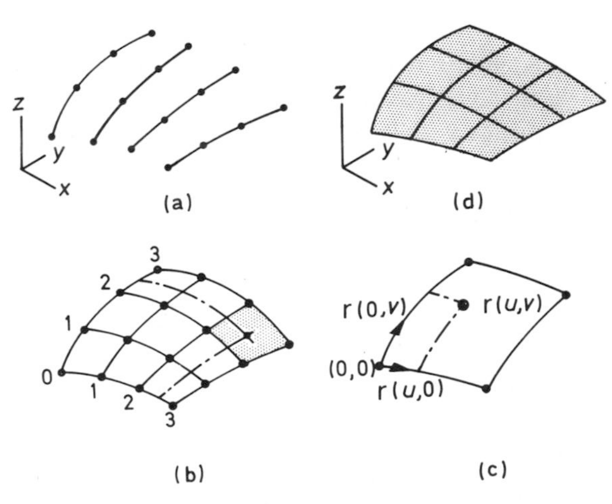

Figure 10.41

The first set of splines are analogous to a series of different directrices. The second set of splines are analogous to a generatrix which changes its shape as it progresses across the directing curves.

If, as in Fig. 10.41(c) we have a twelve node surface then there are six tiles or patches, and in accord with the previous method we compute the surface by letting u and v take values from 0 to 1 across each patch and with a plane curve a different equation is formed for each interval (patch).

If a point P in patch 6, Fig. 10.41(c), has coordinates (2.4, 2.6) in relation to the absolute intrinsic scale then the coordinates of P in relation to the origin of patch 6 are 0.4 and 0.6.

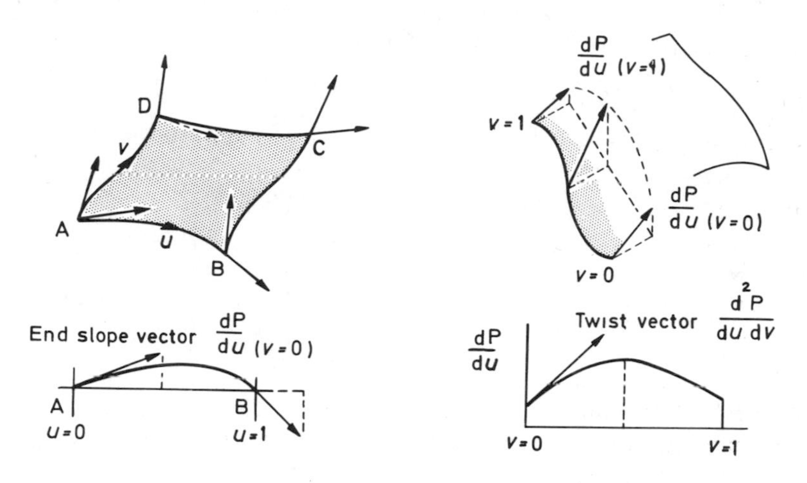

Figure 10.42

In order to provide continuity of position and slope throughout the surface we need to match the slopes in successive patches.

$$
\begin{aligned}
P &= \Sigma\Sigma\ \bar{\mathbf{A}}_{ij}\, u^i\, v^j \\
\frac{\partial P}{\partial u} &= \Sigma\Sigma\ i\ \bar{\mathbf{A}}_{ij}\ u^{i-1}\ v^j \qquad (10.15) \\
\frac{\partial P}{\partial v} &= \Sigma\Sigma\ j\ \mathbf{A}_{ij}\, u^i\, v^{j-1}
\end{aligned}
$$

Where $\partial P/\partial u$ and $\partial P/\partial v$ are tangents to the surface and $\partial^2 P/\partial u \partial v$ is the usual measure of twist. There are thus two tangent vectors and one twist vector to be calculated or specified for each patch corner. The calculation or specification of $\partial P/\partial u$, $\partial P/\partial v$, $\partial^2 P/\partial u \partial v$ determines the magnitude of the coefficients $\mathbf{A}_{ij}$.

The essence of the NMG method described by Hughes and Davies is to define the surface coordinates (x, y, z) in terms of two parameters u and v, in such a way that the y coordinate is independent of u. In the case of a wing surface, for instance, the curves of constant v are the streamwise aerofoil sections, defined by $x = x(u)$, $z = z(u)$ and the surface as a whole being described by

$$x = x(u, v)$$
$$y = y(v)$$
$$z = y(u, v)$$

The spanwise surface may thus be generated so that it passes through a number of equally spaced aerofoil (cross) sections denoted by $(1, z_1), (2, z_2), \ldots (m, z_m)$ with slopes of $\dot{z}_1, \dot{z}_2, \ldots \dot{z}_m$ etc. Where the equation

$$z(u) = z_i f_0(u) + z_{i+1} f_1(u) + \dot{z}_i f_2(u) + \dot{z}_{i+1} f_3(u)$$

represents the curve passing through the interval i to $i + 1$, whence

$$z(u) = [f_0(u)\ f_1(u)\ f_2(u)\ f_3(u)] \begin{bmatrix} z_i \\ z_{i+1} \\ \dot{z}_i \\ \dot{z}_{i+1} \end{bmatrix}$$

A data block of m points would thus contain $2m$ coefficients and this compares well with the $4m - 4$ coefficients needed to resolve the curve $z(u) = a_0 + a_1 u + a_2 u^2 + a_3 u^3$ explicitly.

(N.B. Hughes and Davis adopt the sequence $(z_i, \dot{z}_i, z_{i+1}, \dot{z}_{i+1})$ and denote the intrinsic functions by (f_1, f_2, f_3, f_4).

The equivalent surface definition in which the intrinsic functions are applied in both the u and v directions is given by the matrix representation

$$z(u, v) = [f_0(u)\ f_1(u)\ f_2(u)\ f_3(u)] \begin{bmatrix} & \\ & \\ & \end{bmatrix} \begin{bmatrix} f_0(v) \\ f_1(v) \\ f_2(v) \\ f_3(v) \end{bmatrix} \tag{10.16}$$

where the data block takes the form

$$\left[\begin{array}{c|c} \text{position coordinates} & v \text{ tangents} \\ \hline u \text{ tangents} & uv \text{ twists} \end{array}\right]$$

and an extended data block would follow the following pattern:

$$\left[\begin{array}{cccc|cccc} z_{00} & z_{01} & \cdots & z_{0m} & z^{v}_{00} & z^{v}_{01} & \cdots & z^{v}_{0m} \\ z_{10} & z_{11} & \cdots & z_{1m} & z^{v}_{10} & z^{v}_{11} & \cdots & z^{v}_{1m} \\ & & & & & & & \\ z_{n0} & & & & z^{v}_{n0} & & & \\ \hline z^{u}_{00} & & & & z^{uv}_{00} & z^{uv}_{01} & & \\ z_{10} & & & & & & & \\ & & & & & & & \\ z^{u}_{n0} & & & & z^{uv}_{n0} & & & z^{uv}_{nm} \end{array}\right]$$

Since there are four vectors u, v, $\dot{u}$, $\dot{v}$, associated with each control point, there are $4mn$ vectors in an $m \times n$ array and this compares favourably with the $16(m - 1)(n - 1)$ coefficients which would need to be calculated if the equation

$$\mathbf{Z}(u, v) = \sum_{i=0}^{3} \sum_{j=0}^{3} a_{ij}\, u^{i}\, v^{j}$$

were to be solved without reference to the intrinsic functions.

So far no mention has been made regarding continuity of curvature, however since

$$\begin{aligned} f_0(u) &= 1 - 3u^2 + 2u^3 & \ddot{f}_0(u) &= -6 + 12u \\ f_1(u) &= 3u^2 - 2u^3 & \ddot{f}_1(u) &= 6 - 12u \\ f_2(u) &= u - 2u^2 + u^3 & \ddot{f}_2(u) &= -4 + 6u \\ f_3(u) &= -u^2 + u^3 & \ddot{f}_3(u) &= -2 + 6u \end{aligned} \tag{10.17}$$

we may derive the equation which governs continuity of curvature as follows.

Continuity of curvature is achieved when the incoming and outgoing curvatures at a point are equal and it follows that for the interval $i - 1, i$

$$[\ddot{f}_0(1)\ \ddot{f}_1(1)\ \ddot{f}_2(1)\ \ddot{f}_3(1)] \begin{bmatrix} z_{i-1} \\ z_i \\ \dot{z}_{i-1} \\ \dot{z}_i \end{bmatrix} = [\ddot{f}_0(0)\ \ddot{f}_1(0)\ \ddot{f}_2(0)\ \ddot{f}_3(0)] \begin{bmatrix} z_i \\ z_{i+1} \\ \dot{z}_i \\ \dot{z}_i \end{bmatrix}$$

and by setting $u = 1, u = 0$, in equations 10.17 continuity of curvature is assured,

$$[6\ \ -6\ \ 2\ \ 4]_{u=1} \begin{bmatrix} z_{i-1} \\ z_i \\ \dot{z}_{i-1} \\ \dot{z}_i \end{bmatrix} = [-6\ \ 6\ \ -4\ \ -2]_{u=0} \begin{bmatrix} z_i \\ z_{i+1} \\ \dot{z}_i \\ \dot{z}_{i+1} \end{bmatrix}$$

whence

$$6z_{i-1} - 6z_i + 2\dot{z}_{i-1} + 4\dot{z}_i = -6z_i + 6z_{i+1} - 4\dot{z}_i + 2\dot{z}_{i+1}$$

The condition relating the slopes at $i-1, i, i+1$, and the positions at $i+1, i-1$, is therefore

$$\dot{z}_{i-1} + 4\dot{z}_i + \dot{z}_{i+1} = 3(z_{i+1} - z_{i-1})$$

and this expression must be satisfied at every point if continuity of curvature is required.

Numerical Master Geometry (and similar schemes) are a total computer based commitment. The N.M.G. system was initially developed by BAC (Preston) and BAC (Weybridge) and is based on programs in FORTRAN IV. The system was initially developed to facilitate the interactive design manufacture and interrogation (by stress men and others) of three dimensional surface shapes as currently encountered in aircraft, ships and automobiles, etc. The wide scope of the system does, however, extend its use to the design and manufacture of both external and internal shapes of most practical kinds. Propellers, turbine blades, plenum chambers, shoe lasts, tailors' dummies and *objets d'art* are among items listed by Hitch and Sabin in an early paper.

10.12 THE CONCEPT OF A TOTAL CAE SYSTEM

Whilst the total integration of design analysis, manufacture: estimating, packaging, marketing, after sales service and customer feed-back is regarded by many as a natural development. The short term implementation of any such scheme is clearly subject to many technical, managerial and economic constraints.

With regard to design analysis and manufacture. The requirement is that the total system should integrate both the activity and product of all traditionally separate phases and must provide immediate access to project and other relevant data on an on-going/recall and here-now interactive basis: and yet the problems associated with the protection of RESTRICTED data are not to be underrated.

At the present time many companies have their own N.C. machine tools which can be told (programmed), to varying degrees, to produce a relatively complex component. Whilst a growing number of quantity producers have automated *robots* that do other straightforward jobs.

The A.P.T. (automated programmed tools) system of computer assisted programming of numerically controlled machine tools and draughting machines was pioneered jointly by the Illinois Institute of Technology Research Institute, (N.B. this is the correct U.S. phraseology), and Univac, Sperry Rand Corporation over twenty years ago.

The use of these machines does not in principle entail in depth geometrical knowledge, as the task of directing these machines is largely a matter of programming. To this end a very comprehensive APT dictionary is available.

Early CAD and CAM systems could not provide integration of the complete product development cycle. However, there are now several so called CAE systems on the market. One such system is CAMX.

The following explicit summary of CAMX user facilities figure 10.43 is reproduced here, by kind permission of Ferranti Cetex Graphics Ltd., Livingstone, Scotland.

The current position in respect of computer aided draughting CADD is that we have, or rather can have, if we are prepared to pay, the facility to free-draw onto a visual display unit (typically much smaller than the traditional small double elephant/A.I. drawing board but with the facility to erase and blow-up for detailed work at the press of a button). The free-draw mode may entail the use of a hand held light pen or a joystick device, as in an aeroplane. Diagrams may also be drawn on the screen by inputting coordinate data via a typewriter. All geometrical details are stored within the machine (until they are dumped onto a disc or erased) and so called PRIMITIVES, comprising standard geometrical shapes, forms and annotations are available via a push button menu or via a disc. A so-called HARD COPY of the finished work may be obtained from a flat bed plotter. Microfilm facilities are also available.

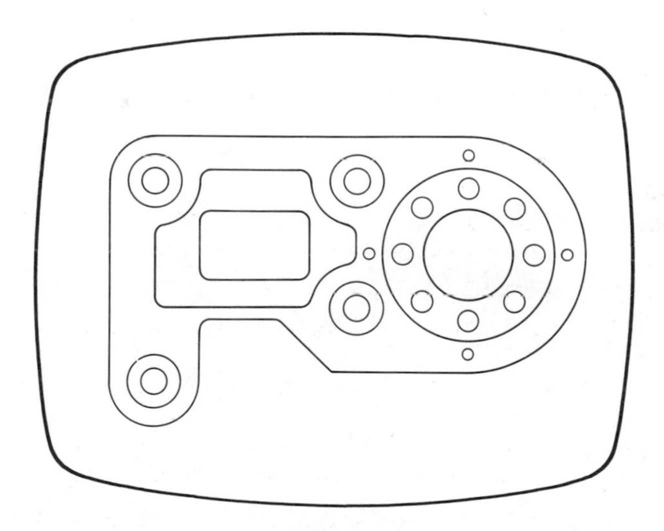

DESIGN

At the CAM-X workstation the designer constructs 2D, 2½D or 3D shapes from conceptual ideas or sketches. 2D views are represented on the graphics display for verification and editing before being transferred to the solid geometric modeller.

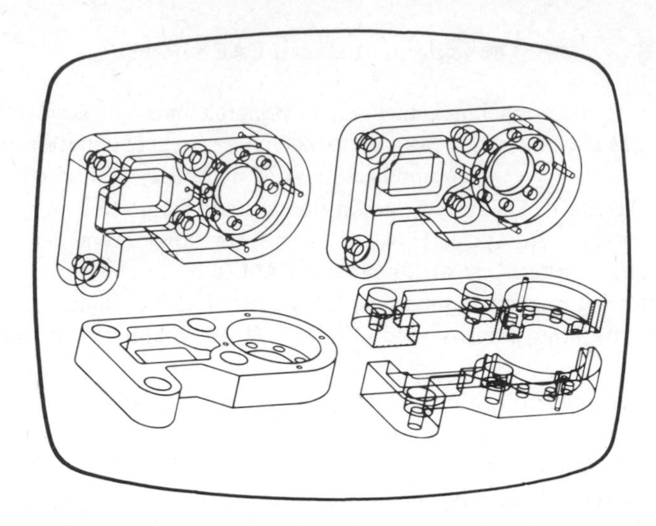

MODELLING

The graphic shapes created at the design stage are combined with non graphic attributes such as material specification and properties to generate solid 2½D and 3D models. Starting with the overall product model, views from any angle and sections through any plane help to verify the product design enabling the designer to establish sub-assemblies and subsequently the detail components.

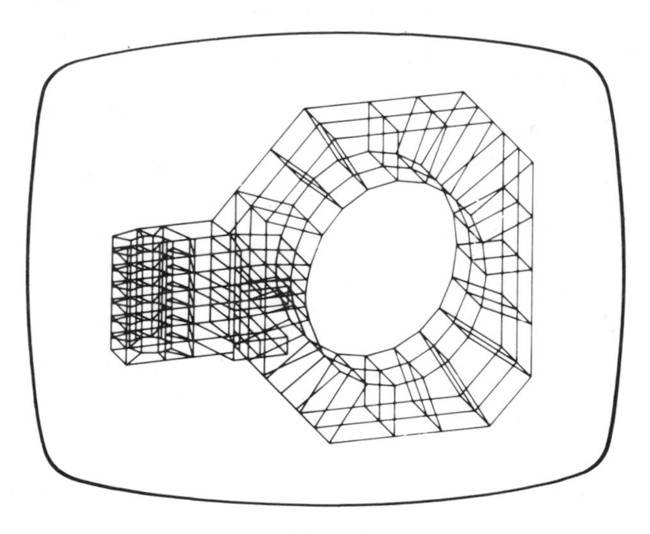

ANALYSIS

The designer who needs to analyse stresses in complex components transmits the solid model to an automatic F.E. mesh generator for subsequent input to any one of the conventional proprietary FE analysis packages. The generated mesh is viewed at the workstation for verification and editing.

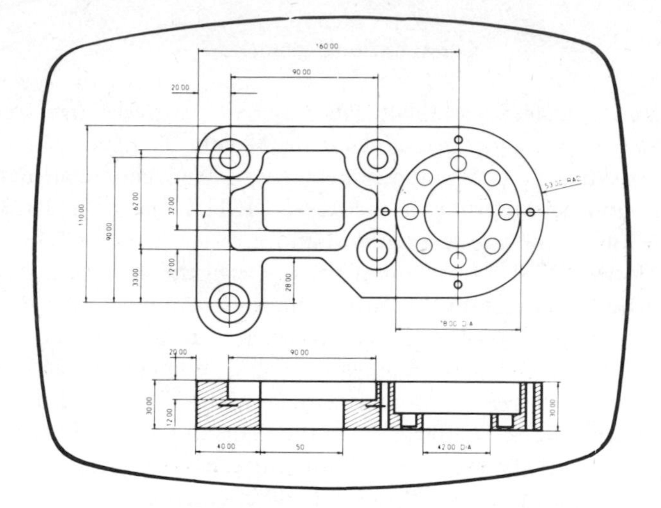

DRAUGHTING

Hardcopy drawings can be produced at any stage in the design, modelling and analysis processes. The designer can generate fully dimensioned drawings, create orthographic drawings, isometric and perspective views of components and assemblies with or without hidden line removal. Sectioned views may be shown through any specified plane.

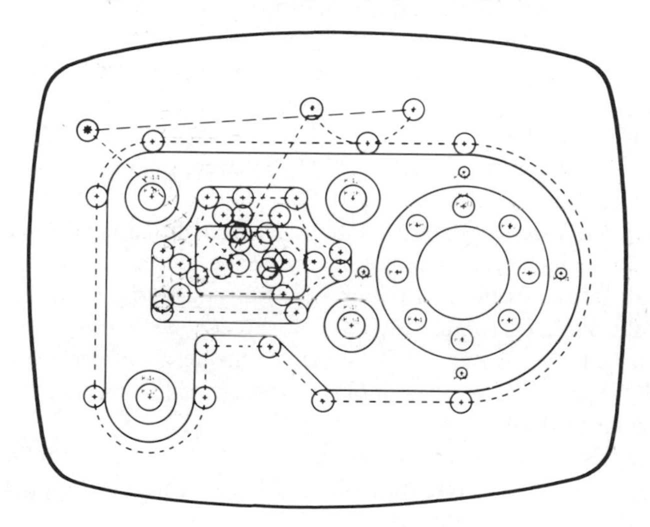

MANUFACTURE

Part descriptions are recalled from the database and machining information is added by the production engineer at the workstation to automatically generate N.C. data for drilling, milling, turning and, following the shape nesting facility, flame cutting and nibbling NC data is available. Animated visualisation of the tool paths gives a rapid verification before automatic post processing to a wide range of N.C. machines.

Figure 10.43

The source of detailed coordinate data may be: a relatively large, true-to-scale orthographic drawing: a physical model: a mathematical model.

When a true-to-scale orthographic drawing is to hand, coordinate data may be picked off using an electronically actuated DIGITIZER. A similar 3D device may also be used in the case of a physical model.

In The Bézier UNISURF system of shape design the designer may prepare a scant set of lines defining his board requirements and opt to perfect these by means of the computational techniques introduced in this chapter. As we have seen the designer defines an open polygon of straight lines which is displayed on a graphics screen. The system responds with a smooth curve which approximates the polygon. By making modifications to the shape of the straight sided polygon the designer can modify the shape of the curve in a predictable way until it satisfies all his requirements and this is a truely interactive process.

The following examples are reproduced by kind permission of Delta Computer Aided Engineering Ltd., Birmingham. They illustrate the type and scope of work currently undertaken.

The computer aided design and manufacturing programmes, code named DELTACAM 1 were developed by Delta C.A.E. Ltd, from three proven sets of software, CNC and POLYSURF, proprietary products of the CAD centre and DUCT from the Wolfson Cambridge Industrial Unit. DUCT and POLYSURF facilitate the 3 dimensional presentation of a component and may also be used for defining surfaces for finite element analysis. Programmes in DELTACAM 1 enable 2, $2\frac{1}{2}$ and 3D components to be produced using almost any machine tool fitted with CNC control.

(N.B. The development of DELTA CAM software has been supported by the Department of Industry since 1977).

DELTA CAM 2 incorporates an interactive draughting facility which allows the designer complete freedom to modify the shape of a component as it is being designed and facilitates the production of a fully dimensional working drawing once the initial design is completed.

DUCT programmes are currently applied to the design of engineering castings and mouldings and as mentioned above are being used to programme numerical controlled machine tools.

Current applications include the design and manufacture of tooling for combustion engine manifold cores, see figure 10.44, duct intersection blends figure 10.45, and so on.

With regard to surface design, DUCT has a powerful interactive facility which enables the designer to develop a complete surface definition of an object and to modify both the original cross sections and spine curve profiles to suit the ongoing requirements of the developing design.

The automatic calculation of cross sectional area surface area and volume is an important adjunct, which greatly speeds the normally laborious process of estimating the amount of material contained in a casting, moulding or forging,

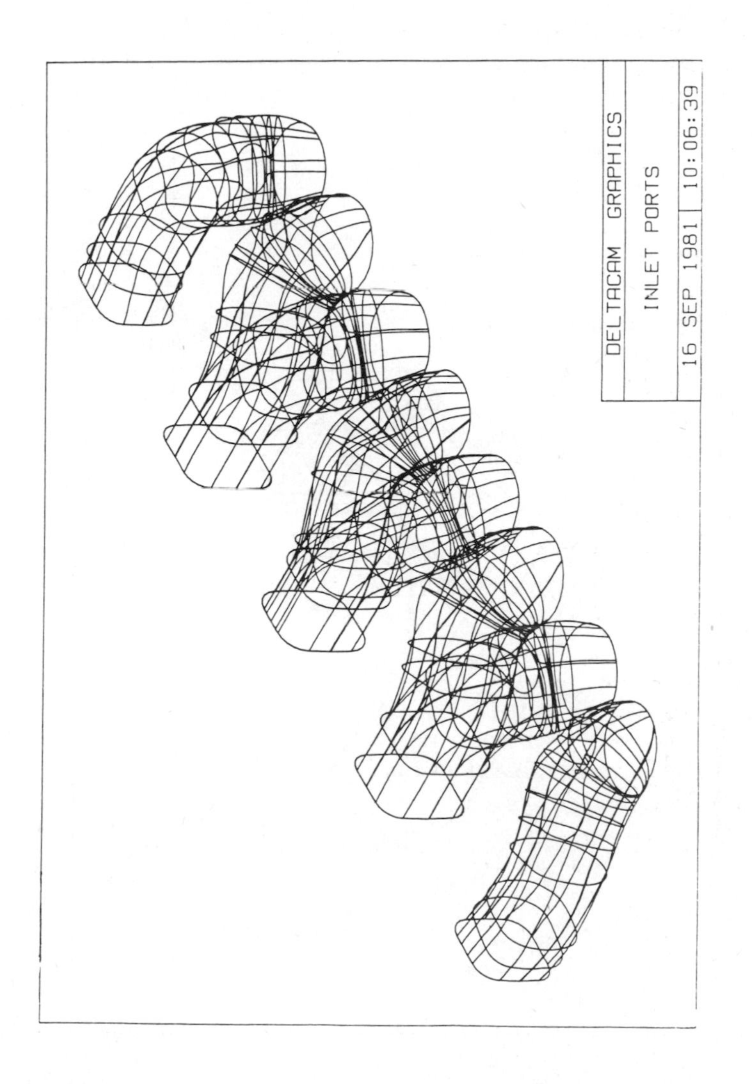

Figure 10.44

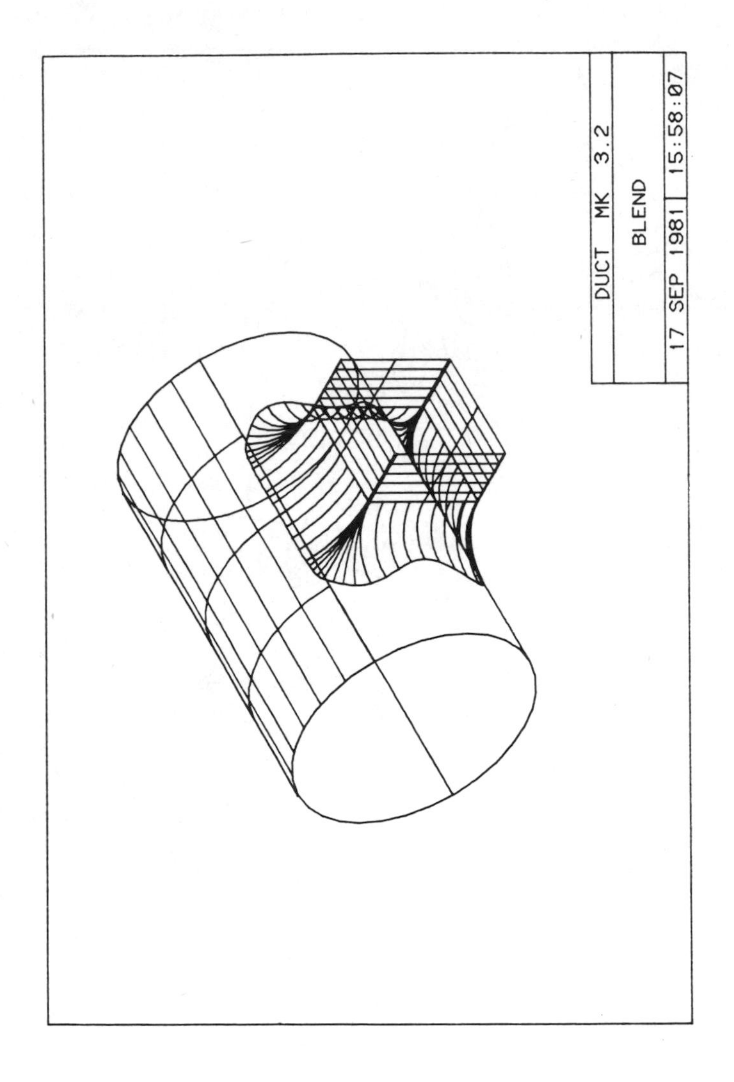

Figure 10.45

or alternatively the interal volume or fluid content of a bottle or container.

Many engineering components, such as the con rod are characterised by interconnected primitive forms (two circular cylinders spaced and joined by a rod piece) which are there after blended.

11 Computer aided graphics

The inventions of science can be used to do the old things more conveniently or cheaply, or to do things that could not be done before.

J. M. Richards

11.1 INTRODUCTION

Throughout this book we have been principally concerned with the derivation and application of equations relating to elemental spatial forms. Emphasis has been placed on the use of the Cartesian coordinate method and it is clear from the few simplistic examples considered that an immense amount of routine calculation is inevitably involved in all but the simplest cases.

In view of the repetitious nature of many of the tasks to be performed, there is much which can be programmed and delegated to the computer. Not only are stresses, strains and natural frequencies, in structures, machines and fluids, calculated in detail and with extreme precision (using the finite element method) but multi-axis machine tools and programmed robots now produce the shapes and forms of many of our designs.

The on-going provision of hardware and software necessary to these operations is, perhaps, the most remarkable achievement ever accomplished by man in the space of a quarter-century. There are many splendid books on the market that treat these issues. We, however, continue to concentrate our attention on the related geometrical problems. One of the more recent developments, much in the news at the moment, is the widespread industrial implementation of new computer aided graphics facilities and it is to this field of endeavour we devote our attention.

11.1.1 Computer aided graphics

There are numerous picture projection techniques used by the architect and engineer and the basis of these has been developed in Chapter 1. We now consider how these same techniques can be executed by a computer aided graphics system.

The first, most obvious, requirement is a means of defining an object in terms that the computer can understand, i.e. all cardinal points on the object are specified by calling all relevant (x, y, z), (r, θ, ϕ) coordinate dimensions, from which the actual sizes and angles, etc. are determined.

When the problem is considered, *ab initio*, a drawing or at least a reasonable sketch of the object is required, from which the coordinate data can be calculated or taken. This initial picture may be produced manually, as in Chapter 1, or it may be produced using the free-draw facility of a modern computer aided graphics device.

When a formal drawing is produced by hand, it is deemed to be of a dimensional accuracy commensurate with the work in hand. Such that all relevant coordinates may be picked off the drawing, by means of a digitizer and stored in a retrievable form. (N.B. Readers needing an introduction to computer aided draughting and manufacturing systems are referred to *Computer Aided Design and Manufacture*, by Besant).

An alternative approach is to input one's requirements using the free-draw facility of a graphics device. The shaky lines one draws with a cumbersome light pen, or touchy joystick, are automatically smoothed by the computer. Or, if one merely points to specific positions, lines and curves can be drawn between and/or through these point positions simply by pressing the appropriate button(s). Coordinate data may, of course, also be loaded using a keyboard or data tape.

11.1.2 Calculation of coordinate data

All who wish to instruct the computer to draw, display and print-out pictures and plans of specific three-dimensional objects must first supply a complete list of all relevant coordinates. In the simple case of a unit cube sited centrally about the origin, with edges parallel or perpendicular to the coordinate axes, is easy enough. Calling the coordinates of a tetrahedron positioned off-zero, with its axis at a compound angle presents additional problems, whilst calling the coordinates of a truncated octahedron is a little more difficult and for those without prior geometrical understanding the calculation of the coordinates of polyhedron with, say, more than twenty nodes (vertices) may well prove an intractable problem.

As is so frequently the case in problem solving the difficulty of solution depends on the approach adopted and the approach adopted in this case is likely to be influenced by one's own prowess at three-dimensional thinking and/or by one's knowledge of the structural configuration of each particular figure.

If one can adopt the right, or rather the most expedient approach, one's troubles are, at the very least, halved. If one is unlucky enough to adopt the wrong approach then one is likely to get bogged down. A popular way of viewing a cubeoctahedron is shown at the top of Fig. 11.1, and no small amount of work would be involved in calculating its nodal coordinates when in this orientation. However, as readers who have learned the lessons of Chapter 1 will know, the calculation and subsequent specification of coordinates can be much more easily achieved by considering the two mutually perpendicular orthographic views in the lower left-hand corner, Fig. 11.1.

The stark simplicity of these two views allows all the coordinates to be written

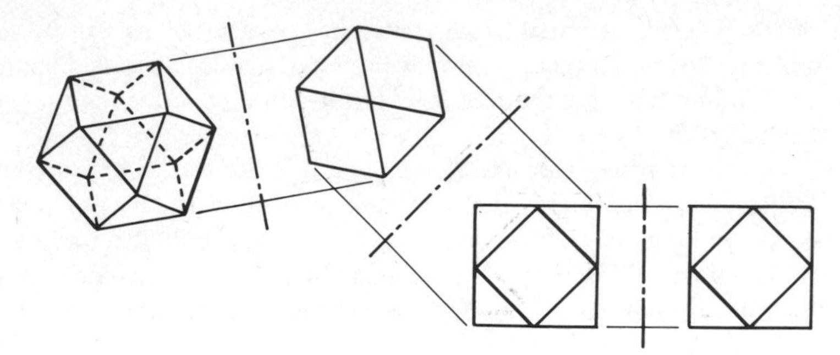

Figure 11.1

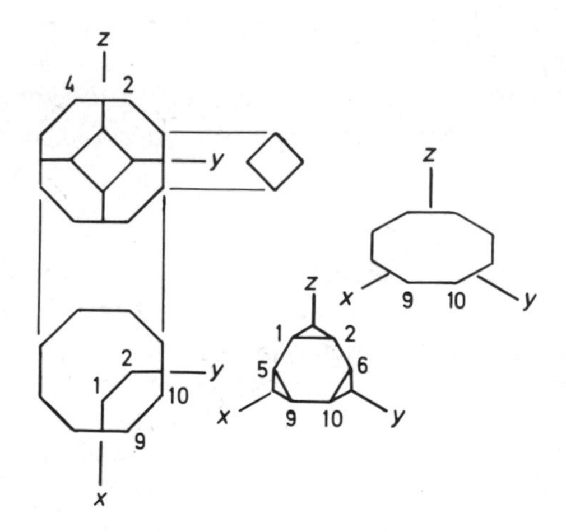

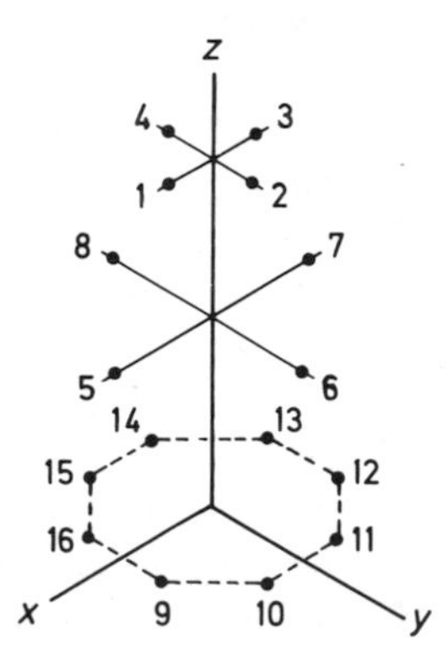

Figure 11.2

directly. Starting with the frontal face the twelve vertices are numbered in counter-clock order.

If local axes are sited as shown in the figure, with *ox, oy, oz* respectively parallel to the ground lines then the call-up of coordinates for a unit edge length is as follows:

V	x	y	z
1	0	0	$\frac{1}{\sqrt{2}}$
2	0	$\frac{1}{\sqrt{2}}$	$\sqrt{2}$
3	0	$\sqrt{2}$	$\frac{1}{\sqrt{2}}$
4	0	$\frac{1}{\sqrt{2}}$	0
5	$\sqrt{2}$	0	$\frac{1}{\sqrt{2}}$
6	$\sqrt{2}$	$\frac{1}{\sqrt{2}}$	$\sqrt{2}$
7	$\sqrt{2}$	$\sqrt{2}$	$\frac{1}{\sqrt{2}}$
8	$\sqrt{2}$	$\frac{1}{\sqrt{2}}$	0
9	$\frac{1}{\sqrt{2}}$	0	$\sqrt{2}$
10	$\frac{1}{\sqrt{2}}$	$\sqrt{2}$	$\sqrt{2}$
11	$\frac{1}{\sqrt{2}}$	$\sqrt{2}$	0
12	$\frac{1}{\sqrt{2}}$	0	0

An alternative approach well suited to the truncated octahedron is to isolate two diametrically opposite squares and regard these as polar faces. The equatorial girth can then be recognised as a regular octagon. Between the top polar face and the equatorial, four other vertices lie in a plane parallel to the equatorial. If the four vertices of the square polar face lie along the $\pm ox$, $\pm oy$ axes, the four sides of the equatorial octagon lie across the $\pm ox$, $\pm oy$ axes, and the four points in the intermediate layer lie on the axes $\pm ox$, $\pm oy$ as shown in Fig. 11.2(a) and this distribution of vertices is repeated below the equator as shown in Fig. 11.2.

Coordinates below the equator are thus numerically the same as corresponding coordinates above the equator except that the sign of the z coordinate is changed. Similar sign asymmetrics are also present in x and y.

The coordinates for the unit edge truncated octahedron, listed below, are easily obtained from the figure

V	x	y	z
1	$\frac{1}{\sqrt{2}}$	0	$\sqrt{2}$
2	0	$\frac{1}{\sqrt{2}}$	$\sqrt{2}$
3	$-\frac{1}{\sqrt{2}}$	0	$\sqrt{2}$
4	0	$-\frac{1}{\sqrt{2}}$	$\sqrt{2}$
5	$\sqrt{2}$	0	$\frac{1}{\sqrt{2}}$
6	0	$\sqrt{2}$	$\frac{1}{\sqrt{2}}$
7	$-\sqrt{2}$	0	$\frac{1}{\sqrt{2}}$
8	0	$-\sqrt{2}$	$\frac{1}{\sqrt{2}}$
9	$\sqrt{2}$	$\frac{1}{\sqrt{2}}$	0
10	$\frac{1}{\sqrt{2}}$	$\sqrt{2}$	0
11	$-\frac{1}{\sqrt{2}}$	$\sqrt{2}$	0
12	$-\sqrt{2}$	$\frac{1}{\sqrt{2}}$	0
13	$-\sqrt{2}$	$-\frac{1}{\sqrt{2}}$	0
14	$-\frac{1}{\sqrt{2}}$	$-\sqrt{2}$	0
15	$\frac{1}{\sqrt{2}}$	$-\sqrt{2}$	0

cont'd

V	x	y	z
16	$\sqrt{2}$	$-\frac{1}{\sqrt{2}}$	0
17	$\sqrt{2}$	0	$-\frac{1}{\sqrt{2}}$
18	0	$\sqrt{2}$	$-\frac{1}{\sqrt{2}}$
19	$-\sqrt{2}$	0	$-\frac{1}{\sqrt{2}}$
20	0	$-\sqrt{2}$	$-\frac{1}{\sqrt{2}}$
21	$\frac{1}{\sqrt{2}}$	0	$-\sqrt{2}$
22	0	$\frac{1}{\sqrt{2}}$	$-\sqrt{2}$
23	$-\frac{1}{\sqrt{2}}$	0	$-\sqrt{2}$
24	0	$-\frac{1}{\sqrt{2}}$	$-\sqrt{2}$

(N.B. In many cases the coordinates of an object may be picked off a scale drawing, or model, by means of an electronic recording device known as a digitiser).

Once the coordinates of an object have been obtained (by whatever means) for one orientation, the transformational methods of Chapter 5 may then be applied to reorientate the figure in any attitude required.

11.2 ORTHOGRAPHIC TRANSFORMATION

Given a set of (x, y, z) coordinates which completely define an object, the orthographic traces may be readily plotted in two or more orthographic planes. Plan, elevation, end view, auxiliary view(s) are produced, one from the other, by means of pure orthogonal rotations.

As in the manual technique the object is set up wholly within the positive octant, formed by the three-way intersection of picture planes, see Chapter 1, article 1.1, and providing one is content with so-called wire frame drawings, that in principle is all there is to it. (N.B. A wire frame drawing is one in which hidden detail has not been removed).

If we are using a conventional computerised plotting table we set up the origin of our picture plot axes to suit the relative size of adjacent orthographic views. In the centre of the screen, or plotting space, if three orthographic traces of the same size are envisaged. (N.B. The home zero on a conventional plotting table and screen is a small distance in from the bottom left-hand corner). Hence **a new XY datum** representing the intersection of the orthographic picture planes is set up as explained.

To avoid the possibility of overlapping views, we position the object as explained in Chapter 1 well out in the positive octant. (N.B. We observe that in a third angle projection the picture plane hinge lines OX, OY, OZ, form a left-handed triad). However, since we are to operate on object coordinates (which form a right-handed triad) and plot the results in a two-dimensional picture space, defined by the axes XOY, all rotations obey our previously defined sign convention.

If the coordinates (1, 3, 4), (5, 3, 4), (5, 6, 4), (1, 6, 4) define one face of a rectangular block, in which the (x, y) coordinates are plotted in XOY and called the plan. Then the elevation is obtained by inducing a positive 90° rotation about the axis OX. (N.B. We make OX and ox coincident in this instance). The outcome of a 90° object rotation about the axis ox is as formulated below

$$\begin{bmatrix} 1 & 3 & 4 & 1 \\ 5 & 3 & 4 & 1 \\ 5 & 6 & 4 & 1 \\ 1 & 6 & 4 & 1 \\ 1 & 3 & 8 & 1 \\ 5 & 3 & 8 & 1 \\ \cdot & \cdot & \cdot & \cdot \\ \cdot & \cdot & \cdot & \cdot \\ 0 & 0 & 0 & 1 \end{bmatrix} \begin{bmatrix} 1 & 0 & 0 & 0 \\ 0 & 0 & 1 & 0 \\ 0 & -1 & 0 & 0 \\ 0 & 0 & 0 & 1 \end{bmatrix} = \begin{bmatrix} 1 & -4 & 3 & 0 \\ 5 & -4 & 3 & 0 \\ 5 & -4 & 6 & 0 \\ 1 & -4 & 6 & 0 \\ 1 & -8 & 3 & 0 \\ 5 & -8 & 3 & 0 \\ \cdot & \cdot & \cdot & \cdot \\ \cdot & \cdot & \cdot & \cdot \\ 0 & 0 & 0 & 1 \end{bmatrix}$$

The plan and elevation defined above is as shown in Fig. 11.3.

We may similarly plot auxiliary views by setting up a new ground line. If we require a view in a direction normal to the line through the points (5, 3) and (1, 6) in plan, then we need to compute the block coordinates for a 90° rotation about the new ground line, defined by the new axis ox'. First we rotate the axis ox through a negative angle of arc tan 3/4. Whence sin = −3/5, cos = +4/5. The coordinates of the plan view relative to the ox' axis are given by transformation R_{oz}

$$\begin{bmatrix} 1 & 3 & 4 & 1 \\ 5 & 3 & 4 & 1 \\ 5 & 6 & 4 & 1 \\ 1 & 6 & 4 & 1 \\ \cdot & \cdot & \cdot & \cdot \\ \cdot & \cdot & \cdot & \cdot \\ 0 & 0 & 0 & 1 \end{bmatrix} \begin{bmatrix} \frac{4}{3} & \frac{3}{5} & 0 & 0 \\ -\frac{4}{3} & \frac{4}{5} & 0 & 0 \\ 0 & 0 & 1 & 0 \\ 0 & 0 & 0 & 1 \end{bmatrix} = \begin{bmatrix} -\frac{5}{5} & \frac{15}{5} & \frac{20}{5} & 0 \\ \frac{11}{5} & \frac{27}{5} & \frac{20}{5} & 0 \\ \frac{2}{5} & \frac{39}{5} & \frac{20}{5} & 0 \\ -\frac{14}{5} & \frac{27}{5} & \frac{20}{5} & 0 \\ \cdot & \cdot & \cdot & \cdot \\ \cdot & \cdot & \cdot & \cdot \\ 0 & 0 & 0 & 1 \end{bmatrix} = \frac{1}{5} \begin{bmatrix} -5 & 15 & 20 & 0 \\ 11 & 27 & 20 & 0 \\ 2 & 39 & 20 & 0 \\ -14 & 27 & 20 & 0 \\ \cdot & \cdot & \cdot & \cdot \\ \cdot & \cdot & \cdot & \cdot \\ 0 & 0 & 0 & 1 \end{bmatrix}$$

An object rotation of 90° about the new axis ox' then produces the auxiliary view

$$\frac{1}{5} \begin{bmatrix} -5 & 15 & 20 & 0 \\ 11 & 27 & 20 & 0 \\ 2 & 39 & 20 & 0 \\ -14 & 27 & 20 & 0 \\ \cdot & \cdot & \cdot & \cdot \\ \cdot & \cdot & \cdot & \cdot \\ 0 & 0 & 0 & 1 \end{bmatrix} \begin{bmatrix} 1 & 0 & 0 & 0 \\ 0 & 0 & 1 & 0 \\ 0 & -1 & 0 & 0 \\ 0 & 0 & 0 & 1 \end{bmatrix} = \frac{1}{5} \begin{bmatrix} -5 & -20 & 15 & 0 \\ 11 & -20 & 27 & 0 \\ 2 & -20 & 39 & 0 \\ -14 & -20 & 27 & 0 \\ \cdot & \cdot & \cdot & \cdot \\ \cdot & \cdot & \cdot & \cdot \\ 0 & 0 & 0 & 1 \end{bmatrix}$$

and picture corresponding to these coordinates is shown in the figure 11.3.

If it had been found that the auxiliary view overlapped the elevation then it could have been moved quite easily, simply by adding a negative constant to the auxiliary oy coordinates. That is, by applying a single axis translation.

If the position of the auxiliary view is acceptable then we may re-express the coordinates in terms of the picture screen axes XOY and this entails bringing the ground line axis ox' into coincidence with the picture axis OX. That is, we induce positive rotation of arc tan 3/4 on the axis ox', whence

$$\frac{1}{5} \begin{bmatrix} -5 & -20 & 15 & 0 \\ 11 & -20 & 27 & 0 \\ 2 & -20 & 39 & 0 \\ -14 & -20 & 27 & 1 \\ \cdot & \cdot & \cdot & \cdot \\ \cdot & \cdot & \cdot & \cdot \\ 0 & 0 & 0 & 1 \end{bmatrix} \begin{bmatrix} \frac{4}{5} & -\frac{3}{5} & 0 & 0 \\ \frac{3}{5} & \frac{4}{5} & 0 & 0 \\ 0 & 0 & 1 & 0 \\ 0 & 0 & 0 & 1 \end{bmatrix} = \frac{1}{25} \begin{bmatrix} -80 & -65 & 15 & 0 \\ -16 & -113 & 27 & 0 \\ -52 & -78 & 39 & 0 \\ -116 & -38 & 27 & 0 \\ \cdot & \cdot & \cdot & \cdot \\ \cdot & \cdot & \cdot & \cdot \\ 0 & 0 & 0 & 1 \end{bmatrix}$$

The point P_3 (in auxillary view) for instance has picture coordinates X = –116/25, Y = –38/25 and this is seen to be so from the figure.

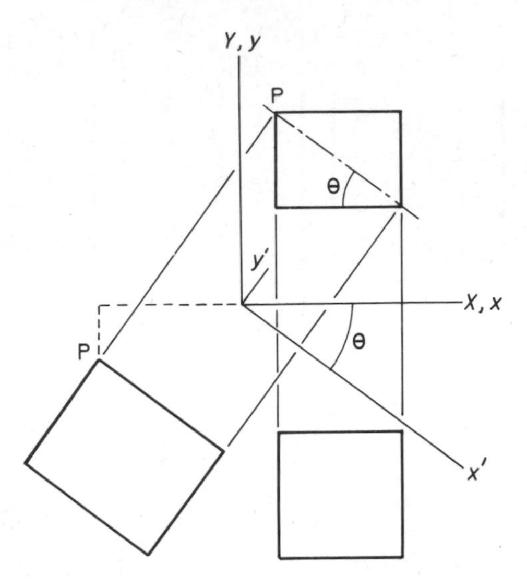

Figure 11.3

11.3 ISOMETRIC TRANSFORMATION

An isometric projection is a picture plane representation of a three-dimensional object rotated and tilted, in 3-space, in such a way that its three principal direction vectors lie at angles (relative to the eye) which make the orthographic traces of these vectors equal in length.

If the plan and elevation of a unit cube are as shown in Fig. 11.4 then a typical isometric representation may be obtained as an end view by direct orthographic projection.

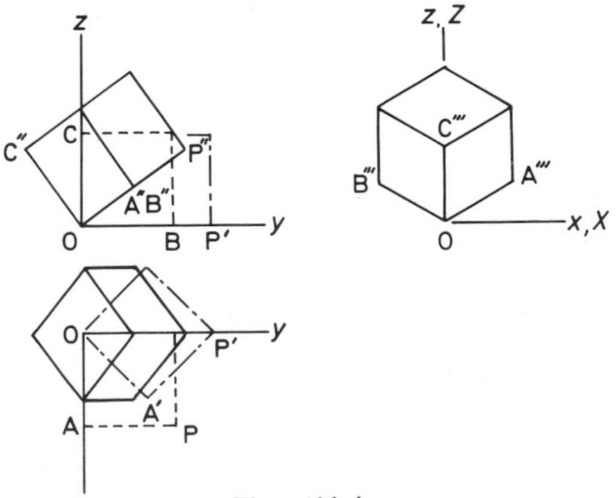

Figure 11.4

The isometric orientation is obtained by first rotating the cube through 45° about the axis *oz* then tilting, that is rotating, the cube through 35.26° about the axis *ox*. (N.B. Proof of these angles is given later).

If the unit cube is initially sited in plan and elevation with its sides OA, OB, OC, aligned with the *ox, oy, oz* axes, then a rotation of 45° about the axis *oz*, followed by a tilt of 35.26° about the axis *ox* brings the cube to the required position. We observe that the orthographic end view is identical to an isometric projection and hence the edges OA and OB are inclined at ±30° to the axis *ox* and the edge OC is vertical. The scale and other features of the end view make it identical with a correctly scaled isometric.

If the orthographic coordinates of the cube corner points are:

O	(0, 0, 0)
A	(1, 0, 0)
P	(1, 1, 0)
B	(0, 1, 0)
C	(0, 0, 1)
	(1, 0, 1)
	(1, 1, 1)
	(0, 1, 1)

We know that an object rotation about the axis *oz* is given by

$$[x' \quad y' \quad z' \quad 1] = [x \quad y \quad z \quad 1] \begin{bmatrix} \cos\theta_z & \sin\theta_z & 0 & 0 \\ -\sin\theta_z & \cos\theta_z & 0 & 0 \\ 0 & 0 & 1 & 0 \\ 0 & 0 & 0 & 1 \end{bmatrix}$$

whence

$$(x', y', z') = (x\cos\theta_z - y\sin\theta_z, x\sin\theta_z + y\cos\theta_z, z)$$

(N.B. It is clear that a rotation about *oz* produces no change in the *z* coordinates, hence $z' = z$.

The transformed coordinates of the point P(1, 1, 0) are $(\cos\theta_z - \sin\theta_z, \sin\theta_z + \cos\theta_z, 0)$. For a rotation of 45° about the axis *oz*, $\cos\theta_z = \sin\theta_z = 1/\sqrt{2}$, hence the coordinates of P(1, 1, 0) are $(0, \sqrt{2}, 0)$ and similarly for other points, details for which are as given in the table.

	Pt	x'	y'	z'
O	(0, 0, 0)	0	0	0
A	(1, 0, 0)	$\frac{1}{\sqrt{2}}$	$\frac{1}{\sqrt{2}}$	0
P	(1, 1, 0)	0	$\sqrt{2}$	0
B	(0, 1, 0)	$-\frac{1}{\sqrt{2}}$	$\frac{1}{\sqrt{2}}$	0
C	(0, 0, 1)	0	0	1
	(1, 0, 1)	$\frac{1}{\sqrt{2}}$	$\frac{1}{\sqrt{2}}$	1
	(1, 1, 1)	0	$\sqrt{2}$	1
	(0, 1, 1)	$-\frac{1}{\sqrt{2}}$	$\frac{1}{\sqrt{2}}$	1

We next rotate (tilt) the cube through 35.26° about the original axis *ox*. The rotary transformation about the axis *ox* is

$$[x'' \quad y'' \quad z'' \quad 1] = [x' \quad y' \quad z' \quad 1] \begin{bmatrix} 1 & 0 & 0 & 0 \\ 0 & \cos\theta_x & \sin\theta_x & 0 \\ 0 & -\sin\theta_x & \cos\theta_x & 0 \\ 0 & 0 & 0 & 1 \end{bmatrix}$$

hence

$$(x'', y'', z'') = (x', \ y' \cos\theta_x - z' \sin\theta_x, y' \sin\theta_x + z' \cos\theta_x)$$

That is point P(1, 1, 0) lies at $(0, \sqrt{2}\cos\theta_x, \sqrt{2}\sin\theta_x)$ when $\theta_x = 35.26°$, as it does for an isometric. The point P(1, 1, 0) transforms to the point (0, 1.15476, 0.816408) and the coordinates of other points are as tabulated.

	Pt	x''	y''	z''
O	(0, 0, 0)	0	0	0
A	(1, 0, 0)	0.7071	0.57738	0.40820
P	(1, 1, 0)	0	1.15476	0.81641
B	(0, 1, 0)	−0.7071	0.57738	0.40820
C	(0, 0, 1)	0	−0.57738	0.81641
	(1, 0, 1)	0.7071	0	1.22468
	(1, 1, 1)	0	0.57738	1.63288
	(0, 1, 1)	−0.7071	0	1.22468

The coordinates (x'', y'', z'') are the 3-space coordinates of a unit cube orientated in one of the most common isometric positions and Fig. 11.4, may thus be plotted using the (x'', y'', z'') coordinates given, or it may be produced by treating the end view as an orthographic projection, in this event an additional orthogonal rotation is applied, as in article 11.2. (N.B. We observe that the ratio z''/y'' point A, for instance, equals 0.40820/0.70711 = 0.57728 = arc tan 30° as it should do.

Having demonstrated the method the two rotation θ_z and θ_x may be combined to form a single matrix.

Thus the contatenation of the two separate rotations

$$\begin{bmatrix} \cos\theta_z & \sin\theta_z & 0 & 0 \\ -\sin\theta_z & \cos\theta_z & 0 & 0 \\ 0 & 0 & 1 & 0 \\ 0 & 0 & 0 & 1 \end{bmatrix} \begin{bmatrix} 1 & 0 & 0 & 0 \\ 0 & \cos\theta_x & \sin\theta_x & 0 \\ 0 & -\sin\theta_x & \cos\theta_x & 0 \\ 0 & 0 & 0 & 1 \end{bmatrix}$$

yields

$$\left[\begin{array}{c|c|c|c} \cos\theta_z + 0 + 0 + 0 & 0 + \sin\theta_z \; \cos\theta_x + 0 + 0 & 0 + \sin\theta_z \; \sin\theta_x + 0 + 0 & 0 \\ -\sin\theta_z + 0 + 0 + 0 & 0 + \cos\theta_z \; \cos\theta_x + 0 + 0 & 0 + \cos\theta_z \; \sin\theta_x + 0 + 0 & 0 \\ 0 \quad + 0 + 0 + 0 & 0 + \quad 0 \; - \sin\theta_x + 0 & 0 + \quad 0 \quad \cos\theta_x + 0 & 0 \\ 0 \quad + 0 + 0 + 0 & 0 + \quad 0 \qquad + 0 & 0 + \quad 0 \qquad + 0 & 1 \end{array}\right]$$

$$\begin{bmatrix} \cos\theta_z & \sin\theta_z \cos\theta_x & \sin\theta_z \sin\theta_x & 0 \\ -\sin\theta_z & \cos\theta_z \cos\theta_x & \cos\theta_z \sin\theta_x & 0 \\ 0 & -\sin\theta_x & \cos\theta_x & 0 \\ 0 & 0 & 0 & 1 \end{bmatrix}$$

For the isometric case in which $\theta_z = 45°$, $\theta_x = 35.26°$, the complete orthographic to isometric transformation is represented by the matrix:

$$\begin{bmatrix} 0.7071 & 0.5774 & 0.4082 & 0 \\ -0.7071 & 0.5774 & 0.4082 & 0 \\ 0 & -0.5774 & 0.8165 & 0 \\ 0 & 0 & 0 & 1 \end{bmatrix}$$

The point P(1, 1, 0) thus transforms from its orthographic position (1, 1, 0) to its isometric position (0.7071 – 0.7071), (0.5774 + 0.5774), (0.4082 + 0.4082) that is (0, 1.1548, 0.8164) and this is confirmed by the previous table.

11.3.1 Derivation of isometric rotation angles

The three sides of the unit cube OA, OB, OC, which initially lie along the *ox, oy, oz*, axes are representative of unit vectors, and since each is homed on zero we see from Fig. 11.4 that

The point A(1, 0, 0) transforms to:

$$(\cos\theta_z, \sin\theta_z\cos\theta_x, \sin\theta_z\sin\theta_x)$$

The point B(0, 1, 0) transforms to:

$$(-\sin\theta_z, \cos\theta_z\cos\theta_x, \cos\theta_z\sin\theta_x)$$

The point C(0, 0, 1) transforms to:

$$(0, -\sin\theta_x, \cos\theta_x)$$

and these points plot to an orthographic end view which corresponds to the isometric scale.

The apparent lengths of OA″, OB″, OC″, are clearly equal and these apparent lengths are of the form $\sqrt{x^2 + z^2}$ hence the unit isometric vectors are

$$OA'' = \sqrt{\cos^2\theta_z + \sin^2\theta_z\sin^2\theta_x}$$

$$OB'' = \sqrt{\sin^2\theta_z + \cos^2\theta_z\sin^2\theta_x}$$

$$OC'' = \sqrt{\cos^2\theta_x} = \cos\theta_x$$

and since (in an isometric) these three lengths are equal we may equate them.

$$\cos^2\theta_z + \sin^2\theta_z\sin^2\theta_x = \sin^2\theta_z + \cos^2\theta_z\sin^2\theta_x = \cos^2\theta_x$$

whence

$$\cos^2\theta_z + \sin^2\theta_z\sin^2\theta_x = \cos^2\theta_x$$

and

$$\sin^2\theta_z + \cos^2\theta_z\sin^2\theta_x = \cos^2\theta_x$$

adding these two equations yields

$$1 + \sin^2\theta_x = 2\cos^2\theta_x$$

hence

$$\cos^2\theta_x = \frac{2}{3} \quad \text{and} \quad \sin^2\theta_x = \frac{1}{3}$$

$$\cos\theta_x = \frac{\sqrt{2}}{\sqrt{3}} \quad \text{whence} \quad \theta_x = 35.26°$$

and on substituting back into the original equation we obtain the corresponding value of θ_z

$$\cos^2\theta_z + \frac{\sin^2\theta_z}{3} = \frac{2}{3}$$

$$2\cos^2\theta_z = 1$$

$$\cos^2\theta_z = \frac{1}{2}$$

$$\cos\theta_z = \frac{1}{\sqrt{2}} \quad \text{whence } \theta_z = 45^\circ$$

$\theta_x = 35.26^\circ, \theta_z = 45^\circ$ are thus the rotational angles required. (N.B. On substituting these angles we obtain an immediate check that the lengths OA″, OB″, OC″, do in fact equal the isometric scale.

We note that

$$\cos^2 45 = \sin^2 45 = \frac{1}{2}, \ \cos^2 35.26 = \frac{1}{3}, \ \sin^2 35.26 = \frac{2}{3}$$

and hence

$$OA'' = \sqrt{\frac{1}{2} + \frac{1}{2}\,\frac{1}{3}} = \sqrt{\frac{2}{3}}$$

$$OB'' = \sqrt{\frac{1}{2} + \frac{1}{2}\,\frac{1}{3}} = \sqrt{\frac{2}{3}}$$

$$OC'' = \sqrt{\frac{2}{3}} = \sqrt{\frac{2}{3}}$$

and $\sqrt{\frac{2}{3}} = 0.8165$ the factor by which the true "orthographic" length is reduced in a pure isometric presentation.

11.4 DIMETRIC TRANSFORMATIONS

All three scales of an isometric projection are equal by virtue of the fact that the three unit scales tilt away or towards the eye at the same angle. In article 11.3 the orthographic cube was brought into the isometric orientation by applying a "symmetrical" rotation of 45° about the axis *oz* followed by a rotation (tilt) of 35.26° about the axis *ox*. If the cube is tilted through an angle other than 35.26° the scale corresponding to *oz*″ is no longer equal to the scales corresponding to *ox*″ and *oy*″. In a dimetric projection two axes are drawn to the same scale, the scale of the third axis being dependent on the angle of tilt. The presentations, Fig. 11.5, are dimetric projections.

If OA, OB, OC, are the sides of a unit cube with sides aligned along the orthographic axes *ox*, *oy*, *oz*, then the points A(1, 0, 0), B(0, 1, 0), C(0, 0, 1)

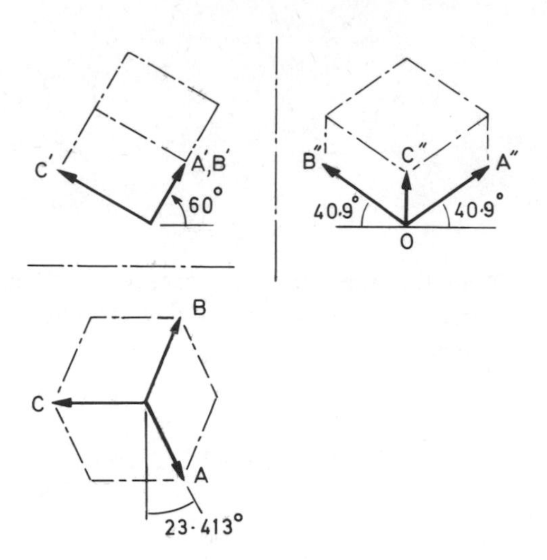

Figure 11.5

transform to the points A″, B″, C″, where the transformed coordinates are given for rotations of θ_z and θ_x by the concatenation matrix, article 11.3.

$$\begin{bmatrix} 1 & 0 & 0 & 0 \\ 0 & 1 & 0 & 0 \\ 0 & 0 & 1 & 0 \\ 0 & 0 & 0 & 1 \end{bmatrix} \begin{bmatrix} \cos\theta_z & \sin\theta_z\cos\theta_x & \sin\theta_z\sin\theta_x & 0 \\ -\sin\theta_z & \cos\theta_z\cos\theta_x & \cos\theta_z\sin\theta_x & 0 \\ 0 & -\sin\theta_x & \cos\theta_x & 0 \\ 0 & 0 & 0 & 1 \end{bmatrix}$$

whence the orthographic coordinates of A″, B″, C″, are:

$$A''\ (\cos\theta_z, \sin\theta_z\cos\theta_x, \sin\theta_z\sin\theta_x)$$
$$B''\ (-\sin\theta_z, \cos\theta_z\cos\theta_x, \cos\theta_z\sin\theta_x)$$
$$C''\ (0,\ -\sin\theta_x,\ \cos\theta_x)$$

If we wish to preserve equal scales along the dimetric axes OA″, OB″ we must clearly rotate the cube through 45° about the axis *oz* and then tilt to an angle other than 35.26° about the axis *ox*.

If for the purpose of example we tilt the cube through an angle of 60° then:

$$\cos\theta_x = \frac{1}{2}, \sin\theta_x = \frac{\sqrt{3}}{2}, \cos\theta_z = \cos\theta_x = \frac{1}{\sqrt{2}}$$

and the transformed coordinates are:

$$A''\left(\frac{1}{\sqrt{2}}, \frac{\sqrt{3}}{4}, \frac{\sqrt{3}}{2\sqrt{2}}\right)$$

$$B''\left(-\frac{1}{\sqrt{2}}, \frac{1}{2\sqrt{2}}, \frac{\sqrt{3}}{2\sqrt{2}}\right)$$

$$C''\left(0, -\frac{\sqrt{3}}{2}, \frac{1}{2}\right)$$

The apparent lengths of OA″, OB″, OC″ (in the *xoz* orthographic plane of Fig. 11.5) are of the form $\sqrt{x^2 + z^2}$ and hence the apparent lengths are

$$OA'' = \sqrt{\frac{1}{2} + \frac{3}{8}} = \sqrt{\frac{7}{8}} = 0.9354$$

$$OB'' = \sqrt{\frac{1}{2} + \frac{3}{8}} = \sqrt{\frac{7}{8}} = 0.9354$$

$$OC'' = \sqrt{0 + \frac{1}{4}} = \sqrt{\frac{1}{4}} = 0.5000$$

The scales corresponding to OA″ and OB″ are therefore equal and due to the tilt of 60°, the apparent length of OC″ is foreshortened. The three scales are OA″ = OB″ = 0.9354, OC″ = 0.5000.

The apparent angle of OA″ and OB″ relative to the orthographic (picture) axis *ox* is obtained from the coordinate table.

$$\text{arc tan } \frac{OA''}{OB''} = \frac{\sqrt{3}}{2\sqrt{2}}\bigg/\frac{1}{\sqrt{2}} = \frac{\sqrt{3}}{2}$$

whence angle A″*ox* = 40° – 54′ and angle B″*ox* = –40° – 54′.

If the end view, Fig. 11.5, is acceptable as a pictorial representation, then given the appropriate angles, scales and coordinates, the dimetric representation may be readily drawn without reference to other orthographic views.

11.5 ENGLISH OBLIQUE TRANSFORMATION

A system of drawing, known as English oblique, in which the frontal plane is drawn true shape/true scale with top and side planes visible, is a conventional method sometimes used to speed the process of manual draughting. It is clear, however, that it is not physically possible to view three sides of a cube when one side is viewed true shape. (N.B. The terms Cavalier and Cabinet projection are used in the United States. (a) to describe an oblique view with scales 1 : 1 : 1 (b) to describe an oblique view with scales 1 : 1 : 0.5 respectively).

If the axis *ox*″ is set horizontal with *oy*″ directed away at an angle then the axis *oz*″ must tilt out of the picture plane and must be foreshortened. Conversely, if the axis oz'' is vertical then the axis ox'' cannot be horizontal and hence it is the axis ox'' which must be foreshortened.

Suppose we ignore the fact that either OA″ or OC″ is foreshortened when the (side) OB″ is visible and let us suppose we have artistic licence to draw OA″ = OC″, even though we know the plane containing OA″ and OC″ is tilted.

If OA″ = OC″, $(x_A^2 + z_A^2) = (x_C^2 + z_C^2)$, and since

$$OA'' = (\cos\theta_z, \sin\theta_z \cos\theta_x, \sin\theta_z \sin\theta_x)$$

$$OB'' = (-\sin\theta_z, \cos\theta_z \cos\theta_x, \cos\theta_z \sin\theta_x)$$

$$OC'' = (0, \ -\sin\theta_x, \ \cos\theta_x)$$

$$\cos^2\theta_z + \sin^2\theta_z \sin^2\theta_x = \cos^2\theta_x = 1 - \sin^2\theta_x$$

and since θ_x and θ_z are necessarily small, $\sin\theta_z \triangleq \tan\theta_x$, hence

$$\sin^2\theta_z \triangleq \tan^2\theta_x \quad \text{and} \quad \cos^2\theta \triangleq 1 - \tan^2\theta_x$$

whence

$$x^* = z^* = \sqrt{1 - \tan^2\theta_x}$$

and since

$$\begin{aligned} y^* &= \sqrt{\sin^2\theta_z + \cos^2\theta_z \sin^2\theta_x} \\ &= \sqrt{\tan^2\theta_x + (1 - \tan^2\theta_x)\sin^2\theta_x} \\ &= \sqrt{\tan^2\theta_x + \sin^2\theta_x - \tan^2\theta_x \sin^2\theta_x} \\ &= \sqrt{\tan^2\theta_x \cos^2\theta_x + \sin^2\theta_x} \\ &= \sqrt{2\sin^2\theta_x} \end{aligned}$$

We see that for small angles of tilt (θ_x up to 10°) $\sqrt{1 - \tan^2\theta_x}$ differs from unity by less than 2% and $\sqrt{2\sin^2 10} = 0.2456 = 0.25$ say. Scales of 1:0.25:1) and angle $y^*ox = 45°$, thus provide a reasonable representation for tilts of 10°.

If we wish to give OB″ more prominence then we could arbitrarily choose to make OB″ = 0.5, such that $y^* = 0.5^2 = 2\sin^2\theta_x$ and this would correspond to a tilt of $\theta_x = 20.7048°$. Under these conditions

$$\cos^2\theta_x = \quad 1 - \frac{0.5^2}{2} \quad = 0.875$$

whence

$$\cos\theta_x = 0.9354$$

In manual draughting we may, in the interests of simplicity, choose to use scales of (1 : 1 : 0.5) and $y^*ox = 45°$, $\theta_x = 20.7048°$, $\theta_z = 22.208°$ without too much artistic licence. In computer graphics, however, it is the computer that does the work and such simplifications are unnecessary.

11.6 AXONOMETRIC TRANSFORMATIONS

An axonometric transformation is one in which all three scales are different. Any sequence of object rotations which do not lead to isometric or dimetric orientations may be used. The choice of two scales is arbitrary, the third scale depends on the other two.

Following the procedure of articles 11.4 and 11.5 we may write the three scales as follows:

The point A(1, 0, 0) transforms to

$$(\cos\theta_z, \sin\theta_z\cos\theta_x, \sin\theta_z\sin\theta_x)$$

The point B(0, 1, 0) transforms to

$$(-\sin\theta_z, \cos\theta_z\cos\theta_x, \cos\theta_z\sin\theta_x)$$

The point C(0, 0, 1) transforms to

$$(0, -\sin\theta_x, \cos\theta_x)$$

and any combination of θ_z and θ_x which does not lead to an isometric or diametric presentation may be used.

By taking a rotation θ_z other than 45°, we release the axes ox'' and oy'' from their previously symmetrical arrangement. If, for the purpose of example, we take $\theta_z = \theta_x = 30°$

A(1, 0, 0) transforms to:

$$(0.8660, 0.4330, 0.2500)$$

B(0, 1, 0) transforms to:

$$(-0.5000, 0.7500, 0.4330)$$

C(0, 0, 1) transforms to:

$$(0, -0.5000, 0.8660)$$

and the apparent axonometric angles are

$$\text{arc tan A}ox = \frac{0.2500}{0.8660} = 0.28868$$

$$\text{angle A}ox = 16.10°$$

$$\text{arc tan B}ox = -\frac{0.4330}{0.5000} = -0.8660$$

$$\text{angle B}ox = 139.11°$$

$$\text{arc tan } Cox = \frac{0.8660}{0} = \infty$$

$$\text{angle } Cox = 90^\circ$$

The axonometric of a unit cube drawn to these scales is shown in Fig. 11.6.

Rider
Throughout the preceding discussions we applied rotations θ_z and θ_x but an infinite number of alternative view points are available – including rotations in θ_y.

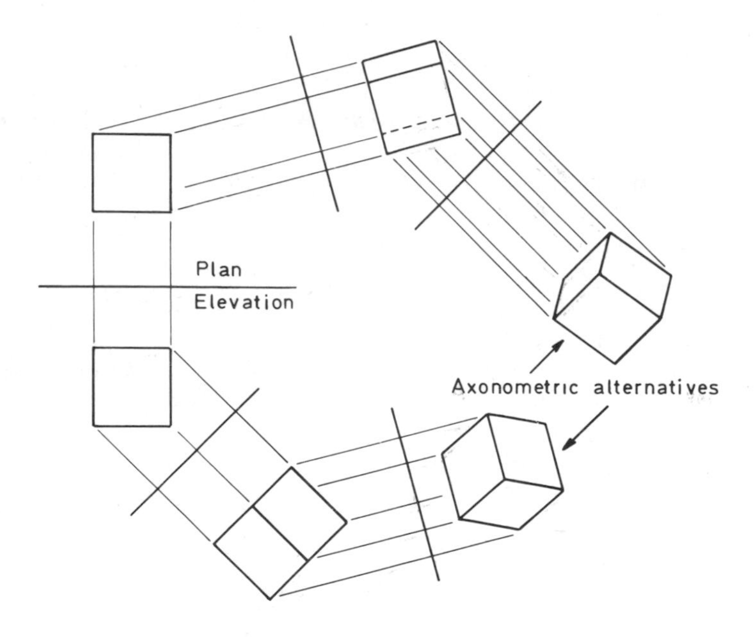

Figure 11.6

11.7 PERSPECTIVE TRANSFORMATIONS

Whilst parallel pictorial projection is adequate for most ordinary engineering purposes, the nature and scale of architectural design and the aesthetic/ergonomic aims of industrial design warrant (true life) perspective illustrations.

Instead of viewing an object from an infinite distance with parallel vision, we may assume a more realistic stance at some lesser distance and make use of our natural divergent vision.

If we consider the path of a single divergent sight line, Fig. 11.7, we see that it intersects the display screen (supposedly placed in front of the object) at a point $P(x_s y_s)$ some finite distance d from the eye.

An object line, such as, PQ which lies parallel to the oy(or ox) axis transforms to a parallel line P*Q* of reduced length, in a perspective presentation and this applies to all lines which are normal to the line of vision and lie parallel to the display screen. Object lines such as P_0P which lie parallel to the viewing direction, i.e. parallel to the axis oz appear as inclined lines on the display screen. The line P_0P thus transforms to a perspective line P_0P^* which is directed towards the vanishing point VP_z. The impression of visual inclination (of the line P_0P^*) is clearly dependent on the position of the vanishing point along oz.

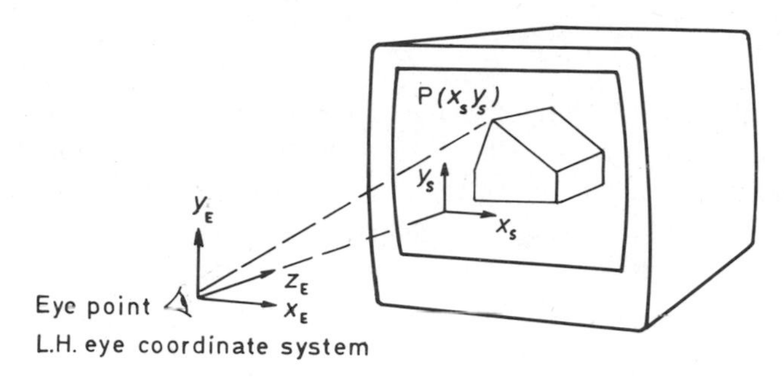

Figure 11.7

It will be observed, from Fig. 11.7, that the points P and Q are transformed to the points P*, Q* in two distinct operations. (1) The orthogonal projection of Q intersects the xoy plane at Q_0. The sight line EQ intersects the xoy plane at Q'_0 and the perspective position of Q* is obtained by projecting Q'_0 orthogonally (parallel to oz), the position of Q* being given by its intersection with the line Q_0VP_z. The perspective coordinates of Q* are shown in broken lines, as are the perspective coordinates of P*.

The perspective coordinates of object points which are subject to a single point perspective transformation may be derived analytically as follows.

A perspective projection of the type, Fig. 11.8, of all object points $P(x, y, z)$ onto the $z = 0$ plane, is given by the following transformation.

$$[X \quad Y \quad Z \quad H] = [x \quad y \quad z \quad 1] \begin{bmatrix} 1 & 0 & 0 & 0 \\ 0 & 1 & 0 & 0 \\ 0 & 0 & 0 & r \\ 0 & 0 & 0 & 1 \end{bmatrix}$$

whence

$$[X \quad Y \quad Z \quad H] = [x, \quad y, \quad o, \quad rz + 1]$$

and hence the perspective screen coordinates are

$$x^* = \frac{x}{rz + 1}, \quad y^* = \frac{y}{rz + 1}$$

The perspective coordinate z^* is given by $z^* = z/(rz + 1)$ in general but $z^* = 0/1 = 0$, when the display screen is at $z = 0$.

The values pertaining to the perspective transformation, Fig. 11.8, are

Pt	x	y	z	r	rz	$rz + 1$
P	2	1	4	$\frac{1}{4}$	1	2
P*	1	$\frac{1}{2}$	2	—	—	—
Q	2	6	4	$\frac{1}{4}$	1	2
Q*	1	3	2	—	—	—

The practical significance of the perspective constant r is apparent from Fig. 11.8, in which it is clear that $+1/r$ is the distance of the vanishing point along oz and $-1/r$ is the distance d to the eye point (in the negative oz direction). If $r = 0$, $\pm 1/r$ is infinite and the perspective construction, Fig. 11.8, degenerates to a parallel projection.

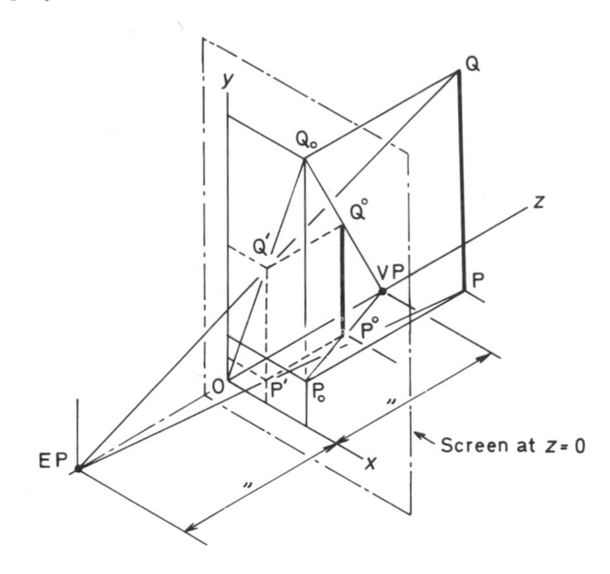

Figure 11.8

For reasons we shall not go into it is not always desirable to site the display screen in the plane $z = 0$ and the plane $z = 1$ is generally to be preferred. (N.B. When $z = 1$ the third column of the perspective transformation matrix reads $(0, 0, 1, 0)$).

The single point presentation described above may be extended to include two and three point perspectives by including vanishing points at $1/p$ and $1/q$ (in the x and y directions) as follows.

$$[X \quad Y \quad Z \quad H] = [x \quad y \quad z \quad 1] \begin{bmatrix} 1 & 0 & 0 & p \\ 0 & 1 & 0 & q \\ 0 & 0 & 1 & r \\ 0 & 0 & 0 & 1 \end{bmatrix}$$

We may clearly produce one and two point perspectives about any chosen axes, simply by specifying the appropriate p, q and/or r.

The general three point perspective with vanishing points at $(1/p, 0, 0)$, $(0, 1/q, 0)$, $(0, 0, 1/r)$ is given by the full complement of perspective p, q and r terms, whence

$$x^* = \frac{x}{px + qy + rz + 1}, \frac{y}{px + qy + rz + 1}, \frac{z}{px + qy + rz + 1}$$

If the orientation of the object is not to our liking then we may rotate or tilt it, to a more acceptable orientation simply by applying the appropriate rotary transformation.

An object rotation about oy, for instance, would be induced by the combined transformation

$$\begin{bmatrix} \cos\theta & 0 & -\sin\theta & 0 \\ 0 & 1 & 0 & 0 \\ \sin\theta & 0 & \cos\theta & 0 \\ 0 & 0 & 0 & 1 \end{bmatrix} \begin{bmatrix} 1 & 0 & 0 & p \\ 0 & 1 & 0 & q \\ 0 & 0 & 1 & r \\ 0 & 0 & 0 & 1 \end{bmatrix}$$

A three point perspective with vanishing points at $(1/p, 0, 0)$, $(0, 1/q, 0)$, $(0, 0, 1/r)$ is given by the full complement of perspective terms.

$$x^* = \frac{x}{px + qy + rz + 1}, \frac{y}{px + qy + rz + 1}, \frac{z}{px + qy + rz + 1}$$

The best practical approach is to represent the horizon by a horizontal line and set one of the axes, oz say, perpendicular to it. We then have freedom to choose the scale of ox^* say, plus the apparent angles made by ox^*, oy and the horizon. The scales for oy^*, oz^* must then be determined.

Example

Determine the picture plane coordinates of the unit cube, the orthographic coordinates of which are given. Make $OV_x = 4$, angle $OV_xV_y = 30°$, angle $OV_yV_x = 45°$.

Solution

Now we know from article 1.11.6, that OV_x is perpendicular to V_yV_z, OV_y is perpendicular to V_xV_y and we already have OV_z perpendicular to V_xV_y, the horizon.

If OV = 4, OH = 2, the scale of oy^* is $2\sqrt{2}$ and since OV_y is perpendicular to V_xV_z the scale of oz^* is $2\sqrt{3} - 2 = 2(\sqrt{3} - 1)$. The perspective constants are therefore

$$p = \frac{1}{4} = 0.25,\ q = \frac{1}{2\sqrt{2}} \simeq 0.354, r = \frac{1}{2\sqrt{3} - 1} \simeq 0.405$$

The x coordinate of A(1, 0, 0) is transformed by the denominator $1/(1 + 0.25)$, whence A*(0.8, 0, 0). The x and y coordinates of P(1, 1, 0) are transformed by the denominator $1/(1 + 0.25 + 0.354) = 1/1.604$ whence P* = (0.623, 0.623, 0) and other coordinates are as listed.

Point	x	y	z	x^*	y^*	z^*
O	0	0	0	0	0	0
A	1	0	0	0.800	0	0
P	1	1	0	0.623	0.623	0
B	0	1	0	0	0.739	0
C	0	0	1	0	0	0.712
	1	0	1	0.604	0	0.604
	1	1	1	0.500	0.500	0.500
	0	1	1	0	0.569	0.569

The perspective cube drawn to these dimensions is shown in Fig. 11.9.

(N.B. All coordinates x^*, y^*, z^*, are of course plotted parallel to the OV_x, OV_y, OV_z axes respectively, but may be transformed to rectangular picture coordinates as previously explained).

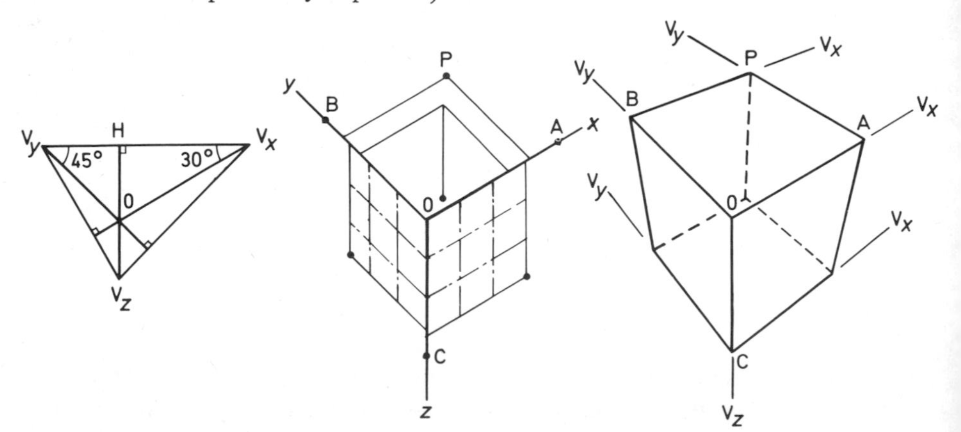

Figure 11.9

11.8 THE EYE COORDINATE SYSTEM

Throughout the preceding discussion it has been our practice to transform the position and orientation of objects about a fixed frame of reference and we now consider the alternative approach of manoeuvring the observer about and around the object. The basic problem is to set up the observer at some distance and angle away from the object so that the object can be viewed, as it would be viewed in real life. The terms pan and tilt are commonly used to denote θ rotations in the horizontal plane and downward ϕ rotations in vertical object space, in accord with the spherical coordinate system. If the axes $oxyz$ are used to define the geometrical properties of the object, and the observer's eye is situated at $(x_E\ y_E\ z_E)$, then the coordinates of the object relative to the eye are $(x - x_E), (y - y_E), (z - z_E)$. When (x_E, y_E, z_E) are given in terms of the spherical coordinate system,

$$x_E = e \cos\phi \sin\theta, y_E = e \sin\phi, z_E = e \cos\phi \cos\theta$$

or since the direction cosines of OE = e are

$$\frac{x_E}{e} = \cos\alpha, \frac{y_E}{e} = \cos\beta, \frac{z_E}{e} = \cos\gamma$$

$$\cos\alpha = \cos\phi \sin\theta, \cos\beta = \sin\phi, \cos\gamma = \cos\phi \cos\theta$$

hence

$$\text{pan angle } \theta = \text{arc tan } \frac{\cos\alpha}{\cos\gamma}$$

$$\text{tilt angle } \phi = \text{arc tan } \frac{\cos\beta}{\sqrt{\cos^2\alpha + \cos^2\gamma}} = \frac{\cos\beta}{\sin\beta}$$

We thus establish a new set of right-handed axes at the eye point. Now since the visual display screen is two dimensional we need to define our picture in terms of two coordinate dimensions and in accord with international practice we choose to look along the axis oz and set the axes ox and oy normal to our line of vision.

We now undertake to transform the axis ox, oy, oz, into a correctly orientated viewing system.

(N.B. The axes ox, oy, oz, are initially coincident with the object. It will be found that five separate geometrical transformations are required to convert these axes into a left-handed triad ox_E, oy_E, oz_E, known as the eye coordinate system, followed by two further transformations which related the physical coordinates to the electronic make-up of the screen).

We have considered the rotation of objects about a fixed frame of reference at length, and we now consider the converse problem of rotating sets of axes about a fixed frame of reference. It is clear that a positive object rotation is akin to a negative axis rotation and since the two operations are complementary we observe that if the matrix for rotating objects is R_0, then the matrix for rotating axes is $R_A = R_0^T$ where R_0^T is the transpose of the matrix R_0.

1. Object rotations. When rotating objects about a fixed frame of reference we use the object transformations

$$\theta_x = \begin{bmatrix} 1 & 0 & 0 \\ 0 & c & s \\ 0 & -s & c \end{bmatrix}, \quad \theta_y = \begin{bmatrix} c & 0 & -s \\ 0 & 1 & 0 \\ s & 0 & c \end{bmatrix}, \quad \theta_z = \begin{bmatrix} c & s & 0 \\ -s & c & 0 \\ 0 & 0 & 1 \end{bmatrix}$$

2. Axis rotations. When rotating axes about a fixed frame of reference we use the axis transformations

$$\theta_x = \begin{bmatrix} 1 & 0 & 0 \\ 0 & c & -s \\ 0 & s & c \end{bmatrix}, \quad \theta_y = \begin{bmatrix} c & 0 & s \\ 0 & 1 & 0 \\ -s & 0 & c \end{bmatrix}, \quad \theta_z = \begin{bmatrix} c & -s & 0 \\ s & c & 0 \\ 0 & 0 & 1 \end{bmatrix}$$

where $c = \cos\theta_x, \cos\theta_y, \cos\theta_z$, and $s = \sin\theta_x, \sin\theta_y, \sin\theta_z$ as appropriate.

11.8.1 The viewing transformation

Now suppose we wish to view an object located at O, from a standing point (eye point) located, at a position $x = 6, y = 8, z = 10$, as shown in Fig. 11.10.

We first set up a set of right-handed axes *oxyz* at the required eye point and this we do by siting the axes at the point (+6, +8, +10), such that the object point O lies at (–6, –8, –10) relative to the new set of axes.

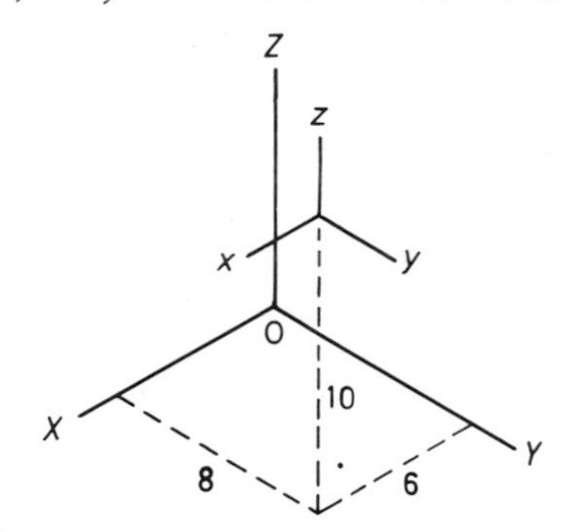

Figure 11.10

The position of the eye point relative to the object datum O is given by the position matrix

$$\begin{bmatrix} 1 & 0 & 0 & 0 \\ 0 & 1 & 0 & 0 \\ 0 & 0 & 1 & 0 \\ 6 & 8 & 10 & 1 \end{bmatrix}$$

and the position of the object datum relative to the eye point datum is given by:

$$\begin{bmatrix} 1 & 0 & 0 & 0 \\ 0 & 1 & 0 & 0 \\ 0 & 0 & 1 & 0 \\ -6 & -8 & -10 & 1 \end{bmatrix}$$

(N.B. Having established a set of right-handed axes at the proposed eye point, we may convert to a left-handed system at the outset (in which case all subsequent positive rotations are reversed in sense) or we may retain the right-handed format for the time being (in which case our positive sign convention is presumed). We will retain right-handed axes for the time being.

Now with a view to setting up screen coordinates $x_s o\, y_s$ we need to direct our viewing axis oz towards the object and this we do in five distinct operations. The first three of which are as follows.

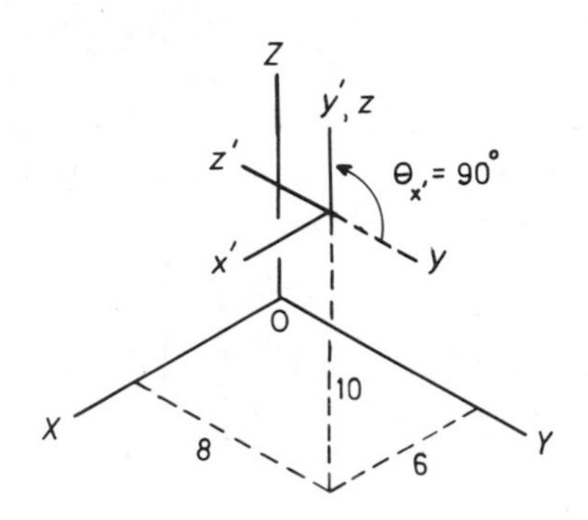

Figure 11.11

A positive rotation of 90° about the axis ox. Fig. 11.11, followed by a negative rotation of arc tan 3/4 about the axis oy', followed by a positive rotation of arc tan 1 about the transformed axis ox'.

We first rotate the triad through 90° about the axis of ox by means of the transformation R^T_{ox}. On substituting $\sin\theta_x = 1$, $\cos\theta_x = 0$ we obtain:

$$\begin{bmatrix} 1 & 0 & 0 & 0 \\ 0 & 1 & 0 & 0 \\ 0 & 0 & 1 & 0 \\ -6 & -8 & -10 & 1 \end{bmatrix} \begin{bmatrix} 1 & 0 & 0 & 0 \\ 0 & 0 & -1 & 0 \\ 0 & 1 & 0 & 0 \\ 0 & 0 & 1 & 1 \end{bmatrix} = \begin{bmatrix} 1 & 0 & 0 & 0 \\ 0 & 0 & -1 & 0 \\ 0 & 1 & 0 & 0 \\ -6 & -10 & 8 & 1 \end{bmatrix}$$

whence the new orientation of axes is as shown in Fig. 11.11.

The first row [1 0 0 0] transforms to [1 0 0 0] and this confirms that there is no change in the direction of ox and hence ox' is directed along ox.

The second row [0 1 0 0] transforms to [0 0 –1 0] and this indicates that oz' is aligned with $-oy$.

The third row [0 0 1 0] transforms to [0 1 0 0] and this indicates that oy' is aligned with oz.

The fourth row [–6 –8 –10 1] transforms to [–6 –10 +8 1] and this confirms that the ox'' axis remains aligned with the ox' axis. The axis oy' points in the direction of oz and hence the object zero O lies at a distance of –10 along the positive oy' axis. The axis oz' points in the direction of $-oy$ and hence the oz' component of the object zero is +8 units. The fourth and final element 1 indicates that the coordinates (–6, –10, +8) are numerically true scale.

The relationships between the new axes ox', oy', oz' is summarised below.

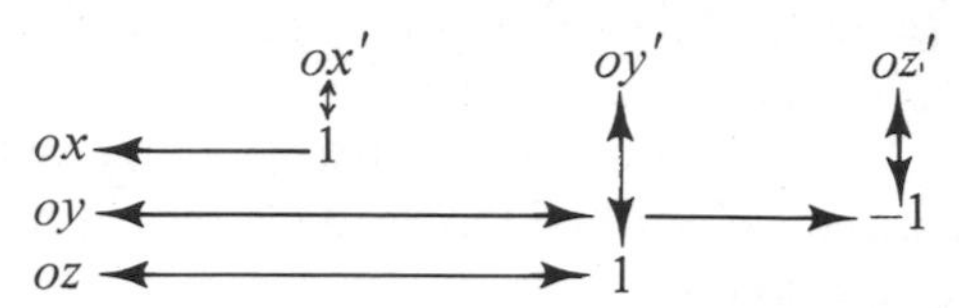

The right-handed triad is next swung about the axis oy' through a negative angle of arc tan 3/4 so that oz'' points towards axis oz, as shown in Fig. 11.14.

The sine of a negative angle of arc tan 3/4 is –3/5, and the cosine is +4/5, and on substituting these values into the rotational transformation R^T_{oy} we reach the orientation, Fig. 11.14.

$$\begin{bmatrix} 1 & 0 & 0 & 0 \\ 0 & 0 & -1 & 0 \\ 0 & 1 & 0 & 0 \\ -6 & -10 & 8 & 1 \end{bmatrix} \begin{bmatrix} \frac{4}{5} & 0 & -\frac{3}{5} & 0 \\ 0 & 1 & 0 & 0 \\ \frac{3}{5} & 0 & \frac{4}{5} & 0 \\ 0 & 0 & 1 & 0 \end{bmatrix} = \begin{bmatrix} \frac{4}{5} & 0 & -\frac{3}{5} & 0 \\ -\frac{3}{5} & 0 & -\frac{4}{5} & 0 \\ 0 & 1 & 0 & 0 \\ 0 & -10 & 10 & 1 \end{bmatrix}$$

The first row [1 0 0 0] transforms to [4/5 0 –3/5 0] and this signifies that the direction ratios of the new axes $ox''y''z''$ are (4/5, 0, –3/5) with respect to the old. The second row [0 0 –1 0] transforms to [–3/5 0 –4/5 0] and this signifies that the direction ratios of the new axes $ox''y''z''$ are (–3/5, 0, –4/5). The third row [0 1 0 0] transforms to [0 1 0 0] and this signifies that there is no change in the direction of oy'.

The fourth row [0 –10 10 1] is computed as follows:

$$-6 \cdot \frac{4}{5} + 8 \cdot \frac{3}{5} = -\frac{24}{5} + \frac{24}{5} = 0, (-10 \, . \, 1) = -10,$$

$$-6 \cdot -\frac{3}{5} + 8 \cdot \frac{4}{5} = \frac{18}{5} + \frac{32}{5} = +10, (1 \, . \, 1) = 1.$$

whence the x'', y'', z'' coordinates of the object zero are (0, –10, +10) as confirmed by the figure.

A positive rotation of amount arc tan 1 about the ox'' axis points oz'' towards O and since $\sin \theta_x = \cos \theta_x = 1/\sqrt{2}$ the transformation R^T_{oz} brings the axes to the chosen viewing direction.

$$\begin{bmatrix} 4/5 & 0 & -3/5 & 0 \\ -3/5 & 0 & -4/5 & 0 \\ 0 & 1 & 0 & 0 \\ 0 & -10 & +10 & 1 \end{bmatrix} \begin{bmatrix} 1 & 0 & 0 & 0 \\ 0 & 1/\sqrt{2} & -1/\sqrt{2} & 0 \\ 0 & 1/\sqrt{2} & 1/\sqrt{2} & 0 \\ 0 & 0 & 0 & 1 \end{bmatrix} = \begin{bmatrix} 4/5 & -3/5\sqrt{2} & -3/5\sqrt{2} & 0 \\ -3/5 & -4/5\sqrt{2} & -4/5\sqrt{2} & 0 \\ 0 & 1/\sqrt{2} & -1/\sqrt{2} & 0 \\ 0 & 0 & 10/\sqrt{2} & 1 \end{bmatrix}$$

Row 3 of this final matrix confirms that the object zero datum 0, lies at $x''' = 0$, $y''' = 0$, $z''' = 10\sqrt{2}$, and since $\sqrt{6^2 + 8^2 + 10^2} = 10\sqrt{2}$, this is clearly a correct observation.

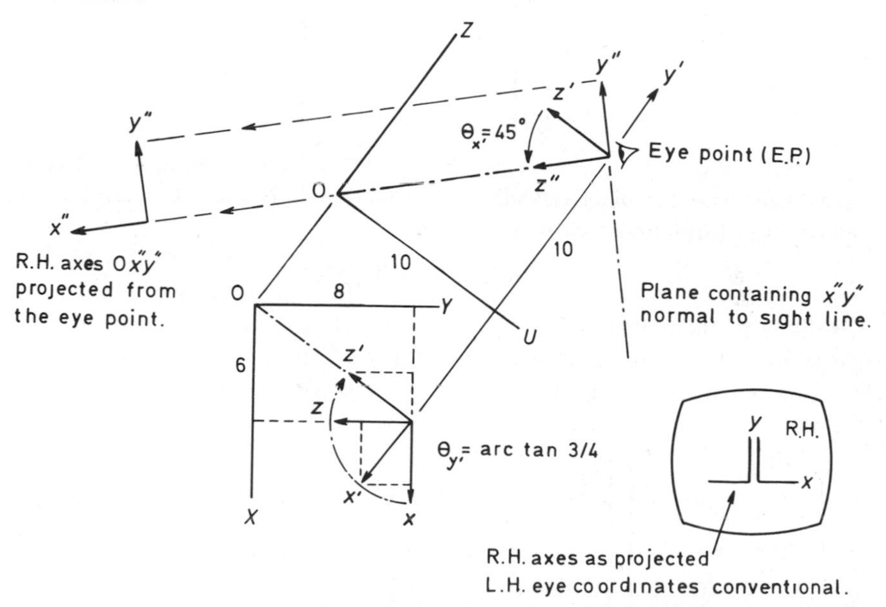

Figure 11.12

We see from Fig. 11.12, that the object axis oz is aligned with the projection of the axis oy''', but since right-handed axes have been used the orientation of the ox''' and oy''' in the picture plane is unconventional. We therefore switch to left-handed axes, so that when the observer looks along the positive oz''' sight line the orientation of the picture screen coordinates is x_s to the right, y_s up is positive. Hence $ox_E = -ox'''$, $oy_E = oy'''$, with oz_E the sight line. The left-handed triad of axes ox_E, oy_E, oz_E is known as the eye coordinate system.

The direction of ox''' is reversed by means of the reflective transformation

$$\begin{bmatrix} 4/5 & -3/5\sqrt{2} & -3/5\sqrt{2} & 0 \\ -3/5 & -4/5\sqrt{2} & -4/5\sqrt{2} & 0 \\ 0 & 1/\sqrt{2} & -1/\sqrt{2} & 0 \\ 0 & 0 & 10\sqrt{2} & 1 \end{bmatrix} \begin{bmatrix} -1 & 0 & 0 & 0 \\ 0 & 1 & 0 & 0 \\ 0 & 0 & 1 & 0 \\ 0 & 0 & 0 & 1 \end{bmatrix} = \begin{bmatrix} -4/5 & -3/5\sqrt{2} & -3/5\sqrt{2} & 0 \\ +3/5 & -4/5\sqrt{2} & -4/5\sqrt{2} & 0 \\ 0 & 1/\sqrt{2} & -1/\sqrt{2} & 0 \\ 0 & 0 & 10\sqrt{2} & 1 \end{bmatrix}$$

(N.B. It would have been equally acceptable to transform the axes ox''', oy''', to the conventional ox_s, oy_s orientation and then reverse the direction of oz''', but the same number of individual transformations would have been involved.)

The physical magnitude of the final coordinate dimensions clearly depends on the absolute size of the object viewed and it is unlikely that these dimensions will fit neatly within the available screen size, and in general the above matrix of picture plane coordinates will need to be scaled by a matrix N of the form

$$N_s = \begin{bmatrix} d/s & 0 & 0 & 0 \\ 0 & d/s & 0 & 0 \\ 0 & 0 & 1 & 0 \\ 0 & 0 & 0 & 1 \end{bmatrix}$$

where d is the distance from eye point to screen and s is approximately half the screen size. The complete viewing transformation V is thus comprised of six separate transformations, whence

$$V = TR_1 R_2 R_3 R_F N$$

Bearing in mind that rotations are non-comutative we may concatenate the matrices R_{ox}, $R_{oy'}$, $R_{ox'}$ by recalling that a rotation of axes about ox followed by a rotating of axes about oy is given by $A \times B = AB$.

$$\begin{bmatrix} 1 & 0 & 0 & 0 \\ 0 & Cx & -Sx & 0 \\ 0 & Sx & Cx & 0 \\ 0 & 0 & 0 & 1 \end{bmatrix} \begin{bmatrix} Cy' & 0 & Sy' & 0 \\ 0 & 1 & 0 & 0 \\ -Sy' & 0 & Cy' & 0 \\ 0 & 0 & 0 & 1 \end{bmatrix} = \begin{bmatrix} Cy' & 0 & Sy' & 0 \\ SxSy' & Cx & -SxCy' & 0 \\ -CxSy' & Sx & CxCy' & 0 \\ 0 & 0 & 0 & 1 \end{bmatrix}$$

and subsequently performing the operation

$$\begin{bmatrix} Cy' & 0 & Sy' & 0 \\ SxSy' & Cx & -SxCy' & 0 \\ -CxSy' & Sx & CxCy' & 0 \\ 0 & 0 & 0 & 1 \end{bmatrix} \begin{bmatrix} 1 & 0 & 0 & 0 \\ 0 & Cx'' & -Sx'' & 0 \\ 0 & Sx'' & Cx'' & 0 \\ 0 & 0 & 0 & 1 \end{bmatrix}$$

$$= \begin{bmatrix} Cy' & Sy'Sx'' & Sy'Cx'' & 0 \\ SxSy' & CxCx'' - SxCy'Sx'' & -CxSx'' - SxCy'Cx'' & 0 \\ -CxSy' & SxCx'' - CxCy'Sx'' & -SxSx'' + CxCy'Cx'' & 0 \\ 0 & 0 & 0 & 1 \end{bmatrix}$$

11.8.2 The screen coordinate system

A small visual display unit has a screen size of about 30 cm (diagonal) and the screen size of a larger unit is about 50 cm (diagonal) with a natural viewing distance s between 25 and 75 cm. When equi-positive and negative values of x_s and y_s are required we set the ox_s, oy_s datum in the centre of the screen. The half sized of a typical screen varies between 15 and 25 cm and hence a scaling ratio within the range $1 < d/s < 5$, is to be expected, and the need to relate the physical dimensions of the image to the electronic dot make-up of the screen is also important. For a screen with a viewable area of 372 mm × 248 mm, comprising 1536 dots × 1024 dots, the screen coordinates in terms of pixle dots are $x_s = 768 . x_c/z_c + 768$ and $y_s = 512 . y_c/z_c + 512$, where x_c/z_c and y_c/z_c are the so-called clipped coordinates given by the transformation $[x_c \; y_c \; z_c \; 1] = [x_E \; y_E \; z_E \; 1] \; [N]$.

11.8.3 Zoom/windowing

The absolute size of current graphics display screens is considerably less than the designer's traditional drawing board and the need for a zoom facility by which small areas of detail may be instantly enlarged and worked upon, is an obvious practical requirement. A zoom/windowing facility and a means to recall the original scale is always provided.

11.8.4 Picture clipping

As we have seen in article 11.7 perspective viewing is characterised by divergent vision and it is clear that the actual magnitudes of the screen coordinates $(x_s \, y_s)$ depend on both the real object size and on the relative distances between eye, screen and object.

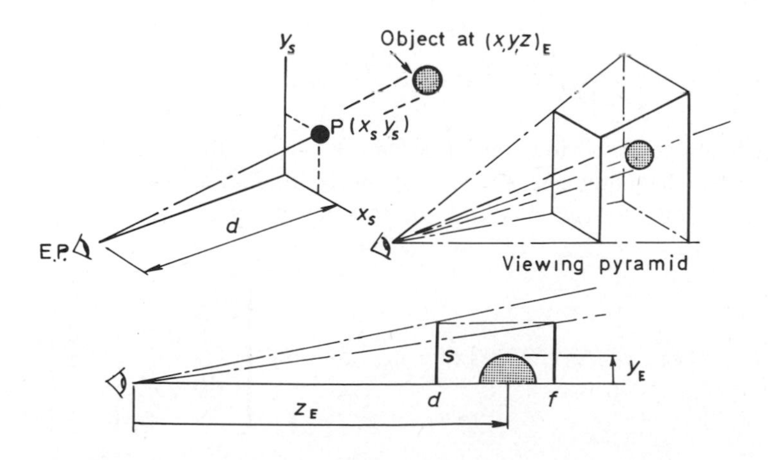

Figure 11.13

It is clear from Fig. 11.13, that the screen coordinates are given by

$$\frac{x_s}{d} = \frac{x_E}{z_E}, \quad \frac{y_s}{d} = \frac{y_E}{z_E}$$

and by introducing the actual half screen size s the screen coordinates may be expressed in non-dimensional form as follows

$$x_s = \frac{d}{s}\,\frac{x_E}{z_E}, \quad y_s = \frac{d}{s}\,\frac{x_E}{z_E}$$

where the ratio d/s is a scaling factor used to fit the picture within the screen. The presence of z_E in the denominator confirms that in a perspective view a near object point measured in $(x_E y_E z_E)$ appears on the screen to a larger scale than does a more distant object point with global and hence eye coordinates $(x_E y_E z_{E+++})$ and this is clearly due to the greater divergence produced by the former. Two real object lines with identical $(x_E, y_E)_s$ but different $(z_E)_s$ clearly transform to different (perspective) lengths on the screen.

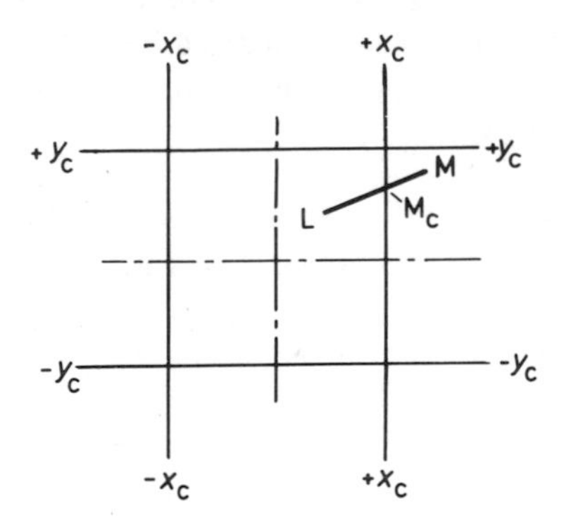

Figure 11.14

It is moreover obvious that any feature, such as the object line LM, which penetrates the bound of the viewing pyramid, Fig. 11.14, will be excluded (clipped) from the picture, and for reasons of computational convenience a clipper matrix

$$[x_c \quad y_c \quad z_c \quad 1] = [x_E \quad y_E \quad z_E \quad 1]\begin{bmatrix} d/s & 0 & 0 & 0 \\ 0 & d/s & 0 & 0 \\ 0 & 0 & 1 & 0 \\ 0 & 0 & 0 & 1 \end{bmatrix}$$

is commonly used to define this bound. The bound $(\pm x_c, \pm y_c)$ thus represents the size of the clipping rectangle sited at $z_c = z_E$, and a computer routine which clips any stray ends, from lines such as LM, Fig. 11.14, is generally provided.

(N.B. Clipping stray detail which lies beyond the bound of the graphics device is important, for if left undone may generate overflow and a hopelessly scrambled picture.)

The (x, y) coordinate of the end point L, Fig. 11.14, are clearly less than the $(x_c y_c)$ bounds and hence points on the line LM near to L are retained and on the grounds that $(x_m y_m)$ exceed $(x_c y_c)$ points near to M are erased, and a new point M′ (defining the intersection of the line LM with $+x_c$) automatically calculated, thus providing a revised definition of the line LM.

11.8.5 Perspective depth

Whilst the screen coordinates $(x_s y_s)$ suffice to map the picture we need (for the purpose of hidden line elimination) a measure of perspective depth and in order to obtain an acceptable level of depth precision we define two cut off planes D and F normal to the eye, such that the object is closely sandwiched between them. These two cut off planes combine with the viewing pyramid to form a truncated viewing pyramid (sometimes called the viewing box) within which the visible object is wholly enclosed. That is $x_s \leqslant \pm 1$, $y_x \leqslant \pm 1$, and $0 \leqslant z_s \leqslant +1$. When the front face of the object is conceptually in contact with the display screen the plane D and the screen S are both at a distance $e = d$ from the eye, and the plane F is at a distance commensurate with the object.

11.8.6 Hidden detail removal

The theory given above is sufficient to enable, so called, wire frame perspectives to be produced. The elimination of hidden detail has not been considered. There are numerous ways in which hidden detail can be removed from a computer graphics picture, but in all but trivial cases the demand on computer time is substantial.

For maximum depth resolution a point on the object at a distance $z_\mathrm{E} = d$ maps to $z_s = 0$ and a point on the object at a distance $z_\mathrm{E} = f$ maps to $z_s = 1$. The expression

$$z_s = \frac{\left(1 - \dfrac{d}{z_\mathrm{E}}\right)}{\left(1 - \dfrac{d}{f}\right)}$$

clearly produces this mapping.

Now for any given situation s/d is constant and since $\omega = sz_E/d$, ω varies linearly with z_E and

$$z_s = \frac{\left(1 - \dfrac{d}{z_E}\right)}{\left(1 - \dfrac{d}{f}\right)} = \frac{s}{\omega}\,\frac{\left(\dfrac{z_E}{d} - 1\right)}{\left(1 - \dfrac{d}{f}\right)}$$

and if $z_s = z/\omega$

$$z = s\,\frac{\left(\dfrac{z_E}{d} - 1\right)}{\left(1 - \dfrac{d}{f}\right)}$$

and we have established all transformations in terms of the parameter ω.

We have already seen that $x_s = \dfrac{d}{s}\,\dfrac{x_E}{z_E}$, $y_s = \dfrac{d}{s}\,\dfrac{y_s}{z_E}$ hence the three transformations are

$$x_s = \frac{x_E}{\omega}, \quad y_s = \frac{y_E}{\omega}, \quad z_s = \frac{s}{\omega}\,\frac{\left(\dfrac{z_E}{d} - 1\right)}{\left(1 - \dfrac{d}{f}\right)}$$

With regard to visibility, points with small z_s have priority over points further away.

11.8.6 Hidden detail removal

The theory given above is sufficient to enable, so called, wire frame perspectives to be produced. The elimination of hidden detail has not been considered. There are numerous ways in which hidden detail can be removed from a computer graphics picture, but in all but trivial cases the demand on computer time is substantial.

The added time in cost and effort must therefore be balanced against the advantages gained. For many applications hidden line removal is not strictly necessary, where as in other applications where greater realism is sought the removal of hidden lines is obligatory. Figure 11.15 which shows a valve core, without hidden lines removal thus contrasts with a broadly similar example of an oil rig structure Figure 11.16 in which all hidden lines have been removed.

The support structure Figure 11.16 and the radar antenna dish Figure 11.17 are produced by PATRAN-G. These figures are reproduced by kind permission of Kins Developments Ltd, Epsom.

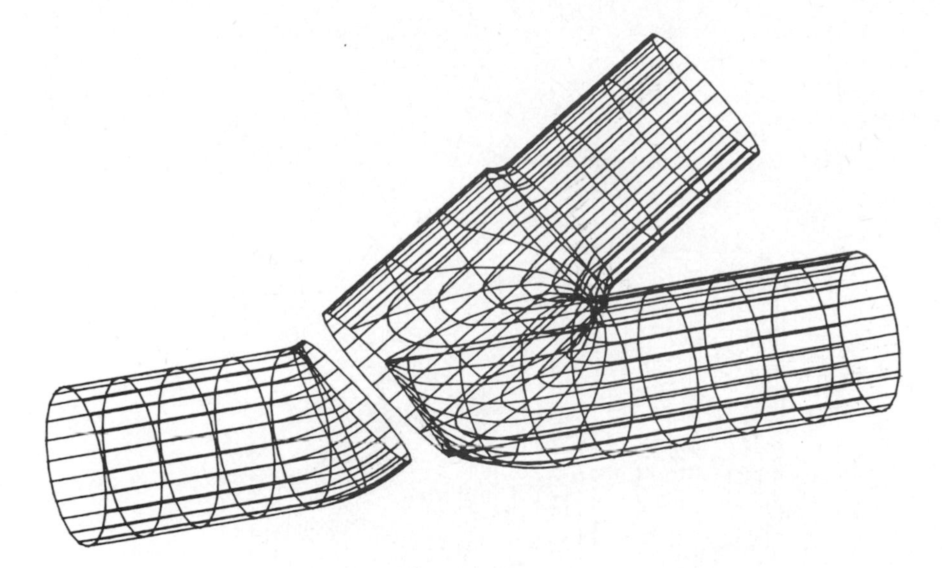

Figure 11.15

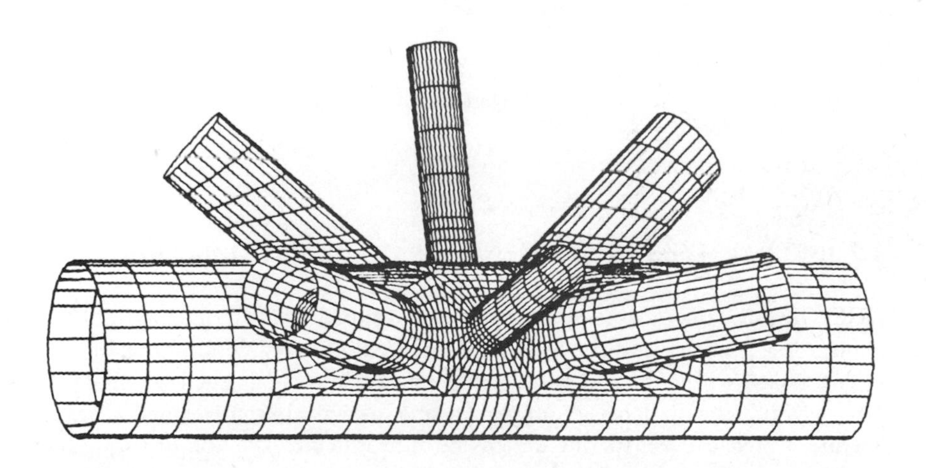

Figure 11.16

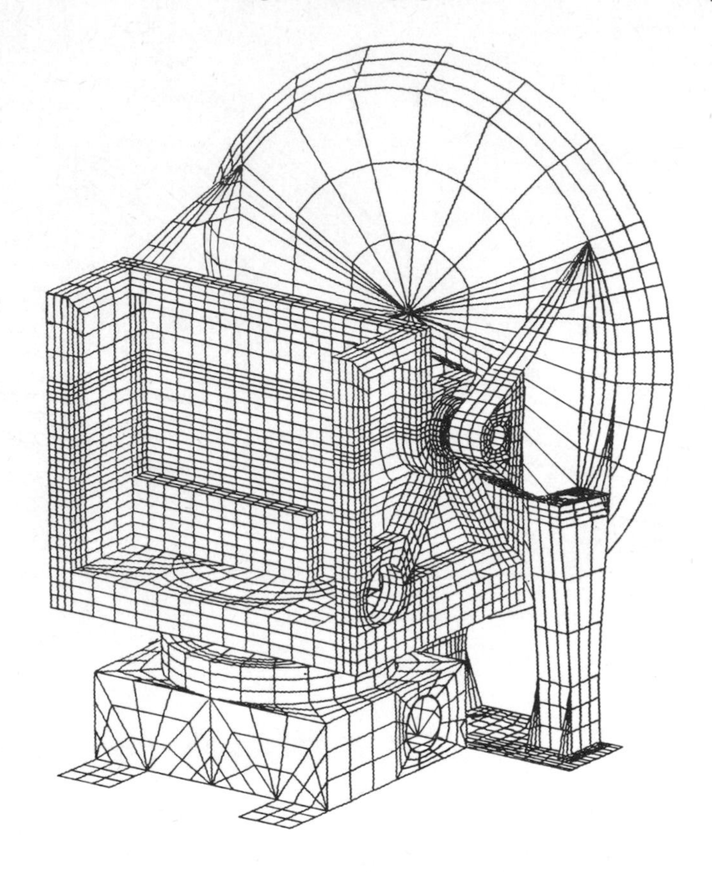

Figure 11.17

11.9 INTERACTIVE THREE DIMENSIONSAL SOLIDS MODELLING

SDRC GEOMOD is an interactive three dimensional solids modelling tool that enables design engineers to construct geometrical concept descriptions of component designs within the computer. Using a powerful system, such as GEOMOD, it is easy for the designer to explore many design alternatives and often to generate a physical feel for the designs he is producing. The ability to produce unrefined 'roughs' as quickly as possible is a very important asset to a project designer and the dual facility to produce roughs very quickly and/or fully refined presentations is an important feature of the SDRC GEOMOD tool. GEOMOD typically produces extremely fine facetted graphic models such as the robot figure 11.18.

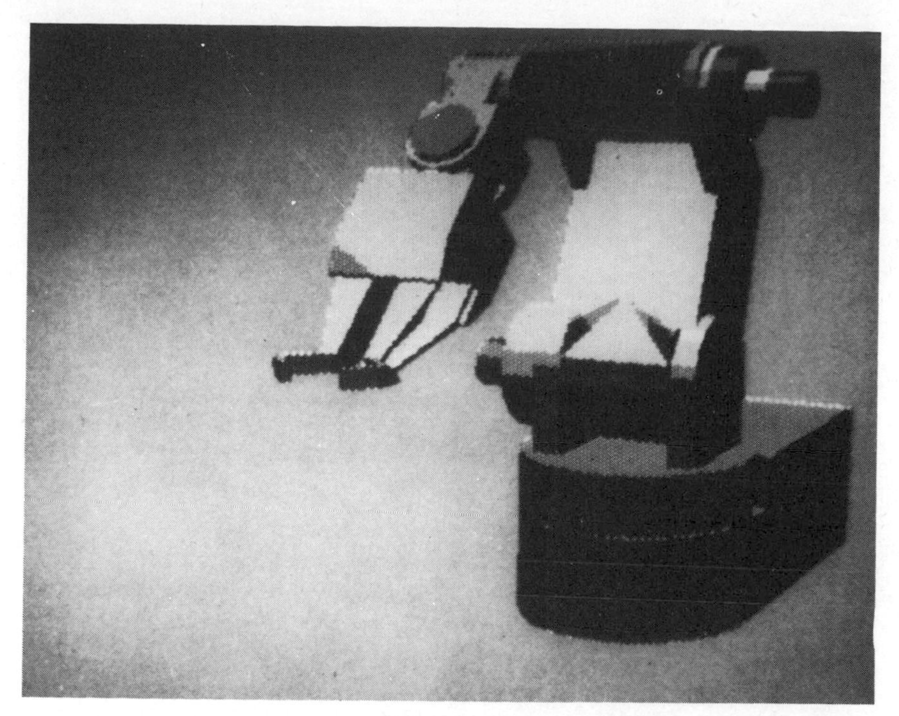

Figure 11.18

Figure 11.19(a)

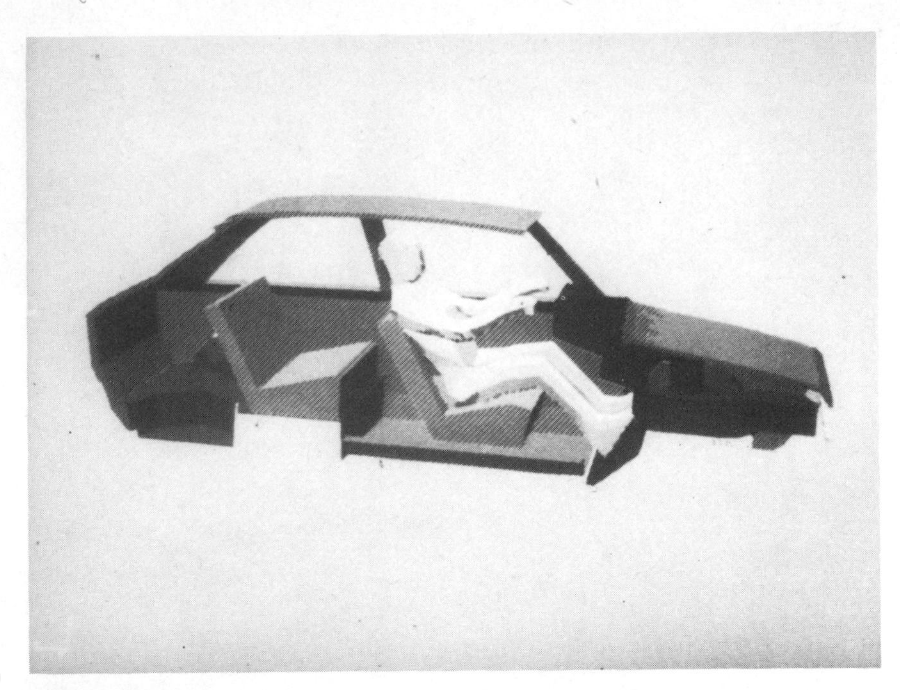

Figure 11.19(b)

Figure 11.19(c)

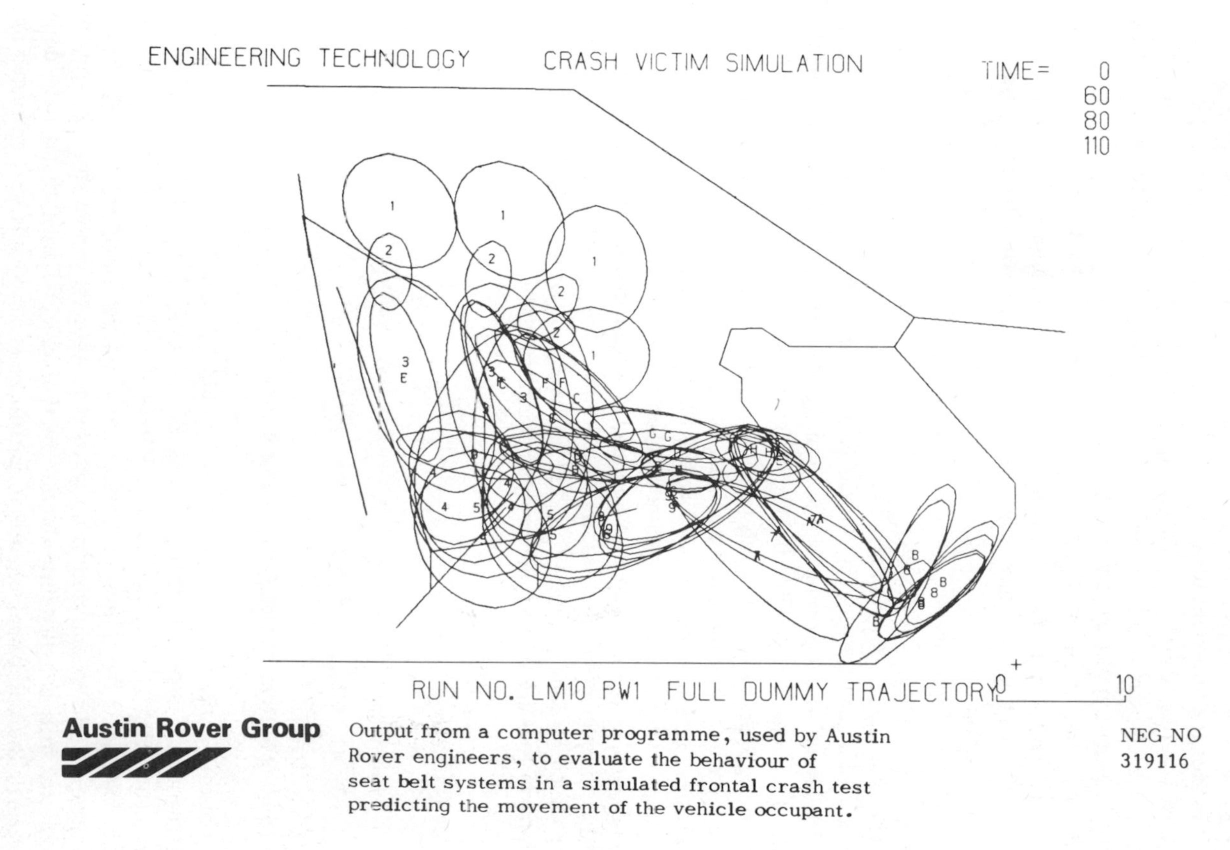

Output from a computer programme, used by Austin Rover engineers, to evaluate the behaviour of seat belt systems in a simulated frontal crash test predicting the movement of the vehicle occupant.

Figure 11.20

With regard to solid modelling the designer may choose from an extensive library of geometrical primatives, comprising cubes, cones, cylinders, hollow cylinders and spheres etc., and these he may freely assemble to produce almost any traditional component form.

A typical turned part may be rapidly built up from a series of co-axial cylinders and a circular or square hole may be bored through a rectangular block using the negation of a cylindrical element.

Boolean operators are available to manoeuvre and blend these shapes into intricate forms and as with all the more powerful programmes the area, volume, mass, centre of gravity and inertia is calculated and if necessary displayed instantaneously.

A particularly impressive example produced by SDRC GEOMOD is the motor vehicle with dummy driver, figure 11.19.

The first figure show the prospective owner out for a simulated test run. In the second figure the side of the car is removed to give one a better idea of seating layout and leg room. In the third figure the computer has removed the dummy and made it possible for the designer (thats you the reader) to occupy the driving seat, so that you can personally observe the existence and impact of blind spots during 'your' computer aided test run.

Figures 11.19 have been reproduced by kind permission of CAE International, Hitchin, to whom my grateful acknowledgement is due.

As a final demonstration of the extent to which the total CAE commitment has been implemented, the readers attention is drawn to figures 11.20 which are reproduced by kind permission of the Austin Rover Group. The original Austin Rover captions are self explanatory, self complimentary.

Concluding Statement and Acknowledgements

In presenting the material kindly contributed by: Austin Rover Group, CAE International, Delta Computer Aided Engineering, Ferranti Cetex Graphics Ltd., and Kins Developments Ltd., I have drawn freely from publicity and other material. I have, however, concentrated on the purely geometrical aspects of these various sytems and in no way have I told the whole story. Bearing in mind that this is a first introduction to the subject I trust that I have done justice to the companies concerned and to personal friends who entrusted me to present their work and companies image fairly and squarely. My personal thanks are due to all concerned.

That is true, Socrates; but these subjects seem to be as yet hardly explored.

Why, yes, I said, and for two reasons: in the first place, no government patronises them, which leads to a want of energy in the study of them, and they are difficult; in the second place, students cannot learn them unless they have a teacher. But then a teacher can hardly be found, and even if he could, as matters now stand, the students, who are very conceited, would not mind him. That, however, would be otherwise if the whole state patronised and honoured these studies; they would find disciples, and there would be continuous and earnest search, and discoveries would be made; since even now, disregarded as they are by the world, and maimed of their fair proportions, and although none of their votaries can tell the use of them, still these studies force their way by their natural charm, and very likely they may emerge into light.

Plato (427-347 BC)

Bibliography

[1] Abbot, W. *Practical Geometry and Engineering Graphics.* Blackie, 1947.

[2] Adler, Irvin. *A New Look at Geometry.* Dobson and Dobson, 1966.

[3] Aitken, A. C. *Determinants and Matrices.* Oliver and Boyd, 1958.

[4] Angell, Ian, O. *A Practical Introduction to Computer Graphics.* Macmillan, 1981.

[5] Apalategui, J.J. and Adams, L. J. *Aircraft Analytic Geometry.* McGraw Hill, 1944.

[6] Ayres, Frank. *Projective Geometry.* McGraw Hill, 1967.

[7] Bell, Robert J. T. *Coordinate Geometry in Three Dimensions.* Macmillan 1928.

[8] Barnhill, Robert, E. and Riesenfeld, Richard, T. *Computer Aided Geometric Design.* Academic Press, 1974.

[9] Benjamin, B. S. *Analysis of Braced Domes.* ASIA Published, 1963.

[10] Berge, Claude. *The Theory of Graphs*. Methuen, 1962.

[11] Besant, C. B. *Computer Aided Design and Manufacture.* Ellis Horwood 2nd, 1983.

[12] Booker, P. J. *A History of Engineering Drawing.* Northgate, 1979.

[13] Bornego, John. *Space Grid Structures.* M.I.T., 1972.

[14] Bostick and Chandler, S. *The Core Course for A Level Mathematics.* Stanley Thornes, 1980.

[15] Bourne, D. E. and Kendall, P. C. *Vector Analysis.* Oldbourne, 1967.

[16] Braid, I. C. and Faux, I. D. *Computer Aided Modelling of Solids.* Ellis Horwood, in prep.

[17] Buckminster-Fuller, R. *Synergetics.* Macmillan, 1975.

[18] Carico, Charles and Drooyan, Irving. *Analytic Geometry.* Wiley, 1980.

[19] Ching, F. D. K. *Architecture, Form, Space and Order.* Design Council, 1980.

[20] Chironis P. Nicholas. *Mechanisms, Linkages and Mechanical Controls.* McGraw Hill, 1965.

[21] Chisholm, J. S. R. *Vectors in Three Dimensional Space.* Cambridge, 1978.

[22] Coates, R. C. and Contie, M. G. *Structural Analysis.* Nelson, 1972.
[23] Coolidge, Julian Lowell. *A History of Conic Sections and Quadric Surfaces.* Dover, 1968.
[24] Cox, H. L. *The Design of Structures of Least Weight.* Pergamon, 1965.
[25] Coxeter, H. S. M. *Introduction to Geometry.* Wiley, 1961.
[26] Coxeter, H. S. M. *Regular Polytypes.* Macmillan Collier, 1963.
[27] Critchlow, Keith. *Order in Space.* Thames and Hudson, 1969.
[28] Cundy, H. M. and Rollett, A. P. *Mathematical Models.* Oxford, 1972.
[29] Day, W. D. *Introduction to Vector Analysis.* Lil, 1966.
[30] Duffy, Joseph. *Analysis of Mechanisms and Robot Manipulators.* Edward Arnold, 1980.
[31] Dunning Davis, J. *Mathematical Methods for Mathematicians, Physical Scientists and Engineers.* Ellis Horwood, 1982.
[32] Faux, I. D. and Pratt, M. J. *Computational Geometry for Design and Manufacture.* Ellis Horwood, 1978.
[33] Fenner, Roger, T. *Finite Element Methods for Engineers.* Macmillan, 1975.
[34] Ferrar, W. L. *Higher Algebra.* Oxford, 1945.
[35] Firby, P. A. and Gardiner, C. F. *Surface Topology.* Ellis Horwood, 1982.
[36] Gardiner, C. F. *Modern Algebra.* Ellis Horwood, 1981.
[37] Gasson, P. C. *Theory of Design.* Batsford, 1974.
[38] Gerard, George. *Structural Stability.* McGraw Hill, 1962.
[39] Gheorghin, Adrian and Dragonep, Virgil. *Geometry of Structural Forms.* Applied Science, 1978.
[40] Gibbs, J. W. *Vector Analysis.* Yale University Press.
[41] Gill, Robert W. *Basic Perspective.* Thames and Hudson, 1974.
[42] Gillespie, R. P. *Partial Differentiation.* Oliver and Boyd, 1954.
[43] Giloi, Wolfgang. *Interactive Computer Graphics.* Prentice Hall, 1978.
[44] Gordon, J. E. *Structures.* Pelican, 1978.
[45] Gordon, William, R. and Sohmer, Bernard. *Intermediate Algebra and Analytic Geometry.* W. H. Allen, 1967.
[46] Goult, R. J. *Applied Linear Algebra.* Ellis Horwood, 1978.
[47] Grace, J. H. *Coordinte Geometry.* University Tutorial Press, 1939.
[48] Rosenberg, F. *Linear Analysis of Frameworks.* Ellis Horwood, 1983.
[49] Gray, Jeremy. *Ideas of Space.* Clarendon, Oxford, 1979.
[50] Gow, Margaret, M. *A Course in Pure Mathematics.* Hodder and Stoughton, 1960.
[51] Hague, B. *An Introduction to Vector Analysis for Scientists and Engineers.* Methuen, 1980.
[52] Hambridge, Jay. *Dynamic Symmetry.* Oxford, 1920.
[53] Hawk, Minor C. *Descriptive Geometry.* McGraw Hill, 1962.
[54] Hemp, W. S. *Optimum Structures.* Clarendon, Oxford, 1973.
[55] Hilson, Barry. *Basic Structural Behaviour via Models.* Wiley (ed Granada).

[56] Hinkle, Rolland. *Kinematics of Machines.* Prentice Hall, 1960.
[57] Holden, Alan. *Shape Space and Symmetry.* Columbia University Press, 1971.
[58] Holt, Michael. *Mathematics in Art.* Studio Vista/van Nostrand Reinholdt, 1971.
[59] Hunt, K. H. *Kinematic Geometry of Mechanisms.* Oxford, 1978.
[60] Kenner, Hugh. *Geodesic Math.* University of California Press, 1976.
[61] Kindle, Joseph, H. *Analytic Geometry.* McGraw Hill, 1950.
[62] Korites, B. J. *Graphics Software for Micros.* Kern, 1983.
[63] Krayszig, E. *Advanced Engineering Mathematics.* Wiley, 1979.
[64] LeCorbusier. *The Modulor.* Faber and Faber, 1964.
[65] Levens, A. S. *Graphical Methods in Research.* Wiley, 1965.
[66] Levitskii, N. I. *Analysis and Synthesis of Mechanisms.* Amerind.
[67] Lipschutz, Martin, M. *Differential Geometry.* McGraw Hill, 1969.
[68] Lockwood, E. H. and Macmillan, R. H. *Geometric Symmetry.* Cambridge, 1978.
[69] Loeb, Arthur, L. *Space Structures.* Addison Wesley, 1976.
[70] Luther, P. and Eisenhart, Fahller. *Coordinate Geometry,* Dover, 1939.
[71] Lynsternix, L. A. *Convex Figures and Polyhedra.* Dover, 1963.
[72] Makowski, Z. S. *Analysis, Design and Construction of Braced Domes.* Pitman, 1983.
[73] Makowski, Z. S. *Steel Space Structures.* Michael Joseph, 1965.
[74] Maxwell, E. A. *Deductive Geometry.* Pergamon, 1962.
[75] Maxwell, E. A. *Geometry by Transformations.* Cambridge, 1975.
[76] Mayall, W. H. *Machines and Perception.* Studio Vista Reinhold, 1968.
[77] Megson, T. H. G. *Aircraft Structures.* Edward Arnold, 1972.
[78] Melzak, Z. A. *Invitation to Geometry.* Wiley, 1983.
[79] Midonick, Henrietta. *The Treasure of Mathematics.* Penguin Books, 1965.
[80] Moffall, William G. and Pearsall, George, W. *The Structure and Properties of Materials, Vol. 1.* Wiley, 1967.
[81] Mold, Josephine. *Workshop Manual Circles Tessellations, Solid Models.* Cambridge, 1974.
[82] Myers, Roy, E. *Microcomputer Graphics.* Addison Wesley, 1982.
[83] Ormerod, Milton, B. *The Architectural Properties of Matter.* Edward Arnold, 1970.
[84] Otto, Frei. *Tensile Structures.* M.I.T., 1973.
[85] Owen, J. B. B. *The Analysis and Design of Lightweight Structures.* Edward Arnold, 1965.
[86] Palmer, Andrew, C. *Structural Mechanics.* Clarendon, Oxford, 1976.
[87] Perry, Owen and Joyce. *New Syllabus Mathematics O/2.* Macmillan, 1979.
[88] Popko, Edward. *Geodesics.* University of Detroit Press.

[89] Pugh, Anthony. *Polyhedra.* University of California Press, 1976.
[90] Pugh Anthony. *Tensegrity.* University of California Press, 1976.
[91] Redcliffe, C. W. and Thomas, George, B. *Calculus and Analytic Geometry.* Addison Wesley.
[92] Reuleaux, Franz. *The Kinematics of Machinery.* Dover, 1963.
[93] Rich, Barnett. *Plane Geometry and Coordinate Geometry.* McGraw Hill, 1963.
[94] Riegels, Friedrick Wilhelm. *Aerofoil Sections.* Butterworths, 1961.
[95] Rogers, David, F. and Adams, J. Alan. *Mathematical Elements for Computer Graphics.* McGraw Hill, 1976.
[96] Roland, Conrad. *Frei Otto Structures.* Longman, 1970.
[97] Rosenauer, N. and Willis, A. N. *Kinematics of Mechanisms.* Dover, 1967.
[98] Rutherford, D. E. *Vector Methods.* Oliver and Boyd, 1957.
[99] Rusinoff, S. E. *Descriptive Geometry.* American Technical Press, 1948.
[100] Ryan, Daniel, L. *Computer Aided Graphics and Design.* Marcel Dekker, 1979.
[101] Ryan, Daniel, L. *Computer Aided Kinematics for Machine Design.* 1981.
[102] Slaby, Steve, M. *Three Dimensional Descriptive Geometry.* Wiley, 1965.
[103] Spiegel, Murray, R. *Vector Analysis.* McGraw Hill, 1959.
[104] Stevens, Peter, S. *Patterns in Nature.* Penguin, 1974.
[105] Stevens, Peter, S. *Handbook of Regular Patterns.* M.I.T. 1980.
[106] Stevens, Karl. *Statics and Strength of Materials.* Prentice Hall, 1979.
[107] Street, W. E. *Technical Descriptive Geometry.* Van Nostrand, 1966.
[108] Suh, C. H. *Kinematics and Machines Design.* Wiley, 1978.
[109] Sweet, M. V. *Algebra Geometry and Trigonometry for Science Students.* Ellis Horwood, in prep.
[110] Teague, Walter Dorwin. *Design This Day.* Studio, 1946.
[111] Thompson, Darcy. *On Growth and Form.* Cambridge, 1969.
[112] Thring, M. W. *Robots and Telechirs.* Ellis Horwood, in prep.
[113] Todhunter, I. *Spherical Trigonometry.* Cambridge, 1871.
[114] Turtle, S. B. *Mechanisms for Engineering Design.* Wiley, 1967.
[115] Waite, Mitchell. *Computer Graphics Primer.* Howard W. Sams, 1980.
[116] Walker, B.S., Gurd, J. R. and Drawneek, E. A. *Interactive Computer Graphics.* Edward Arnold, 1975.
[117] Wen-H-Siung Li. *Engineering Analysis.* Prentice Hall, 1960.
[118] Wenninger, Magnas, J. *Polyhedral Models.* Cambridge, 1974.
[119] Wenninger, Magnas, J. *Spherical Models.* Cambridge, 1979.
[120] Willmore, Floyd., Barr, Donald and Voils, Donald. *Analytic Geometry: a Vector Approach.* Allyn and Bacon, 1971.
[121] Yates, Robert, C. *Curves and Their Properties.* The National Council of Teachers, 1974.
[122] Conference Papers. *Curved Surfaces in Engineering.* I.P.C., 1972.

Index

C

D

E

F

G

L

M

N

O

P

Q

R

S

V

W

X

Y

Z